Universitext

Universitext

Universitext is a series of textbooks that presents material from a wide variety of mathematical disciplines at master's level and beyond. The books, often well class-tested by their author, may have an informal, personal even experimental approach to their subject matter. Some of the most successful and established books in the series have evolved through several editions, always following the evolution of teaching curricula, to very polished texts.

Thus as research topics trickle down into graduate-level teaching, first textbooks written for new, cutting-edge courses may make their way into *Universitext*.

More information about this series at http://www.springer.com/series/223

Igor Chueshov

Dynamics of Quasi-Stable Dissipative Systems

 Springer

Igor Chueshov
Department of Mechanics and Mathematics
Karazin Kharkov National University
Kharkov, Ukraine

ISSN 0172-5939   ISSN 2191-6675   (electronic)
Universitext
ISBN 978-3-319-22902-7   ISBN 978-3-319-22903-4   (eBook)
DOI 10.1007/978-3-319-22903-4

Library of Congress Control Number: 2015949651

Mathematics Subject Classification (2010): 37L05, 37L30, 35B41, 34K25

Springer Cham Heidelberg New York Dordrecht London

Printed on acid-free paper

Springer International Publishing AG Switzerland is part of Springer Science+Business Media (www.
springer.com)

Preface

The main goal of this book is to present background material and recently developed mathematical methods in the study of infinite-dimensional evolutionary models, taking into account dissipativity and stability properties of various forms and origins.

The main feature of dissipative systems is the presence of an energy reallocation mechanism with decaying in higher modes. This mechanism can lead to the appearance of complicated limit regimes and structures in the system, which are stable in a certain sense. It is commonly recognized that the general theory of dissipative systems was significantly stimulated in the 1980s with attempts to find adequate mathematical models to explain turbulence phenomena. By now, significant progress in the study of infinite-dimensional dissipative dynamics has been made (see, e.g., the monographs BABIN/VISHIK [9], CHEPYZHOV/VISHIK [31], CHUESHOV [39], HALE [116], LADYZHENSKAYA [142], ROBINSON [195], SELL/YOU [206], and TEMAM [216] and the references therein).

The main feature of this book in comparison with the sources mentioned is that we systematically present, develop, and use the quasi-stability method originally designed for second order in time models with nonlinear damping in collaboration with Irena Lasiecka in CHUESHOV/LASIECKA [56, 58] (see also our recent survey [60]). Here we extend this method substantially. New classes of second order evolutions, parabolic-type models, and PDE systems with delay are included for consideration.

We hope that this book will be useful not only to mathematicians interested in the general theory of dynamical systems, but also to physicists and engineers interested in both the mathematical background and methods for the asymptotic analysis of infinite-dimensional dissipative systems that arise in continuum mechanics.

Our presentation is based on general and abstract models and covers several important classes of nonlinear PDEs, which generate infinite-dimensional dissipative systems. These classes include heat and reaction-diffusion models, a wide spectrum of models arising in two-dimensional hydrodynamics for studying turbulence phenomena and plate and wave models with nonlinear state-dependent damping. We also consider the nonlinearly damped wave Kirchhoff model and some classes of parabolic and hyperbolic delay problems.

Much of the analysis in this book is devoted to the stability of dynamics and a rigorous reduction of infinite-dimensional systems to some finite-dimensional structures, which are described only by finitely many degrees of freedom. These finite-dimensional structures should be of interest to application-oriented scientists, who pursue the mathematical simulation of real infinite-dimensional phenomena.

The book contains a large number of exercises. As in the famous monograph by Dan Henry [123], they are an integral part of the book. Most of them are placed strategically within the text, rather than at the end of a section. Some of the exercises are routine, while others are general comments and remarks written in "exercise form." This allows us to make the narrative shorter and avoid extra refinement.

The book can be used as a textbook for courses in dissipative dynamics at the graduate level. It is sufficient to know the basic concepts and facts from functional analysis and ordinary differential equations to understand this book. In fact, many parts of the book were already used in advanced undergraduate and beginning graduate courses given by the author at the Kharkov University.

Acknowledgments

It is a great pleasure to express my thanks to all my colleagues who have contributed to my understanding of the nature of dissipative dynamics. My warmest thanks are due to Irena Lasiecka for numerous very stimulating discussions on nonlinear PDE dynamics and for enjoyable collaborations. I thank Alexander Rezounenko for his comments and fruitful discussions on state-dependent delay models. I am grateful to my son Gennadiy for his generous help in plotting all of the illustrations and to my wife, Galina, for her permanent encouragement of my work on this book. I am also indebted to the Springer editorial staff (in particular, to Donna Chernyk) for their interest and much appreciated assistance with this project. During its realization starting from the very first draft of Spring 2014 I received 11 (eleven!) reviews from quite a few anonymous reviewers. I thank all of them for their valuable comments and suggestions.

Kharkov, Ukraine   Igor Chueshov
June 2015

Contents

Introduction

The general theory of dynamical systems originated from the qualitative theory of ordinary differential equations, the foundations of which were laid by H. Poincaré (1854–1912) and A.M. Lyapunov (1857–1918). A very important contribution to the theory was made by G.D. Birkhoff (1884–1944). He was an inventor of the term "dynamical system" and he developed the theory at the abstract level using topological methods to a great extent. The notion of a dynamical system is the mathematical formalization of the general scientific concept of evolution (time-dependent) processes. These processes can be of quite different natures. Dynamical systems naturally arise in the study of many physical, chemical, biological, ecological, economical, and even social phenomena. The notion of a dynamical system includes a set of its possible states (state space) and a law of the state evolution in time. Thus, the term "dynamical system" covers a wide class of models which may describe arbitrary objects evolving in time and also time-dependent processes. For instance, this class of objects and processes includes models generated by nonlinear evolutionary partial differential equations (PDEs) arising in continuum mechanics and mathematical physics. These models require infinite-dimensional spaces for the representation of a variety of possible states. In this book we concentrate on (infinite-dimensional) systems which demonstrate various types of relocation and dissipation of energy. It seems (see, e.g., the discussion in HALE [116], RAUGEL [188], TEMAM [216]) that for the first time these effects were formalized in the paper LEVINSON [150], where the notion of (dynamical) dissipativity was introduced in its modern (mathematical) form; see also BILLOTI/LASALLE [13], CODDINGTON/LEVINSON [75], PLISS [182, 183]. Dissipativity means that the limiting dynamics becomes localized in the phase space. This can be expressed as a statement on the existence of a bounded absorbing set. In the case of systems with a finite number of degrees of freedom this localization allows us to select limiting objects such as attractors, which carry important information concerning the qualitative behavior of the system. The situation is quite different for infinite-dimensional systems. To single out the corresponding limiting regimes we need additional compactness properties of evolutions. This makes the theory much more complicated in infinite dimensions.

Nevertheless, by now several important aspects of the theory of infinite-dimensional systems have been developed with a concentration on different classes of PDE models (see, e.g., the monographs BABIN/VISHIK [9], CHEPYZHOV/VISHIK [31], CHUESHOV [39], CHUESHOV/LASIECKA [56, 58], HALE [116], LADYZHENSKAYA [142], ROBINSON [195], SELL/YOU [206], TEMAM [216] and the surveys BABIN [7], MIRANVILLE/ZELIK [166], RAUGEL [188]).

This book focuses on the dynamics of infinite-dimensional dissipative systems. In order to achieve a reasonable level of generality, our consideration is fairly abstract and tuned to general classes of evolutions, which are defined on abstract spaces. Our aim is to present general methods and abstract results pertaining to fundamental dynamical system properties related to long-time behavior. Our main tool is based on quasi-stability properties of the corresponding dissipative system. Roughly speaking, the quasi-stability means that we are able to control the divergence of two trajectories by decomposing their difference into convergent and compact parts.

The main features of the book (comparative to other sources) are the following:

- We present, develop, and illustrate an approach to the compactness of dynamics which is based on the relatively recent observation made in KHANMAMEDOV [134] (see also CHUESHOV/LASIECKA [56, 58]) and has proved to be very useful for the study of problems with critical nonlinearities. This approach appeared as a method of compensated compactness by means of potential energy for the second order in time evolution equations and was already applied in many situations. In fact, this approach represents a weak form of quasi-stability.

- To study problems related to the finite dimensionality of attractors and their smoothness properties, we suggest and develop a new version of the quasi-stability method originally introduced in CHUESHOV/LASIECKA [51] (see also CHUESHOV/LASIECKA [56, 58] and the recent survey CHUESHOV/LASIECKA [60]) for some classes of evolution equations of second order in time. The main advantage of this method is minimal requirements concerning the initial smoothness of the dynamics.

- In our presentation we are strongly oriented on the application of the theory to infinite-dimensional systems, which have their roots in continuum mechanics and mathematical physics. However, to make the abstract schemes more transparent and to present different possible scenarios of complicated behavior we use low-dimensional ODE examples intensively. Some of them are low-mode approximations of the real-world PDE models.

- We provide the basic concepts of the theory of dynamical systems in modern form adapted to the infinite-dimensional (locally noncompact) case. We present some material in the form of exercises. Some of them contain additional information about the objects under consideration. This makes the text more concentrated. In fact, many of the exercises can be changed into plain text by substituting the title "Exercise" with words like "easy to see." However, we prefer to keep them in this "quantized" form and believe that this makes the text more user-friendly for the reader.

The book is organized as follows.

Chapter 1 deals with the basic notation and definitions of the abstract theory of dynamical systems. Here we explain such notions as trajectories and ω-limit sets. We also present several notions of stability (Lagrange, Poisson, and Lyapunov). We discuss possible types of behaviors of individual trajectories such as wandering and nonwandering, recurrent, almost recurrent, and almost periodic motions. We mainly follow the classical sources such as NEMYTSKII/STEPANOV [171] and SIBIRSKY [212]. In this chapter we also present a complete theory of 1D continuous dynamical systems and discuss the Poincaré-Bendixson theory presenting the main types of qualitative dynamics in 2D systems with continuous time. In conclusion, by means of examples we consider elements of bifurcation theory.

Chapters 2 and 3 are central to the book. They pertain to the long-time behavior of (infinite-dimensional) dynamical systems and involve quasi-stability ideas in different forms.

Chapter 2 starts with the foundations of the general theory of dissipative systems. Although in many considerations in this and the next chapters we deal with both discrete and continuous time systems, our main point of interest is continuous systems with infinite-dimensional phase spaces. First we introduce several notions of dissipativity and present a certain useful and rather general criterion for dissipativity. The next topic is asymptotic compactness. Analysis of the existing literature shows that there are two popular equivalent forms of this notion, which are important from the point of view of the PDE applications. One of them is due to Olga Ladyzhenskaya, and the second one was suggested and explored by Jack Hale. Relying on the properties of the Kuratowski measure of noncompactness, we suggest several convenient criteria for asymptotic compactness which demonstrate some weak forms of quasi-stability. Then we present the central result of this chapter and the whole theory of dissipative systems. It is a theorem stating that dissipativity and asymptotic compactness are necessary and sufficient conditions for the existence of a compact global attractor. We also discuss various forms of stability of global attractors and the reduction principle, which states the possibility to reduce dynamics to smaller phase spaces. The chapter concludes with the considerations devoted to a rather wide and important class of gradient systems. This class assumes the existence of a Lyapunov-type function on the phase space. The main features of these systems are the possibilities (i) to avoid the dissipativity property in explicit form in the proof of the existence of a global attractor, and (ii) to describe a structure of the attractor via unstable manifolds.

In **Chapter 3** we continue to present the theory of dissipative systems, concentrating on finite-dimensional behavior in the infinite-dimensional case. Here we introduce and discuss the notions of Hausdorff and fractal (box-counting) dimensions. The main result concerning dimension which we present in the book is a generalization of the celebrated Ladyzhenskaya theorem on dimension of invariant sets for (locally) Lipschitz mappings in Banach spaces. We compare this approach with the volume contraction method (see, e.g., BABIN/VISHIK [9], CHEPYZHOV/VISHIK, [31], TEMAM [216]), which requires C^1 smoothness of evolutions. It is known from examples in CHUESHOV/LASIECKA [56] that this gap between Lipschitz continuity and C^1 smoothness can be critical for several classes of systems. The version of the generalization presented is also an extension of schemes suggested earlier in CHUESHOV/LASIECKA [56, 58] and proved to be useful in the case of models with critical nonlinearities. Motivated by this generalization and also by ideas presented in CHUESHOV/LASIECKA [56, 58], we introduce the notion of a quasi-stable system in a more general form than the one that was discussed in CHUESHOV/LASIECKA [58]. This new version of quasi-stability allows us to cover not only the second order in time models (as in CHUESHOV/LASIECKA [58]) but also several types of parabolic systems. Delay perturbations are also included in the scheme. The notion of quasi-stability is rather natural from the point of view of long-time behavior. It pertains to decomposition of the flow into exponentially stable and compact parts. However, in contrast with the standard "splitting" method (see BABIN/VISHIK [9] or TEMAM [216]), the quasi-stable decomposition refers to a difference of two trajectories and is related to different forms of dynamical squeezing. For quasi-stable systems we show (i) the existence of compact global attractors (with the help of the new asymptotic compactness criterion presented in the previous chapter), and (ii) finiteness of the fractal dimension of these attractors (relying on our extension of the Ladyzhenskaya-type result). For these systems it is also possible to establish the existence of fractal exponential attractors, which are finite-dimensional forward invariant sets attracting trajectories with an exponential speed. These objects were introduced in the 1990s (see EDEN ET AL. [92] and the references therein). We also discuss other consequences of quasi-stability, such as determining functionals and regularity properties of trajectories from the attractor. Thus, quasi-stability provides important tools, which automatically deliver a set of important properties of long-time dynamics.

The rest of the book, **Chapters 4–6**, is devoted to applications of the general abstract theory presented in Chapters 2 and 3 and demonstrates capabilities of the quasi-stability method. These applications deal with different classes of evolution equations of the form

$$u_t = F(u), \quad t > 0, \quad u\big|_{t=0} = u_0, \tag{*}$$

in a Hilbert space X. These classes include parabolic- and hyperbolic-type models and also their delay perturbations. The main goal is to demonstrate how the general theory developed can be implemented in the studies of particular systems of the form (*). Usually this kind of implementation follows some standard scheme and requires a realization of several steps.

- *Step 1: Generation.* We need to check whether the equation in (*) generates a dynamical system. For this we need to prove global well-posedness of the Cauchy problem in (*), i.e., to establish the existence and uniqueness statement and show continuous dependence of solutions on initial data u_0. This makes it possible to introduce the evolution operator S_t which maps u_0 in the solution taken at moment t. Thus, we can construct a dynamical system with the phase space X and make an attempt to apply the general theory.

- *Step 2: Basic qualitative properties.* This step usually assumes answers on the questions about dissipativity and related energy balance equalities. These equalities are usually the main tool in the proof of dissipativity. In the case of gradient systems an explicit form of dissipativity can be avoided. Instead the set $\mathcal{N} = \{u \in X : F(u) = 0\}$ of stationary (time-independent) solutions should be studied.
- *Step 3: Long-time dynamics.* The main goal is to establish the existence of a global attractor. For this we need asymptotic compactness, which can be established by application of either smoothening properties of the evolution operator (in the case of parabolic-type models) or the compensated compactness method mentioned above (in the case of second order in time equations). Another way is to show that the system is quasi-stable, even in some weak form.
- *Step 4: Other features of asymptotic behavior.* This step includes issues related to the finite dimension of the attractor and its other similar properties. Studies of exponential attractors and determining functionals also provide important information about the long-time dynamics. All related results can be obtained in one shot in the case when we manage to show that the system is quasi-stable. Qualitative model-dependent methods can also be applied at this stage.

We demonstrate all these steps for several types of models in Chapters 4–6.

Chapter 4 deals with abstract evolution models of parabolic type. We discuss two types of models here. The first one is a general abstract parabolic-type equation which models reaction-diffusion processes. First we prove several (local and global) well-posedness statements, which rely on the notion of a mild solution and are motivated by perturbation-type results presented in PAZY [181]. Next we deal with dissipativity and compactness. It is remarkable that for this class of models dissipativity implies the compactness of the system (i.e., the existence of a compact absorbing set). Then using the standard multipliers technique we establish what is called the Ladyzhenskaya squeezing property. This property allows us to show that the system is quasi-stable and thus to apply all the techniques presented in Chapter 3. The squeezing property also provides an approach to study the data assimilation problem for the parabolic models considered. This problem is a question on how to incorporate available observation data in computational schemes to improve the quality of the future evolution predictions of the corresponding dynamical system. This problem has a long history and was studied by many authors at different levels (see the references in Chapter 4).

A similar program is realized for a certain abstract class of models which are motivated by 2D hydrodynamics. In contrast with the first type of models, the studies of this class are based on the notion of a weak solution of variational type and involve the Galerkin method. This class contains a wide variety of 2D hydro-dynamical models, including the Navier-Stokes equations, magnetohydrodynamic (MHD) equations, the Boussinesq model for Bénard convection, the 2D magnetic Bénard problem, and also some models of turbulence.

Chapter 5 specializes in the direction of second order systems. It develops and applies material presented in Chapters 2 and 3 for this particular type of system.

We deal with a second order in time equation with damping and source terms of different structure whose abstract form is the following Cauchy problem in a separable Hilbert space:

$$u_{tt} + K(u)u_t + Au + B(u) = 0, \quad t > 0; \quad u|_{t=0} = u_0, \quad u_t|_{t=0} = u_1. \qquad (**)$$

This model represents nonlinear wave dynamics with the damping (operator) coefficient $K(u)$ which depends on displacement u (but not on velocity u_t). Formally, we can rewrite the equation in (**) as a first order equation of the form (*) and apply the scheme described above. The main achievements in this chapter deal with well-posedness and the existence of a compact finite-dimensional attractor for different situations. We prove that the corresponding system is asymptotically quasi-stable, and then we apply general theorems on properties of quasi-stable systems. This allows us to establish the existence of a fractal exponential attractor and give the conditions that guarantee the existence of a finite number of determining functionals. In the case when the set of equilibria is finite and hyperbolic, we show that every trajectory is attracted by some equilibrium at an exponential rate. By means of an example we also consider the case when the main elliptic part A is nonlinear. The motivation for this is the Kirchhoff wave equation in a bounded domain in $\mathbb{R}^d$ which demonstrates the interplay of a nonlinear state-dependent nonlocal damping and nonlinear stiffness. Again, the main method is quasi-stability.

In **Chapter 6** we consider the qualitative dynamics of abstract evolution equations containing delay terms. We start with delay perturbations of the parabolic-type models considered in Chapter 4. The corresponding perturbations are Lipschitz. For these equations we prove well-posedness and study the long-time dynamics. We also consider a parabolic problem with a singular delay term. This allows us to include some population dynamics models with state-dependent delays in the scope of the theory developed. Then we deal with a class of second order in time nonlinear evolution equations with state-dependent delays. This class covers several important PDE models arising in the theory of nonlinear plates. We prove well-posedness in a certain space of functions which are C^1 in time. The solutions constructed generate a dynamical system in a C^1-type space over a delay time interval. The main result shows that this dynamical system possesses compact global and exponential attractors of finite fractal dimension. To obtain this result we adapt the developed method of quasi-stability estimates.

The **Appendix** provides necessary background and preliminary material used throughout the book. In particular, here we describe basic properties of vector-valued function spaces and quote several compactness theorems for them. We also present some extensions of the standard Gronwall inequality and provide some material on calculus in infinite-dimensional spaces. We discuss several important issues related to ordinary differential equations and the Orlicz-type result of genericity of uniqueness in the case of abstract parabolic models. The monotonicity method for 2D hydrodynamical models is also presented here.

As we mentioned, the book can be used by beginners wanting first to learn the *basic* theory of dissipative systems and also by specialists wanting to prepare a lecture for graduate students. In this case we would recommend at the first stage of reading (or teaching) to restrict yourselves to the first four sections of Chapter 1 with reference to the examples in Sections 1.7-1.9 to get an idea of how the dynamics may depend on the dimension of the system. We then recommend switching to Chapter 2. In this chapter some material can also be omitted at the first reading (Subsections 2.2.3 and 2.3.4, for instance). Next we recommend reading the first two subsections in Section 3.1 and then switching to Section 3.4 on quasi-stability to complete the basic general theory portion. For a basic overview of applications we recommend concentrating on Sections 4.1, 4.2 and on the first part of Section 4.3. Then one could go to Section 6.1 on delay systems. Another possibility is, after reading Section 4.1 on operators with a discrete spectrum, to then switch to the second order in time models considered in the first two sections of Chapter 5. Alternatively, after Section 4.1, the readers can also switch to the 2D hydrodynamical models presented in the second part of Chapter 4.

Chapter 1
Basic Concepts

This chapter collects basic definitions, notions and also the simplest illustrating statements from the general theory of dynamical systems. We also describe all possible dynamical scenarios in 1D and 2D continuous systems and, by means of examples, discuss the principal bifurcation pictures. Our intention in the latter materials is to give the reader some feeling on what kind of dynamics can arise for low-dimensional (1 or 2) continuous time evolutions.

We mainly follow the presentation given in NEMYTSKII/STEPANOV [171] and SIBIRSKY [212] and also rely on the classical ODE sources; see CODDINGTON/ LEVINSON [75], HARTMAN [120], LEFSCHETZ [148] and also BAUTIN /LEONTOVICH [11], REISSING/SANSONE/CONTI [189]).

1.1 Evolution operators and dynamical systems

As already mentioned in the Introduction, the notion of a dynamical system includes a set of its possible states (state space) and a law of the evolution of the state in time. Below we take a complete metric space X as a set of possible states. We denote by $\mathbb{T}_+$ all non-negative elements on $\mathbb{T}$, where $\mathbb{T}$ is either $\mathbb{R}$ or $\mathbb{Z}$ and represents the time.

Definition 1.1.1. A family $\{S_t\}_{t \in \mathbb{T}_+}$ of continuous mappings of X into itself is said to be an *evolution operator* (or evolution semigroup, or semiflow) if it satisfies the semigroup property:

$$S_0 = Id, \quad S_{t+\tau} = S_t \circ S_\tau \quad \text{for all } t, \tau \geq 0.$$

In the case when $\mathbb{T} = \mathbb{R}$ we assume in addition that the mapping $t \mapsto S_t x$ is continuous from $\mathbb{R}_+$ into X for every $x \in X$. The pair (X, S_t) is said to be a *dynamical system* with the phase (or state) space X and the evolution operator S_t.

© Springer International Publishing Switzerland 2015
I. Chueshov, *Dynamics of Quasi-Stable Dissipative Systems*, Universitext,
DOI 10.1007/978-3-319-22903-4_1

If $\mathbb{T} = \mathbb{Z}$, then the evolution operator (and dynamical system) is called discrete (or with discrete time). If $\mathbb{T} = \mathbb{R}$, then S_t (resp. (X, S_t)) is called an evolution operator (resp. dynamical system) with continuous time. If a notion of dimension can be defined for the phase space X (e.g., if X is a linear space), the value $\dim X$ is called a dimension of the dynamical system.   ∎

The following examples illustrate Definition 1.1.1.

Example 1.1.2 (Ordinary differential equations). Let $F : \mathbb{R}^d \mapsto \mathbb{R}^d$ be a (nonlinear) mapping. Consider the equation

$$\frac{du(t)}{dt} = F(u(t)), \quad t \geq 0, \quad u(0) = u_0 \in \mathbb{R}^d. \tag{1.1.1}$$

If this problem has a unique solution for every initial data $u_0 \in \mathbb{R}^d$ which continuously depends on u_0, then it generates an evolution semigroup S_t in $X = \mathbb{R}^d$ by the formula $S_t u_0 = u(t, u_0)$, where $u(t, u_0)$ is the solution to problem (1.1.1). Thus, we have a dynamical system (X, S_t) with the phase space $X = \mathbb{R}^d$.   ∎

Example 1.1.3 (Mappings). Let X be a complete metric space. Consider a mapping $F : X \mapsto X$. Let $n \in \mathbb{Z}_+$. Then the n-fold composition $S_n \equiv F \circ \cdots \circ F$ of the mapping F provides us with an evolution family. If the mapping F is continuous, then we obtain a discrete time dynamical system (X, S_n). Therefore, the pair (X, F) completely determinates this (discrete time) dynamical system. This is why a pair (X, F) consisting of the space X and the (one-step) mapping F is also often called a dynamical system.   ∎

The following example shows how a single mapping can generate a dynamical system with continuous time.

Example 1.1.4 (Continuous time systems from mappings). As in the previous example, let X be a complete metric space and $F : X \mapsto X$ be a continuous mapping. Consider the difference equation with continuous argument

$$u(t + 1) = F(u(t)), \quad t \in \mathbb{R}_+.$$

Any solution of this equation can be easily constructed from data $\phi(\xi)$ defined on $[0, 1]$ by the formula

$$u(t) = S^n(\phi(t - n)), \quad n \leq t < n + 1, \quad n \in \mathbb{Z}_+,$$

where $S_n \equiv F \circ \cdots \circ F$. This function u is continuous on $\mathbb{R}_+$ when

$$\phi \in Y = \{\phi \in C([0, 1], X) : \phi(1) = F(\phi(0))\},$$

where $C([0, 1], X)$ is the space of continuous functions on $[0, 1]$ with values in X. Now we can define a continuous time evolution operator in Y by the formula

$$S_t : \phi(\xi) \mapsto F^{[t+\xi]}(\phi(\{t + \xi\})), \quad \xi \in [0, 1], \quad t \in \mathbb{R}_+,$$

where $[\xi]$ is the integer part of ξ and $\{\xi\}$ is its fractional part. Thus, we arrive at a continuous time system (Y, S_t). This kind of system (mainly in the case when X is an interval in $\mathbb{R}$) was intensively studied in SHARKOVSKY/MAISTRENKO/ROMA-NENKO [209]; see also SHARKOVSKY ET AL. [208]. Features of the dynamics in the system (Y, S_t) provide a motivation for the recently introduced and developed notion of *ideal turbulence*; see SHARKOVSKY [207]. ∎

Example 1.1.5 (Bebutov dynamical system). Let $X = C(\mathbb{R})$ be the space of all continuous functions on $\mathbb{R}$ equipped with the Bebutov metric:

$$\operatorname{dist}(\psi, \phi) = \sup_{r > 0} \min \left\{ \sup_{|x| \le r} |\psi(x) - \phi(x)|, \ \frac{1}{r} \right\}.$$

In this case X becomes a complete metric space, and convergence with respect to this metric is equivalent to uniform convergence on bounded sets (see, e.g., SIBIRSKY [212]). As an evolution operator S_t we take the left shift operator

$$(S_t f)(x) = f(x + t), \ f \in X, \ t \ge 0.$$

This system (X, S_t) is called the Bebutov (shift) dynamical system. It is convenient to demonstrate different types of dynamics of individual trajectories with the help of this system (see, e.g., NEMYTSKII/STEPANOV [171], SIBIRSKY [212] and the references therein). ∎

Remark 1.1.6 (Closed evolutions[1]). Many general dynamical properties can be established without assuming the continuity of evolution operators S_t. This can be important in the study of some infinite-dimensional PDE models. Instead of continuity, following PATA/ZELIK [179] we can assume that evolution operators S_t are *closed* in the corresponding space X. This means that for every $t > 0$ the properties $x_n \to x$ and $S_t x_n \to y$ for some $x, y \in X$ as $n \to \infty$ imply that $S_t x = y$. It is clear that continuity of mappings S_t implies their closeness. However, the inverse statement is valid under some additional conditions only. Namely, one can show that if S_t is closed and maps any compact set into a relatively compact set, then $x \mapsto S_t x$ is continuous. Indeed, let $x_n \to x$ as $n \to \infty$. For every $t > 0$ we can choose a subsequence $\{n_m\}$ such that $S_t x_{n_m} \to y$ for some $y \in X$. By the closeness of S_t this implies that $S_t x = y$. Moreover, one can see that the sequence $\{S_t x_n\}$ cannot have other limiting points except $y = S_t x$. This means continuity of S_t. On the other hand, the mapping $f : \mathbb{R}_+ \mapsto \mathbb{R}_+$ given by the formula

$$f(x) = \begin{cases} (1 - x)^{-1} & \text{if } 0 \le x < 1; \\ x & \text{if } x \ge 1, \end{cases}$$

---

[1] We recommend omitting of this remark at the first reading.

provides us with a closed evolution operator which is not continuous (for other examples we refer to PATA/ZELIK [179]). Closed evolutions also arise in the study of the long-time dynamics of a class of PDE systems with state-dependent delay (see Section 6.2 in Chapter 6).

We also note that operator closedness is a well-known concept in the theory of linear (unbounded) operators, see, e.g., DUNFORD/SCHWARTZ [88, Chapter 2] or YOSIDA [229, Chapter 2]. To our best knowledge, in the context of evolution operators this notion appeared in BABIN/VISHIK [9] as a (weak) closedness of an evolution (strongly continuous) semigroup (see also CHUESHOV [39] and Theorem 2.3.18 below) and in PATA/ZELIK [179] for the general case.   ∎

In the study of qualitative behavior a notion of equivalence of dynamical systems plays an important role. This equivalence relation allows us to divide wide collections of dynamical systems into classes of systems with very similar behaviors.

Definition 1.1.7 (Topological equivalence). Two dynamical systems (X, S_t) and $(\tilde{X}, \tilde{S}_t)$ are said to be *topologically equivalent* (or isomorphic) if there exists a homeomorphism h from X onto $\tilde{X}$ such that $h(S_t x) = \tilde{S}_t h(x)$ for all $x \in X$ and $t \in \mathbb{T}_+$. In this case the evolution operators S_t and $\tilde{S}_t$ are called *topologically conjugate*.   ∎

The following exercise illustrates this definition.

Exercise 1.1.8. Let $\alpha, \beta > 0$ and $\alpha, \beta \neq 1$. Then two discrete systems $(\mathbb{R}_+, \alpha x)$ and $(\mathbb{R}_+, \beta x)$ are topologically equivalent if and only if either $\{\alpha, \beta > 1\}$ or else $\{\alpha, \beta < 1\}$. Hint: To prove the sufficient part look for a homeomorphism h of $\mathbb{R}_+$ of the form $h(x) = x^\gamma$ with some $\gamma > 0$; the necessary part can be proved by the contradiction argument.   ∎

1.2 Trajectories, invariant sets, and equilibria

Now we recall several well-known notions from the theory of dynamical systems (see, e.g., BABIN/VISHIK [9], CHUESHOV [39], NEMYTSKII/STEPANOV [171], SIBIRSKY [212], TEMAM [216] and the references cited in these monographs).

Let S_t be an evolution semigroup in X. A set $D \subset X$ is said to be *forward* (or positively) *invariant* (with respect to S_t) if $S_t D \subseteq D$ for all $t \geq 0$. It is *backward* (or negatively) *invariant* if $S_t D \supseteq D$ for all $t \geq 0$. The set D is said to be *invariant* (or strictly invariant) if it is both forward and backward invariant; that is, $S_t D = D$ for all $t \geq 0$.

Some properties of invariant sets are listed in the following exercise.

Exercise 1.2.1. Prove the following statements.

(A) The union of an arbitrary collection of forward invariant sets is also forward invariant (the same is true concerning backward and strict invariance).

(B) The nonempty intersection of an arbitrary collection of forward invariant sets is also forward invariant. Show by means of examples that this cannot be true in general for backward invariant sets. Hint: Consider the discrete dynamical system (X, f) with $X = [0, 1]$ and

$$f(x) = \begin{cases} 3x & \text{if } 0 \leq x < 1/3; \\ 1 & \text{if } 1/3 \leq x < 2/3; \\ 3(1-x) & \text{if } 2/3 \leq x \leq 1, \end{cases}$$

Show that $f(A_1 \cap A_2) \not\supseteq A_1 \cap A_2$ when $A_1 = [0, 2/3]$ and $A_2 = [1/3, 1]$.

(C) Let S_t be surjective, i.e., $S_t X = X$ for every $t > 0$. If B is a forward invariant set, then the complement $X \setminus B$ is backward invariant.

(D) If S_t is injective, i.e., for each t the equality $S_t x = S_t y$ implies $x = y$, the complement $X \setminus B$ is a forward invariant set for every backward invariant set B.

(E) Let S_t be a one-to-one mapping. Then the complement $X \setminus B$ of every invariant set B is also invariant.

(F) If B is forward invariant, then the closure $\overline{B}$ of B is also forward invariant. The same is true for backward invariance if we assume that the closure $\overline{B}$ of B is a compact set. Make sure that the compactness of $\overline{B}$ is essential for its backward (and strict) invariance. Hint: Consider the discrete system $(\mathbb{R}_+, f)$ with $f(x) = (1+x)^{-1} + x \sin^2 x$ and show that $f(\mathbb{R}_+) = f(\text{int } \mathbb{R}_+) = \text{int } \mathbb{R}_+$, where $\text{int } \mathbb{R}_+ = \{x \in \mathbb{R}_+ : x > 0\}$.

(G) Let S_t and $\tilde{S}_t$ be two topologically conjugate semiflows in X and $\tilde{X}$. Let $h : X \mapsto \tilde{X}$ be the corresponding homeomorphism. Then a set D is invariant (resp. forward or backward invariant) if and only if $\tilde{D} = h(D)$ is invariant (resp. forward or backward invariant).

∎

Let S_t be an evolution operator in X. For any $D \subset X$ we denote by

$$\gamma_D^t \equiv \bigcup_{\tau \geq t} S_\tau D$$

the *tail* (from the moment t) of the trajectories emanating from D. It is clear that $\gamma_D^t = \gamma_{S,D}^0 \equiv \gamma_{S,D}^+$. If $D = \{v\}$ is a single point set, then $\gamma_v^+ \equiv \gamma_v^0$ is said to be a *positive semitrajectory (or semiorbit)* emanating from v. A curve $\gamma \equiv \{u(t) : t \in \mathbb{T}\}$ in X is said to be a *full trajectory* iff $S_t u(\tau) = u(t + \tau)$ for any $\tau \in \mathbb{T}$ and $t \geq 0$. For every $v \in X$ there exists a positive semitrajectory which contains v. Since S_t is not necessarily an invertible operator, this is not true for a full trajectory. Positive semitrajectories are forward invariant sets. Full trajectories are invariant sets. The set $\gamma^{\tau_1, \tau_2} = \{u(t) : \tau_1 \leq t \leq \tau_2\}$ is called a *segment* (or a piece) with time interval $[\tau_1, \tau_2]$ of the trajectory γ. A trajectory $\gamma = \{u(t) : t \in \mathbb{T}\}$ is called a *periodic* trajectory (or periodic orbit, or a cycle) if there exists $T > 0$, such that

$u(t + T) = u(t)$ for all $t \in \mathbb{T}$. In this case any point on γ is called *periodic* (or *T*-periodic). The minimal positive number T possessing this property is called a period of a trajectory. An element $v_0 \in X$ is called a *fixed point* of an evolution operator if $S_t v_0 = v_0$ for all $t \geq 0$ (synonyms: equilibrium, stationary point, rest point).

Exercise 1.2.2. Let S_t and $\tilde{S}_t$ be two topologically conjugate semiflows in X and $\tilde{X}$. Let $h : X \mapsto \tilde{X}$ be the corresponding homeomorphism. Show that

(A) A point v is fixed for S_t if and only if $h(v)$ is a fixed point for $\tilde{S}_t$.
(B) γ is a periodic orbit for S_t if and only if $h(\gamma)$ is a periodic orbit for $\tilde{S}_t$ with the same period.

∎

Exercise 1.2.3. Prove that the set of all fixed points is closed.   ∎

Exercise 1.2.4. If there exists the limit $v = \lim_{t \to +\infty} S_t w$ for some $w \in X$, then v is a fixed point. Thus, semitrajectories can converge to fixed points only.   ∎

The Schauder fixed point theorem[2] makes it possible to prove the following assertion on the existence of a fixed point.

Theorem 1.2.5. *Let (X, S_t) be a continuous dynamical system on some complete metric space X such that the mapping $(t; x) \mapsto S_t x$ is continuous. Then every forward invariant set M which is homeomorphic to a compact convex set in some Banach space contains a fixed point.*

Proof. We use the same argument as in SIBIRSKY [212].

Let $\{t_n\}$ be a sequence of positive numbers such that $t_n \to 0$ and h be a homeomorphism which maps M onto $\tilde{M} = h(M)$ which is a convex compact set in some Banach space. Let

$$R_n \equiv h \circ S_{t_n} \circ h^{-1} : \tilde{M} \mapsto \tilde{M}$$

By the Schauder theorem there exists $y_n \in \tilde{M}$ such that $R_n y_n = y_n$. The sequence $\{y_n\}$ is compact. This allows us to find an element $x_* \in M$ and sequences $\{x_n\} \subset X$ and $\{\tau_n > 0\} \subset \mathbb{R}$ such that

$$S_{\tau_n} x_n = x_n, \ n = 1, 2, \ldots, \quad \text{and} \quad x_n \to x_*, \ \tau_n \to 0 \ \text{as} \ n \to \infty. \tag{1.2.1}$$

By continuity of $S_t x$ with respect to $(t; x)$ we have that for every $\varepsilon > 0$ there exist $t_\varepsilon > 0$ and $n_\varepsilon > 0$ such that

$$\text{dist}\,(S_t x_n, x_*) < \varepsilon \ \text{for all} \ t \in [0, t_\varepsilon], \ n \geq n_\varepsilon.$$

---

[2]The Schauder theorem (see, e.g., ZEIDLER [231, Volume I, Chapter 2]) states that any continuous mapping from a convex compact set in a Banach space into itself has a fixed point.

Since $\tau_n \to 0$, this yields

$$\text{dist}\,(S_t x_n, x_*) < \varepsilon \ \text{ for all } \ t \in [0, \tau_n], \ n \geq \bar{n}_\varepsilon.$$

Since x_n is a τ_n-periodic point, this implies that the relation above is valid for all $t \in \mathbb{R}_+$, i.e.,

$$\text{dist}\,(S_t x_n, x_*) < \varepsilon \ \text{ for } \ t \in \mathbb{R}_+, \ n \geq \bar{n}_\varepsilon.$$

In the limit $n \to \infty$ we obtain that

$$\text{dist}\,(S_t x_*, x_*) < \varepsilon \ \text{ for every } \ t \in \mathbb{R}_+ \text{ and } \varepsilon > 0,$$

which implies that $S_t x_* = x_*$ for all $t \in \mathbb{R}_+$ and thus completes the proof. $\qquad \square$

We apply Theorem 1.2.5 in Section 1.8 to prove a version of the Poincaré-Bendixson theorem for 2D dynamical systems.

Remark 1.2.6. Under conditions of Theorem 1.2.5 the relations in (1.2.1) mean that every vicinity of a point $x_* \in X$ contains periodic points for (X, S_t) with arbitrary small periods. Exactly this property allowed us to show that x_* is an equilibrium for this system. However, bear in mind that we cannot guarantee that the periods τ_n in (1.2.1) are minimal. Thus, we do not know whether periodic points with arbitrary small periods arise. Moreover, it is known from YORKE [228] (see also Theorem 1.8.8 below) that there are some restrictions from below on possible periods of solutions to finite-dimensional autonomous ODEs with smooth nonlinearities. We also refer to ROBINSON/VIDAL-LÓPEZ [197, 198] (see also ROBINSON [196]) for similar restrictions in the case of parabolic PDE models. $\qquad \blacksquare$

1.3 Omega-limit sets

To describe asymptotic behavior it is convenient (see, e.g., BIRKHOFF [14], LEFSCHETZ [148], NEMYTSKII/STEPANOV [171]) to use the concept of an ω-limit set. The set

$$\omega(D) \equiv \bigcap_{t>0} \overline{\gamma_D^t} = \bigcap_{t>0} \overline{\bigcup_{\tau \geq t} S_\tau D} \qquad (1.3.1)$$

is called the *ω-limit set* of the trajectories emanating from D (the bar over a set means the closure).

Exercise 1.3.1. If $\omega(D) \neq \emptyset$, then $\omega(D)$ is closed and $\omega(S_t D) = \omega(D)$ for every $t > 0$. $\qquad \blacksquare$

Exercise 1.3.2. If v is a fixed point, then $\omega(v) = \{v\}$; if γ is a periodic orbit, then $\omega(\gamma) = \gamma$.   ∎

In the following well-known assertion we provide an alternative (to (1.3.1)) description of ω-limit sets.

Proposition 1.3.3. *Let S_t be an evolution operator in a complete metric space X and $D \subset X$. Then $x \in \omega(D)$ if and only if there exist sequences $t_n \to +\infty$ and $x_n \in D$ such that $S_{t_n} x_n \to x$ as $n \to \infty$.*

Proof. If $x \in \omega(D)$, then

$$x \in \overline{\bigcup_{\tau \geq n} S_\tau D} \quad \text{for every } n = 0, 1, \ldots$$

Therefore, there exist $t_n \geq n$ and $x_n \in D$ such that $\mathrm{dist}(x, S_{t_n} x_n) \leq 1/n$ and thus $S_{t_n} x_n \to x$ as $n \to \infty$.

Now we assume that $x = \lim_{n \to \infty} y_n$, where $y_n = S_{t_n} x_n$, for some $t_n \to +\infty$ and $x_n \in D$. It is obvious that

$$y_n \in \bigcup_{\tau \geq t} S_\tau D \subset \overline{\bigcup_{\tau \geq t} S_\tau D} \quad \text{for all } n \geq n_0,$$

where n_0 is defined such that $t_n \geq t$ for all $n \geq n_0$. Therefore,

$$x = \lim_{n \to \infty} y_n \in \overline{\bigcup_{\tau \geq t} S_\tau D} \quad \text{for all } t \geq 0.$$

Thus, $x \in \omega(D)$.   □

Exercise 1.3.4. Let S_t and $\tilde{S}_t$ be two topologically conjugate semiflows in X and $\tilde{X}$ with the homeomorphism $h : X \mapsto \tilde{X}$. Then $h(\omega(D)) = \omega(h(D))$.   ∎

Exercise 1.3.5. Any ω-limit set (if it exists) is forward invariant.   ∎

If $\gamma = \{u(t) : t \in \mathbb{T}\}$ is a full trajectory, we can define both ω- and α-limit sets of γ by the formulas

$$\omega(\gamma) = \bigcap_{t>0} \overline{\bigcup \{u(\tau) : \tau \geq t\}} \quad \text{and} \quad \alpha(\gamma) = \bigcap_{t<0} \overline{\bigcup \{u(\tau) : \tau \leq t\}}. \quad (1.3.2)$$

Exercise 1.3.6. Let $\gamma = \{u(t) : t \in \mathbb{T}\}$ be a full trajectory. Show that

$$x \in \omega(\gamma) \Leftrightarrow \left\{ x : \exists t_n \to +\infty \text{ such that } x = \lim_{n \to \infty} u(t_n) \right\},$$

$$x \in \alpha(\gamma) \Leftrightarrow \left\{ x : \exists t_n \to -\infty \text{ such that } x = \lim_{n \to \infty} u(t_n) \right\}.$$

The sets $\omega(\gamma)$ and $\alpha(\gamma)$ (if they exist) are forward invariant.   ∎

The following assertion shows that any semitrajectory spends arbitrary large time intervals in an arbitrary neighborhood of any forward invariant subset in $\omega(v)$.

Proposition 1.3.7. *Let (X, S_t) be a dynamical system. Assume that $(t; x) \mapsto S_t x$ is continuous from $\mathbb{R}_+ \times X$ into X (in the case when $\mathbb{T} = \mathbb{R}$). Let $\omega(v) \neq \emptyset$ for some $v \in X$ and A is a forward invariant subset of $\omega(v)$ (the equality $A = \omega(v)$ is allowed). Then for any $\varepsilon > 0$, $T > 0$ and $t_* \geq 0$ there exists $\bar{t} \geq t_*$ such that*

$$S_t v \in \mathcal{O}_\varepsilon(A) \quad \text{for all } t \in [\bar{t}, \bar{t} + T], \tag{1.3.3}$$

where $\mathcal{O}_\varepsilon(A) = \{y \in X : \text{dist}(y, A) < \varepsilon\}$ is the ε-neighborhood of the set A.

Proof. Let $q \in A$. Then by the continuity property of $(t; x) \mapsto S_t x$ for any $\varepsilon > 0$ and $T > 0$ there exists $\delta > 0$ such that

$$\sup\{\text{dist}(S_t x, S_t q) : t \in [0, T]\} \leq \varepsilon \quad \text{provided } \text{dist}(x, q) \leq \delta.$$

Since $q \in \omega(v)$, by Proposition 1.3.3 we can choose arbitrary large $\bar{t}$ such that $\text{dist}(S_{\bar{t}} v, q) \leq \delta$. Thus,

$$\sup\{\text{dist}(S_t v, A) : t \in [\bar{t}, \bar{t} + T]\} \leq \sup\{\text{dist}(S_t S_{\bar{t}} v, S_t q) : t \in [0, T]\} \leq \varepsilon,$$

hence (1.3.3) is valid. $\qquad\square$

Now we present a condition under which a given point from $\omega(v)$ is either fixed or periodic.

Proposition 1.3.8. *Let the hypotheses of Proposition 1.3.7 be in force. Let $w \in \omega(v)$ and $w = \lim_{n \to \infty} S_{t_n} v$ for some sequence $t_n \to +\infty$ (such a sequence exists by Proposition 1.3.3). If this sequence $\{t_n\}$ can be chosen such that the differences $t_{n+1} - t_n$ are uniformly bounded, then the point w is either fixed or periodic.*

Proof. We can assume that the sequence $\{t_n\}$ is increasing and there exist positive k and K such that $k \leq t_{n+1} - t_n \leq K$ for $n = 1, 2 \ldots$. Indeed, if necessary, we can choose the subsequence $\tilde{t}_{n+1} = \min\{t_m : t_m \geq \tilde{t}_n + k\}$ with $\tilde{t}_1 = t_1$. Next we can choose a subsequence $\{n_m\}$ such that $t_{n_m+1} - t_{n_m} \to t_*$ for some $t_* > 0$ when $m \to \infty$. Now using the continuity $(t; x) \mapsto S_t x$ we obtain that

$$w = \lim_{m \to \infty} S_{t_{n_m+1}} v = \lim_{m \to \infty} S_{t_{n_m+1} - t_{n_m}} S_{t_{n_m}} v = S_{t_*} w.$$

This means that w is either fixed or periodic. $\qquad\square$

Below we provide conditions under which $\omega(D)$ is nonempty. In the next section we discuss this issue in the case when D is a single point set.

1.4 Limiting properties of individual trajectories

We start with the following notion (see NEMYTSKII/STEPANOV [171] or SIBIRSKY [212]), which is important not only in the studies of long-time dynamics of individual trajectories but also for the existence of global minimal attractors; see Section 2.3.

Definition 1.4.1 (Lagrange stability). A semitrajectory $\gamma_v^+ = \{S_t v : t \in \mathbb{T}_+\}$ (and its initial point v) is said to be *Lagrange stable* if the closure $\overline{\gamma_v^+}$ of γ_v^+ is compact in X. ∎

Exercise 1.4.2. The set of all Lagrange stable points is forward invariant. ∎

Exercise 1.4.3. Let S_t and $\tilde{S}_t$ be two topologically conjugate semiflows on X and $\tilde{X}$ with the homeomorphism $h : X \mapsto \tilde{X}$. Then a point v is Lagrange stable (with respect to S_t) if and only if $h(v)$ is Lagrange stable (with respect to $\tilde{S}_t$). ∎

Exercise 1.4.4. Let $X = C(\mathbb{R})$ and (X, S_t) be the corresponding Bebutov system (see Example 1.1.5). Show that every bounded uniformly continuous function from $C(\mathbb{R})$ is a Lagrange stable point for the system (X, S_t). ∎

The following assertion contains criteria for Lagrange stability.

Theorem 1.4.5. *A semitrajectory* $\gamma_v^+ = \{S_t v : t \in \mathbb{T}_+\}$ *is Lagrange stable if and only if the following two conditions*[3] *are satisfied:*

(i) *the ω-limit set $\omega(v)$ emanating from v is a nonempty compact set;*
(ii) $\mathrm{dist}_X(S_t v, \omega(v)) \to 0$ *as* $t \to +\infty$.

Proof. Let γ_v^+ be Lagrange stable. Then $\overline{\gamma_v^\tau}$ is a compact set for every $\tau > 0$. Since $\overline{\gamma_v^\tau}$ is a decreasing sequence of compact sets, by (1.3.1) $\omega(v)$ is a nonempty compact set. To prove the convergence property in (ii) we use the contradiction argument. Assume that there exists a sequence $\{t_n \to +\infty\}$ such that

$$\mathrm{dist}_X(S_{t_n} v, \omega(v)) \geq \delta > 0 \quad \text{for all} \quad n = 1, 2, \ldots \tag{1.4.1}$$

By the compactness of $\overline{\gamma_v^+}$ there exist an element $z \in X$ and a subsequence $\{t_{n_m}\}$ such that $S_{t_{n_m}} v \to z$ as $m \to \infty$. Moreover, by Proposition 1.3.3, $z \in \omega(v)$. This contradicts the property in (1.4.1).

Assume now that the conditions in Theorem 1.4.5 are satisfied. To prove Lagrange stability of γ_v^+ we need to show that any sequence of the form $\{S_{t_n} v\}$ contains a convergent subsequence. If the sequence $\{t_n\}$ contains an (infinite) bounded subsequence, then the conclusion follows from the continuity of the evolution operator. Thus, we need to consider the case when $t_n \to \infty$. In this case by

---

[3]If the space X is locally compact, then the second condition can be omitted. See, e.g., SIBIRSKY [212, Theorem 2.8].

(ii) there exists a sequence $\{z_n\}$ in $\omega(v)$ such that $\text{dist}_X(S_{t_n}v, z_n) \to 0$ as $n \to \infty$. By (i) we can choose a subsequence $\{z_{n_m}\}$ which converges to some element $z \in \omega(v)$. It is clear that $S_{t_{n_m}}v \to z$ as $m \to \infty$. Thus, the sequence $\{S_{t_n}v\}$ is relatively compact and hence γ_v^+ is Lagrange stable. $\square$

Proposition 1.4.6. *Let a semitrajectory* $\gamma_v^+ = \{S_t v : t \in \mathbb{T}_+\}$ *be Lagrange stable. Then* $\omega(v)$ *is a (strictly) invariant set.*

Proof. By Exercise 1.3.5, $S_t\omega(v) \subset \omega(v)$. To obtain reverse inclusion we write an element $z \in \omega(v)$ as

$$z = \lim_{n \to \infty} S_{t_n} v = \lim_{n \to \infty} S_t S_{t_n - t} v.$$

By Lagrange stability the sequence $\{S_{t_n - t}v\}$ is relatively compact for each $t > 0$ and thus contains a subsequence $\{S_{t_{n_m} - t}v\}$ such that $S_{t_{n_m} - t}v \to w_t$ for some element $w_t \in X$. It is clear from Proposition 1.3.3 that $w_t \in \omega(v)$. Hence $z = S_t w_t$. Thus $\omega(v) \subset S_t\omega(v)$. $\square$

Theorem 1.4.7. *Let a semitrajectory* $\gamma_v^+ = \{S_t v : t \in \mathbb{T}_+\}$ *be Lagrange stable. Then the* ω-*limit set* $\omega(v)$ *emanating from* v *is connected.*

Proof. By Theorem 1.4.5, $\omega(v)$ is a nonempty compact set. Assume that $\omega(v)$ is not connected, i.e., $\omega(v) = K \cup K_*$, where K and K_* are two nonempty disjoint compact sets such that $\text{dist}(K, K_*) = 2\delta > 0$. Take $k \in K$ and $k_* \in K_*$. By Proposition 1.3.3 there exist sequences $\{t_n\}$ and $\{t_n^*\}$ such that $t_n, t_n^* \to \infty$ and

$$\lim_{n \to \infty} S_{t_n} v = k \quad \text{and} \quad \lim_{n \to \infty} S_{t_n^*} v = k_*.$$

Moreover, we can assume that $t_n < t_n^*$ and also

$$\text{dist}(S_{t_n} v, K) < \delta \quad \text{and} \quad \text{dist}(S_{t_n^*} v, K) > \delta.$$

Since $\phi(t) = \text{dist}(S_t v, K)$ is a continuous function, this implies that there exists $\tau_n \in [t_n, t_n^*]$ such that $\text{dist}(S_{\tau_n} v, K) = \delta$. The Lagrange stability of γ_v^+ implies that the sequence $\{S_{\tau_n} v\}$ is relatively compact; i.e., there exist $\{n_m\}$ and $z \in X$ such that $S_{\tau_{n_m}} v \to z$ as $m \to \infty$. By Proposition 1.3.3, $z \in \omega(v)$. This contradicts the relation $\text{dist}(z, K) = \delta$. $\square$

The following example shows that *without* assuming Lagrange stability the ω-limit set can be non-connected.

Example 1.4.8 (Non-connected limit set). We present an analytic realization of the example given in SIBIRSKY [212, p. 39] in the graphic form. See Figure 1.1.

Let $\alpha, \omega > 0$ and $H(s) = \arctan s$, $s \in \mathbb{R}$. In the strip $X = \{(x; y) : x \in \mathbb{R}, |y| \leq \frac{\pi}{2}\}$ we define an evolution operator by the formula

$$S_t(x_0; y_0) = (x(t, x_0, y_0); x(t, x_0, y_0)),$$

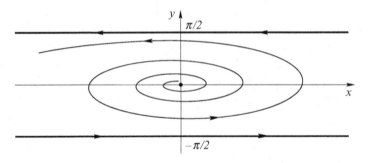

Fig. 1.1 Dynamics near non-connected ω-limit set

where for $|y_0| < \frac{\pi}{2}$ we suppose that

$$x(t, x_0, y_0) = \mathrm{Re}\left[(x_0 + iH^{-1}(y_0)) \exp\{(\alpha + i\omega)t\}\right]$$

and

$$y(t, x_0, y_0) = H\left(\mathrm{Im}\left[(x_0 + iH^{-1}(y_0)) \exp\{(\alpha + i\omega)t\}\right]\right)$$

(here $i = \sqrt{-1}$). In the case $|y_0| = \frac{\pi}{2}$ we take

$$x(t, x_0, y_0) = x_0 - t\,\mathrm{sign}\,y_0, \quad y(t, x_0, y_0) = y_0.$$

One can see that S_t is an evolution operator with continuous semitrajectories. All nonzero semitrajectories are unbounded and thus Lagrange unstable. The calculations based on Proposition 1.3.3 show that

$$\omega(v) = \left\{x \in \mathbb{R}, \ y = -\frac{\pi}{2}\right\} \cup \left\{x \in \mathbb{R}, \ y = \frac{\pi}{2}\right\}$$

for any $v = (x_0; y_0) \neq (0; 0)$ with $|y_0| < \frac{\pi}{2}$. See Figure 1.1.   ∎

Using properties of ω-limit sets it is possible to suggest some classification of individual trajectories and to introduce an important notion of Poisson stability.

Definition 1.4.9 (Poisson stability).   If the set $\omega(v)$ is empty, then the point v and the semitrajectory γ_v^+ are called *departing*. In the case when $\omega(v) \neq \emptyset$ but $\omega(v) \cap \gamma_v^+ = \emptyset$ the point v and the semitrajectory γ_v^+ are called *asymptotic*. If $\omega(v) \cap \gamma_v^+ \neq \emptyset$, then the point v and the semitrajectory γ_v^+ are called *Poisson stable*.   ∎

The simplest examples of Poisson stable trajectories are stationary points and periodic trajectories.

Exercise 1.4.10. Prove the following statements.

(A) The point v and the semitrajectory γ_v^+ are Poisson stable if and only if there exists $t_* \geq 0$ such that $\gamma_v^{t_*} \subset \omega(v)$. If $\omega(v)$ is strictly invariant and the evolution operator possesses the backward uniqueness property ($S_t u = S_t v$ for some $t > 0$ implies $u = v$), then we can take $t_* = 0$.

(B) The set of all Poisson stable points is strictly invariant; i.e., (i) if v is Poisson stable, then $S_t v$ is also Poisson stable for every $t > 0$ and (ii) if $S_t v$ is Poisson stable for some $t > 0$, then v is Poisson stable.

∎

The following theorem gives a topological criterion for the Poisson stability.

Theorem 1.4.11. *Let (X, S_t) be a dynamical system with continuous time. Then the point v and the semitrajectory γ_v^+ are Poisson stable if and only if γ_v^+ is not homeomorphic to the semi-axis $\mathbb{R}_+$.*

Proof. Let γ_v^+ be a Poisson stable semitrajectory which contains neither equilibrium nor periodic orbit. In this case the mapping $t \mapsto \Psi(t) \equiv S_t v \in \gamma_v^+$ is one-to-one and continuous from $\mathbb{R}_+$ on γ_v^+. However, the inverse mapping $\Psi^{-1} : \gamma_v^+ \mapsto \mathbb{R}_+$ is not continuous. Indeed, there exists $t_* \geq 0$ such that $v_* = S_{t_*} v \in \omega(v)$. Therefore, by Proposition 1.3.3 we can find a sequence $t_n \to +\infty$ such that $y_n := S_{t_n} v \to v_*$ as $n \to \infty$. In this case $\Psi^{-1}(y_n) = t_n$ goes to $+\infty$ and not to $\Psi^{-1}(v_*) = t_*$.

To continue let us assume that γ_v^+ is homeomorphic to $\mathbb{R}_+$ and $\Phi : \gamma_v^+ \mapsto \mathbb{R}_+$ is the corresponding homeomorphism. Then the mapping $G = \Phi \circ \Psi$ maps $\mathbb{R}_+$ onto itself and is one-to-one and continuous. Thus, it is strictly increasing. This allows us to show that G is a homeomorphism. Then $\Psi = \Phi^{-1} \circ G$ is also a homeomorphism as it is a composition of two homeomorphisms. Thus, γ_v^+ cannot be Poisson stable.

Now assume that γ_v^+ is not Poisson stable. In this case the mapping $t \mapsto \Psi(t) \equiv S_t v \in \gamma_v^+$ remains one-to-one and continuous from $\mathbb{R}_+$ on γ_+^v. Moreover, the inverse mapping $\Psi^{-1} : \gamma_v^+ \mapsto \mathbb{R}_+$ is also continuous. Indeed, if some sequence $\{x_n\} \subset \gamma_v^+$ converges to some point $x \in \gamma_v^+$, then $t_n = \Psi^{-1}(x_n)$ is bounded (otherwise $x \in \omega(v)$). Let $\tilde{t}$ be a limiting point for $\{t_n\}$. Then it is obvious that $x = S_{\tilde{t}} v$. This means that $\tilde{t} = \Psi^{-1}(x)$. Thus, $x_n \to x$ implies that $\Psi^{-1}(x_n) \to \Psi^{-1}(x)$; i.e., Ψ is a homeomorphism between γ_+^v and $\mathbb{R}_+$. This completes the proof. □

1.5   Recurrent properties of trajectories

We continue the study of qualitative properties of individual trajectories. Our main goal in this section is to show what kinds of scenarios are possible in the dynamics of individual trajectories. The realization of this or other types of recurrent behaviors in a concrete system is not a simple task and lies beyond the general theory. However (see NEMYTSKII/STEPANOV [171], SIBIRSKY [212] and the references therein), all motions described below can be demonstrated in the Bebutov system (see Example 1.1.5).

1.5.1 Wandering and nonwandering points

We introduce the following concept (see BIRKHOFF [14]).

Definition 1.5.1 (Wandering and nonwandering points). Let S_t be an evolution operator on X. A point $v \in X$ and semitrajectory γ_v^+ are said to be *nonwandering* (in X) if for any $t_* \in \mathbb{T}_+$ and any neighborhood $\mathscr{O}(v) \subset X$ of v there exists a moment of time $t \geq t_*$ such that $\mathscr{O}(v) \cap S_t \mathscr{O}(v) \neq \emptyset$. A point $y \in X$ and semitrajectory γ_y^+ are called *wandering* (in X) if they are not nonwandering; i.e. there exists a number $t_* \in \mathbb{T}_+$ and a neighborhood $\mathscr{O}(y) \subset X$ such that $\mathscr{O}(y) \cap S_t \mathscr{O}(y) = \emptyset$ for all $t \geq t_*$. ∎

Exercise 1.5.2. The set of all wandering points is open and thus the set of all nonwandering points is closed. ∎

Proposition 1.5.3. *The set of all nonwandering points is forward invariant.*

Proof. By continuity of S_t for any $\varepsilon > 0$ and $\tau \geq 0$ there exists $\delta = \delta_\varepsilon > 0$ such that $S_\tau \mathscr{O}_\delta(v) \subset \mathscr{O}_\varepsilon(S_\tau v)$, where $\mathscr{O}_\varepsilon(w)$ is the ε-neighborhood of the point w. Therefore

$$\mathscr{O}_\varepsilon(S_\tau v) \cap S_t \mathscr{O}_\varepsilon(S_\tau v) \supset S_\tau \mathscr{O}_\delta(v) \cap S_\tau S_t \mathscr{O}_\delta(v) \supset S_\tau \left[\mathscr{O}_\delta(v) \cap S_t \mathscr{O}_\delta(v) \right].$$

The latter set is not empty provided $\mathscr{O}_\delta(v) \cap S_t \mathscr{O}_\delta(v) \neq \emptyset$. This implies the conclusion. □

Exercise 1.5.4. Let D be a closed forward invariant set in X. The point $v \in D$ is said to be nonwandering in D, if it is nonwandering for the restriction of S_t on D endowed with the induced topology. Prove that if $v \in D$ is nonwandering in D, then v is nonwandering in X. ∎

Proposition 1.5.5. *Every point from $\omega(v)$ is nonwandering.*

Proof. Let $q \in \omega(v)$ for some $v \in X$. Then by Proposition 1.3.3, for any neighborhood $\mathscr{O}(q)$ there exists $w \in \gamma_v^+$ such that $w \in \mathscr{O}(q)$. Since $\omega(w) = \omega(v)$ (see Exercise 1.3.1), we have $q \in \omega(w)$. Applying Proposition 1.3.3 again, we obtain that for any $t_* \geq 0$ there exists $t > t_*$ such that $S_t w \in \mathscr{O}(q)$. □

Exercise 1.5.6. Assume that there exists a Lagrange stable trajectory γ_v^+. Then the set of nonwandering points is nonempty. ∎

1.5.2 Center of attraction

Let S_t be an evolution operator with either discrete or continuous time on some complete metric space X. As in NEMYTSKII/STEPANOV [171] and SIBIRSKY [212], we introduce some characteristics which describe the amount of time spent by the trajectory near a given set.

Let E be a Borel set in X and $\chi_E(x)$ be the corresponding characteristic function ($\chi_E(x) = 1$ for $x \in E$ and $\chi_E(x) = 0$ for $x \notin E$). We define the *time of occurrence* of the point v in the set E during the time interval $[0, T]$ by the formulas

$$\tau(v, T, E) = \int_0^T \chi_E(S_t v)dt \quad \text{(the continuous time case, } \mathbb{T} = \mathbb{R})$$

and

$$\tau(v, T, E) = \sum_{n=0}^T \chi_E(S_n v) \quad \text{(the discrete time case, } \mathbb{T} = \mathbb{Z}).$$

If the limit

$$\mathbb{P}(E, v) = \lim_{T \to +\infty} \frac{\tau(v, T, E)}{T}$$

exists, then we call it the *relative time* spent by the trajectory starting in v in the set E. It is also convenient to introduce *lower* $\underline{\mathbb{P}}$ and *upper* $\overline{\mathbb{P}}$ relative times by the formulas

$$\underline{\mathbb{P}}(E, v) = \liminf_{T \to +\infty} \frac{\tau(v, T, E)}{T} \quad \text{and} \quad \overline{\mathbb{P}}(E, v) = \limsup_{T \to +\infty} \frac{\tau(v, T, E)}{T},$$

which exist for *every* Borel set E (due to the fact that $0 \leq \tau(v, T, E)/T \leq 1$ for all $T > 0$).

Exercise 1.5.7. Let $\mathbb{P}_*(E)$ be either $\underline{\mathbb{P}}(E, v)$ or $\overline{\mathbb{P}}(E, v)$. Show that (a) $\mathbb{P}_*(E) \leq \mathbb{P}_*(F)$ when $E \subset F$ and (b) $\overline{\mathbb{P}}(E \cup F) \leq \overline{\mathbb{P}}(E) + \overline{\mathbb{P}}(F)$.   ∎

Exercise 1.5.8. Let $\alpha \in \mathbb{R}$. Consider the one-dimensional dynamical system (X, S_t) with $X = \mathbb{R}$ and $S_t x = x e^{\alpha t}$. Calculate the time of occurrence and the relative time as functions of α for the set $E = [-1, 1]$ and for every initial point $v \in \mathbb{R}$, $v \neq 0$. Make sure that $\mathbb{P}(E, v) = 0$ if $\alpha > 0$ and $\mathbb{P}(E, v) = 1$ when $\alpha < 0$.   ∎

Exercise 1.5.9 (Relative times for rotations). Let $(\mathbb{R}^2, S_t)$ be a dynamical system in $\mathbb{R}^2$ with evolution operator given (in the complex form) by the relation

$$S_t(x + iy) = e^{i\omega t}(x + iy) \quad \text{for } x + iy \in \mathbb{C} = \mathbb{R} + i\mathbb{R}.$$

Show that the relative time for the set $E = \{(x; y) : 0 \leq x < +\infty, \ |y| \leq h\}$ has the form

$$\mathbb{P}(E, (x; y)) = \begin{cases} 1/2, & \text{if } x^2 + y^2 \leq h^2; \\ \frac{1}{\pi} \arcsin \frac{h}{\sqrt{x^2 + y^2}}, & \text{if } x^2 + y^2 > h^2. \end{cases}$$

   ∎

The notion of a center of attraction for an individual trajectory plays an important role in the study of asymptotic dynamics (see, e.g., SIBIRSKY [212] and the references therein).

Definition 1.5.10 (Center of attraction). A forward invariant closed set $V \subset X$ is said to be a *center of attraction* for a semitrajectory γ_v^+ if we have that $\mathbb{P}(\mathscr{O}_\varepsilon(V), v) = 1$ for any $\varepsilon > 0$, where $\mathscr{O}_\varepsilon(V)$ is an ε-neighborhood of the set V. If the set V does not contain a proper invariant subset with the same property, then this set is called the *minimal* center of attraction.   ∎

Exercise 1.5.11. Let (X, S_t) be the dimensional dynamical system described in Exercise 1.5.8. Show that in the case when $\alpha < 0$ every interval $[-\beta, \beta]$ is a center of attraction for every point $v \in \mathbb{R}$ and $\{0\}$ is the minimal center of attraction. If $\alpha > 0$ the set $\mathbb{R} \setminus (-\beta, \beta)$ is a center of attraction for every point $v \neq 0$ and there is no minimal center of attraction for this v.   ∎

The following result concerning the center of attraction was established in SIBIRSKY [212].

Theorem 1.5.12. *Let S_t be an evolution operator in X and γ_v^+ be a Lagrange stable semitrajectory. Then the set*

$$W_v = \left\{ x \in X : \limsup_{T \to +\infty} \frac{\tau(v, T, \mathscr{O}_\delta(x))}{T} > 0 \text{ for every } \delta > 0 \right\} \tag{1.5.1}$$

is a nonempty closed set lying in the ω-limit set $\omega(v)$ such that

$$\mathbb{P}(\mathscr{O}_\varepsilon(W_v), v) \equiv \lim_{T \to +\infty} \frac{\tau(v, T, \mathscr{O}_\varepsilon(W_v))}{T} = 1 \text{ for every } \varepsilon > 0. \tag{1.5.2}$$

The set W_v is forward invariant. Moreover, W_v is the minimal center of attraction for γ_v^+.

Proof. We split the proof into several steps.

Step 1: W_v is nonempty. Indeed, let $\varepsilon > 0$ be fixed and

$$K = \begin{cases} \overline{\gamma_v^+} \setminus \mathscr{O}_\varepsilon(W_v), & \text{if } W_v \neq \emptyset; \\ \overline{\gamma_v^+}, & \text{if } W_v = \emptyset. \end{cases}$$

If $K = \emptyset$, we obviously have that $W_v \neq \emptyset$. Let $K \neq \emptyset$. For every $q \in K$ we have that $q \notin W_v$ and thus there exists $\delta = \delta_q > 0$ such that

$$\mathbb{P}(\mathscr{O}_\delta(q), v) \equiv \lim_{T \to +\infty} \frac{\tau(v, T, \mathscr{O}_\delta(q))}{T} = 0 \text{ for every } q \in K. \tag{1.5.3}$$

Since K is compact, there exists a finite collection $\{q_i : i = 1, \ldots, m\} \subset K$ such that $K \subset \cup_{i=1}^m \mathscr{O}_{\delta_{q_i}}(q_i)$. Therefore, by Exercise 1.5.7 it follows from (1.5.3) that $\mathbb{P}(K, v) = 0$. On the other hand, we obviously have that $\mathbb{P}(\overline{\gamma}_v^+, v) = 1$. Thus K is a *proper* subset of $\overline{\gamma}_v^+$ and thus $W_v \neq \emptyset$.

Step 2: Relation (1.5.2) *holds.* Indeed,

$$\overline{\gamma}_v^+ \subset \left(\overline{\gamma}_v^+ \setminus K\right) \cup K \subset \mathscr{O}_\varepsilon(W_v) \cup K.$$

Thus $\tau(v, T, \overline{\gamma}_v^+) \leq \tau(v, T, \mathscr{O}_\varepsilon(W_v)) + \tau(v, T, K)$ and hence

$$1 = \lim_{T \to \infty} \frac{\tau(v, T, \overline{\gamma}_v^+)}{T} \leq \limsup_{T \to \infty} \frac{\tau(v, T, \mathscr{O}_\varepsilon(W_v))}{T} \leq 1.$$

This implies (1.5.2).

Step 3: W_v *belongs to* $\omega(v)$. Indeed, let $q \notin \omega(v)$. By Theorem 1.4.5, $\omega(v)$ is a compact set. Therefore, there exist a neighborhood $\mathscr{O}_\eta(q)$ of q and a neighborhood $\mathscr{O}_\eta(\omega(v))$ of the set $\omega(v)$ such that $\mathscr{O}_\eta(q) \cap \mathscr{O}_\eta(\omega(v)) = \emptyset$. By Theorem 1.4.5, $S_t v$ converges to $\omega(v)$. Thus $S_t v \in \mathscr{O}_\eta(\omega(v))$ for all t large enough. This implies that $\overline{\mathbb{P}}(\mathscr{O}_\eta(q), v) = 0$ and thus $q \notin W_v$.

Step 4: W_v *is closed.* Let $q \in X \setminus W_v$. Then there exists a neighborhood $\mathscr{O}_\eta(q)$ such that $\overline{\mathbb{P}}(\mathscr{O}_\eta(q), v) = 0$. Thus for any $p \in \mathscr{O}_\eta(q)$ there exists $\delta > 0$ such that $p \in \mathscr{O}_\delta(p) \subset \mathscr{O}_\eta(q)$ and hence $\overline{\mathbb{P}}(\mathscr{O}_\delta(p), v) \leq \overline{\mathbb{P}}(\mathscr{O}_\eta(q), v) = 0$. Therefore, $\mathscr{O}_\eta(q) \subset X \setminus W_v$. Thus W_v is closed.

Step 5: W_v *is forward invariant.* Let $q \in W_v$. By the continuity of S_t, for every $\varepsilon > 0$ there is $\delta > 0$ such that $S_t \mathscr{O}_\delta(q) \subset \mathscr{O}_\varepsilon(S_t q)$. Thus,

$$\tau(v, T, \mathscr{O}_\varepsilon(S_t q)) \geq \tau(v, T, S_t \mathscr{O}_\delta(q)) = \int_0^T \chi_{S_t \mathscr{O}_\delta(q)}(S_r v) dr$$

(for the definiteness we consider the continuous time case only). One can see that

$$\{r : S_r v \in S_t \mathscr{O}_\delta(q)\} \supset \{r : S_{r-t} v \in \mathscr{O}_\delta(q)\}$$

in the case when $r \geq t$. Hence,

$$\chi_{S_t \mathscr{O}_\delta(q)}(S_r v) \geq \chi_{\mathscr{O}_\delta(q)}(S_{r-t} v) \quad \text{for all } r \geq t.$$

Consequently,

$$\tau(v, T, \mathscr{O}_\varepsilon(S_t q)) \geq \int_t^T \chi_{\mathscr{O}_\delta(q)}(S_{r-t} v) dr \geq \int_0^T \chi_{\mathscr{O}_\delta(q)}(S_r v) dr - t.$$

This implies that $\overline{\mathbb{P}}(\mathscr{O}_\varepsilon(S_t q), v) > 0$ for any $\varepsilon > 0$. Thus W_v is forward invariant.

Step 6: W_v is a minimal center of attraction. To see this, it is sufficient to prove the following lemma.

Lemma 1.5.13. *Let the hypotheses of Theorem 1.5.12 be in force. Then for every closed set V with property $\mathbb{P}(\mathscr{O}_\varepsilon(V), v) = 1$, for any $\varepsilon > 0$ we have that $W_v \subseteq V$.*

Proof. Suppose $q \notin V$. Then there exists $\eta > 0$ such that $\mathscr{O}_\eta(q) \subset X \setminus \mathscr{O}_\eta(V)$. Since $\mathbb{P}(\mathscr{O}_\eta(V), v) = 1$, we have $\mathbb{P}(\mathscr{O}_\eta(q), v) = 0$, which means that $q \notin W_v$.   □

This completes the proof of Theorem 1.5.12.   □

1.5.3   Almost recurrent and recurrent trajectories

In the class of nonwandering trajectories we can extract a class with stronger recurrence properties (see NEMYTSKII/STEPANOV [171] and SIBIRSKY [212]).

Definition 1.5.14 (Almost recurrent trajectory). A semitrajectory γ_v^+ uniformly approximates a set $Q \subseteq X$ if for any $\varepsilon > 0$ there exists $T > 0$ such that every segment $\gamma_v^{\tau, \tau+T}$ approximates the set Q within ε, i.e.,

$$Q \subseteq \mathscr{O}_\varepsilon(\gamma_v^{\tau, \tau+T}) \quad \text{for any } \tau \geq 0.$$

A semitrajectory γ_v^+ and the point v are said to be *almost recurrent* if γ_v^+ uniformly approximates the point v; i.e., for any $\varepsilon > 0$ there exists $T > 0$ such that $v \in \mathscr{O}_\varepsilon(\gamma_v^{\tau, \tau+T})$ for any $\tau \geq 0$.   ■

Exercise 1.5.15. Every almost recurrent point is nonwandering.   ■

Proposition 1.5.16. *If γ_v^+ is an almost recurrent semitrajectory, then $\omega(v)$ is not empty, $v \in \omega(v)$, and hence γ_v^+ is Poisson stable.*

Proof. By Definition 1.5.14 applied to $\varepsilon = 1/n$ we have that

$$\text{dist}(v, \gamma_v^{n, n+T_n}) \leq 1/n, \quad n = 1, 2, \ldots$$

for some sequence $T_n > 0$. Thus there exists $t_n \in [n, n + T_n]$, $t_n \to \infty$, such that

$$\text{dist}(v, S_{t_n} v) \leq 2/n \quad \text{for all } n = 1, 2, \ldots.$$

Thus $v = \lim_{n \to \infty} S_{t_n} v$. Therefore, by Proposition 1.3.3, $v \in \omega(v)$.   □

Proposition 1.5.17. *The set of all almost recurrent points is forward invariant.*

Proof. Let v be an almost recurrent point and $q = S_t v$ for some $t > 0$. By the continuity of S_t, for every $\varepsilon > 0$ there is $\delta > 0$ such that

$$\text{dist}(q, S_t p) \leq \varepsilon \quad \text{provided } \text{dist}(v, p) \leq \delta. \tag{1.5.4}$$

On the other hand, since v is almost recurrent, there is $T = T(\delta)$ such that

$$\text{dist}(v, \gamma_v^{\tau,\tau+T}) \le \delta \quad \text{for every } \tau \ge 0.$$

Therefore (1.5.4) yields

$$\text{dist}(q, \gamma_q^{\tau,\tau+T}) = \text{dist}(q, S_t \gamma_v^{\tau,\tau+T}) \le \varepsilon \quad \text{for every } \tau \ge 0.$$

Thus $q = S_t v$ is an almost recurrent point.   □

As in BIRKHOFF [14] (see also NEMYTSKII/STEPANOV [171] and SIBIRSKY [212]), we introduce the following subclass of almost recurrent motions.

Definition 1.5.18 (Recurrent trajectory). A semitrajectory γ_v^+ and the point v are said to be *recurrent* if γ_v^+ uniformly approximates itself; i.e., for any $\varepsilon > 0$ there exists $T > 0$ such that $\gamma_v^+ \subset \mathcal{O}_\varepsilon(\gamma_v^{\tau,\tau+T})$ for any $\tau \ge 0$.   ■

Exercise 1.5.19. Every periodic or fixed point is recurrent.   ■

Exercise 1.5.20. The set of all recurrent points is forward invariant.   ■

Exercise 1.5.21. If a semitrajectory γ_v^+ is recurrent, then γ_v^+ is bounded.   ■

Proposition 1.5.22. *Let γ_v^+ be a recurrent semitrajectory. Then this semitrajectory γ_v^+ is Lagrange stable.*

Proof. By definition, for every $\varepsilon > 0$ there exists $T = T_\varepsilon > 0$ such that γ_v^+ lies in $\mathcal{O}_{\varepsilon/4}(\gamma_v^{0,T})$. Therefore $\overline{\gamma_v^+} \subset \mathcal{O}_{\varepsilon/2}(\gamma_v^{0,T})$. Since $K = \overline{\gamma_v^{0,T}}$ is compact for every fixed $T > 0$, there exists a finite $\varepsilon/2$-net for K, i.e., a finite set $\{p_n : n = 1, \ldots, m\}$ in K such that

$$K \subset \cup_{n=1}^m \mathcal{O}_{\varepsilon/2}(p_n).$$

In this case $\overline{\gamma_v^+} \subset \cup_{n=1}^m \mathcal{O}_\varepsilon(p_n)$. Thus, for every $\varepsilon > 0$ there exists a finite ε-net for $\overline{\gamma_v^+}$. Therefore (see, e.g., DIEUDONNÉ [85, Section 3.17] or Proposition A.3.4 in the Appendix), the set $\overline{\gamma_v^+}$ is compact.   □

Below we use the notion of *minimal* set which we define as a nonempty forward invariant closed set which does not contain a proper nonempty forward invariant closed subset.

Proposition 1.5.23. *Any compact minimal set Σ is strictly invariant.*

Proof. Take $v \in \Sigma$. Then $\overline{\gamma_v^+} \subset \Sigma$. Thus the semitrajectory γ_v^+ is Lagrange stable. Hence by Proposition 1.4.6, $\omega(v)$ is a strictly invariant subset of Σ. By minimality $\Sigma = \omega(v)$. Thus Σ is strictly invariant.   □

Exercise 1.5.24. Show that every semitrajectory γ^+ from a compact minimal set Σ is dense in Σ.   ■

Now we return to the recurrence. In fact, the following properties of recurrent trajectories were established in BIRKHOFF [14].

Theorem 1.5.25 (Birkhoff). *Let (X, S_t) be a dynamical system. Then*

- *Any semitrajectory from a compact minimal set is recurrent.*
- *If we assume in addition that $(t, x) \mapsto S_t x$ is continuous from $\mathbb{R}_+ \times X$ into X (in the continuous time case), then the closure of any recurrent semitrajectory is a compact minimal set.*

Proof. Let Σ be a minimal set and $\gamma_v^+ \subset \Sigma$ be not recurrent. Then there exists ε_0 such that for any $T > 0$ we have

$$\gamma_v^+ \not\subset \mathscr{O}_{\varepsilon_0}(\gamma_v^{\tau, \tau+T}) \quad \text{for some} \ \ \tau = \tau(\varepsilon_0, T) \geq 0.$$

Thus, there exist sequences $\{T_n \to +\infty\}$ and $\{p_n = S_{\tau_n} v\}$ such that

$$\gamma_v^+ \not\subset \mathscr{O}_{\varepsilon_0}(\gamma_{p_n}^{0, T_n}) \quad \text{for all} \ \ n = 1, 2, \ldots.$$

Therefore, we can find a sequence $\{q_n\} \subset \gamma_v^+$ such that

$$\text{dist}(q_n, S_t p_n) \geq \varepsilon_0 \quad \text{for all} \ \ t \in [0, T_n], \ \ n = 1, 2, \ldots. \tag{1.5.5}$$

Since Σ is compact, by an appropriate choice of subsequences, we can assume that

$$q_n \to q \in \overline{\gamma_v^+} \quad \text{and} \quad p_n \to p \in \Sigma \quad \text{as} \ n \to \infty.$$

Thus (1.5.5) implies that $\text{dist}(q, S_t p) \geq \varepsilon_0$ for every $t > 0$. Hence the closure $\overline{\gamma_p^+}$ of γ_p^+ is a forward invariant (see Exercise 1.2.1(F)) proper closed subset of Σ, which is impossible.

To prove the second part of Theorem 1.5.25 we first establish the following lemma concerning almost recurrent semitrajectories.

Lemma 1.5.26. *Assume that $(t, x) \mapsto S_t x$ is continuous from $\mathbb{R}_+ \times X$ into X (in the continuous time case). Let γ_v^+ be an almost recurrent trajectory. Then the omega limit set $\omega(v)$ is minimal. By Proposition 1.5.16 this implies that $\overline{\gamma_v^+} = \omega(v)$ and thus is also minimal.*

Proof. Let A be a closed forward invariant proper subset of $\omega(v)$. Then $v \notin A$. Indeed, if $v \in A$, then $\overline{\gamma_v^+} \subset A$ and thus $\omega(v) \subset A$, which contradicts our assumption concerning A. Let $d = \text{dist}(v, A)$ and $\varepsilon < d/2$. By almost recurrence of v there exists $T > 0$ such that

$$v \in \mathscr{O}_\varepsilon(\gamma_v^{\tau, \tau+T}) \quad \text{for every} \ \ \tau \geq 0.$$

On the other hand, by Proposition 1.3.7 we have that

$$\gamma_v^{\bar{\tau}, \bar{\tau}+T} \subset \mathscr{O}_\varepsilon(A) \quad \text{for some} \ \ \bar{\tau} \geq 0.$$

This implies that $v \in \mathscr{O}_{2\varepsilon}(A)$. This is impossible because $2\varepsilon < \text{dist}(v, A)$. $\qquad \square$

To conclude the proof of Theorem 1.5.25 we note that if γ_v^+ is recurrent, then by Proposition 1.5.22 γ_v is Lagrange stable; i.e., $\overline{\gamma_v^+}$ is a compact set which is minimal according to Lemma 1.5.26. □

1.5.4 Almost periodic trajectories

Definition 1.5.27. A full trajectory $\gamma = \{u(t) : t \in \mathbb{R}\}$ is said to be *almost periodic* if for any $\varepsilon > 0$ there exists $T = T(\varepsilon) > 0$ such that any time interval of the length T contains a number τ such that $\text{dist}(u(t + \tau), u(t)) \le \varepsilon$ for every $t \in \mathbb{R}$. ∎

Exercise 1.5.28. Show that equilibria and periodic orbits are almost periodic. ∎

Exercise 1.5.29. A full trajectory $\gamma = \{u(t) : t \in \mathbb{R}\}$ is almost periodic if and only if for every $\varepsilon > 0$ there exists T such that

$$\text{dist}(u(t), \gamma^{\tau,\tau+T}) \le \varepsilon \quad \text{for all } t, \tau \in \mathbb{R}, \tag{1.5.6}$$

where $\gamma^{a,b} \equiv \{u(t) : a \le t \le b\}$ denotes a segment of the trajectory γ. Moreover, (1.5.6) can be written in the form:

$$\gamma \subset \mathcal{O}_\varepsilon(\gamma^{\tau,\tau+T}) \quad \text{for every } \tau \in \mathbb{R}. \tag{1.5.7}$$

∎

We note that (1.5.7) implies that every almost periodic trajectory is recurrent not only in the forward time direction ($\tau > 0$), as described in Definition 1.5.18, but also possesses a similar property in the backward ($\tau < 0$) time direction.

Proposition 1.5.30. *Let $\gamma = \{u(t) : t \in \mathbb{R}\}$ be an almost periodic trajectory. Then*

- *The closure $\overline{\gamma}$ of γ is a compact set and $\overline{\gamma} = \omega(\gamma) = \alpha(\gamma)$.*
- *For every $p \in \overline{\gamma}$ there exists a full almost periodic trajectory $\gamma_* = \{w(t) : t \in \mathbb{R}\} \subset \overline{\gamma}$ such that $w(0) = p$. Moreover, under the conditions of Lemma 1.5.26 by the minimality property (see Lemma 1.5.26) we have $\overline{\gamma_*} = \overline{\gamma}$.*

Proof. It follows from (1.5.7) that any semitrajectory $\gamma^{\tau,+\infty}$ is recurrent for every τ. By Propositions 1.5.16 and 1.5.22, this implies that $\overline{\gamma}$ is compact and $\overline{\gamma} \subset \omega(\gamma)$. It follows from Exercise 1.3.6 that $\omega(\gamma) \subset \overline{\gamma}$. Thus $\overline{\gamma} = \omega(\gamma)$. Equality $\overline{\gamma} = \alpha(\gamma)$ follows by the same arguments as Propositions 1.5.16 and 1.5.22 applied to the backward direction of time.

To prove the second statement we note that for every $p \in \overline{\gamma} = \omega(\gamma)$ there is a sequence $\{t_n \to +\infty\}$ such that $u(t_n) \to p$ as $n \to \infty$. Since $\overline{\gamma}$ is compact and

$$w_n(t) \equiv u(t_n + t) \in \gamma \quad \text{for every } t \in \mathbb{R}, \tag{1.5.8}$$

the sequence $\{z_n = u(t_n + m)\}$ is relatively compact for every $m \in \mathbb{Z}$. Hence, by the standard diagonal procedure there exist a sequence $\{n_l\}$ and elements $w_m \in X$ such that $w_0 = p$ and

$$u(t_{n_l} + m) \to w_m \text{ as } l \to \infty \text{ and } S_1 w_m = w_{m+1}, \quad m \in \mathbb{Z}.$$

Therefore, there exists a full trajectory $\gamma_* = \{w(t) : t \in \mathbb{R}\} \subset \overline{\gamma}$ such that $w(m) = w_m$, $w(0) = p$, and

$$u(t_{n_l} + t) \to w(t) \text{ as } l \to \infty \text{ for every } t \in \mathbb{R}.$$

It follows from Definition 1.5.27 that for every $\varepsilon > 0$ there exists T such that for every $t_* \in \mathbb{R}$ we can find $\tau \in [t_*, t_* + T]$ such that $\text{dist}(u(t), u(t + \tau)) \leq \varepsilon$ for all $t \in \mathbb{R}$. Thus, substituting $t_{n_l} + t$ instead of t yields

$$\text{dist}(u(t_{n_l} + t), u(t_{n_l} + t + \tau)) \leq \varepsilon \quad \text{for all } t \in \mathbb{R}, \quad l \in \mathbb{Z}_+.$$

Therefore, after the limit transition $l \to \infty$ we obtain that

$$\text{dist}(w(t), w(t + \tau)) \leq \varepsilon \quad \text{for all } t \in \mathbb{R}.$$

Thus γ_* is almost periodic.   □

For more details concerning recurrent and chaotic properties of individual trajectories we refer to BIRKHOFF [14], GUCKENHEIMER/HOLMES [114], KATOK/ HASSELBLATT [132], NEMYTSKII/STEPANOV [171], SHARKOVSKY ET AL. [208], SIBIRSKY [212] and the references therein.

1.6   Equilibria and Lyapunov stability

There are many sources which discuss the notion of Lyapunov stability; see the monographs CODDINGTON/LEVINSON [75], HARTMAN [120], LEFSCHETZ [148], SIBIRSKY [212], for instance. Our main goal in this section is to present the result stating that this kind of stability for an equilibrium is equivalent to the existence of some function with specific properties defined on a neighborhood of this equilibrium.

We first recall the general concept of Lyapunov stability.

Definition 1.6.1 (Lyapunov stability). A point v and the semitrajectory γ_v^+ are called *Lyapunov stable* if for any $\varepsilon > 0$ there exists $\delta > 0$ such that $\text{dist}(S_t v, S_t w) \leq \varepsilon$ for all $t \geq 0$ and $w \in X$ with the property $\text{dist}(v, w) \leq \delta$.   ■

Specifying this definition for fixed points, we arrive at the following.

Definition 1.6.2 (Lyapunov stability of fixed points). A fixed point v is said to be *Lyapunov stable* if for any $\varepsilon > 0$ there exists $\delta > 0$ such that $\mathrm{dist}(v, S_t w) \leq \varepsilon$ for all $t \geq 0$ and $w \in X$ with the property $\mathrm{dist}(v, w) \leq \delta$. If, moreover, $\mathrm{dist}(v, S_t w) \to 0$ as $t \to +\infty$, then v is said to be *asymptotically* Lyapunov stable.[4] ∎

Definition 1.6.3 (Local Lyapunov function). A non-negative real-valued function $V(x)$ defined on some neighborhood $\mathscr{O}_\eta(v)$ of a fixed point v is called a (local) *Lyapunov function* for the point v when the following conditions hold:

- $V(w_n) \to 0$ if and only if $w_n \to v$ as $n \to \infty$;
- if for some $t_* > 0$ and $w \in X$ we have that $S_t w \in \mathscr{O}_\eta(v)$ for all $t \in [0, t_*]$, then $V(S_t w) \leq V(w)$ for $t \in [0, t_*]$.

∎

Exercise 1.6.4. Let v_0 be a Lyapunov stable fixed point. Show that

$$\cap_{\delta > 0} \overline{\cup \{\gamma_v^+ : \mathrm{dist}(v, v_0) \leq \delta\}} = \{v_0\}$$

∎

The following exercise shows that a local Lyapunov function is not unique.

Exercise 1.6.5. Assume that $V(x)$ is a local Lyapunov function for v defined on $\mathscr{O}_\eta(v)$. Let f be a strictly increasing continuous scalar function defined on the closure of $\mathrm{range}\,(V) = \{V(x) : x \in \mathscr{O}_\eta(v)\}$ and $f(0) = 0$. Show that $W(x) = f(V(x))$ is also a local Lyapunov function. ∎

Theorem 1.6.6. *A fixed point $v \in X$ is Lyapunov stable if and only if there exists a local Lyapunov function for v.*

Proof of Theorem 1.6.6. We use the same argument as in SIBIRSKY [212].

Let a fixed point $v \in X$ be Lyapunov stable and $\varepsilon_0 > 0$. Then there exists a $\mathscr{O}_{\delta_0}(v)$ such that $S_t x \in \mathscr{O}_{\varepsilon_0}(v)$ for all $x \in \mathscr{O}_{\delta_0}(v)$. Thus, we can define a function V by the formula

$$V(x) = \sup_{t \geq 0} \mathrm{dist}(S_t x, v) \quad \text{for all } x \in \mathscr{O}_{\delta_0}(v). \tag{1.6.1}$$

Let us prove that $V(x)$ is a Lyapunov function for v.

By the stability of v, for any $\varepsilon > 0$ there exists $0 \leq \delta \leq \delta_0$ such that

$$\mathrm{dist}(S_t q, v) \leq \varepsilon \quad \text{for all } t \geq 0 \text{ provided } \mathrm{dist}(q, v) \leq \delta. \tag{1.6.2}$$

---

[4]There are examples showing that the property $\mathrm{dist}(v, S_t w) \to 0$ as $t \to +\infty$ does not imply the Lyapunov stability. See, e.g., TESCHL [217, p. 168] and also Example 1.9.6 with $\mu = 0$ in Section 1.9.

Thus, if $w_n \to v$ as $n \to \infty$, then there exists n_0 such that $\mathrm{dist}(S_t w_n, v) \le \varepsilon$ for all $t \ge 0$ and $n \ge n_0$. Hence $V(w_n) \to 0$ as $n \to \infty$.

By (1.6.1) we have that

$$\mathrm{dist}(w_n, v) \le \sup_{t \ge 0} \ \mathrm{dist}(S_t w_n, v) = V(w_n)$$

and thus $w_n \to v$ as $n \to \infty$ provided $V(w_n) \to 0$ as $n \to \infty$.

The monotonicity of $V(S_t w)$ follows from the semigroup property of S_t. Indeed, we have that

$$V(S_t w) = \sup_{\tau \ge 0} \ \mathrm{dist}(S_{\tau + t} w, v) = \sup_{\tau \ge t} \ \mathrm{dist}(S_\tau w, v) \le \sup_{\tau \ge 0} \ \mathrm{dist}(S_\tau x, v) = V(w).$$

Now we assume that there exists a local Lyapunov function $V(x)$ for a fixed point $v \in X$ which is defined on $\mathscr{O}_\eta(v)$. We take $\varepsilon < \eta$ and set

$$\lambda = \inf_x \{ V(x) \ : \ \varepsilon \le \mathrm{dist}(x, v) < \eta \}.$$

By the first requirement concerning V in Definition 1.6.3 we have that $\lambda > 0$ and there exists $0 < \delta < \varepsilon$ such that $V(x) < \lambda$ provided $\mathrm{dist}(x, v) < \delta$. Now we show that for these ε and δ the relation in (1.6.2) holds. Indeed, if (1.6.2) is not true for some $q \in \mathscr{O}_\delta(v)$, then by the continuity of $t \mapsto S_t q$ there exists $t_* > 0$ such that $\mathrm{dist}(S_t q, v) < \varepsilon$ for all $t \in [0, t_*)$ and $\mathrm{dist}(S_{t_*} q, v) = \varepsilon$, which implies that $V(S_{t_*} q) \ge \lambda$. However, by the second requirement in Definition 1.6.3 we have that $V(S_{t_*} q) \le V(q) < \lambda$. Thus, we arrive at a contradiction.

This completes the proof of Theorem 1.6.6. $\square$

Theorem 1.6.7. *Let $v \in X$ be a fixed point of some dynamical system (X, S_t). Then v is asymptotically Lyapunov stable if and only if there exists a local Lyapunov function for v on $\mathscr{O}_\eta(v)$ possessing the property*

$$V(S_t w) \to 0 \ \text{ as } \ t \to +\infty \ \text{ provided } \ S_t w \in \mathscr{O}_\eta(v) \ \text{ for all } \ t \ge 0. \tag{1.6.3}$$

Proof. If v is asymptotically Lyapunov stable, then by Theorem 1.6.6 there exists a Lyapunov function $V(x)$ for v. We have that $S_t w \to v$ for all w from some neighborhood of v. By the first property in Definition 1.6.3 this implies (1.6.3).

Suppose that there exists a Lyapunov function $V(x)$ possessing property (1.6.3). Then by Theorem 1.6.6, v is a Lyapunov stable fixed point. This means that for any $0 < \varepsilon \le \eta$ there exists $\delta > 0$ such that $\mathrm{dist}(v, S_t w) \le \varepsilon$ for all $t \ge 0$ and $w \in X$ with the property $\mathrm{dist}(v, w) \le \delta$. In this case we can apply (1.6.3) to conclude that $V(S_t w) \to 0$ as $t \to +\infty$. Thus, the first property in Definition 1.6.3 yields that $S_t w \to 0$. Hence, v is asymptotically Lyapunov stable. $\square$

1.7 Complete theory of 1D continuous systems

Now we describe all possible scenarios in one-dimensional systems with continuous time. We start with several simple observations by including them in the following exercises.

Exercise 1.7.1. Describe the dynamics in the models generated by the following equations: (a) $\dot{x} = x$, (b) $\dot{x} = -x$, (c) $\dot{x} = |x|$, (d) $\dot{x} = 1$. Show that the corresponding dynamical systems cannot be topologically equivalent. ∎

Exercise 1.7.2. Any continuous dynamical system $(S_t, \mathbb{R})$ on $\mathbb{R}$ is monotone, i.e., if $x \leq y$, then $S_t x \leq S_t y$ for all $t \in \mathbb{R}_+$. ∎

Exercise 1.7.3. Show that a continuous dynamical system on $\mathbb{R}$ cannot contain nontrivial periodic orbits. ∎

Exercise 1.7.4. Show that any semitrajectory $\gamma_+ = \{S_t x : t \in \mathbb{R}_+\}$ of a continuous dynamical system on $\mathbb{R}$ is a graph of a monotone function, i.e., $t \mapsto S_t x$ is either non-decreasing or non-increasing. ∎

Exercise 1.7.5. Consider the Cauchy problem

$$\dot{x}(t) = -x(t)^3, \quad x(t) = x_0.$$

(A) Show that this equation generates the dynamical system $(\mathbb{R}, S_t)$ with the evolution operator S_t given by the formula

$$S_t x_0 = x_0 (1 + 2tx_0^2)^{-1/2}, \quad x_0 \in \mathbb{R}.$$

(B) Make sure that there are no nonzero semitrajectories which can be extended to a full trajectory.
(C) Show that the zero equilibrium is asymptotically Lyapunov stable.
(D) Show that the formula in (1.6.1) gives us a Lyapunov function of the form $V(x) = |x|$ for the zero equilibrium. Thus by Exercise 1.6.5, $V_\lambda(x) = |x|^\lambda$ is also a Lyapunov function for every $\lambda > 0$.

∎

We can say more about 1D systems on $\mathbb{R}$ in the case when the evolution operator is invertible; i.e., S_t is a one-parameter continuous *group* of continuous mapping. We call this system a dynamical system with continuous reversible time. In this case any point x belongs to some full trajectory (cf. Exercise 1.7.5).

We start with the following simple observations.

Exercise 1.7.6. Let $(\mathbb{R}, S_t)$ be a dynamical system on the real line $\mathbb{R}$ with continuous reversible time. Assume that the system has no equilibrium points. Show that any full trajectory $\gamma = \{S_t x : t \in \mathbb{R}\}$ is a graph of a strictly monotone function with the range $\mathbb{R}$. ∎

Exercise 1.7.7. Let $(\mathbb{R}, S_t)$ and $(\mathbb{R}, \tilde{S}_t)$ be two dynamical systems on the real line $\mathbb{R}$ with continuous reversible time. Assume that both systems have no equilibrium points. Show that these systems are topologically equivalent. Hint: Use the result of Exercise 1.7.6 and show that the mapping $S_t 0 \mapsto \tilde{S}_t 0$ gives the desired homeomorphism.   ∎

This exercise means that in the absence of equilibria any system on $\mathbb{R}$ with continuous reversible time is topologically equivalent with the system of right shifts on the real line: $S_t x = x + t$, $x, t \in \mathbb{R}$.

Since the set $\mathcal{N}$ of equilibrium points for $(\mathbb{R}, S_t)$ is closed (see Exercise 1.2.3), we have that

$$\mathbb{R} \setminus \mathcal{N} = \cup_i(a_i, b_i), \quad \text{for some } a_i, b_i \in \{-\infty\} \cup \mathcal{N} \cup \{+\infty\}.$$

Every interval (a_i, b_i) is called adjacent to $\mathcal{N}$.

Exercise 1.7.8. Let $(\mathbb{R}, S_t)$ be a dynamical system on $\mathbb{R}$. Show that any adjacent interval (a, b) to $\mathcal{N}$ is a strictly invariant set. Moreover, if $(\mathbb{R}, S_t)$ has continuous reversible time, then $t \mapsto S_t x$ is a strictly monotone continuous mapping of $\mathbb{R}$ onto (a, b) for each $x \in (a, b)$.   ∎

In the following theorem it is important to bear in mind that any homeomorphism of real line $\mathbb{R}^1$ onto itself is represented by a strictly monotone function.

Theorem 1.7.9. *Let $(\mathbb{R}, S_t)$ and $(\mathbb{R}, \tilde{S}_t)$ be two dynamical systems on the real line $\mathbb{R}$ with continuous reversible time. We denote by $\mathcal{N}$ and $\tilde{\mathcal{N}}$ the corresponding sets of equilibrium points (one/both of them can be empty). These systems are conjugate if and only if there exists a homeomorphism ψ on $\mathbb{R}$ such that $\psi(\mathcal{N}) = \tilde{\mathcal{N}}$ and the directions of motions on the corresponding intervals adjacent to $\mathcal{N}$ and $\tilde{\mathcal{N}}$ are compatible (i.e., they coincide if $\psi(x)$ increases and are oppositely directed if $\psi(x)$ decreases).*

Proof. We need only to prove the sufficient part. For this we use the same argument as in SIBIRSKY [212].

Let ψ be a homeomorphism with the properties formulated in the theorem. We construct a conjugate mapping h as follows. If $x \in \mathcal{N}$, we set $h(x) = \psi(x)$. Then inside every adjacent interval $\mathcal{I} = (a, b)$ for $\mathcal{N}$ we fix a point x_{ab}. The end points of this interval are mapped into end points of some adjacent interval $\tilde{\mathcal{I}}$ for $\tilde{\mathcal{N}}$. We fix a point $\tilde{x}_{ab}$ inside $\tilde{\mathcal{I}}$. Now we define the mapping h on (a, b) as

$$h(S_t x_{ab}) = \tilde{S}_t \tilde{x}_{ab} \quad \text{for all } t \in \mathbb{R}.$$

One can see that the mapping we have defined is a homeomorphism with the properties

$$h(S_t x) = \tilde{S}_t h(x) \quad \text{for all } x, t \in \mathbb{R}.$$ □

We also have the following assertion.

Theorem 1.7.10. *Every dynamical system* $(\mathbb{R}, S_t)$ *with continuous reversible time is conjugate with a system generated by some ordinary differential equation.*

Proof. We use the argument given in SIBIRSKY [212].

Let us set $f(x) = 0$ for every $x \in \mathcal{N}$. Then inside every adjacent interval $\mathscr{I} = (a, b)$ for $\mathcal{N}$ we fix a point x_{ab} and denote

$$f(x) = (x - a)(x - b)\operatorname{sign}[x_{ab} - S_1 x_{ab}], \quad x \in (a, b).$$

If $a = -\infty$ we take

$$f(x) = (x - b)\operatorname{sign}[x_{ab} - S_1 x_{ab}], \quad x \in (a, b).$$

We perform this similarly in the case $b = \infty$. One can see that the equation $\dot{x} = f(x)$ generates a dynamical system with continuous reversible time. By Theorem 1.7.9 this system is topologically equivalent to $(S_t, \mathbb{R})$. $\square$

We do not know whether a similar result holds in higher dimensions.

Exercise 1.7.11. Assume that 1D dynamical systems $(\mathbb{R}, S_t)$ and $(\mathbb{R}, \tilde{S}_t)$ with reversible time have finite numbers of equilibria

$$\mathcal{N} = \{x_1 < x_2 < \ldots < x_N\} \quad \text{and} \quad \tilde{\mathcal{N}} = \{\tilde{x}_1 < \tilde{x}_2 < \ldots < \tilde{x}_{\tilde{N}}\}.$$

We also denote $x_0 = \tilde{x}_0 = -\infty$ and $x_{N+1} = \tilde{x}_{\tilde{N}+1} = +\infty$. Show that $(\mathbb{R}, S_t)$ and $(\mathbb{R}, \tilde{S}_t)$ are equivalent if $N = \tilde{N}$ and the motions of S_t and $\tilde{S}_t$ on the intervals (x_i, x_{i+1}) and $(\tilde{x}_i, \tilde{x}_{i+1})$ have the same directions for all $i = 0, \ldots, \tilde{N}$. ∎

The following exercise demonstrates that the condition of time reversibility is important in Theorem 1.7.9 and Exercise 1.7.11.

Exercise 1.7.12. Show that the system $(\mathbb{R}, S_t)$ considered in Exercise 1.7.5 is not equivalent to the system generated by the equation $\dot{x} = -x$. Hint: The equivalence would imply that in the system discussed in Exercise 1.7.5 every point belongs to a full trajectory. ∎

The following property of 1D systems makes it possible to study their dynamics in detail.

Exercise 1.7.13 (Comparison principle for 1D ODE). Let $x(t)$ and $y(t)$ be solutions to the 1D equations

$$\dot{x} = f(x) \quad \text{and} \quad \dot{y} = g(y)$$

on some interval $[0, T]$ with $f, g \in C^1(\mathbb{R})$. Assume that

$$x(0) \geq y(0) \quad \text{and} \quad f(x(t)) \geq g(x(t)) \quad \text{for } t \in [0, T].$$

Prove that $x(t) \geq y(t)$ for $t \in [0, T)$. Hint: Show that $z(t) = x(t) - y(t)$ satisfies the linear equation $\dot{z} = a(t)z(t) + b(t)$ for some non-negative $b(t)$.   ∎

As an application of the comparison principle we suggest the following exercise.

Exercise 1.7.14 (Lyapunov exponent). Let $(\mathbb{R}, S_t)$ be a system generated by some 1D equation

$$\dot{x} = f(x), \quad t > 0, \quad x(0) = x_0 \in \mathbb{R}, \tag{1.7.1}$$

where $f \in C^1(\mathbb{R})$. Let x_* be an (isolated) equilibrium and $f'(x_*) < 0$. Show that x_* is asymptotically stable for $(\mathbb{R}, S_t)$ and there exists a vicinity $\mathcal{O}_\delta(x_*)$ such that

$$\lim_{t \to +\infty} \frac{\ln |S_t x|}{t} = f'(x_*) \quad \text{for all } x \in \mathcal{O}_\delta(x_*). \tag{1.7.2}$$

Hint: Use the comparison principle for $\dot{x} = f(x)$ and the linear equation

$$\dot{y} = [f'(x_*) \pm \varepsilon]y$$

near the equilibrium x_*. We note that the limit on the left-hand side of (1.7.2) is called the Lyapunov exponent and provides the exact rate of the convergence to the stable equilibrium x_*.   ∎

We conclude this section with several facts related to non-uniqueness and blow-up phenomena for 1D ODEs.

We start with the standard non-uniqueness example (see, e.g., HARTMAN [120]), which shows that the uniqueness statement for the 1D ODE in (1.7.1) cannot be true without the Lipschitz assumption for f.

Example 1.7.15. The functions $x(t) = t^2$ and $x(t) \equiv 0$ solve the Cauchy problem in $\mathbb{R}$: $\dot{x} = 2\sqrt{|x|}$, $t > 0$ and $x(0) = 0$.   ∎

In relation to this example, it is interesting to mention the following simple assertion concerning uniqueness for 1D equations.

Proposition 1.7.16. *Let $f : \mathbb{R} \mapsto \mathbb{R}$ be a continuous function. Then a Cauchy problem*

$$\dot{x} = f(x), \quad t > 0, \quad x(0) = x_0 \in \mathbb{R}, \tag{1.7.3}$$

has a unique (local) solution in a neighborhood of the point x_0, provided

- *either f is Lipschitz in some neighborhood of x_0,*
- *or else $f(x_0) \neq 0$.*

Proof. We note that the existence of local solutions to (1.7.3) is well known (see, e.g., CODDINGTON/LEVINSON [75] or HARTMAN [120] and also Theorem A.1.2 in the Appendix). The same theorem yields the uniqueness in the Lipschitz case. Thus, we need to consider the case $f(x_0) \neq 0$ only. In this case there is a neighborhood of the point x_0 such that $1/f(x)$ is continuous and preserves the sign. This implies that every solution to (1.7.3) in this neighborhood satisfies the equation

$$\frac{d}{dt}F(x(t)) = 1 \text{ with } F(x) = \int_{x_0}^x \frac{d\xi}{f(\xi)}.$$

Moreover, $x(t)$ solves (1.7.3) if and only if it solves the functional equation $F(x(t)) = t$ for all $t > 0$ small enough. Since F is strictly monotone near x_0, we have that $x(t) = F^{-1}(t)$ for small $t > 0$ and thus the local solution $x(t)$ is locally unique. □

Remark 1.7.17. Proposition 1.7.16 implies that any solution $x(t)$ to the equation considered in Example 1.7.15 is locally unique for every initial datum $x_0 \neq 0$. We note that this fact does not mean that any extension of $x(t)$ outside a small neighborhood of this x_0 is also unique. For instance, the function

$$x(t) = \begin{cases} -(1-t)^2, & 0 \leq t \leq 1; \\ 0, & 1 < t \leq a; \\ (t-a)^2, & a < t < \infty, \end{cases}$$

for every $a \geq 1$ solves the equation $\dot{x} = 2\sqrt{|x|}$, $t > 0$, with the initial datum $x(0) = -1$. This solution is unique until it reaches the branching point $x_* = 0$ at the time $t_* = 1$. ∎

The following example shows that the smoothness of the function f is not sufficient for global existence. As can be seen from Exercise 1.7.20, the behavior of the right-hand side $f(x)$ as $|x| \to \infty$ is responsible for this.

Exercise 1.7.18. Show that any solution $x(t)$ to the following Cauchy problem:

$$\dot{x} = x^2, \quad t > 0, \quad x(0) = x_0 \in \mathbb{R},$$

has the form $x(t) = x_0(1 - x_0 t)^{-1}$, which blows up[5] for each initial datum $x_0 > 0$ at the time $T_* = 1/x_0$. ∎

The model below demonstrates more complicated types of behaviors of local solutions.

Exercise 1.7.19. Let $\lambda, \varkappa > 0$, $\mu \in \mathbb{R}$. Show that any solution $x(t)$ to the problem:

$$\dot{x} + \lambda x + \mu |x|^\varkappa x = 0, \quad t > 0, \quad x(0) = x_0 \in \mathbb{R}, \tag{1.7.4}$$

has the form

---

[5]This means that $|x(t)| \to \infty$ as $t \to T_*$ from the left.

$$x(t) = e^{-\lambda t} x_0 \left[1 + \frac{\mu}{\lambda} \left(1 - e^{-\lambda \varkappa t} \right) |x_0|^\varkappa \right]^{-1/\varkappa}. \tag{1.7.5}$$

Make sure that problem (1.7.4) has a global solution if and only if either $\mu \geq 0$ or else $\mu < 0$ and $|\mu| \lambda^{-1} |x_0|^\varkappa \leq 1$.

Show that in the case $\lambda = 0$ problem (1.7.4) has a global solution if and only if either $\mu \geq 0$ or else $\mu < 0$ and $x_0 = 0$. Hint: In the limit $\lambda \to 0$ (1.7.5) gives $x(t) = x_0 [1 + \mu \varkappa t |x_0|^\varkappa]^{-1/\varkappa}$.   ∎

To conclude this section, we mention the following result on global existence for 1D ODEs with a continuous right-hand side (see, e.g., CODDINGTON/LEVINSON [75]).

Exercise 1.7.20. Let f be a continuous function on $\mathbb{R}$. Then for every $x_0 \in \mathbb{R}$ any local solution to problem (1.7.3) can extended on the whole time semi-axis $\mathbb{R}_+$, provided there exists a continuous function $\psi(r)$ on $\mathbb{R}_+$ such that

$$\forall x \in \mathbb{R}: \quad |f(x)| \leq \psi(|x|) \quad \text{and} \quad \exists \delta \geq 0: \quad \int_\delta^\infty \frac{dr}{\psi(r)} = \infty.$$

Hint: Apply the non-explosion criterion of Theorem A.1.2.   ∎

1.8   Possible types of qualitative behaviors in 2D systems

The theory of continuous 2D systems is much more complicated than 1D theory. Nevertheless, due to the fact that trajectories in 2D systems separate the phase space into two parts and cannot intersect each other, it is still possible to develop a rather deep and complete theory of 2D continuous systems (see, e.g., the monographs CODDINGTON/LEVINSON [75], HARTMAN [120], LEFSCHETZ [148], NEMYTSKII/STEPANOV [171] and also BAUTIN/LEONTOVICH [11], REISSING/SANSONE/CONTI [189]). In this section we will discuss possible scenarios of dynamical behavior in continuous systems on the plane $\mathbb{R}^2$.

1.8.1   General facts and Poincaré-Bendixson theory

We start with the following assertion, which gives a complete description of possible structures of ω-limit sets of individual trajectories (for the proof we refer to CODDINGTON/LEVINSON [75], LEFSCHETZ [148], NEMYTSKII/STEPANOV [171]).

Theorem 1.8.1. *Let $\gamma^+(v) = \{S_t v : t \geq 0\}$ be a semitrajectory of a continuous 2D dynamical system $(\mathbb{R}^2, S_t)$. Assume that $\gamma^+(v)$ is Lagrange stable (see Defini-*

tion 1.4.1). In this case the ω-limit set $\omega(v)$ is a compact connected strictly invariant set (see Section 1.4). There are only two possibilities for this set:

- *either $\omega(v)$ is a cycle (periodic trajectory);*
- *or $\omega(v)$ consists of some subset $\mathscr{N}_v$ of the set $\mathscr{N}$ of equilibria and (possibly) some set of full trajectories $\gamma = \{u(t) : t \in \mathbb{R}\}$ such that $u(t) \to \mathscr{N}_v$ as $t \to \pm\infty$ (this means that the α-limit $\alpha(\gamma)$ and ω-limit $\omega(\gamma)$ sets belong to $\mathscr{N}_v$).*

If the set $\mathscr{N}$ of equilibria is finite, then under the conditions of Theorem 1.8.1 we have only one of the following possibilities:

- $\omega(v)$ is a single equilibrium;
- $\omega(v)$ is a single cycle;
- $\omega(v)$ consists of a number of equilibrium points and full trajectories connecting these points.

We illustrate these types of behaviors in the following exercises.

Exercise 1.8.2. Consider the systems in $\mathbb{R}^2$ generated by the equations

$$\dot{x}_1 = -\lambda x_1, \quad \dot{x}_2 = -\mu x_2, \quad \lambda, \mu > 0.$$

Show that the equilibrium $(0,0)$ is the ω-limit set for every semitrajectory of the system. ∎

Exercise 1.8.3. Let $(\mathbb{R}^2, S_t)$ be the system generated by the equations

$$\begin{cases} \dot{x}_1 = -\beta x_2 + x_1 - x_1(x_1^2 + x_2^2), \\ \dot{x}_2 = \beta x_1 + x_2 - x_2(x_1^2 + x_2^2). \end{cases}$$

Show that in the case $\beta \neq 0$, the circle $\gamma = \{(x_1, x_2) : x_1^2 + x_2^2 = 1\}$ is the ω-limit set for *every* nonzero semitrajectory. If $\beta = 0$, then every point of this circle is an equilibrium which is an ω-limit set for *some* semitrajectory. Hint: Use the polar coordinates $(x_1 = \varrho \cos\varphi, x_2 = \varrho \sin\varphi)$ to simplify equations. ∎

Exercise 1.8.4 (see Chueshov [39], Chapter 1). Let us consider the following quasi-Hamiltonian system in $\mathbb{R}^2$:

$$\begin{cases} \dot{q} = \frac{\partial H}{\partial p} - \mu H \frac{\partial H}{\partial q}, \\ \dot{p} = -\frac{\partial H}{\partial q} - \mu H \frac{\partial H}{\partial p}, \end{cases}$$

with $H(p,q) = \frac{1}{2}p^2 + q^4 - q^2$ and $\mu > 0$. Show that these equations generate a dynamical system in $\mathbb{R}^2$ and the separatrix $\Gamma = \{(q,p) : H(p,q) = 0\}$ is the ω-limit set for any trajectory starting in the domain $\{(q,p) : H(p,q) > 0\}$. This set Γ consists of the equilibrium point $(0,0)$ and also two (homoclinic) trajectories

$$\gamma_+ = \{(q_+(t), p_+(t)) : t \in \mathbb{R}^2\} \text{ and } \gamma_- = \{(q_-(t), p_-(t)) : t \in \mathbb{R}^2\}$$

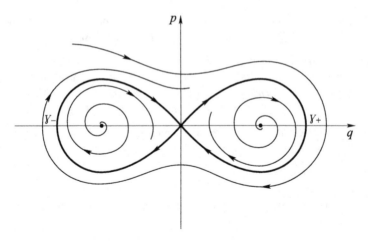

Fig. 1.2 ω-limit set consisting of two homoclinics and unstable equilibrium

such that $q_+(t) > 0$, $q_-(t) < 0$, and $(q_\pm(t), p_\pm(t)) \to (0,0)$ as $t \to \pm\infty$. The corresponding dynamics is shown in Figure 1.2.   ∎

There are several important facts related to the Poincaré-Bendixson theory which provides criteria for the existence of equilibria and periodic orbits. For instance, Theorem 1.8.1 implies the following assertion.

Corollary 1.8.5. *Let B be a bounded closed forward invariant set for* $(\mathbb{R}^2, S_t)$. *Assume that B does not contain equilibria. Then there is a periodic orbit inside B.*

Proof. Any point $v \in B$ is Lagrange stable in B. Thus $\omega(v)$ lies in B and thus does not contain equilibria. Therefore, by Theorem 1.8.1, $\omega(v)$ is a cycle.   □

We also mention the following result due to Bendixson (see the reference given in CODDINGTON/LEVINSON [75]).

Theorem 1.8.6 (Bendixson). *Let a 2D system* $(\mathbb{R}^2, S_t)$ *possess a periodic trajectory* γ. *Then there is at least one equilibrium inside the domain bounded by* γ.

Proof. Since γ is a simple closed curve, then the domain D bounded by γ is homeomorphic to the unit disc. The set D is definitely forward invariant. Therefore, we can apply Theorem 1.2.5.   □

Theorem 1.8.6 gives us the following assertion.

Corollary 1.8.7. *A simply connected domain D in* $\mathbb{R}^2$ *does not contain periodic orbits provided there are no equilibria inside D.*

This corollary means that the situation of Corollary 1.8.5 can be realized in the case when B is not simply connected only.

1.8.2 Lower bounds for cycle periods

In spite of the several criteria mentioned above, the problem of existence of (non-constant) periodic solutions of *autonomous* equations is much more difficult than the existence of fixed points. An autonomous system does not contain any a priori exact information about the period of a possible periodic solution. In this context it is interesting to mention the following result due to YORKE [228], which provides the lower bounds for possible periods.

Theorem 1.8.8 (Yorke, 1969). *Any (nontrivial) periodic orbit of the equation $\dot{x} = f(x)$ ($x \in \mathbb{R}^d$), where f is globally Lipschitz with the constant L on $\mathbb{R}^d$, $d \geq 2$, has period $T \geq 2\pi/L$.*

Exercise 1.8.9. Show that all nontrivial solutions of the system in $\mathbb{R}^2$ generated by

$$\dot{x}_1 = -\omega x_2, \quad \dot{x}_2 = \omega x_1,$$

are periodic with the minimal period $T = 2\pi/\omega$. Thus, the estimate for the period in Theorem 1.8.8 is sharp. ∎

The model borrowed from FARKAS [97] shows that without the global Lipschitz property a system may possess orbits of arbitrary minimal period.

Exercise 1.8.10 (Farkas, [97]). Let $\alpha > 0$. Show that for every $T \in (0, \infty)$ in the system $(\mathbb{R}^2, S_t)$ generated by the equations

$$\dot{x}_1 = -x_2(x_1^2 + x_2^2)^\alpha, \quad \dot{x}_2 = x_1(x_1^2 + x_2^2)^\alpha,$$

there is a periodic orbit with minimal period T. Hint: In the polar coordinates $(x_1 = \varrho \cos \varphi, x_2 = \varrho \sin \varphi)$ the equations have the form $\dot{\varrho} = 0$ and $\dot{\varphi} = \varrho^{2\alpha}$. ∎

The globally Lipschitz condition in Theorem 1.8.8 can be relaxed in the following way.

Exercise 1.8.11. Assume that $\dot{x} = f(x)$ generates a dynamical system in $\mathbb{R}^d$ which possesses a (bounded) forward invariant set B. Show that there are no nontrivial periodic solutions inside B with period less than $2\pi/L_B$, where L_B is the Lipschitz constant for f on B. ∎

Proof of Theorem 1.8.8. We follow the line of the argument given in the short note BUSENBERG/FISHER/MARTELLI [21] and rely on the following Poincaré-Wirtinger inequality:

$$\int_0^T |u(t)|^2 dt \leq \frac{T^2}{4\pi^2} \int_0^T |\dot{u}(t)|^2 dt \tag{1.8.1}$$

for every scalar T-periodic function possessing the properties

$$\dot{u} \in L_2(0,T), \quad \int_0^T u(t)dt = 0.$$

The inequality in (1.8.1) easily follows from calculations with Fourier series.

Let $x(t) = (x_1(t), \dots x_d(t))$, be a periodic orbit with the (minimal) period T. Let

$$v(t) \equiv (v_1(t), \dots, v_d(t)) = x(t) - x(t - \tau)$$

with a fixed $\tau > 0$. By periodicity we obviously have that

$$\int_0^T v_i(s)ds = 0 \text{ for every } \tau > 0, \; i = 1, \dots, d.$$

Therefore by (1.8.1),

$$\int_0^T |v_i(t)|^2 dt \le \frac{T^2}{4\pi^2} \int_0^T |\dot{v}_i(t)|^2 dt = \frac{T^2}{4\pi^2} \int_0^T |f_i(x(t)) - f_i(x(t-\tau))|^2 dt$$

This implies that

$$\int_0^T |v(t)|_{\mathbb{R}^d}^2 dt = \sum_{i=1}^d \int_0^T |v_i(t)|^2 dt$$

$$\le \frac{L^2 T^2}{4\pi^2} \int_0^T |x(t) - x(t-\tau)|_{\mathbb{R}^d}^2 dt = \frac{L^2 T^2}{4\pi^2} \int_0^T |v(t)|_{\mathbb{R}^d}^2 dt.$$

If $LT < 2\pi$, then $v(t) = x(t) - x(t-\tau) \equiv 0$ for $t \in [0,T]$ and for every $\tau > 0$. Thus $x(t)$ is a stationary point.   □

We note that the argument above does not use the fact that the equation is finite-dimensional and can be applied to ODEs with globally Lipschitz right-hand sides in arbitrary Hilbert spaces BUSENBERG/FISHER/MARTELLI [21]. Some results are available in Banach spaces; see BUSENBERG/FISHER/MARTELLI [21] and also the recent paper NIEUWENHUIS/ROBINSON/STEINERBERGER [172] and the references therein. The same idea was already applied in ROBINSON [196] and ROBINSON/VIDAL-LÓPEZ [197, 198] to obtain lower bounds for periods of solutions to some classes of semilinear parabolic equations.

1.8.3   Example of ω-limit set with three unstable equilibria

To give an additional illustration of possible structures of ω-limit sets in 2D systems we consider the following coupled equations:

$$\dot{x} = (x - \lambda)y, \tag{1.8.2a}$$

$$\dot{y} = \lambda x + \frac{1}{2}(x^2 - y^2), \tag{1.8.2b}$$

with $\lambda > 0$. This system was considered before in GUCKENHEIMER/HOLMES [114, Section 18].

One can see that for any initial data $(x_0; y_0) \in \mathbb{R}^2$ the system in (1.8.2) has a unique local solution; i.e., it generates a local semiflow.

Exercise 1.8.12. Show that the system in (1.8.2) can be written in the Hamiltonian form:

$$\dot{x} = \frac{\partial H}{\partial y}, \quad \dot{y} = -\frac{\partial H}{\partial x},$$

where

$$H(x, y) = -\frac{\lambda}{2}(x^2 + y^2) + \frac{1}{2}\left(xy^2 - \frac{x^3}{3}\right) + \frac{2}{3}\lambda^3$$
$$= \frac{1}{2}(x - \lambda)\left(y^2 - \frac{1}{3}(x + 2\lambda)^2\right).$$

Thus $H(x, y)$ is a constant on solutions.   ∎

Exercise 1.8.13. Show that

$$Y_0 = (0; 0), \quad Y_1 = (-2\lambda; 0), \quad Y_\pm = (\lambda; \pm\sqrt{3}\lambda)$$

are equilibria for (1.8.2). These points can be seen in Figure 1.3.   ∎

Exercise 1.8.14. Using (1.8.2a) show that the value $z(t) = x(t) - \lambda$ preserves its sign in time evolution. The same is true for $z_\pm(t) = y(t) \pm \frac{1}{\sqrt{3}}(x(t) + 2\lambda)$. Moreover, the lines (see Figure 1.3)

$$I_0 = \{x - \lambda = 0\} \text{ and } I_\pm = \{y \pm \frac{1}{\sqrt{3}}(x + 2\lambda) = 0\}$$

are invariant sets. Hint: Using the Hamiltonian representation with H written as $H(x, y) = \frac{1}{2}(x - \lambda)z_+z_-$ one can see that z_+ satisfies the equation

$$\dot{z}_+ = \frac{1}{2}[2(x - \lambda)/\sqrt{3} - z_-]z_+$$

and a similar relation for z_-.   ∎

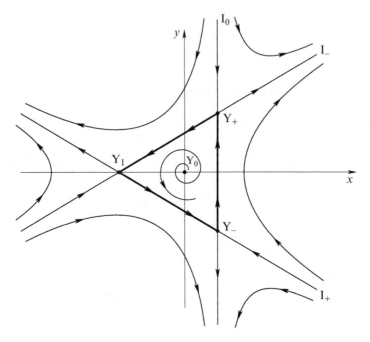

Fig. 1.3 Qualitative dynamics of quasi-Hamiltonian 2D system (1.8.3)

Exercise 1.8.15. On the vertical line $I_0 = \{x = \lambda\}$ the dynamics of system (1.8.2) is described by the equation

$$\dot{y} = \frac{3}{2}\lambda^2 - \frac{1}{2}y^2, \ t > 0, \quad y(0) = y_0.$$

Describe the qualitative behavior of trajectories on I_0. Make sure that solutions starting with $y_0 < -\sqrt{3}\lambda$ blow up. ∎

Exercise 1.8.16. Describe the dynamics on the lines $I_\pm = \{y \pm \frac{1}{\sqrt{3}}(x + 2\lambda) = 0\}$. Hint: See Figure 1.3. ∎

Exercise 1.8.17. Using the result of Exercise 1.8.14, make sure that the triangle

$$\Delta = \left\{(x; y) \ : \ -2\lambda < x < \lambda, \ y^2 < \frac{1}{3}(x + 2\lambda)^2\right\}$$

and its closure $\bar{\Delta}$ are forward invariant sets and the (local) semiflow restricted on $\bar{\Delta}$ is global. The triangle Δ is shown in Figure 1.3. ∎

Now we consider a quasi-Hamiltonian modification of system (1.8.2) of the form

$$\dot{x} = \frac{\partial H}{\partial y} - \mu H \frac{\partial H}{\partial x}, \quad \dot{y} = -\frac{\partial H}{\partial x} - \mu H \frac{\partial H}{\partial y}, \qquad (1.8.3)$$

where $H(x, y)$ is the same as in Exercise 1.8.12 and $\mu > 0$ is a parameter. It is clear that equations (1.8.3) generate a local semiflow. Moreover, one can see that

- The points Y_0, Y_1, and $Y_\pm$ defined in Exercise 1.8.13 are equilibria for the quasi-Hamiltonian system in (1.8.3).
- The lines I_0 and $I_\pm$ defined in Exercise 1.8.14 are invariant with respect to the semiflow generated by (1.8.3).
- The triangle Δ (see Exercise 1.8.17) and its closure $\bar{\Delta}$ are invariant with respect to the dynamics governed by (1.8.3). Moreover, for every initial data $(x_0; y_0) \in \bar{\Delta}$ there exists a *global* solution to (1.8.3). Thus, equations (1.8.3) generate a dynamical system in the triangle $\bar{\Delta}$.

Using the obvious relation

$$\frac{d}{dt} H = -\mu H \left[\left(\frac{\partial H}{\partial x} \right)^2 + \left(\frac{\partial H}{\partial y} \right)^2 \right]$$

on solutions to (1.8.3), one can also see that for $\mu > 0$ the ω-limit set for every nonzero point inside Δ is the boundary $\partial \Delta$, which consists of three (unstable) equilibria (vertexes of the triangle) and connecting them full trajectories (sides of the triangle).[6] The qualitative behavior of the system generated by (1.8.3) is shown in Figure 1.3. For other illustrations of possible dynamics in 2D systems we refer to the 2D examples in Section 1.9.

1.8.4 On 2D systems generated by a second order equation

An important class of 2D systems arises from the Newton laws of dynamics of a material point which lead to the following equation:

$$\ddot{x} = f(x, \dot{x}), \quad x|_{t=0} = x_0, \quad \dot{x}|_{t=0} = x_1.$$

We can write this equation as a first order 2D system

$$\dot{x} = y, \quad \dot{y} = f(x, y), \quad x|_{t=0} = x_0, \quad y|_{t=0} = x_1,$$

and apply the standard Peano-Carathéodory result (see Theorem A.1.2 in the Appendix) to guarantee the local existence of solutions when $f(x, y)$ is a continuous function. To show that the model above generates a dynamical system, we need additional hypotheses. We discuss these conditions, concentrating mainly on models which are important from an applications point of view.

---

[6]In the Hamiltonian case ($\mu = 0$) every nonzero trajectory γ starting in Δ is periodic, i.e., $\gamma = \omega(\gamma) = \alpha(\gamma)$.

Example 1.8.18 (Lienard equation, I). We start with a model with a state-dependent damping coefficient. Let $k(x)$ be a locally Lipschitz function on $\mathbb{R}$ and $U(x) \in C^1(\mathbb{R})$ with locally Lipschitz derivative $U'(x)$. Assume also that

$$\inf_{x \in \mathbb{R}} k(x) > -\infty \quad \text{and} \quad \inf_{x \in \mathbb{R}} U(x) > -\infty.$$

One can see that the problem

$$\ddot{x} + k(x)\dot{x} + U'(x) = 0, \quad x|_{t=0} = x_0, \quad \dot{x}|_{t=0} = x_1, \tag{1.8.4}$$

has a unique solution $x(t)$ which generates a dynamical system on $\mathbb{R}^2$ with the evolution operator S_t given by the formula $S_t(x_0; x_1) = (x(t); \cdot x(t))$. Moreover, any solution satisfies the energy balance equation

$$\frac{1}{2}\dot{x}(t)^2 + U(x(t)) + \int_0^t k(x(\tau))\dot{x}(\tau)^2 d\tau = \frac{1}{2}x_1^2 + U(x), \quad t > 0.$$

∎

Remark 1.8.19. Some authors (see, e.g., LEFSCHETZ [148]) called (1.8.4) the Cartwright-Littlewood equation. In the case when

$$k(x) = -k_0(1 - x^2), \ k_0 > 0, \quad \text{and} \quad U(x) = \frac{a}{2}x^2, \ a > 0,$$

the model in (1.8.4) is called the van der Pol equation. If

$$k(x) \equiv k_0 \geq 0 \quad \text{and} \quad U(x) = \frac{a}{4}x^4 + \frac{b}{3}x^3 + \frac{c}{2}x, \ a > 0, \ b, c \in \mathbb{R},$$

then (1.8.4) is called the Duffing equation. ∎

Example 1.8.20 (Lienard equation, II). In this model we deal with nonlinear damping depending on velocity only. Let $U(x) \in C^1(\mathbb{R})$ with locally Lipschitz derivative $U'(x)$ and $\inf_{x \in \mathbb{R}} U(x) > -\infty$. The problem

$$\ddot{x} + g(\dot{x}) + U'(x) = 0, \quad x|_{t=0} = x_0, \quad \dot{x}|_{t=0} = x_1 \tag{1.8.5}$$

generates a dynamical system on $\mathbb{R}^2$ if one of the following conditions holds:

(a) $g(s) \in C(\mathbb{R})$ and there exists $c \in \mathbb{R}_+$ such that $g(s) + as$ is not decreasing;
(b) $g(s)$ is locally Lipschitz and $g(s)s \geq 0$.

This result can be derived from Theorem A.1.2 and relies on the corresponding energy balance relation which allows us to show the absence of blow-up phenomena.

∎

Exercise 1.8.21. Let $(\mathbb{R}^2, S_t)$ be a dynamical system generated by the Duffing equation

$$\ddot{x} + x^3 - x = 0.$$

Calculate the relative times (see the definition in Section 1.5.2) which different trajectories spend near $(0; 0)$. ∎

Exercise 1.8.22 (Krasovskii example [136]). Consider the equation

$$\ddot{x} + k(x^2 + \dot{x}^2)\dot{x} + x = 0$$

with the damping coefficient function $k(r)$ possessing the properties

$$k \in Lip_{loc}(\mathbb{R}_+), \quad \inf_{r \in \mathbb{R}_+} k(r) > -\infty.$$

Show that (i) this equation generates a dynamical system in $\mathbb{R}^2$ and (ii) for each root ϱ_0 of the function $k(r)$ the set $\{(x; \dot{x}) : x^2 + \dot{x}^2 = \varrho_0^2\}$ is a periodic orbit of period $T = 2\pi$. ∎

1.9 Bifurcation theory by means of examples

If we deal with a family of dynamical systems (S_t^μ, X) depending on the parameter μ, in principle, we can observe different types of qualitative behaviors for different values of the parameter μ. Moreover, it is well known (see, e.g., GUCKENHEIMER/HOLMES [114], HALE/KOCAK [117], KUZNETSOV [139]) that small changes in the parameters can produce large changes in the qualitative behavior of trajectories. According to KUZNETSOV [139] the appearance of a topologically nonequivalent dynamical behavior under variation of parameters is called a *bifurcation* and the goal of bifurcation theory is to produce bifurcation diagrams that divide the parameter space into regions of topologically equivalent systems.

In this section by means of examples we demonstrate several types of bifurcations (for a general bifurcation theory we refer to GUCKENHEIMER/HOLMES [114], HALE/KOCAK [117], KUZNETSOV [139]) and the references therein). The examples presented here have the dimension 1 or 2. However, all of them can be used to produce similar pictures of qualitative behavior of infinite-dimensional (PDE) models; see the discussion in Section 4.2.5.

Example 1.9.1 (Pitchfork bifurcation). We consider a 1D system generated by the equation

$$\dot{x} = \mu x - x^3.$$

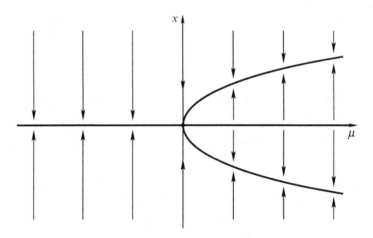

Fig. 1.4 Pitchfork bifurcation: loss of stability of the zero equilibrium

For $\mu \leq 0$ the system has one stable fixed point $x = 0$. This point becomes unstable and splits off two stable fixed points $\pm\sqrt{\mu}$ when $\mu > 0$. The point $x = 0$ remains fixed but becomes unstable. For each $\mu \in \mathbb{R}$ this qualitative behavior is presented in Figure 1.4. The bold line and curve show equilibria. The arrows demonstrate dynamics of trajectories for each $\mu \in \mathbb{R}$. ∎

Example 1.9.2 (Transcritical bifurcation).   The system generated by

$$\dot{x} = \frac{\mu x - 2x^2}{1 + x^2}$$

has two fixed points for $\mu \neq 0$ which collide and exchange stability at $\mu = 0$ ($x = 0$ is stable when $\mu < 0$ and unstable for $\mu > 0$). The qualitative behavior is presented in Figure 1.5. The bold lines are equilibria. The arrows demonstrate stability/instability effects for a fixed μ. ∎

Example 1.9.3 (Saddle-node (fold) bifurcation).   We consider on $\mathbb{R}^2$ a dynamical system generated by the equations

$$\dot{x} = \frac{\mu - x^2}{1 + x^2}, \quad \dot{y} = -y. \tag{1.9.1}$$

We observe the following *bifurcation* behavior (see Figure 1.6):

- $\mu < 0$: there are no equilibrium points (for every initial data $(x_0; y_0)$ and we have that $S_t(x_0; y_0) = (x(t); y(t))$ tends to $(-\infty; 0)$ as $t \to \infty$), see Figure 1.6(a).
- $\mu = 0$: a non-hyperbolic unstable equilibrium arises at $(0; 0)$ and we have the following picture (Figure 1.6(b)):

  - if $x_0 < 0$, then $S_t(x_0; y_0) = (x(t); y(t))$ tends to $(-\infty; 0)$ as $t \to \infty$;
  - if $x_0 \geq 0$, then $S_t(x_0; y_0) \to (0; 0)$ as $t \to \infty$.

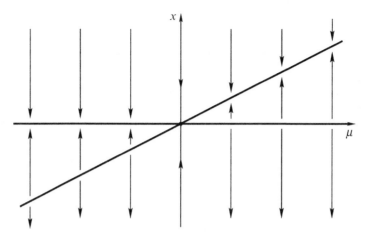

Fig. 1.5 Transcritical bifurcations: exchange of stability

- $\mu > 0$: we have two equilibrium points $(\pm\sqrt{\mu}; 0)$ and the following picture:
    - if $x_0 \geq \sqrt{\mu}$, then $S_t(x_0; y_0) = (x(t); y(t))$ tends to $(\sqrt{\mu}; 0)$ as $t \to \infty$;
    - if $|x_0| < \sqrt{\mu}$, then $S_t(x_0; y_0) \to (\sqrt{\mu}; 0)$ as $t \to \infty$;
    - if $x_0 < -\sqrt{\mu}$, then $S_t(x_0; y_0) \to (-\infty; 0)$ as $t \to \infty$.

    Hence we have two equilibria; one of them is a stable node and another is a saddle (there are both stable and unstable directions), see Figure 1.6(c).

Thus, when μ becomes positive we observe a generation of two equilibria connected by a heteroclinic trajectory from the regular picture (without any rest points). ∎

Example 1.9.4 (Andronov-Hopf bifurcation).   We consider a family $(S_t^\mu, \mathbb{R}^2)$ of dynamical systems generated by the following equations:

$$\begin{cases} \dot{x}_1 = \mu x_1 - x_2 - x_1(x_1^2 + x_2^2), \\ \dot{x}_2 = x_1 + \mu x_2 - x_2(x_1^2 + x_2^2) \end{cases} \tag{1.9.2}$$

In polar coordinates $(x_1 = \varrho\cos\varphi, x_2 = \varrho\sin\varphi)$ the problem in (1.9.2) can be written as

$$\begin{cases} \dot{\varrho} = \varrho(\mu - \varrho^2), \\ \dot{\varphi} = 1. \end{cases} \tag{1.9.3}$$

Therefore, we observe the following bifurcation picture:

- $\mu < 0$: unique exponentially stable equilibrium (focus), see Figure 1.7(a);
- $\mu = 0$: unique (non-exponential) stable focus, see Figure 1.7(a);

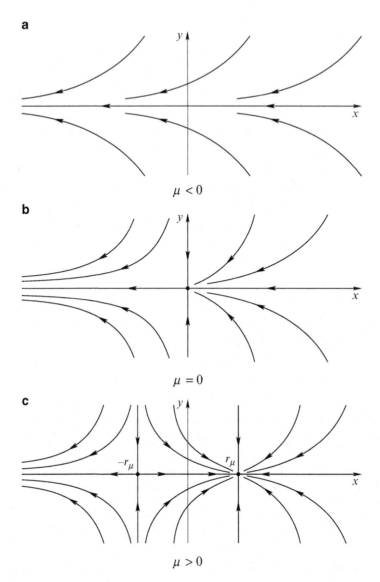

Fig. 1.6 Saddle-node bifurcation, generation of two equilibria from regular picture: (a) no equilibria, (b) saddle-node equilibrium, (c) saddle $(-r_\mu; 0)$ and node $(r_\mu; 0)$ equilibria, $r_\mu = \sqrt{\mu}$

- $\mu > 0$: unstable focus at zero and stable periodic orbit (the circle with center at 0 and radius $r_\mu = \sqrt{\mu}$). The dynamics is shown in Figure 1.7(b).

Thus, a periodic orbit arises from zero equilibria. The period does not depend on μ. The size of the orbit is small for small $\mu > 0$. We observe the "soft" regime of the cycle appearance. This means (see, e.g., KUZNETSOV [139]) that small changes

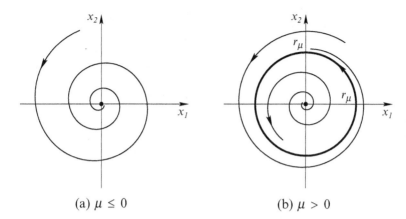

Fig. 1.7 Andronov-Hopf bifurcation: generation of periodic orbit from equilibrium: (a) stable focus, (b) unstable focus and stable periodic orbit, $r_\mu = \sqrt{\mu}$

of the bifurcation parameter μ cannot produce large changes in the dynamics of an individual trajectory whose initial data do not depend on μ. In other words, if we change μ back and forth, the dynamics of the trajectories changes continuously. ∎

All bifurcations above are local. They can be detected by looking at small neighborhoods of equilibrium (fixed) points. Now we give two examples of global (nonlocal) bifurcations.

Example 1.9.5 (Generation of periodic orbit from infinity). This is a small modification of the system presented in Example 1.9.4. We consider the following equations:

$$\begin{cases} \dot{x}_1 = x_1 - x_2 - \mu x_1(x_1^2 + x_2^2), \\ \dot{x}_2 = x_1 + x_2 - \mu x_2(x_1^2 + x_2^2) \end{cases} \tag{1.9.4}$$

In polar coordinates this problem can be written as

$$\dot{\varrho} = \varrho(1 - \mu\varrho^2), \quad \dot{\varphi} = 1.$$

Therefore, we observe the following bifurcation picture (see Figure 1.8):

- $\mu \leq 0$: unique unstable focus, Figure 1.8(a);
- $\mu > 0$: unstable focus at zero and stable periodic orbit (the circle with center at 0 and radius $1/\sqrt{\mu}$), Figure 1.8(b).

Thus, a periodic orbit arises from infinity. The size of the orbit goes to infinity as $\mu \to +0$. In this limit the orbit disappears at infinity. ∎

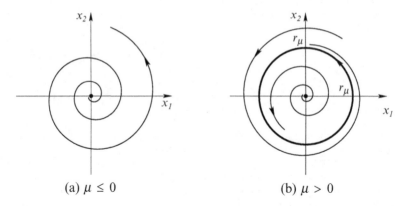

(a) $\mu \leq 0$ (b) $\mu > 0$

Fig. 1.8 Generation of periodic orbit from infinity: (a) unstable focus, (b) unstable focus and stable cycle, $r_\mu = 1/\sqrt{\mu}$

Example 1.9.6 (Saddle-node homoclinic bifurcation).   As in KUZNETSOV [139, p. 59] we consider following equations on the plane $\mathbb{R}^2$:

$$\begin{cases} \dot{x}_1 = x_1(1 - x_1^2 - x_2^2) - x_2(1 + \mu + x_1), \\ \dot{x}_2 = x_1(1 + \mu + x_1) + x_2(1 - x_1^2 - x_2^2) \end{cases} \tag{1.9.5}$$

In polar coordinates ($x_1 = \varrho \cos\varphi$, $x_2 = \varrho \sin\varphi$) the problem in (1.9.5) has the form

$$\begin{cases} \dot{\varrho} = \varrho(1 - \varrho^2), \\ \dot{\varphi} = 1 + \mu + \varrho \cos\varphi. \end{cases} \tag{1.9.6}$$

For all $\mu \in \mathbb{R}$ the unit circle $\{(\varrho; \varphi) : \varrho = 1\}$ is an invariant set of the corresponding dynamical system $(S_t^\mu, \mathbb{R}^2)$. At $\mu = 0$, there is a (non-hyperbolic) equilibrium point of the system: $x^* = (\varrho^*; \varphi^*)) = (1; \pi)$. For small positive values of μ the equilibrium on the circle disappears (Figure 1.9(c)), while for small negative μ it splits into a saddle and a node connected by (heteroclinic) orbits (saddle-node bifurcation on the circle); see Figure 1.9(a). Thus, for $\mu > 0$ a stable limit cycle appears in the system coinciding with the unit circle. This circle is always an invariant set in the system, but for $\mu \leq 0$ it contains equilibria and thus does not represent a periodic orbit. So we observe a generation of a periodic orbit. We note that for $\mu = 0$ there is exactly one orbit that is homoclinic to the non-hyperbolic equilibrium x^*. Thus, we observe a generation of a periodic orbit from a homoclinic trajectory. We also note that this example with $\mu = 0$ demonstrates the effect that a globally attracting equilibrium can be unstable.   ∎

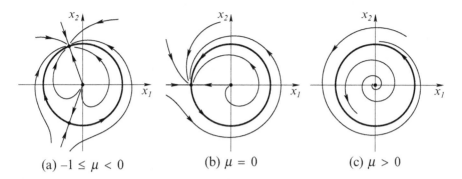

Fig. 1.9 Saddle-node homoclinic bifurcation, generation of periodic orbit from saddle and node via homoclinic trajectory: (a) saddle and node with connecting heteroclinic orbits, (b) non-hyperbolic equilibrium on circle with homoclinic orbit, (c) unstable focus and stable periodic orbit

For the theory and further examples of possible bifurcation scenarios we refer to GUCKENHEIMER/HOLMES [114], HALE/KOCAK [117], KUZNETSOV [139] and the references therein. As we have already mentioned, it is well to bear in mind that all scenarios represented by ODE systems can be realized in evolution PDE models. See the discussion in Section 4.2.5.

Chapter 2
General Facts on Dissipative Systems

In this chapter we deal with the qualitative theory pertinent to (infinite-dimensional) dissipative systems. Our presentation is based mainly on some new criteria for asymptotic compactness which rely on a certain weak form of quasi-stability. We also emphasize a role of gradient systems for the existence of global attractors. A similar approach was discussed earlier in CHUESHOV/LASIECKA [56, 58] in a short form without many details. For other possible approaches to the topic we refer to the monographs BABIN/VISHIK [9], CHUESHOV [39], HALE [116], HENRY [123], LADYZHENSKAYA [142], ROBINSON [195], SELL/YOU [206], TEMAM [216] and the surveys BABIN [7] and RAUGEL [188].

Our main focus is on questions such as the existence of global attractors and their structure. We present ideas and methods which are applicable to the systems generated by nonlinear partial differential equations. We discuss these applications in Chapters 4–6 in detail for several PDE classes. In the current chapter we illustrate general results on long-time dynamics by means of finite-dimensional ODE examples only. Questions related to dimensions and smoothness of attractors for infinite-dimensional systems are considered in Chapter 3.

2.1 Dissipative dynamical systems

The main topic of this book is that of dissipative dynamical systems. As already mentioned in the Introduction, from a physical point of view, dissipative systems are characterized by relocation and dissipation of energy. This means that the energy of higher modes is dissipated and relocated to low modes. The interaction of these two mechanisms can lead to the appearance of complicated limit regimes and structures in the system that are stable in a suitable sense.

© Springer International Publishing Switzerland 2015
I. Chueshov, *Dynamics of Quasi-Stable Dissipative Systems*, Universitext,
DOI 10.1007/978-3-319-22903-4_2

We start with a description of several concepts which present both dissipation and relocation on a formal level.

Definition 2.1.1. Let S_t be an evolution operator on a complete metric space X and (X, S_t) be the corresponding dynamical system.

- A closed set $B \subset X$ is said to be *absorbing* for S_t if for any bounded set $D \subset X$ there exists $t_0(D)$ such that $S_t D \subset B$ for all $t \geq t_0(D)$.
- S_t is said to be *(bounded) dissipative* if it possesses a bounded absorbing set B. If the phase space X of a dissipative evolution operator S_t is a Banach space, then the radius of a ball containing an absorbing set is called a *radius of dissipativity* of S_t.
- S_t is said to be *point dissipative* if there exists a bounded set $B_0 \subset X$ such that for any $x \in X$ there is $t_0(x)$ such that $S_t x \in B_0$ for all $t \geq t_0(x)$.

We apply the same terminology to the corresponding dynamical system (X, S_t). ∎

The following criterion of dissipativity covers many cases which are important from an applications point of view.

Theorem 2.1.2 (Criterion of dissipativity). *Let (X, S_t) be a continuous dynamical system in some Banach space X. Assume that*

- *there exists a continuous function $U(x)$ on X possessing the properties*

$$\phi_1(\|x\|) \leq U(x) \leq \phi_2(\|x\|), \quad \forall x \in X, \tag{2.1.1}$$

where ϕ_i are continuous functions on $\mathbb{R}_+$ such that $\phi_i(r) \to +\infty$ as $r \to +\infty$;
- *there exist a derivative $\frac{d}{dt}U(S_t y)$ for every $t > 0$ and $y \in X$, a positive function[1] $\alpha(r)$ on $\mathbb{R}_+$, and a positive number ϱ such that*

$$\frac{d}{dt}U(S_t y) \leq -\alpha(\|y\|) \quad provided \ \|S_t y\| > \varrho. \tag{2.1.2}$$

Then the dynamical system (X, S_t) is dissipative with an absorbing set of the form

$$B_* = \{x : \|x\| \leq R_*\}, \tag{2.1.3}$$

where the constant R_* depends on the functions ϕ_1 and ϕ_2 and the constant ϱ only.

Proof. The argument involves some kind of "barrier method"; see, e.g., REISSING/SANSONE/CONTI [189] for a discussion in the ODE case.

Let us choose $R_0 > \varrho$ such that $\phi_1(r) > 0$ for all $r \geq R_0$. Let

$$L = \sup\{\phi_2(r) : r \leq 1 + R_0\}.$$

---

[1] This function $\alpha(r)$ may tend to zero as $r \to +\infty$.

We show that the ball B in (2.1.3) is absorbing provided $R_* \geq R_0 + 1$ is chosen such that $\phi_1(r) > L$ for $r \geq R_*$. This choice is definitely possible and R_* can be taken dependent on ϕ_1, ϕ_2, and ϱ only.

Our argument consists of two steps.

Step 1. First we show that

$$\|S_t y\| \leq R_* \quad \text{for all } t \geq 0 \text{ and } \|y\| \leq R_0. \tag{2.1.4}$$

Indeed, if this is not true, then for some $y \in X$ such that $\|y\| \leq R_0$ there exists a time $\bar{t} > 0$ possessing the property $\|S_{\bar{t}} y\| > R_*$. By the continuity of $S_t y$ this implies that there exists $0 < t' < \bar{t}$ such that $\|S_{t'} y\| = 1 + R_0 > \varrho$. Let

$$t_0 = \sup\{\tau < \bar{t} : \|S_\tau y\| = 1 + R_0\}. \tag{2.1.5}$$

It is clear that $\|S_{t_0} y\| = 1 + R_0 > \varrho$. Therefore, equation (2.1.2) implies that

$$\phi_1(\|S_t y\|) \leq U(S_t y) \leq U(S_{t_0} y) \leq L \quad \text{for } t \in [t_0, t_1],$$

where

$$t_1 = \sup\{t : \|S_\tau y\| \geq \varrho \text{ for all } t_0 \leq \tau \leq t\}.$$

This means that $\|S_t y\| \leq R_*$ for all $t \in [t_0, t_1]$. Since $\|S_{\bar{t}} y\| > R_*$ we have that $t_0 < t_1 < \bar{t}$. Moreover, it is clear that $\|S_{t_1} y\| = \varrho$. Thus, there exists $t_2 \in (t_1, \bar{t})$ (hence $t_2 > t_0$) such that $\|S_{t_2} y\| = 1 + R_0$. This contradicts the definition of t_0 in (2.1.5) and thus (2.1.4) is proved.

Step 2. Let us assume now that B is an arbitrary bounded set in X that lies outside the closed ball with the radius R_0. Then equation (2.1.2) implies that

$$U(S_t y) \leq U(y) - \alpha(\|y\|)t \leq L_B - \alpha_B t \quad \text{for } t \in [0, \tilde{t}], \; y \in B, \tag{2.1.6}$$

where $\tilde{t} = \sup\{t : \|S_\tau y\| \geq \varrho \text{ for all } 0 \leq \tau \leq t\}$ and

$$L_B = \sup\{U(x) : x \in B\}, \quad \alpha_B = \inf\{\alpha(x) : x \in B\}.$$

We can assume that $L_B > L$. If $\tilde{t} \leq t_B \equiv (L_B - L)/\alpha_B$, then, since $\|S_{\tilde{t}} y\| = \varrho$, by (2.1.4) we have that $\|S_t y\| \leq R_*$ for all $t \geq t_B$. If $\tilde{t} > t_B$, then by (2.1.6) and (2.1.1)

$$\phi_1(\|S_t y\|) \leq U(S_t y) \leq L \quad \text{for } t \in [t_B, \tilde{t}]$$

and hence $\|S_t y\| \leq R_*$ for $t \in [t_B, \tilde{t}]$. Since $\|S_{\tilde{t}} y\| = \varrho$, by (2.1.4) we have that $\|S_t y\| \leq R_*$ for all $t \geq \tilde{t}$. Consequently, the set B_* given by (2.1.3) is absorbing.

$\square$

We conclude this section with several exercises which illustrate the notion of dissipativity.

Exercise 2.1.3. Show that the hypothesis (2.1.2) in Theorem 2.1.2 can be replaced by the requirement

$$\frac{d}{dt}U(S_t y) + \phi_3(\|S_t y\|) \le \beta, \qquad (2.1.7)$$

where $\phi_3(r)$ is a continuous function such that

$$\liminf_{r \to \infty} \phi_3(r) > \beta$$

and β is a positive constant. In particular, (2.1.7) is true with $\phi_3(r) = \alpha\phi_1(r)$ if we assume that

$$\frac{d}{dt}U(S_t y) + \alpha U(S_t y) \le \beta, \qquad (2.1.8)$$

where α and β are positive constants.   ∎

Exercise 2.1.4. Let (2.1.8) be in force. Solving the inequality in (2.1.8), show that the set

$$\{x \in X : U(x) \le R\}$$

is a forward invariant absorbing set provided $R > \beta/\alpha$.   ∎

Exercise 2.1.5. Show that the dynamical system generated in $\mathbb{R}$ by the differential equation $\dot{x} + f(x) = 0$ (see Section 1.7 in Chapter 1) is dissipative, provided the function $f(x)$ possesses the additional property: $xf(x) \ge \delta x^2 - c$, where $\delta > 0$ and c are constants. Hint: Take $U(x) = x^2$ and use the result of Exercise 2.1.3. Find an upper estimate for the minimal radius of dissipativity.   ∎

Exercise 2.1.6. Let (X, S_t) be a dissipative system and let B_0 be a bounded absorbing set. Show that there exists $t_* \ge 0$ such that the set $B_* = \cup\{S_t B_0 : t \ge t_*\}$ is a bounded forward invariant absorbing set for (X, S_t).   ∎

Exercise 2.1.7. Consider a discrete dynamical system $(\mathbb{R}, f)$, where f is a continuous function on $\mathbb{R}$. Show that the system is dissipative, provided there exist $\rho > 0$ and $0 < \alpha < 1$ and such that $|f(x)| \le \alpha|x|$ when $|x| \ge \rho$.   ∎

Exercise 2.1.8 (Duffing equation). Consider a dynamical system in $\mathbb{R}^2$ generated (see Remark 1.8.19 and Example 1.8.18) by the Duffing equation

$$\ddot{x} + \gamma\dot{x} + x^3 - ax = b,$$

where a and b are real numbers and $\gamma > 0$. Using the properties of the function

$$U(x, \dot{x}) = \dot{x}^2 + \frac{1}{2}x^4 + ax^2 + v[2x\dot{x} + \gamma x^2],$$

where $v > 0$ is small enough, show that this dynamical system is dissipative. ∎

Exercise 2.1.9 (Lorenz system). Consider the Lorenz system arising as a three-mode Galerkin approximation in the problem of convection in a thin layer of liquid:

$$\begin{cases} \dot{x} = -\sigma x + \sigma y, \\ \dot{y} = rx - y - xz, \\ \dot{z} = -bz + xy. \end{cases}$$

Here σ, r, and b are positive numbers. Prove the dissipativity of the dynamical system generated by these equations in $\mathbb{R}^3$. Hint: Consider the function

$$V(x, y, z) = x^2 + y^2 + (z - r - \sigma)^2$$

on the trajectories of the system. ∎

Exercise 2.1.10 (Krasovskii equation). Consider the system generated by the equation

$$\ddot{x} + k(x^2 + \dot{x}^2)\dot{x} + x = 0,$$

in $\mathbb{R}^2$, see Exercise 1.8.22. Show that this system is dissipative under the conditions

$$k \in L_\infty(\mathbb{R}_+) \cap Lip_{loc}(\mathbb{R}_+), \quad \liminf_{r \to +\infty} k(r) > 0.$$

Hint: Consider the function $V(x, \dot{x}) = x^2 + \dot{x}^2 + vx\dot{x}$ with $v > 0$ small enough on the trajectories. ∎

2.2 Asymptotic compactness and smoothness

To study the long-time dynamics of infinite-dimensional systems we also need some properties of asymptotic compactness. There are several ways to formulate these properties depending on the structure of the corresponding model (see, e.g., the monographs BABIN/VISHIK [9], HALE [116], LADYZHENSKAYA [142], TEMAM [216] and the references therein). Below we mainly concentrate on the approaches suggested by LADYZHENSKAYA [142] and HALE [116]. We also refer to HARAUX [118], where some concepts of asymptotic compactness were used for the first time.

2.2.1 Basic definitions and facts

We start with several notions of compactness of an evolution operator.

Definition 2.2.1. Let S_t be an evolution operator on a complete metric space X.

- S_t is said to be *compact* if it possesses a compact absorbing set.
- S_t is said to be *conditionally compact* if for any bounded set D such that $S_t D \subset D$ for $t > 0$ there exist $t_D > 0$ and a compact set K in the closure $\overline{D}$ of D, such that $S_t D \subset K$ for all $t \geq t_D$.
- S_t is said to be *asymptotically compact* if the following Ladyzhenskaya condition (see LADYZHENSKAYA [142] and the references therein) holds: for any bounded set B in X such that the tail $\gamma^\tau(B) := \cup_{t \geq \tau} S_t B$ is bounded for some $\tau \geq 0$ we have that any sequence of the form $\{S_{t_n} x_n\}$ with $x_n \in B$ and $t_n \to \infty$ is relatively compact.
- An evolution operator S_t is said to be *asymptotically smooth* if the following Hale condition (see, e.g., HALE [116]) is valid: for every bounded set D such that $S_t D \subset D$ for $t > 0$ there exists a compact set K in the closure $\overline{D}$ of D, such that $S_t D$ converges uniformly to K in the sense that

$$\lim_{t \to +\infty} d_X\{S_t D \,|\, K\} = 0, \quad \text{where } d_X\{A|B\} = \sup_{x \in A} \text{dist}_X(x, B). \tag{2.2.1}$$

We apply the same terminology in the case of dynamical systems. Below we also use the notation $S_t D \rightrightarrows K$ as $t \to \infty$ in the case when (2.2.1) holds. ∎

Exercise 2.2.2. If S_t is a compact evolution operator, then S_t is conditionally compact. The latter property implies that S_t is asymptotically compact and asymptotically smooth. ∎

Exercise 2.2.3. Show that any dissipative conditionally compact system is compact. Hint: By Exercise 2.1.6 there exists a bounded forward invariant absorbing set. ∎

The proposition below shows that asymptotic smoothness is equivalent to asymptotic compactness.

Proposition 2.2.4. *An evolution operator S_t in some metric space X is asymptotically compact if and only if it is asymptotically smooth.*

Proof. We start with the following key lemma, which is also important in further considerations.

Lemma 2.2.5. *Let an evolution operator S_t be asymptotically compact on X and D be a bounded set. Assume the tail $\gamma^\tau(D)$ is bounded for some $\tau \geq 0$. Then the ω-limit set[2] $\omega(D)$ is a nonempty compact strictly invariant set such that $S_t D \rightrightarrows \omega(D)$ as $t \to \infty$.*

---

[2]We recall that the notions of a tail and an ω-limit set were introduced in Chapter 1.

Proof. It follows from asymptotic compactness that for any $t_n \to \infty$ and $x_n \in D$ the sequence $\{S_{t_n}x_n\}$ is relatively compact. Therefore, there exist a subsequence $\{n_m\}$ and an element $z \in X$ such that $S_{t_{n_m}}x_{n_m} \to z$ as $m \to \infty$. By Proposition 1.3.3,

$$\omega(D) = \left\{ z \in X \: : \: z = \lim_{t \to \infty} S_{t_n}x_n \text{ for some } t_n \to +\infty, \ x_n \in D \right\}. \qquad (2.2.2)$$

Hence $\omega(D)$ contains the element z, at least, and thus $\omega(D)$ is not empty.

To prove compactness of $\omega(D)$ we note that by (2.2.2) for any sequence $\{z_n\}$ in $\omega(D)$ there exist $t_n \to \infty$ and $x_n \in D$ such that $\mathrm{dist}_X(S_{t_n}x_n, z_n) \leq 1/n$. By asymptotic compactness there exist a subsequence $\{n_m\}$ and an element $\hat{z}$ such that $S_{t_{n_m}}x_{n_m} \to \hat{z} \in \omega(D)$ as $m \to \infty$. Thus, we have that $z_{n_m} \to \hat{z}$. This means that $\omega(D)$ is relatively compact. In the same way, if $z_n \to \bar{z}$ as $n \to \infty$, then $\bar{z} = \hat{z} \in \omega(D)$, i.e., $\omega(D)$ is closed.

Now we prove invariance of $\omega(D)$. Let $z \in \omega(D)$ and $z = \lim_{n \to \infty} S_{t_n}x_n$. Then $S_t z = \lim_{n \to \infty} S_{t+t_n}x_n$. Thus, due to (2.2.2) $\omega(D)$ is forward invariant. To prove backward invariance we consider the sequence $\{S_{t_n-t}x_n\}$ for some fixed $t > 0$ and n such that $t_n > t$. By asymptotic compactness this sequence is relatively compact. Thus, there exist a sequence $\{n_m\}$ and an element $v \in \omega(D)$ such that $y_m \equiv S_{t_{n_m}-t}x_{n_m} \to v$. We also have that $S_t y_m \to z$. Thus $z = S_t v$ and hence $S_t \omega(D) \supset \omega(D)$, i.e., $\omega(D)$ is backward invariant.

Assume that $S_t D \rightrightarrows \omega(D)$ is not true. Then there exist $\delta > 0$ and sequences $t_n \to \infty$ and $x_n \in D$ such that $\mathrm{dist}_X(S_{t_n}x_n, \omega(D)) \geq \delta$ for all n. As above, $\{S_{t_n}x_n\}$ is relatively compact. Therefore, $S_{t_{n_m}}x_{n_m} \to z \in \omega(D)$ for some subsequence $\{n_m\}$. This contradicts the relation $\mathrm{dist}_X(S_{t_n}x_n, \omega(D)) \geq \delta$. $\qquad\square$

Now we return to the proof of Proposition 2.2.4.

Let S_t be asymptotically compact and $B \subset X$ be an invariant bounded set. By Lemma 2.2.5, $\omega(B)$ is a compact set which attracts B. Thus, the Hale condition (see Definition 2.2.1) holds.

Let S_t be asymptotically smooth and $B \subset X$ be a bounded set such that the tail $\gamma^\tau(B) = \cup_{t \geq \tau} S_t B$ is bounded for some $\tau \geq 0$. Since $B_* \equiv \gamma^\tau(B)$ is forward invariant, by the Hale condition $S_t B_*$ converges uniformly to a compact set K. Thus $S_{t_n}x_n \to K$ for any sequences $x_n \in B$ and $t_n \to \infty$. Hence $\{S_{t_n}x_n\}$ is relatively compact. $\qquad\square$

The following exercise provides some sufficient conditions of asymptotic compactness of semiflows. They were established and applied by many authors (see, e.g., TEMAM [216, Chapter 1] and also LADYZHENSKAYA [142] and RAUGEL [188]).

Exercise 2.2.6. An evolution operator S_t in some metric space X is asymptotically compact provided one of the following conditions is valid:

(A) There exists a compact set K such that $S_t B \rightrightarrows K$ as $t \to \infty$ for every bounded set B in X.

(B) For any bounded set B there exists a compact set K_B such that $S_t B \rightrightarrows K_B$ as $t \to \infty$.

(C) X is a Banach space and there exists a decomposition $S_t = S_t^{(1)} + S_t^{(2)}$, where $S_t^{(1)}$ is uniformly compact for large t; that is, for any bounded set B there exists $t_0 = t_0(B)$ such that the set $\gamma^{(1)}(B; t_0) := \bigcup_{\tau \geq t_0} S_\tau^{(1)} B$ is relatively compact in X and $S_t^{(2)}$ is uniformly stable in the sense that

$$r_B(t) = \sup \left\{ \|S_t^{(2)} x\|_X \ : \ x \in B \right\} \to 0 \quad \text{as} \quad t \to \infty. \tag{2.2.3}$$

Hint: **(C)** implies **(B)** with $K_B = \text{Closure}_X \left\{ \gamma^{(1)}(B; t_0) \right\}$. Statement **(B)** applied to a bounded sequence $B = \{x_n\}$ yields the convergence of $S_{t_n} x_n$ to a compact set as $t_n \to \infty$. ∎

2.2.2 Kuratowski's measure of noncompactness

To obtain effective criteria of asymptotic compactness, it is convenient to use Kuratowski's α-measure of noncompactness (see, e.g., AKHMEROV ET AL. [2] and the references therein). The latter is defined by the formula

$$\alpha(B) = \inf\{d \ : \ B \text{ has a finite cover by } \textit{open} \text{ sets of diameter } < d\}$$

on bounded sets of a complete metric space X. We recall that the diameter of the set is defined by the relation $\text{diam}\, B = \sup \{\text{dist}(x, y) \ : \ x, y \in B\}$.

Some elementary properties of the α-measure are collected in the following exercises.

Exercise 2.2.7. Let X be a complete metric space.

(A) Show that in the definition of α-measure we can consider arbitrary coverings, i.e.,

$$\alpha(B) = \inf\{d \ : \ B \text{ has finite cover by (arbitrary) sets of diameter } < d\}.$$

This implies that $\alpha(B) \leq \text{diam}\, B$.
(B) Show that if $K_1 \subset K_2$, then $\alpha(K_1) \leq \alpha(K_2)$ (monotonicity).
(C) Show that $\alpha(K) = \alpha(\overline{K})$, where $\overline{K}$ is the closure of K.
(D) Show that $\alpha(A \cup B) \leq \max\{\alpha(A), \alpha(B)\}$ (semi-additivity).
(E) Show that $\alpha(K) = 0$ if and only if the closure $\overline{K}$ of K is compact.
(F) Show that the set B is bounded if and only if $\alpha(B) < \infty$. ∎

Exercise 2.2.8. Let X be a Banach space. Show that

(A) $\alpha(\lambda B) = |\lambda| \alpha(B)$ for any $\lambda \in \mathbb{R}$, where $\lambda B = \{\lambda x : x \in B\}$ (homogeneity).

(B) $\alpha(y + B) = \alpha(B)$ for any $y \in X$, where $y + B = \{y + x : x \in B\}$ (invariance under translations).

∎

It is known (see, e.g., AKHMEROV ET AL. [2]) that $\alpha(B_R(y)) = 2R$ for every ball $B_R(y) = \{x \in X : \|x - y\|_X < R\}$ in an infinite-dimensional Banach space X. The following propositions are also important (see HALE [116] and SELL/YOU [206]) in the study of asymptotic smoothness of evolution operators.

Proposition 2.2.9. *Let A and B be bounded sets in a Banach space X. Then*

$$\alpha(A + B) \leq \alpha(A) + \alpha(B), \tag{2.2.4}$$

where $A + B = \{x + y : x \in A, \ y \in B\}$

Proof. Take arbitrary $\varepsilon > 0$. Let $\{\mathscr{O}_i^A\}$ and $\{\mathscr{O}_j^B\}$ be coverings of A and B with diameters less than $\alpha(A) + \varepsilon$ and $\alpha(B) + \varepsilon$. Then $\{\mathscr{O}_i^A + \mathscr{O}_j^B\}$ is a covering for $A + B$. It is clear that

$$\mathrm{diam}\{\mathscr{O}_i^A + \mathscr{O}_j^B\} \leq \mathrm{diam}\{\mathscr{O}_i^A\} + \mathrm{diam}\{\mathscr{O}_j^B\} \leq \alpha(A) + \alpha(B) + 2\varepsilon.$$

This implies (2.2.4). □

Proposition 2.2.10. *Let X be a complete metric space and $U_1 \supset U_2 \supset U_3 \ldots$ be nonempty closed sets in X. If $\alpha(U_n) \to 0$ as $n \to \infty$, then $\cap_{n \geq 1} U_n$ is nonempty and compact.*

Proof. For each n take $u_n \in U_n$ and consider the sequence $\{u_n\}$. For every $\varepsilon > 0$ we can find N such that $\alpha(U_N) < \varepsilon$. Thus,

$$K \equiv \{u_n : n = 1, \ldots\} \subset \{u_n : n = 1, \ldots, N - 1\} \bigcup U_N.$$

Hence by Exercise 2.2.7(B,D), $\alpha(K) \leq \alpha(U_N) < \varepsilon$ for every $\varepsilon > 0$. Therefore, $\alpha(K) = 0$ and thus $\bar{K}$ is compact. Thus, there exist $u \in X$ and a subsequence $\{n_m\}$ such that

$$u_{n_m} \in U_{n_m} \text{ and } u_{n_m} \to u \text{ as } m \to \infty.$$

Since U_n is closed for every n, we have that $u \in U_n$ for every n, i.e., $U = \cap_{n \geq 1} U_n$ is not empty. It is clear that U is closed. Since $\alpha(U) \leq \alpha(U_n)$ for $n = 1, 2, \ldots$, we have $\alpha(U) = 0$ and thus by Exercise 2.2.7(E) U is compact. □

Exercise 2.2.11. Prove the continuous analog of Proposition 2.2.10: if $\alpha(U_t) \to 0$ as $t \to \infty$ for some decreasing family $\{U_t\}$ of nonempty closed sets, then $\cap_{t \geq 0} U_t$ is nonempty and compact. ∎

Exercise 2.2.12. Let $B_1 \supset B_2 \supset \ldots$ be a sequence of closed sets in a complete metric space X. Assume that $\mathrm{diam}\, B_n \to 0$ as $n \to \infty$. Show that there exists a unique element $x \in X$ such that $x \in B_n$ for all n (in the case when $\{B_k\}$ are balls in a Banach space, this fact is known as the principle of nested balls). Hint: Apply Proposition 2.2.10 and also the observation made in Exercise 2.2.7(A).   ∎

The following assertion allows us to reformulate the asymptotic smoothness/compactness in the terms of Kuratowski's α-measure.

Proposition 2.2.13. *An evolution operator S_t is asymptotically smooth if and only if for any bounded forward invariant set B we have that $\alpha(S_t B) \to 0$ as $t \to \infty$.*

Proof. Let B be a bounded forward invariant set for S_t.

If S_t is an asymptotically smooth evolution operator, then there exists a compact set K_B such that $S_t B \rightrightarrows K_B$ as $t \to \infty$. By the compactness of K_B, for any $\varepsilon > 0$ there exists a finite set $\{x_k : k = 1, \ldots N_\varepsilon\}$ in K_B such that

$$K_B \subset \bigcup_{k=1}^{N_\varepsilon} \mathscr{B}_k, \quad \text{where} \quad \mathscr{B}_k = \{x \in X : \text{dist}_X(x_k, x) < \varepsilon\}.$$

Since $S_t B \rightrightarrows K_B$, there exists $t_\varepsilon > 0$ such that $S_t B \subset \cup_{k=1}^{N_\varepsilon} \mathscr{B}_k$ for all $t \geq t_\varepsilon$. Thus $\alpha(S_t B) < 2\varepsilon$ for all $t \geq t_\varepsilon$. This implies that $\alpha(S_t B) \to 0$ as $t \to \infty$.

Assume now that $\alpha(S_t B) \to 0$ as $t \to \infty$. Then we can apply the result of Exercise 2.2.11 to a family of the sets $U_t = \overline{S_t B}$ and conclude that

$$\omega(B) = \bigcap_{t>0} \overline{S_t B} \quad \text{is a nonempty compact set.}$$

Thus, it is sufficient to show that $S_t B \rightrightarrows \omega(B)$. If this is not true, then there exist $\delta > 0$ and sequences $t_n \to \infty$ and $x_n \in B$ such that $\text{dist}_X(S_{t_n} x_n, \omega(B)) \geq \delta$ for all n. One can see that for any $t > 0$ there exists N_t such that

$$\{S_{t_n} x_n : n = 1, 2, \ldots\} \subset \{S_{t_n} x_n : n = 1, 2, \ldots, N_t\} \bigcup \overline{S_t B}.$$

Thus $\alpha(\{S_{t_n} x_n : n = 1, 2, \ldots\}) \leq \alpha(S_t B)$, which implies that $\alpha(\{S_{t_n} x_n\}) = 0$. Hence $\{S_{t_n} x_n\}$ is relatively compact. Therefore, $S_{t_{n_m}} x_{n_m} \to z \in \omega(B)$ for some subsequence $\{n_m\}$. This contradicts the relation $\text{dist}_X(S_{t_n} x_n, \omega(B)) \geq \delta$.   □

Using Proposition 2.2.13 we can prove the next assertion, which is a slight modification of the statement proved earlier in HALE [116, Lemma 3.2.3] by another method.

Proposition 2.2.14. *Let S_t be an evolution operator in a Banach space X. Assume that for each $t > 0$ there exists a decomposition $S_t = S_t^{(1)} + S_t^{(2)}$, where $S_t^{(2)}$ is a mapping in X satisfying (2.2.3) and $S_t^{(1)}$ is compact in the sense that for each $t > 0$ the set $S_t^{(1)} B$ is a relatively compact set in X for every $t > 0$ large enough and for every bounded forward invariant set B in X. Then S_t is asymptotically smooth.*

We note that this proposition improves the statement of Exercise 2.2.6(C), because we do not assume compactness of $\gamma^{(1)}(B; t_0)$ here. The size of $S_t^{(1)}B$ may be unbounded as $t \to +\infty$.

Proof. For any bounded forward invariant set B we have that $S_t B \subset S_t^{(1)}B + S_t^{(2)}B$. Therefore, Proposition 2.2.9 (see also Exercise 2.2.7) yields

$$\alpha(S_t B) \leq \alpha(S_t^{(1)}B) + \alpha(S_t^{(2)}B) \leq \alpha(S_t^{(2)}B) \leq \operatorname{diam}\{S_t^{(2)}B\} \leq 2 \sup_{y \in B} \|S_t^{(2)}y\|$$

for all t large enough. Thus by (2.2.3), $\alpha(S_t B) \to 0$ as $t \to \infty$. Hence by Proposition 2.2.13, S_t is asymptotically smooth. □

Keeping in mind Proposition 2.2.13, it is convenient to introduce the following notion (see HALE [116]).

Definition 2.2.15. A (nonlinear) operator V on a complete metric space X is said to be an α-contraction if there exists $0 \leq \kappa < 1$ such that $\alpha(VB) \leq \kappa\alpha(B)$. ∎

The following simple result connects this notion with dynamics.

Exercise 2.2.16. A dynamical system (X, S_t) is asymptotically smooth if there exists $t_* > 0$ such that S_{t_*} is an α-contraction. Hint: For every forward invariant set D we have that $S_t D \subset S_{nt_*}D$, where n is the integer part of t/t_*. ∎

For more discussion of the α-measure from the point of view of dynamical systems we refer to HALE [116] and the references therein; see also SELL/YOU [206, Lemma 22.2].

2.2.3 Criteria of asymptotic compactness via weak quasi-stability

We conclude this section with several assertions that give convenient criteria for asymptotic smoothness/compactness of evolution operators and dynamical systems. These criteria generalize the corresponding statements known due to KHANMAMEDOV [134], MA/WANG/ZHONG [156] and CERON/LOPES [28]. A posteriori they can be treated as some weak forms of quasi-stability discussed in Chapter 3. Roughly speaking, this *weak* quasi-stability means that the difference of two trajectories can be made small for large moments of time modulo some functional which demonstrates some (rather weak) compactness behavior (see, e.g., (2.2.5) below).

We start with the criterion which relies on the idea presented in KHANMAMEDOV [134] and provides more flexibility with respect to more standard methods (see, e.g., the discussion in CHUESHOV/LASIECKA [56, 58] and also the references cited therein).

Theorem 2.2.17. *Let S_t be an evolution operator on a complete metric space X. Assume that for any bounded forward invariant set B in X and for any $\epsilon > 0$ there exists $T \equiv T(\epsilon, B)$ such that*

$$\text{dist}(S_T y_1, S_T y_2) \leq \epsilon + \Psi_{\epsilon, B, T}(y_1, y_2), \quad y_i \in B, \tag{2.2.5}$$

where $\Psi_{\epsilon, B, T}(y_1, y_2)$ is a functional defined on $B \times B$ such that

$$\liminf_{m \to \infty} \liminf_{n \to \infty} \Psi_{\epsilon, B, T}(y_n, y_m) = 0 \quad \text{for every sequence } \{y_n\} \subset B. \tag{2.2.6}$$

Then S_t is an asymptotically smooth evolution operator.

The result stated in Theorem 2.2.17 is an abstract version of Theorem 2 in KHAN-MAMEDOV [134] and can be derived from the arguments given in KHANMAMEDOV [134]. Our proof is shorter and can be easily derived from the following assertion.[3]

Proposition 2.2.18. *Let S_t be an evolution operator on a complete metric space X. Assume that for any bounded positively invariant set B in X and for any $\epsilon > 0$ there exists $T \equiv T(\epsilon, B)$ such that*

$$\liminf_{m \to \infty} \liminf_{n \to \infty} \text{dist}(S_T y_n, S_T y_m) \leq \epsilon \quad \text{for every sequence } \{y_n\} \subset B. \tag{2.2.7}$$

Then S_t is an asymptotically smooth evolution operator.

Proof. By Proposition 2.2.13 it is sufficient to prove that

$$\lim_{t \to \infty} \alpha(S_t B) = 0,$$

where $\alpha(B)$ is Kuratowski's α-measure of noncompactness.

Because $S_{t_1} B \subset S_{t_2} B$ for $t_1 > t_2$, the function $\alpha(t) \equiv \alpha(S_t B)$ is non-increasing. Therefore, it is sufficient to prove that for any $\varepsilon > 0$ there exists $T > 0$ such that $\alpha(S_T B) \leq \varepsilon$. If this is not true, then there is $\varepsilon_0 > 0$ such that $\alpha(S_T B) \geq 5\varepsilon_0$ for all $T > 0$. For this ε_0 we choose T_0 such that (2.2.7) holds. The relation $\alpha(S_{T_0} B) \geq 5\varepsilon_0$ implies that there exists an infinite sequence $\{y_n\}_{n=1}^{\infty}$ such that

$$\text{dist}(S_{T_0} y_n, S_{T_0} y_m) \geq 2\varepsilon_0 \text{ for all } n \neq m, \ n, m = 1, 2, \ldots \tag{2.2.8}$$

If such a sequence does not exist, then we can use the following construction: take arbitrary $y_1 \in B$ and choose $y_2 \in B$ such that $\text{dist}(S_{T_0} y_1, S_{T_0} y_2) \geq 2\varepsilon_0$. Then we take $y_3 \in B$ such that $\text{dist}(S_{T_0} y_3, S_{T_0} y_i) \geq 2\varepsilon_0$ for $i = 1, 2$, and so on. If this procedure stops, we obtain a finite $2\varepsilon_0$-net for $S_{T_0} B$. This means that $\alpha(S_{T_0} B) \leq 4\varepsilon_0$ and contradicts the relation $\alpha(S_{T_0} B) \geq 5\varepsilon_0$. Thus (2.2.8) holds true. This contradicts (2.2.7). □

---

[3]In many cases we can use Proposition 2.2.18 directly. Theorem 2.2.17 is formulated mainly due to priority and historical reasons.

Proposition 2.2.18 can also be used to obtain the following criterion.

Proposition 2.2.19. *Let S_t be an evolution operator on a reflexive Banach space X. Assume that for any bounded forward invariant set B in X and any $\varepsilon > 0$ there exist $T > 0$ and a compact operator K such that*

$$\|(I - K)S_T y\| \le \varepsilon, \quad \forall\, y \in B. \tag{2.2.9}$$

Then the evolution operator S_t is asymptotically smooth.

This proposition was proved in MA/WANG/ZHONG [156] for the case when K is a finite-dimensional projector. Now the relation in (2.2.9) with a projector is known as the "flattening" property (see the discussion in CARVALHO/LANGA/ROBINSON [26] and KLOEDEN/RASMUSSEN [135]).

Proof. By (2.2.9) we have that

$$\|S_T y_1 - S_T y_2\| \le \|(I - K)S_T y_1\| + \|(I - K)S_T y_2\| + \|K(S_T y_1 - S_T y_2)\|$$
$$\le 2\varepsilon + \|K(S_T y_1 - S_T y_2)\|, \quad \forall\, y_1, y_2 \in B.$$

Let $\{y_n\} \subset B$. Since $\{S_T y_n\} \subset B$ is a bounded sequence, there exists a weakly convergent subsequence $\{S_T y_{n_k}\}$. By the compactness of K, we have that

$$\lim_{k,m \to \infty} \|K(S_T y_{n_k} - S_T y_{n_m})\| = 0$$

which implies that

$$\liminf_{m \to \infty} \liminf_{n \to \infty} \|K(S_T y_n - S_T y_m)\| = 0 \quad \text{for every sequence } \{y_n\} \subset B.$$

Thus, we can apply Proposition 2.2.18. $\qquad\square$

The following exercise presents another asymptotic smoothness criterion in reflexive Banach spaces.

Exercise 2.2.20. Let S_t be an evolution operator on a Hilbert space. Assume that S_t is weakly continuous for every $t > 0$; i.e., the condition $x_n \to x$ weakly in X implies that $S_t x_n \to S_t x$ weakly. Show that the evolution operator S_t is asymptotically smooth provided that for any bounded forward invariant set B and for any $\varepsilon > 0$ there exists $T \equiv T(\varepsilon, B)$ such that

$$\limsup_{n \to \infty} \|S_T y_n\| \le \|S_T y\| + \varepsilon \tag{2.2.10}$$

for every sequence $\{y_n\} \subset B$ such that $y_n \to y$ weakly. Hint: Prove first that

$$\limsup_{n \to \infty} \|S_T y_n - S_T y\| \le \varepsilon,$$

then apply Proposition 2.2.18. $\qquad\blacksquare$

The following assertion is a generalization of the results presented in HALE [116] and CERON/LOPES [28] (see also CHUESHOV/LASIECKA [56, 58] where this fact is established by a different method).

Theorem 2.2.21. *Let S_t be an evolution operator on a complete metric space X. Assume that for any bounded forward invariant set B in X there exist $T > 0$, a continuous nondecreasing function $g : \mathbb{R}_+ \mapsto \mathbb{R}_+$, and a pseudometric ϱ_B^T on the set B such that*

(i) *$g(0) = 0$; $g(s) < s$, $s > 0$.*
(ii) *The pseudometric ϱ_B^T is precompact (with respect to the topology of X) in the sense that any sequence $\{x_n\} \subset B$ has a subsequence $\{x_{n_k}\}$ which is Cauchy with respect to ϱ_B^T.*
(iii) *The following estimate holds for every $y_1, y_2 \in B$:*

$$\text{dist}_X(S_T y_1, S_T y_2) \le g\,(\text{dist}_X(y_1, y_2)) + \varrho_B^T(y_1, y_2). \tag{2.2.11}$$

Then the evolution operator S_t is asymptotically smooth.

Remark 2.2.22. The difference between pseudometrics and metrics is that a pseudometric can be degenerate. In our case this means that the property $\varrho_B^T(y_1, y_2) = 0$ does not imply $y_1 = y_2$. We also know that instead of (2.2.11) one may also assume that

$$\text{dist}_X(S_T y_1, S_T y_2) \le g\left(\text{dist}_X(y_1, y_2) + \varrho_B^T(y_1, y_2)\right),$$

(pseudometric inside g); see some details in [56, Chapter 2].   ∎

Proof. We use Proposition 2.2.18.

Let B be a bounded forward invariant set in X with diameter L. One can see that for any $\varepsilon > 0$ we can choose N such that $g^N(L) \le \varepsilon$, where g^N denotes the composition $g \circ \cdots \circ g$. Iterating (2.2.11) we have that

$$\text{dist}_X(S_T^N y_1, S_T^N y_2) \le g\left(\text{dist}_X(S_T^{N-1} y_1, S_T^{N-1} y_2)\right) + \varrho_B^T(S_T^{N-1} y_1, S_T^{N-1} y_2)$$
$$\le g(g(\cdots g(g\,(L)) + \varrho_B^T(y_1, y_2)))$$
$$+ \varrho_B^T(S_T y_1, S_T y_2)) \cdots) + \varrho_B^T(S_T^{N-1} y_1, S_T^{N-1} y_2).$$

The right-hand side of the relation above is a continuous function of L and the expressions of the form

$$\varrho_B^T(S_T^m y_1, S_T^m y_2), \quad m = 1, \dots, N - 1.$$

Since the pseudometric ϱ_B^T is precompact, any sequence $\{x_n\} \subset B$ has a subsequence $\{\hat{x}_{n_k}\}$ such that

$$\lim_{p,q \to \infty} \varrho_B^T(S_T^m \hat{x}_{n_p}, S_T^m \hat{x}_{n_q}) = 0, \quad \forall\, m = 1, \ldots, N-1.$$

This implies that

$$\liminf_{k \to \infty} \liminf_{n \to \infty} \text{dist}_X(S_T^N x_n, S_T^N x_k) \le g^N(L) \le \varepsilon.$$

By Proposition 2.2.18 this implies that S_t is asymptotically smooth. $\qquad\square$

Theorem 2.2.21 implies the following result which was proved earlier in the paper of CERON/LOPES [28].

Proposition 2.2.23. *Let (X, S_t) be a dynamical system in a Banach space X. Assume that for any bounded forward invariant set B in X there exist functions $C_B(t) \ge 0$ and $K_B(t) \ge 0$ such that $\lim_{t \to \infty} K_B(t) = 0$, a time $t_0 = t_0(B)$, and a precompact pseudometric ϱ on X such that*

$$\|S_t y_1 - S_t y_2\| \le K_B(t) \cdot \|y_1 - y_2\| + C_B(t) \cdot \varrho(y_1, y_2), \quad t \ge t_0, \tag{2.2.12}$$

for every $y_1, y_2 \in B$. Then (X, S_t) is an asymptotically smooth dynamical system.

Proof. We apply Theorem 2.2.21 with $g(s) = K_B(T) \cdot s$, where T is chosen such that $K_B(T) < 1$. $\qquad\square$

2.3 Global attractors

The main objects arising in the analysis of the long-time behavior of dissipative dynamical systems are attractors. Their study makes it possible to answer a number of fundamental questions on the properties of limit regimes that can arise in the system. There are several general approaches and methods that allow us to study attractors for a large class of dynamical systems generated by nonlinear partial differential equations (see, e.g., BABIN/VISHIK [9], CHUESHOV [39], HALE [116], LADYZHENSKAYA [142], TEMAM [216] and the references listed therein). In this section we present the main general tools which are usually involved in the theory of infinite-dimensional dissipative systems.

2.3.1 Existence and basic properties

Several definitions of an attractor are available (see, e.g., the discussion in CHUES-HOV [39, Section 1.3]). From the point of view of infinite-dimensional systems, the most convenient concept is a global attractor.[4]

Definition 2.3.1 (Global attractor). Let S_t be an evolution operator on a complete metric space X. A bounded closed set $\mathfrak{A} \subset X$ is said to be a *global attractor* for S_t if

(i) $\mathfrak{A}$ is an invariant set; that is, $S_t \mathfrak{A} = \mathfrak{A}$ for $t \geq 0$.
(ii) $\mathfrak{A}$ is uniformly attracting; that is, for all bounded set $D \subset X$

$$\lim_{t \to +\infty} d_X\{S_t D \,|\, \mathfrak{A}\} = 0 \quad \text{for every bounded set } D \subset X, \tag{2.3.1}$$

where $d_X\{A|B\} = \sup_{x \in A} \text{dist}_X(x, B)$ is the Hausdorff semidistance.

∎

In many sources (see, e.g., BABIN [7], CHEPYZHOV [31], HALE [116], TEMAM [216]) the definition of a global attractor requires this to be a compact set. We do not assume this property because, hypothetically, situations when a global attractor is not compact are possible for systems with degenerate damping mechanisms. See, e.g., Section 5.3.3 in Chapter 5.

Exercise 2.3.2. Show that if a global attractor exists, then it is unique. ∎

Exercise 2.3.3. Show that any backward invariant bounded set belongs to the global attractor. In particular, every stationary point lies in the attractor. ∎

Exercise 2.3.4. Show that

(A) A full trajectory $\gamma = \{u(t) : t \in \mathbb{R}\}$ belongs to the global attractor if and only if γ is a bounded set.
(B) For any x from the attractor $\mathfrak{A}$ there exists a full trajectory $\gamma = \{u(t) : t \in \mathbb{R}\}$ such that $u(0) = x$ and $\gamma \subset \mathfrak{A}$. Hint: The strict invariance property of the attractor implies that there exists a sequence $\{x_{-n} : n = 1, 2, \ldots\} \subset \mathfrak{A}$ such that $S_1 x_{-n} = x_{-(n-1)}$ for all $n = 1, 2, \ldots$ with $x_0 = x$.

Thus, the global attractor can be described as a set of all bounded full trajectories.

∎

The main result on the existence of global attractors is the following assertion.

---

[4] Below we use the Fraktur (Gothic) "A" for notation of global attractors because the Latin version of this letter is overloaded, especially in Chapters 4–6.

Theorem 2.3.5. *Let* (X, S_t) *be a dissipative asymptotically compact dynamical system on a complete metric space* X. *Then* S_t *possesses a unique compact global attractor* $\mathfrak{A}$ *such that*

$$\mathfrak{A} = \omega(B_0) = \bigcap_{t>0} \overline{\bigcup_{\tau \geq t} S_\tau B_0} \tag{2.3.2}$$

for every bounded absorbing set B_0 *and*

$$\lim_{t \to +\infty} (d_X\{S_t B_0 \mid \mathfrak{A}\} + d_X\{\mathfrak{A} \mid S_t B_0\}) = 0, \tag{2.3.3}$$

where as above $d_X\{A|B\} = \sup_{x \in A} \mathrm{dist}_X(x, B)$. *Moreover, if there exists a connected absorbing bounded set,*[5] *then* $\mathfrak{A}$ *is connected.*

Property (2.3.3) states that $\mathfrak{A}$ attracts bounded absorbing sets in the Hausdorff metric which is defined by the formula

$$\mathrm{dist}_H\{A|B\} = d_X\{A|B\} + d_X\{B|A\}$$

for all bounded sets A and B. The convergence in the Hausdorff metric means that for any $\varepsilon > 0$ and for any absorbing set B there exists $t_\varepsilon > 0$ such that $S_t B \subset \mathscr{O}_\varepsilon(A)$ and $A \subset \mathscr{O}_\varepsilon(S_t B)$ for all $t \geq t_\varepsilon$. Here $\mathscr{O}_\varepsilon(D)$ denotes the ε-vicinity of the set D.

We note that in finite-dimensional systems for the existence of a global attractor we need the dissipativity property only. This observation implies that the Duffing (Exercise 2.1.8) and Lorenz (Exercise 2.1.9) systems possess global attractors.

Exercise 2.3.6. Show that the 1D system generated by the equation $\dot{x} + x^3 - x = 0$ on $\mathbb{R}$ possesses a global attractor $\mathfrak{A}$ and $\mathfrak{A}$ is the interval $[-1, 1]$. Hint: See Exercise 2.1.5 for dissipativity; also, make use of the fact that the attractor is a connected set containing the rest points $x = \pm 1$. ∎

Further applications of Theorem 2.3.5 will be presented later.

Proof of Theorem 2.3.5. Since S_t is dissipative, there exists a bounded absorbing set B_0. This implies that for every bounded set D the tail γ_D^t lies in B_0 for all $t \geq t_D$. Therefore, using the asymptotic compactness of (X, S_t), by Lemma 2.2.5 we conclude that $\omega(B_0)$ is a nonempty compact strictly invariant set such that (2.3.1) holds. Thus, the formula in (2.3.2) gives a global attractor.

To prove (2.3.3) we need to show that

$$\lim_{t \to +\infty} \sup \{d_X(x, S_t B_0) : x \in \mathfrak{A}\} = 0.$$

---

[5] We can assume instead that X is a connected space in the sense that every two points from X can be connected by a continuous path.

This follows from the fact that $\mathfrak{A} \subset B_0$, which implies that

$$\mathfrak{A} = S_t\mathfrak{A} \subset S_tB_0 \text{ for all } t > 0.$$

To prove connectedness we use (2.3.3) and the contradiction argument.

Let B_0 be connected. Assume that $\mathfrak{A}$ is not connected, i.e., $\mathfrak{A} = K \cup K_*$, where K and K_* are two nonempty disjoint compact sets such that $\text{dist}(K, K_*) = 3\delta > 0$. By (2.3.3) we have that

$$S_tB_0 \subset \mathcal{O}_\delta(\mathfrak{A}) \equiv \{x \in X : \text{dist}_X(x, \mathfrak{A}) < \delta\} \tag{2.3.4}$$

for all t large enough. Obviously S_tB_0 is connected for each t. Thus, by (2.3.4) we have that $S_tB_0 \subset \mathcal{O}_\delta(\tilde{K})$, where $\tilde{K}$ is either K or K_*, say $\tilde{K} = K$. Using (2.3.3) again we have that

$$K_* \subset \mathfrak{A} \subset \mathcal{O}_\delta(S_tB_0) \subset \mathcal{O}_{2\delta}(\tilde{K})$$

for all t large enough. This is impossible because $\text{dist}(K, K_*) = 3\delta > 0$. $\square$

It is clear that if an evolution operator possesses a compact global attractor, then it is dissipative and asymptotically compact. Thus, Theorem 2.3.5 implies that a dynamical system (X, S_t) has a compact global attractor *if and only if* it is dissipative and asymptotically compact (or asymptotically smooth).

Exercise 2.3.7. Show that under the hypotheses and the notation of Theorem 2.3.5 we have that

$$\mathfrak{A} = \bigcap_{n \geq N} S_{nT}B_0 \text{ for every } N \in \mathbb{Z}_+ \text{ and } T > 0. \tag{2.3.5}$$

Hint: $\mathfrak{A} \subset B_0$ and thus $\mathfrak{A} = S_{nT}\mathfrak{A} \subset S_{nT}B_0$ for every $n \in \mathbb{Z}_+$ and $T > 0$.   ∎

Exercise 2.3.8. Let a system (X, S_t) be dissipative and $V = S_{t_*}$ be an α-contraction for some $t_* > 0$ (see Definition 2.2.15). Then (X, S_t) possesses a compact global attractor which can be written in the form (2.3.5). Hint: See Exercise 2.2.16.   ∎

In some cases it is convenient to use the condition of point dissipativity instead of (bounded) dissipativity. The following assertion can be found in HALE [116] and RAUGEL [188]; see also CARVALHO/LANGA/ROBINSON [26].

Theorem 2.3.9. *An evolution semigroup S_t on some complete metric space X possesses a compact global attractor if and only if*

(i) *S_t is point dissipative;*
(ii) *for every bounded set B there exists $\tau > 0$ such that the tail $\gamma_B^\tau = \cup_{t \geq \tau}S_tB$ is bounded;*
(iii) *S_t is asymptotically smooth.*

Proof. Due to Theorem 2.3.5 it is sufficient to prove that under the conditions above the system (X, S_t) is (bounded) dissipative. To show this, we use the same idea as in RAUGEL [188].

We first establish the following "locally compact" dissipativity property. Namely, we show that there exists a bounded forward invariant set B_* possessing the property: for every compact set K

$$\exists \epsilon = \epsilon_K > 0, \ \exists t_K \geq 0 : \ S_t \mathcal{O}_\epsilon(K) \subset B_* \text{ for all } t \geq t_K, \tag{2.3.6}$$

where $\mathcal{O}_\epsilon(K)$ is the ϵ-neighborhood of K. Indeed, since S_t is point dissipative, there exists a bounded set B_0 such that

$$\forall x_0 \in X, \ \exists t_{x_0} \geq 0 : \ S_t x_0 \in B_0 \text{ for all } t \geq t_{x_0}.$$

We can assume that B_0 is open. In this case, by the continuity of $S_{t_{x_0}}$ there is $\epsilon = \epsilon_{x_0} > 0$ such that

$$S_{t_{x_0}} \mathcal{O}_{\epsilon_{x_0}}(x_0) \subset B_0.$$

Let τ_0 be such that $B_* \equiv \gamma_{B_0}^{\tau_0}$ is bounded. In this case,

$$S_{\tau + t_{x_0}} \mathcal{O}_{\epsilon_{x_0}}(x_0) \subset \gamma_{B_0}^{\tau_0} = B_* \text{ for all } \tau \geq \tau_0.$$

If K is a compact set, then we can find a finite set $\{x_i\}$ in K such that

$$K \subset \mathcal{U} \equiv \cup \mathcal{O}_{\epsilon_{x_i}}(x_i).$$

It is clear that

$$S_t \mathcal{U} = \cup S_t \mathcal{O}_{\epsilon_{x_i}}(x_i) \subset B_* \text{ for all } t \geq \tau_0 + \max_i t_{x_i}.$$

Since $\mathcal{U}$ is open, we can find $\epsilon = \epsilon_K > 0$ such that $\mathcal{O}_\epsilon(K) \subset \mathcal{U}$. Thus (2.3.6) is established.

To conclude the proof we note that for every bounded set B there exists $\tau = \tau_B$ such that γ_B^τ is bounded and forward invariant. Thus, by asymptotic smoothness, there is a compact set K such that

$$\forall \epsilon > 0, \ \exists t_\epsilon \geq 0 : \ S_t [\gamma_B^\tau] \subset \mathcal{O}_\epsilon(K) \text{ for all } t \geq t_\epsilon.$$

Hence the locally compact dissipativity property in (2.3.6) implies the desired conclusion. □

The study of the structure of the global attractors is an important problem from the point of view of applications. There are no universal approaches to this problem. It is well known that even in finite-dimensional cases an attractor can possess an extremely complicated structure. However, some sets that belong to the attractor can be easily pointed out. For example, every stationary point and every bounded full trajectory belong to the global attractor (see Exercises 2.3.3 and 2.3.4). The global attractor also contains unstable motions which can be introduced by the following definition (see, e.g., BABIN/VISHIK [9], CHUESHOV [39], TEMAM [216]).

Definition 2.3.10. Let $\mathcal{N}$ be the set of stationary points of a dynamical system (X, S_t):

$$\mathcal{N} = \{v \in X \,:\, S_t v = v \text{ for all } t \geq 0\}\,.$$

We define the *unstable manifold* $\mathcal{M}^u(\mathcal{N})$ emanating from the set $\mathcal{N}$ as a set of all $y \in X$ such that there exists a full trajectory $\gamma = \{u(t) \,:\, t \in \mathbb{R}\}$ with the properties

$$u(0) = y \text{ and } \lim_{t \to -\infty} \text{dist}_X(u(t), \mathcal{N}) = 0. \tag{2.3.7}$$

∎

Exercise 2.3.11. Show that $\mathcal{M}^u(\mathcal{N})$ is a (strictly) invariant set.   ∎

The following assertion can be found in BABIN/VISHIK [9], CHUESHOV [39], or TEMAM [216], for instance.

Proposition 2.3.12. *Let $\mathcal{N}$ be the set of stationary points of a dynamical system (X, S_t) possessing a global attractor $\mathfrak{A}$. Then $\mathcal{M}^u(\mathcal{N}) \subset \mathfrak{A}$.*

Proof. Let $y \in \mathcal{M}^u(\mathcal{N})$ and $\gamma = \{u(t) \,:\, t \in \mathbb{R}\}$ be the trajectory possessing property (2.3.7). Then there exists $s \leq 0$ such that the set

$$\gamma_s \equiv \{u(t) \,:\, -\infty < t \leq s\} \subset \{z \,:\, \text{dist}(z, \mathcal{N}) \leq 1\}$$

Thus γ_s is bounded. It is also clear that γ_s is backward invariant, i.e., $\gamma_s \subset S_t \gamma_s$ for every $t > 0$. Therefore, the result of Exercise 2.3.3 implies that $\gamma_s \subset \mathfrak{A}$. Since $y \in S_{-s} \gamma_s$, this implies the desired conclusion.   □

In some cases (see Section 2.4 below) it is possible to show that the unstable manifold coincides with the attractor; that is, $\mathcal{M}^u(\mathcal{N}) = \mathfrak{A}$.

To exclude unstable motions from consideration, it is convenient to use the concept of a global *minimal* attractor (see LADYZHENSKAYA[142]). This concept is also useful for a description of the long-time behavior of individual trajectories.

Definition 2.3.13 (Global minimal attractor). Let S_t be an evolution operator on a complete metric space X. A bounded closed set $\mathfrak{A}_{\min} \subset X$ is said to be a *global minimal attractor* for S_t if the following properties hold.

(i) $\mathfrak{A}_{\min}$ is a positively invariant set; that is, $S_t \mathfrak{A}_{\min} \subseteq \mathfrak{A}_{\min}$ for $t \geq 0$;
(ii) $\mathfrak{A}_{\min}$ attracts every point x from X; that is,

$$\lim_{t \to +\infty} \text{dist}_X(S_t x, \mathfrak{A}_{\min}) = 0 \text{ for any } x \in X;$$

(iii) $\mathfrak{A}_{\min}$ is minimal; that is, $\mathfrak{A}_{\min}$ has no proper closed subsets possessing (i) and (ii). ∎

One can prove the following assertion.

Theorem 2.3.14. *Let S_t be an evolution operator on a complete metric space X. Assume that S_t is point dissipative (see Definition 2.1.1). If any semitrajectory γ_v^+ is Lagrange stable (see Definition 1.4.1), then S_t possesses a (unique) global minimal attractor $\mathfrak{A}_{\min}$. Moreover, the attractor $\mathfrak{A}_{\min}$ has the representation*

$$\mathfrak{A}_{\min} = \overline{\bigcup \{\omega(x) \ : \ x \in X\}}. \tag{2.3.8}$$

Proof. Since any positive semitrajectory is Lagrange stable, then by Theorem 1.4.5 and Proposition 1.4.6 each ω-limit set $\omega(x)$ is a strictly invariant compact set which attracts $S_t x$. Thus,

$$A_{\min} = \bigcup \{\omega(x) \ : \ x \in X\} \tag{2.3.9}$$

is a strictly invariant set attracting all semitrajectories. Due to point dissipativity this set $A_{\min}$ is bounded. Now using the continuity of S_t one can see that the closure $\overline{A_{\min}}$ of $A_{\min}$ is forward[6] invariant, and thus $\mathfrak{A}_{\min} \equiv \overline{A_{\min}}$ is a minimal global attractor. □

Exercise 2.3.15. Assume that a system (X, S_t) possesses a *compact* global minimal attractor $\mathfrak{A}_{min}$. Show that in this case any semitrajectory γ_v^+ is Lagrange stable, and thus $\mathfrak{A}_{min}$ has form (2.3.8). ∎

The following assertion (see DE [83]) shows how global and global minimal attractors are related.

Theorem 2.3.16. *Assume that an evolution operator S_t on a complete metric space X possesses a compact global attractor $\mathfrak{A}$. Then there exists a global minimal attractor $\mathfrak{A}_{\min}$ which is a compact subset of $\mathfrak{A}$ and has the form (2.3.8). Moreover, $\mathfrak{A}_{\min}$ is strictly invariant and*

$$\mathfrak{A} = \omega(\mathscr{O}_\delta(\mathfrak{A}_{\min})) = \bigcap_{t>0} \overline{\bigcup_{\tau \geq t} S_\tau(\mathscr{O}_\delta(\mathfrak{A}_{\min}))} \text{ for every } \delta > 0, \tag{2.3.10}$$

where $\mathscr{O}_\delta(D)$ denotes the δ-neighborhood of the set D. Thus, any small neighborhood of $\mathfrak{A}_{\min}$ "generates" the global attractor $\mathfrak{A}$.

Proof. It is clear that we can apply Theorem 2.3.14 and show that $\mathfrak{A}_{\min}$ given by (2.3.8) is a global minimal attractor. Since $\omega(x) \subset \mathfrak{A}$ for every x, we have that $\mathfrak{A}_{\min} \subset \mathfrak{A}$ and thus it is compact.

The set $A_{\min}$ given by (2.3.9) is strictly invariant. Therefore, we can apply Exercise 1.2.1(F) to show that $\mathfrak{A}_{\min} = \overline{A_{\min}}$ is strictly invariant.

---

[6] In general we cannot guarantee the strict invariance of this closure, see Exercise 1.2.1(F).

To prove (2.3.10) we note that for every $\delta > 0$ the set $B_0 = \mathscr{O}_\delta(\mathfrak{A}_{\min})$ is point-absorbing, i.e.,

$$\forall\, x \in X, \ \exists\, t_x > 0 : \ S_t x \in B_0, \ \forall\, t \geq t_x.$$

Thus, we can apply the same argument as in the proof of Theorem 2.3.9 and show that there exist $\varepsilon_0 > 0$ and $t_0 > 0$ such that

$$S_{t_0} \mathscr{O}_{\varepsilon_0}(\mathfrak{A}) \subset \gamma_{B_0}^{t_*} = \bigcup_{\tau \geq t_*} S_\tau B_0$$

for some $t_* \geq 0$. Therefore,

$$\mathfrak{A} = S_{t+t_0}\mathfrak{A} \subset S_t\left(S_{t_0}\mathscr{O}_{\varepsilon_0}(\mathfrak{A})\right) \subset \gamma_{B_0}^{t+t_*} \subset \gamma_{B_0}^{t} \ \text{ for every } t \geq 0.$$

Thus,

$$\mathfrak{A} \subset \bigcap_{t>0} \overline{\gamma_{B_0}^t} \equiv \omega(B_0).$$

This completes the proof of Theorem 2.3.16.   □

For some further discussions of properties of global minimal attractors we refer to LADYZHENSKAYA [142] and DE [83].

2.3.2 Weak global attractor

The most restrictive assumption guaranteeing the existence of a global attractor is asymptotic compactness of the corresponding dynamical system (see Theorem 2.3.5). However, in some cases it is possible to get rid of this requirement. For this we need the notion of a global *weak* attractor.

Definition 2.3.17 (Global weak attractor). Let S_t be an evolution operator in a reflexive Banach space X. A bounded weakly closed set $\mathfrak{A}$ in X is called a *global weak attractor* if (i) it is invariant ($S_t\mathfrak{A} = \mathfrak{A}$ for all $t \geq 0$) and (ii) it is uniformly attracting in the weak topology: for any weak vicinity $\mathscr{O}$ of the set $\mathfrak{A}$ and for every bounded set $B \subset X$ there exists $t_* = t(\mathscr{O}, B) > 0$ such that $S_t B \subset \mathscr{O}$ for all $t \geq t_*$.   ∎

It is clear that if a global attractor exists and is weakly closed, then it is also weak. Thus, in the finite-dimensional case they are the same.

Theorem 2.3.18. *Let S_t be an evolution semigroup on a separable reflexive Banach space X. Assume that S_t is weakly closed; i.e., for every $t > 0$ the weak convergence properties $x_n \to x$ and $S_t x_n \to y$ imply that $y = S_t x$. If this semigroup S_t is dissipative, then it possesses a weak global attractor.*

The argument for the proof relies on weak compactness of bounded sets in a separable reflexive Banach space. For details we refer to BABIN/VISHIK [9] or CHUESHOV [39]. Here we give an alternative argument by showing that the situation can be reduced to Theorem 2.3.5.

Proof. Let D be an absorbing bounded set for S_t. We can suppose that D is weakly closed. Let $t_* \geq 0$ be such that $S_t D \subset D$ for all $t \geq t_*$ and

$$D_* = \gamma_D^{t_*} \equiv \bigcup_{t \geq t_*} S_t D.$$

One can see that D_* is a bounded forward invariant set. Since S_t is weakly closed, the weak closure D_*^w of D_* possesses the same properties. Moreover (see DUNFORD/SCHWARTZ [88, Chapter 5, Section 5]), this set D_*^w endowed with weak topology is a *compact* metric space with respect to the distance

$$\varrho(f, g) = \sum_{n=1}^{\infty} \frac{|l_n(f - g)|}{1 + |l_n(f - g)|}, \quad f, g \in D_*^w,$$

where $\{l_n\}$ is a complete set of functionals on X. Thus, the evolution operator S_t is automatically (asymptotically) compact, and we can apply Theorem 2.3.5. □

We note that sometimes it is also convenient to use not only strong or weak convergences but also other topologies in the definition of global attractors. We refer to BABIN/VISHIK [9] (see also the recent survey BABIN [7]) for the theory of attractors involving two phase spaces with different topologies. We also refer to CHESKIDOV/FOIAS [33], CHESKIDOV [32], FOIAS/ROSA/TEMAM [106] and the literature cited there for some development of the theory of weak attractors with application to 3D hydrodynamics.

2.3.3 Stability properties and reduction principle

In order to describe the stability properties of attractors, we need the following notions.

Definition 2.3.19 (Lyapunov stability of invariant sets). A forward invariant set M is said to be *stable (in the Lyapunov sense)* if for any vicinity $\mathcal{O}$ of the closure $\overline{M}$ of M there exists an open set $\mathcal{O}'$ such that $\overline{M} \subset \mathcal{O}' \subset \mathcal{O}$ and $S_t \mathcal{O}' \subset \mathcal{O}$ for all $t \geq 0$. The set M is *asymptotically stable* iff it is stable and $S_t x \to M$ as $t \to \infty$ for every $x \in \mathcal{O}'$. This set is *uniformly asymptotically stable* if it is stable and

$$\lim_{t \to +\infty} \sup_{x \in \mathcal{O}'} \text{dist}_X(S_t x, M) = 0.$$

We note that when $M = \gamma_v^+ = \{S_t v : t \geq 0\}$ is a semitrajectory, the stability of M as an invariant set follows from its stability as a trajectory (see Definition 1.6.1). However, as we can see in the following exercise, the inverse statement is not true.

Exercise 2.3.20. Show that any (nontrivial) trajectory in the system described in Exercise 1.8.10 is stable as an invariant set but unstable as a trajectory. ∎

To distinguish these two types of (Lyapunov) stability, the stability of a trajectory as an invariant set is often called the *orbital stability* of the trajectory.

The following stability property of compact global attractors is important in many situations (see, e.g., BABIN/VISHIK [9] or CHUESHOV [39]).

Theorem 2.3.21. *Let (X, S_t) be a dynamical system in a complete metric space X possessing a compact global attractor $\mathfrak{A}$. Assume that there exists a bounded vicinity $\mathcal{U}$ of $\mathfrak{A}$ such that the mapping $(t; x) \mapsto S_t x$ is continuous on $\mathbb{R}_+ \times \mathcal{U}$. Then $\mathfrak{A}$ is uniformly asymptotically stable.*

Proof. Let $\mathcal{O}$ be a vicinity of $\mathfrak{A}$. Then there exists $T > 0$ such that $S_t \mathcal{U} \subset \mathcal{O}$ for all $t \geq T$. Now we show that there exists a vicinity $\mathcal{O}^*$ of the attractor such that $S_t \mathcal{O}_* \subset \mathcal{O}$ for all $t \in [0, T]$. If this is not true, then there exist sequences $\{u_n\} \subset X$ and $\{t_n\} \subset [0, T]$ such that $\operatorname{dist}\{u_n, \mathfrak{A}\} \to 0$ and $S_{t_n} u_n \notin \mathcal{O}$. Since $\mathfrak{A}$ is compact, we can choose a subsequence $\{n_k\}$ such that $u_{n_k} \to u \in \mathfrak{A}$ and $t_{n_k} \to t \in [0, T]$. Therefore, the continuity property of the function $(t; x) \mapsto S_t x$ gives us that $S_{t_{n_k}} u_{n_k} \to S_t u \in \mathfrak{A}$. This contradicts the equation $S_{t_n} u_n \notin \mathcal{O}$. Thus, there exists an $\mathcal{O}^* \supset \mathfrak{A}$ such that $S_t \mathcal{O}_* \subset \mathcal{O}$ for $t \in [0, T]$. This implies that $S_t(\mathcal{O}_* \cap \mathcal{U}) \subset \mathcal{O}$ for all $t \in \mathbb{R}_+$. Therefore, the attractor $\mathfrak{A}$ is stable. Thus, by the global attraction property the attractor $\mathfrak{A}$ is uniformly asymptotically stable. □

In certain situations the following *reduction principle* enables us to significantly decrease the number of degrees of freedom in the problem. This is important in the study of infinite-dimensional systems.

Theorem 2.3.22 (Reduction principle). *Let (X, S_t) be a dissipative dynamical system in a complete metric space X. Assume that there exists a positively invariant locally compact[7] closed set M possessing the property of uniform attraction:*

$$\lim_{t \to +\infty} \sup_{x \in D} \operatorname{dist}_X(S_t x, M) = 0 \text{ for every bounded set } D. \qquad (2.3.11)$$

If $\mathfrak{A}$ is a global attractor of the restriction (M, S_t) of the system (X, S_t) on M, then $\mathfrak{A}$ is also a global attractor for (X, S_t).

Proof. We use the same method as in CHUESHOV [39, Chapter 1].
It is sufficient to verify that

$$\lim_{t \to +\infty} \sup_{x \in D} \operatorname{dist}_X(S_t x, \mathfrak{A}) = 0 \qquad (2.3.12)$$

---

[7]In the sense that every bounded subset of that set is relatively compact.

for any bounded set D in X. Assume that there exists a bounded closed set $B \subset X$ such that (2.3.12) does not hold. Then there exist sequences $\{y_n\} \subset B$ and $\{t_n : t_n \to +\infty\}$ such that

$$\text{dist}_X(S_{t_n} y_n, \mathfrak{A}) \geq \delta \quad \text{for some} \quad \delta > 0. \tag{2.3.13}$$

Let B_0 be a bounded absorbing set for (X, S_t). We choose a time t_* such that

$$\sup_{x \in M \cap B_0} \text{dist}_X(S_{t_*} x, \mathfrak{A}) \leq \delta/2 \tag{2.3.14}$$

This choice is possible because $\mathfrak{A}$ is a global attractor for (M, S_t). Equation (2.3.11) implies that

$$\text{dist}_X(S_{t_n - t_*} y_n, M) \to 0, \quad n \to +\infty.$$

The dissipativity of (X, S_t) gives us that $S_{t_n - t_*} y_n \in B_0$ for all n large enough. Therefore, local compactness of the set M guarantees the existence of an element $z \in M \cap B_0$ and a subsequence $\{n_k\}$ such that $z = \lim_{k \to \infty} S_{t_{n_k} - t_*} y_{n_k}$. This implies that $S_{t_{n_k}} y_{n_k} \to S_{t_*} z$. Therefore, equation (2.3.13) gives us that $\text{dist}_X(S_{t_*} z, \mathfrak{A}) \geq \delta$, which contradicts (2.3.14). This completes the proof of Theorem 2.3.22.   □

Example 2.3.23. Consider the following system of ODEs:

$$\begin{cases} \dot{y} + y^3 - \lambda y = yz^2, & t > 0, \quad y\big|_{t=0} = y_0, \\ \dot{z} + z(1 + y^2) = 0, & t > 0, \quad z\big|_{t=0} = z_0, \end{cases} \tag{2.3.15}$$

where $\lambda \in \mathbb{R}$. One can see that for any initial data the problem in (2.3.15) has a unique solution on some semi-interval $[0, t_*)$, where $t_* \leq \infty$ depends on $(y_0; z_0)$. If we multiply the first equation by $y(t)$ and the second equation by $z(t)$, then after taking the sum we obtain

$$\frac{1}{2}\frac{d}{dt}\left[y^2 + z^2 \right] + y^4 - \lambda y^2 + z^2 = 0, \quad 0 < t < t_*.$$

This implies that the function $V(y, z) = y^2 + z^2$ possesses the property

$$\frac{d}{dt} V(y(t), z(t)) + 2V(y(t), z(t)) \leq \frac{(1 + \lambda)^2}{2}, \quad 0 < t < t_*.$$

Therefore,

$$V(y(t), z(t)) \leq V(y_0, z_0) e^{-2t} + \frac{(1 + \lambda)^2}{4}(1 - e^{-2t}), \quad 0 < t < t_*.$$

This implies that any solution to problem (2.3.15) can be extended to the whole semi-axis $\mathbb{R}_+$ and the dynamical system $(\mathbb{R}^2, S_t)$ generated by (2.3.15) is dissipative. Obviously, the set $M = \{(y; 0) : y \in \mathbb{R}\}$ is positively invariant. Moreover, the second equation in (2.3.15) yields that

$$\frac{1}{2} \frac{d}{dt} [z(t)]^2 + [z(t)]^2 \leq 0, \quad t > 0.$$

on the solutions. Hence, $|z(t)| \leq |z_0| e^{-t}$ for all $t > 0$. Thus, the set M exponentially attracts all bounded sets from $\mathbb{R}^2$. Consequently, Theorem 2.3.22 yields that the global attractor of the dynamical system (M, S_t) is also the attractor of the system $(\mathbb{R}^2, S_t)$.

On the set M, equations (2.3.15) are reduced to the problem

$$\dot{y} + y^3 - \lambda y = 0, \quad t > 0, \quad y\big|_{t=0} = y_0. \tag{2.3.16}$$

Thus, the global attractors of the dynamical systems generated by equations (2.3.15) and (2.3.16) coincide, and the study of the dynamics on the plane is reduced to the investigation of properties of a certain one-dimensional dynamical system. ∎

Exercise 2.3.24. Using the same idea as in Exercise 2.3.6, show that the global attractor $\widehat{\mathfrak{A}}$ of the system $(\mathbb{R}, \widehat{S}_t)$ generated by (2.3.16) is the interval $[-\sqrt{\lambda_+}, \sqrt{\lambda_+}]$ in $\mathbb{R}$, where $\lambda_+ = \max\{0, \lambda\}$. Therefore, by Theorem 2.3.22 the global attractor $\mathfrak{A}$ of the dynamical system $(\mathbb{R}^2, S_t)$ generated by (2.3.15) has the form $\mathfrak{A} = \{(y; z) : -\sqrt{\lambda_+} \leq y \leq \sqrt{\lambda_+}, z = 0\}$. ∎

Another example of a model with the reduction possibility is described in the following exercise.

Exercise 2.3.25 (Two-mode plasma equation). This model arises as the lowest mode approximation of some equations arising in plasma physics (see, e.g., CHUE-SHOV/SHCHERBINA [70, 71] and the references therein). We consider the following system of ODEs:

$$\begin{cases} \ddot{y} + \gamma \dot{y} + y^3 - y = |z|^2, \quad t > 0, \quad y\big|_{t=0} = y_0, \ \dot{y}\big|_{t=0} = y_1, \\ i\dot{z} - z(1+y) + i\delta z = 0, \quad t > 0, \quad z\big|_{t=0} = z_0, \end{cases} \tag{2.3.17}$$

where y is a real and z is a complex unknown function. Assume that γ and δ are positive parameters. Prove the following statements:

(A) Equations (2.3.17) generate a dynamical system (X, S_t) in $X = \mathbb{R}^2 \times \mathbb{C}$. Hint: One can see that for any initial data the problem in (2.3.17) has a unique solution on some semi-interval $[0, t_*)$, where $t_* \leq \infty$ depends on $(y_0; y_1; z_0)$. If we multiply the second equation by $\bar{z}(t)$ and take the imaginary part, then we get the relation

$$\frac{d}{dt}|z(t)|^2 + 2\delta|z(t)|^2 = 0$$

on the existence time interval. This allows us to apply the non-explosion criterion (see Theorem A.1.2).

(B) The subspace $X_0 = \{U = (y_0; y_1; 0) : (y_0; y_1) \in \mathbb{R}^2\}$ is (strictly) invariant, and the restriction $(X_0; S_t)$ of (X, S_t) on X_0 is generated by the Duffing equation

$$\ddot{y} + \gamma\dot{y} + y^3 - y = 0, \quad t > 0, \quad y|_{t=0} = y_0, \quad \dot{y}|_{t=0} = y_1. \tag{2.3.18}$$

(C) The subspace $X_0 = \{U = (y_0; y_1; 0) : (y_0; y_1) \in \mathbb{R}^2\}$ is an exponentially attracting set for S_t.

(D) Show that system (2.3.17) is dissipative. Hint: Make use of the same Lyapunov function as in Exercise 2.1.8 for the first equation and the fact that $|z(t)| \leq |z_0| \exp\{-\delta t\}$.

(E) Using the reduction principle, describe the global attractor for the system (X, S_t).

∎

We can also formulate a reduction principle with exponential convergence properties (see FABRIE ET AL. [94] and also CHUESHOV [39, Lemma 1.9.6]).

Theorem 2.3.26. *Let (X, S_t) be a dissipative dynamical system in a complete metric space X. In addition to the hypotheses of Theorem 2.3.22, we assume:*

- *There is an absorbing set B_0 and constants $K, \alpha > 0$ such that*

$$\text{dist}_X(S_t x, S_t y) \leq Le^{\alpha t}\text{dist}_X(x, y) \quad \text{for any } x, y \in B_0. \tag{2.3.19}$$

- *The convergence in (2.3.11) holds with exponential rate, i.e., there exist $K, \gamma > 0$ such that*

$$\sup_{x \in B_0} \text{dist}_X(S_t x, M) \leq Ke^{-\gamma t}, \quad t > 0. \tag{2.3.20}$$

- *The attractor $\mathfrak{A}$ is exponential in M, i.e., for any bounded set D in M there exist positive constants K_D and γ_D and time t_D such that*

$$\sup_{x \in D} \text{dist}_X(S_t x, \mathfrak{A}) \leq K_D e^{-\gamma_D t}, \quad t \geq t_D. \tag{2.3.21}$$

Then $\mathfrak{A}$ is an exponential attractor for (X, S_t), i.e., for any bounded set B in X there exist positive constants K_B and γ_D and time t_B such that

$$\sup_{x \in B} \text{dist}_X(S_t x, \mathfrak{A}) \leq K_B e^{-\gamma_B t}, \quad t \geq t_B. \tag{2.3.22}$$

Proof. We first prove the following lemma.

Lemma 2.3.27 (Fabrie et al. [94]). *Let (X, S_t) be a dynamical system in a complete metric space X. Assume that there exist $L, \alpha > 0$ such that*

$$\text{dist}_X(S_t x, S_t y) \leq L e^{\alpha t} \text{dist}_X(x, y) \quad \text{for any } x, y \in X. \tag{2.3.23}$$

Let M_i, $i = 0, 1, 2$, be subsets in X such that $S_t M_i$ converges to M_{i+1} with exponential speed, $i = 0, 1$. This means that

$$\sup_{x \in M_0} \text{dist}_X(S_t x, M_1) \leq K_1 e^{-\gamma_1 t} \quad \text{and} \quad \sup_{x \in M_1} \text{dist}_X(S_t x, M_2) \leq K_2 e^{-\gamma_2 t} \tag{2.3.24}$$

for some positive constants K_i and γ_i. Then $S_t M_0 \to M_2$ exponentially, i.e.,

$$\sup_{x \in M_0} \text{dist}_X(S_t x, M_2) \leq (L K_1 + K_2) e^{-\gamma t} \quad \text{with } \gamma = \frac{\gamma_1 \gamma_2}{\alpha + \gamma_1 + \gamma_2}. \tag{2.3.25}$$

Proof. Let $x \in M_0$ and $z \in M_2$. Then

$$\text{dist}_X(S_t x, z) \leq \text{dist}_X(S_{\kappa t} S_{(1-\kappa)t} x, S_{\kappa t} w) + \text{dist}_X(S_{\kappa t} w, z)$$

for any $0 \leq \kappa \leq 1$ and $w \in M_1$. By (2.3.23),

$$\text{dist}_X(S_t x, z) \leq L e^{\alpha \kappa t} \text{dist}_X(S_{(1-\kappa)t} x, w) + \text{dist}_X(S_{\kappa t} w, z).$$

Therefore,

$$\text{dist}_X(S_t x, M_2) \leq L e^{\alpha \kappa t} \text{dist}_X(S_{(1-\kappa)t} x, w) + \sup_{y \in M_1} \text{dist}_X(S_{\kappa t} y, M_2)$$

for any $0 \leq \kappa \leq 1$ and $w \in M_1$. This implies that

$$\sup_{x \in M_0} \text{dist}_X(S_t x, M_2) \leq L K_1 e^{[\alpha \kappa - \gamma_1(1-\kappa)]t} + K_2 e^{-\gamma_2 \kappa t}$$

for any $0 \leq \kappa \leq 1$. Taking $\kappa = \gamma_1 (\alpha + \gamma_1 + \gamma_2)^{-1}$ we obtain (2.3.25). $\square$

To conclude the proof of Theorem 2.3.26, we apply Lemma 2.3.27 with

$$X = \hat{B} \equiv \overline{\bigcup_{t \geq \hat{t}} S_t B_0} = M_0, \quad M_1 = M \cap \hat{B}, \quad M_2 = \mathfrak{A},$$

where $\hat{t}$ is chosen such that $\hat{B} \subset B_0$ is a forward invariant absorbing set. $\square$

Exercise 2.3.28. Show that any solution to the 1D ODE in (2.3.16) with $\lambda > 0$ and with initial data $|y_0| \geq \lambda$ has the form

$$y(t) = \sqrt{\lambda} y_0 \left[y_0^2 - (y_0^2 - \lambda) e^{-2\lambda t} \right]^{-1/2}, \quad t \geq 0.$$

Using this formula, prove that the attractor $\hat{\mathfrak{A}}$ (see Exercise 2.3.24) for the system generated by (2.3.16) is exponential. Then apply Theorem 2.3.26 to show that the global attractor $\mathfrak{A}$ of the dynamical system $(\mathbb{R}^2, S_t)$ generated by (2.3.15) is also exponential. ∎

Exercise 2.3.29. Show that the attractor $\mathfrak{A}$ for (2.3.15) for $\lambda = 0$ is not exponential. Hint: Use the formula from Exercise 1.7.5. ∎

2.3.4 Stability of attractors with respect to parameters

We next deal with the stability of attractors with respect to perturbations of a dynamical system. For this we consider a family of dynamical systems (X, S_t^λ) with the same phase space X and with evolution operators S_t^λ depending on a parameter λ from a complete metric space Λ.

We start with the following simple assertion (see, e.g., ROBINSON [195, Theorem 10.16]). We also refer to BABIN/VISHIK [9] and HALE [115] for similar results on semicontinuity.

Proposition 2.3.30. *Let X be a complete metric space and S_t^λ be a family of evolution semigroups on X possessing global attractors $\mathfrak{A}^\lambda$ for $\lambda \in \Lambda$. Assume that*

- *the attractors $\mathfrak{A}^\lambda$ are uniformly bounded, i.e., there exists a bounded set B_0 such that $\mathfrak{A}^\lambda \subset B_0$;*
- *there exists $t_0 \geq 0$ such that $S_t^\lambda x \to S_t^{\lambda_0} x$ as $\lambda \to \lambda_0$ for each $t \geq t_0$ uniformly with respect to $x \in B_0$, i.e.,*

$$\sup_{x \in B_0} \mathrm{dist}_X\left(S_t^\lambda x, S_t^{\lambda_0} x\right) \to 0 \quad as \ \lambda \to \lambda_0. \tag{2.3.26}$$

Then the family $\{\mathfrak{A}^\lambda\}$ of attractors is upper semicontinuous at the point λ_0, i.e.,

$$d_X\left\{\mathfrak{A}^\lambda \mid \mathfrak{A}^{\lambda_0}\right\} \equiv \sup\left\{\mathrm{dist}_X(x, \mathfrak{A}^{\lambda_0}) : x \in \mathfrak{A}^\lambda\right\} \to 0 \ as \ \lambda \to \lambda_0.$$

Proof. Given $\varepsilon > 0$ there exists $t > t_0$ such that $S_t^{\lambda_0} B_0 \subset \mathscr{O}_\varepsilon(\mathfrak{A}^{\lambda_0})$. We also have that

$$\mathrm{dist}_X\left(S_t^\lambda x, S_t^{\lambda_0} B_0\right) \leq \sup_{y \in B_0} \mathrm{dist}_X\left(S_t^\lambda y, S_t^{\lambda_0} y\right), \quad \forall\, x \in B_0.$$

Thus,

$$\exists \delta > 0 : \ S_t^\lambda B_0 \subset \mathscr{O}_{2\varepsilon}(\mathfrak{A}^{\lambda_0}) \ \text{ as soon as } \ \mathrm{dist}_\Lambda(\lambda, \lambda_0) < \delta.$$

Consequently,

$$\forall \varepsilon > 0: \quad \mathfrak{A}^\lambda = S_t^\lambda \mathfrak{A}^\lambda \subset S_t^\lambda B_0 \subset \mathcal{O}_{2\varepsilon}(\mathfrak{A}^{\lambda_0}) \quad \text{when} \ \ \text{dist}_\Lambda(\lambda, \lambda_0) < \delta.$$

This implies the conclusion.   □

We illustrate the statements of this proposition and the next theorem in Exercises 2.3.34 and 2.3.35 below.

The following assertion, which was proved by KAPITANSKY/KOSTIN [130] (see also BABIN/VISHIK [9] and CHUESHOV [39]), assumes a much weaker hypothesis concerning convergence of semigroups. This can be critical in singularly perturbed evolutions; see examples in KAPITANSKY/KOSTIN [130]. However, in contrast with the previous assertion, some uniform compactness property of the attractors is assumed.

Theorem 2.3.31 (Kapitansky-Kostin [130]). *Assume that a dynamical system (X, S_t^λ) in a complete metric space X possesses a compact global attractor $\mathfrak{A}^\lambda$ for every $\lambda \in \Lambda$. Assume that the following conditions hold.*

(i) There exists[8] a compact $K \subset X$ such that $\mathfrak{A}^\lambda \subset K$.
(ii) If $\lambda_k \to \lambda_0$, $x_k \to x_0$, and $x_k \in \mathfrak{A}^{\lambda_k}$, then

$$S_\tau^{\lambda_k} x_k \to S_\tau^{\lambda_0} x_0 \quad \text{for some } \tau > 0. \tag{2.3.27}$$

Then the family $\{\mathfrak{A}^\lambda\}$ of attractors is upper semicontinuous at the point λ_0; that is,

$$d_X\{\mathfrak{A}^\lambda \mid \mathfrak{A}^{\lambda_0}\} = \sup\{\text{dist}_X(x, \mathfrak{A}^{\lambda_0}) : x \in \mathfrak{A}^\lambda\} \to 0 \ \text{ as } \lambda \to \lambda_0. \tag{2.3.28}$$

Moreover, if (2.3.27) holds for every $\tau > 0$, then the upper limit $\mathfrak{A}(\lambda_0, \Lambda)$ of the attractors $\mathfrak{A}^\lambda$ at λ_0 defined by the formula

$$\mathfrak{A}(\lambda_0, \Lambda) = \bigcap_{\delta > 0} \overline{\bigcup \{\mathfrak{A}^\lambda : \lambda \in \Lambda, \ 0 < \text{dist}(\lambda, \lambda_0) < \delta\}} \tag{2.3.29}$$

is a nonempty compact strictly invariant set lying in the attractor $\mathfrak{A}^{\lambda_0}$ and possessing the property

$$d_X\{\mathfrak{A}^\lambda \mid \mathfrak{A}(\lambda_0, \Lambda)\} \to 0 \ \text{ as } \lambda \to \lambda_0. \tag{2.3.30}$$

Proof. Assume that equation (2.3.28) does not hold. Then there exists a sequence $\lambda_m \to \lambda_0$ such that $d_X\{\mathfrak{A}^{\lambda_m} \mid \mathfrak{A}^{\lambda_0}\} \geq 2\delta$ for all $m = 1, 2, \ldots$ and for some $\delta > 0$.

---

[8]This property can be relaxed, see Exercise 2.3.33 below.

Thus we can find a sequence $x_m \in \mathfrak{A}^{\lambda_m}$ such that $\mathrm{dist}_X(x_m, \mathfrak{A}^{\lambda_0}) \geq \delta$. But this sequence $\{x_m\}$ lies in the compact K. Therefore, without loss of generality we can assume that $x_m \to x_0$ for some $x_0 \in K$ such that $x_0 \notin \mathfrak{A}^{\lambda_0}$. We show now that this fact leads to a contradiction.

Let $\gamma_m = \{u_m(t) : t \in \mathbb{R}\} \subset \mathfrak{A}^{\lambda_m}$ be a full trajectory of the dynamical system $(X, S_t^{\lambda_m})$ passing through the element x_m ($u_m(0) = x_m$). Using the standard diagonal process, one can see that there exist a subsequence $\{m_n\}$ and a sequence of elements $u_N \in K$ such that

$$\lim_{n \to \infty} u_{m_n}(-N\tau) = u_N \quad \text{for all} \ \ N = 0, 1, \ldots,$$

with $u_0 = x_0$, where τ is the same as in (2.3.27). The condition (ii) also implies that

$$u_{N-L} = \lim_{n \to \infty} u_{m_n}(-(N-L)\tau) = \lim_{n \to \infty} S_{L\tau}^{\lambda_{m_n}} u_{m_n}(-N\tau) = S_{L\tau}^{\lambda_0} u_N$$

for all $N = 1, 2, \ldots$ and $L = 1, \ldots, N$. Therefore, the function

$$u(t) = \begin{cases} S_t^{\lambda_0} u_0, & \text{for} \ t > 0; \\ S_{t+N\tau}^{\lambda_0} u_N, & \text{for} \ -\tau N \leq t < -\tau(N-1), \ \ N = 1, 2, \ldots, \end{cases}$$

gives a full trajectory γ of $(X, S_t^{\lambda_0})$ passing through x_0. It is obvious that this trajectory is bounded. Therefore, by Exercise 2.3.4(A), $\gamma \subset \mathfrak{A}_0^{\lambda}$. This contradicts the relation $x_0 \notin \mathfrak{A}^{\lambda_0}$ and thus completes the proof of (2.3.28).

To prove the assertion concerning the set $\mathfrak{A}(\lambda_0, \Lambda)$ given by (2.3.29), we first note that, by the assumption in (i),

$$\mathfrak{A}^\delta(\lambda_0, \Lambda) = \overline{\bigcup \{\mathfrak{A}^\lambda : \lambda \in \Lambda, \ 0 < \mathrm{dist}(\lambda, \lambda_0) < \delta\}}$$

is a compact set for each $\delta > 0$ and $\mathfrak{A}^\delta(\lambda_0, \Lambda) \supset \mathfrak{A}^{\delta'}(\lambda_0, \Lambda)$ for every $\delta \geq \delta'$. Thus

$$\mathfrak{A}(\lambda_0, \Lambda) = \bigcap_{\delta > 0} \mathfrak{A}^\delta(\lambda_0, \Lambda)$$

is a nonempty compact set. By (2.3.28) we have that $\mathfrak{A}(\lambda_0, \Lambda) \subseteq \mathfrak{A}^{\lambda_0}$. The (strict) invariance of $\mathfrak{A}(\lambda_0, \Lambda)$ follows from the obvious relation

$$x \in \mathfrak{A}(\lambda_0, \Lambda) \quad \text{if and only if} \quad \left\{ \exists \lambda_n \to \lambda_0, \ \exists x_n \in \mathfrak{A}^{\lambda_n} : x = \lim_{n \to \infty} x_n \right\}.$$

$$(2.3.31)$$

By (2.3.27) with $\tau > 0$ arbitrary we obtain that

$$S_t^{\lambda_0} x = \lim_{n \to \infty} S_t^{\lambda_n} x_n, \quad \forall t \geq 0.$$

Since $S_t^{\lambda_n} x_n \in \mathfrak{A}^{\lambda_n}$, the criterion in (2.3.31) implies that $S_t^{\lambda_0} x \in \mathfrak{A}^{\lambda_0}$ for every $t \geq 0$, i.e., $S_t^{\lambda_0} \mathfrak{A}(\lambda_0, \Lambda) \subset \mathfrak{A}(\lambda_0, \Lambda)$. To prove the backward invariance of $\mathfrak{A}(\lambda_0, \Lambda)$ we note that by invariance of the attractors $\mathfrak{A}^{\lambda_n}$ there exists a sequence $y_n \in \mathfrak{A}^{\lambda_n}$ such that $x_n = S_t^{\lambda_n} y_n$. Due to the assumption in (i) and the criterion in (2.3.31), we can choose a subsequence $\{n_k\}$ such that $y_{n_k} \to y \in \mathfrak{A}(\lambda_0, \Lambda)$. Thus,

$$x = \lim_{k \to \infty} x_{n_k} = \lim_{k \to \infty} S_t^{\lambda_{n_k}} y_{n_k} = S_t^{\lambda_0} y,$$

which implies that $S_t^{\lambda_0} \mathfrak{A}(\lambda_0, \Lambda) \supset \mathfrak{A}(\lambda_0, \Lambda)$. Relation (2.3.30) follows from (2.3.31). This completes the proof of Theorem 2.3.31.   □

In the following two exercises we suggest that the reader make sure that condition (i) in Theorem 2.3.31 concerning uniform compactness can be relaxed.

Exercise 2.3.32. Let $\{B_n\}$ be a sequence of bounded sets in a complete metric space X. Assume that there exists a compact set K such that

$$d_X \{B_n \,|\, K\} = \sup \{\text{dist}_X(x, K) \,:\, x \in B_n\} \to 0 \text{ as } n \to \infty.$$

Then every sequence $\{x_n\}$ with $x_n \in B_n$ contains a subsequence $\{x_{n_k}\}$ such that $x_{n_k} \to z$ as $k \to \infty$ for some $z \in K$.   ■

Exercise 2.3.33. Using the result of the previous exercise, show that condition (i) in Theorem 2.3.31 can be changed to the following one: there exists a compact set K_{λ_0} such that

$$d_X \{\mathfrak{A}^\lambda \,|\, K_{\lambda_0}\} \to 0 \text{ as } \lambda \to \lambda_0$$

(if condition (i) holds, then this property is definitely true with $K_{\lambda_0} = K$).   ■

The situation with the (full) continuity of attractors $\mathfrak{A}^\lambda$ with respect to λ is more complicated. In general the family $\{\mathfrak{A}^\lambda\}$ is not lower semicontinuous at the point λ_0; that is, the property $d_X \{\mathfrak{A}^{\lambda_0} \,|\, \mathfrak{A}^{\lambda_k}\} \to 0$ as $\lambda_k \to \lambda_0$ does not hold. The corresponding examples (borrowed from BABIN [7] and RAUGEL [188]) are given in the following exercises.

Exercise 2.3.34 (Raugel [188]). We consider a dynamical system generated in $\mathbb{R}$ by the following equation:

$$\dot{x} = (1 - x)(x^2 - \lambda), \quad t > 0, \quad x(0) = x_0 \in \mathbb{R}.$$

Prove that for each value of the parameter $\lambda \in [-1, 1]$ this dynamical system possesses a global attractor $\mathfrak{A}^\lambda$. Show that

$$\mathfrak{A}^\lambda = \begin{cases} [-\sqrt{\lambda}, 1] & \text{for } \lambda \geq 0; \\ \{1\}, & \text{for } \lambda < 0. \end{cases}$$

Thus $d_X \{ \mathfrak{A}^{\lambda_k} \mid \mathfrak{A}^{\lambda_0} \} \to 0$ as $\lambda_k \to \lambda_0$ for every $\lambda_0 \in [-1, 1]$ and $d_X \{ \mathfrak{A}^0 \mid \mathfrak{A}^{\lambda_k} \} = 1$ as $\lambda_k \to -0$, which means that $\mathfrak{A}^\lambda$ is not (fully) continuous at $\lambda = 0$. Moreover,

$$\mathfrak{A}(0, [-1, 0]) = \{1\} \neq \mathfrak{A}^0,$$

where $\mathfrak{A}(0, [-1, 0])$ is the upper limit defined according to (2.3.29).   ∎

A similar idea is realized in the next exercise.

Exercise 2.3.35 (Babin [7]). Let $(\mathbb{R}, S_t^\lambda)$ be a dynamical system generated by the equation

$$\dot{x} = -x \big[(|x| - 1)^2 - \lambda \big], \quad t > 0, \quad x(0) = x_0 \in \mathbb{R}.$$

Prove that for each value of the parameter $\lambda \in \mathbb{R}$ the system $(\mathbb{R}, S_t^\lambda)$ possesses a global attractor $\mathfrak{A}_\lambda$ and

$$\mathfrak{A}_\lambda = \begin{cases} [-1 - \sqrt{\lambda}, 1 + \sqrt{\lambda}] & \text{for } \lambda \geq 0; \\ \{0\}, & \text{for } \lambda < 0. \end{cases}$$

Thus, $\mathfrak{A}_\lambda$ is continuous with respect to λ for every $\lambda \neq 0$ and is not lower semicontinuous at $\lambda = 0$.   ∎

We note that in order to prove lower semicontinuity under the hypotheses of Theorem 2.3.31 some additional assumptions should be imposed (see BABIN/VISHIK [9]). However, the lower semicontinuity property is generic under simple compactness assumptions (see the discussion in the surveys BABIN [7], RAUGEL [188] and also the recent note HOANG/OLSON/ROBINSON [124]). In particular, one can prove the following result (see HOANG/OLSON/ROBINSON [124] for the details).

Theorem 2.3.36 (Full continuity of attractors). *Let (X, S_t^λ) be a collection of dynamical systems on a complete metric space X. We suppose that the set Λ of parameters is also a complete metric space. Assume that the following conditions hold.*

(i) (X, S_t^λ) possesses a compact global attractor $\mathfrak{A}^\lambda$ for every $\lambda \in \Lambda$;
(ii) there exists a compact set $K \subset X$ such that $\mathfrak{A}^\lambda \subset K$ for every $\lambda \in \Lambda$;
(iii) for each $t > 0$ the function $\lambda \mapsto S_t^\lambda x$ is continuous uniformly for x in compact subsets of X.

Then the family $\{ \mathfrak{A}^\lambda \}$ of attractors is continuous in λ with respect to the Hausdorff distance

$$d_H \{ A \mid B \} \equiv \sup \{ \mathrm{dist}_X(x, B) \; : \; x \in A \} + \sup \{ \mathrm{dist}_X(x, A) \; : \; x \in B \}$$

at every point λ_0 from some residual set. Thus, the full continuity of $\lambda \mapsto \{ \mathfrak{A}^\lambda \}$ is a generic property.

We recall (see, e.g., BOURBAKI [16]) that in the metric space Λ a *residual* set is the complement of a meager set. A subset D of Λ is said to be *meager* (or a first category set in the Baire sense), if it is contained in a countable union of closed nowhere dense subsets of Λ. A set K is said to be *nowhere dense* if its closure contains no open sets. By the Baire categories theorem (see, e.g., BOURBAKI [16]) any residual set is dense. A property $\mathscr{P}$ is said to be *generic* in Λ if $\mathscr{P}$ holds in some residual set of Λ.

In conclusion, we emphasize that the results presented in this subsection deal with stability of attractors with respect to parameters in the Hausdorff (semi)distance and do not consider issues related to uniform stability of individual perturbed trajectories on large time intervals. In this connection we point out the method of finite-dimensional composed trajectories for global tracking of trajectories of a perturbed system which was developed by BABIN/VISHIK [9, Chapters 7 and 8] (see also a short survey in BABIN [7] and the references therein).

2.4   Gradient systems

In this section we consider gradient systems. The main features of these systems are that (i) in the proof of the existence of a global attractor we can avoid a dissipativity property in explicit form, and (ii) the structure of the attractor can be described via unstable manifolds.

2.4.1   Lyapunov function

We start with the following definition.

Definition 2.4.1. Let $Y \subseteq X$ be a forward invariant set of a dynamical system (X, S_t).

- A continuous functional $\Phi(y)$ defined on Y is said to be a *Lyapunov function* on Y for the dynamical system (X, S_t) if $t \mapsto \Phi(S_t y)$ is a non-increasing function for any $y \in Y$.
- The Lyapunov function $\Phi(y)$ is said to be *strict* on Y if the equation $\Phi(S_t y) = \Phi(y)$ for *all* $t > 0$ and for some $y \in Y$ implies that $S_t y = y$ for all $t > 0$; that is, y is a stationary point of (X, S_t).
- The dynamical system (X, S_t) is said to be *gradient* if there exists a strict Lyapunov function for (X, S_t) on the whole phase space X. This Lyapunov function is usually called *global*.

∎

The simplest examples of Lyapunov functions are given in the following exercises.

Exercise 2.4.2. Let $F : \mathbb{R}^d \mapsto \mathbb{R}$ be a C^2 function such that $F(x) \to +\infty$ as $|x| \to \infty$. Show that the ordinary differential equation

$$\dot{x} = -\nabla F(x), \quad x \in \mathbb{R}^d, \ t > 0,$$

generates a dynamical system $(\mathbb{R}^d, S_t)$ which possesses a strict Lyapunov function $\Phi(x) = F(x)$ on $\mathbb{R}^d$. ∎

Exercise 2.4.3. Consider the second order in time ordinary differential equation

$$\ddot{y} + \gamma\dot{y} + U'(y) = 0, \quad t > 0, \quad y|_{t=0} = y_0, \ \dot{y}|_{t=0} = y_1,$$

where $\gamma > 0$ and $U(y)$ is a C^2 function on $\mathbb{R}$ bounded from above. Show that this equation generates a dynamical system $(\mathbb{R}^2, S_t)$ which possesses a strict Lyapunov function

$$\Phi(y, \dot{y}) = \frac{1}{2}\dot{y}^2 + U(y), \quad (y; \dot{y}) \in \mathbb{R}^2.$$

Hint: See Example 1.8.18 and Remark 1.8.19. ∎

Example 2.4.4. Using the result of Exercise 2.4.3, one can see that the system generated by the plasma equation in (2.3.17) has a strict Lyapunov function on the attractor $\mathfrak{A}$. This is true due to the reduction principle, which shows that the dynamics on $\mathfrak{A}$ can be described by the Duffing equation in (2.3.18). We do not know whether the system generated by (2.3.17) possesses a *global* Lyapunov function, i.e., whether it is gradient. The same effect can be seen in the model considered in Example 2.3.23. ∎

2.4.2 Geometric structure of the attractor

The following result on the structure of a global attractor is known from many sources, including BABIN/VISHIK [9], CHUESHOV [39], HALE [116], HENRY [123], LADYZHENSKAYA [142], TEMAM [216].

Theorem 2.4.5. *Let a dynamical system (X, S_t) possess a compact global attractor $\mathfrak{A}$. Assume that there exists a strict Lyapunov function on $\mathfrak{A}$. Then $\mathfrak{A} = \mathcal{M}^u(\mathcal{N})$, where $\mathcal{M}^u(\mathcal{N})$ denotes the unstable manifold emanating from the set $\mathcal{N}$ of stationary points (see Definition 2.3.10). Moreover, the global attractor $\mathfrak{A}$ consists of full trajectories $\gamma = \{u(t) : t \in \mathbb{R}\}$ such that*

$$\lim_{t \to -\infty} \text{dist}_X(u(t), \mathcal{N}) = 0 \quad \text{and} \quad \lim_{t \to +\infty} \text{dist}_X(u(t), \mathcal{N}) = 0. \tag{2.4.1}$$

Proof. It is known from Proposition 2.3.12 that $\mathscr{M}^u(\mathscr{N}) \subset \mathfrak{A}$. Thus, we need only prove that $\mathfrak{A} \subset \mathscr{M}^u(\mathscr{N})$.

Let $y \in \mathfrak{A}$. By Exercise 2.3.4(B) there exists a full trajectory $\gamma = \{u(t) : t \in \mathbb{R}\}$ passing through y, $u(0) = y$. Since $\gamma \subset \mathfrak{A}$, the set γ is compact. This implies that the α-limit set

$$\alpha(\gamma) = \bigcap_{\tau < 0} \overline{\cup\{u(t) : t \le \tau\}}$$

of the trajectory γ is a nonempty compact set. One can see that the set $\alpha(\gamma)$ is invariant: $S_t\alpha(\gamma) = \alpha(\gamma)$. This follows from its compactness and the description given in Exercise 1.3.6.

Let us show that the Lyapunov function $\Phi(x)$ is a constant on $\alpha(\gamma)$. Indeed, if $u \in \alpha(\gamma)$, then there exists a sequence $\{t_n\}$ such that $t_n \to -\infty$ and $u(t_n) \to u$ as $n \to \infty$ (see Exercise 1.3.6). Consequently,

$$\Phi(u) = \lim_{n\to\infty} \Phi(u(t_n)).$$

By the monotonicity of Φ along trajectories, we have

$$\Phi(u) = \sup_{\tau < 0} \Phi(u(\tau)).$$

Therefore, the limit above does not depend on a sequence $\{u_n\}$ and the function $\Phi(u)$ is a constant on $\alpha(\gamma)$. Hence by invariance of $\alpha(\gamma)$ we have that $\Phi(S_t u) = \Phi(u)$ for all $t > 0$ and $u \in \alpha(\gamma)$. This means that $\alpha(\gamma)$ lies in the set $\mathscr{N}$ of stationary points. Now we prove that

$$\lim_{t\to-\infty} \text{dist}(u(t), \alpha(\gamma)) = 0. \tag{2.4.2}$$

If (2.4.2) is not true, then there exists a sequence $\{t_n \to -\infty\}$ such that

$$\text{dist}_X(u(t_n), \alpha(\gamma)) \ge \delta > 0 \quad \text{for all} \quad n = 1, 2, \ldots \tag{2.4.3}$$

By the compactness of $\overline{\gamma}$ there exist an element $z \in X$ and a subsequence $\{t_{n_m}\}$ such that $u(t_{n_m}) \to z$ as $m \to \infty$. Moreover, by Exercise 1.3.6, $z \in \alpha(\gamma)$. This contradicts the property in (2.4.3) and thus (2.4.2) holds.

Since $\alpha(v) \subset \mathscr{N}$, equation (2.4.2) implies the first relation in (2.4.1) and hence $y \in \mathscr{M}^u(\mathscr{N})$ and $\mathfrak{A} = \mathscr{M}^u(\mathscr{N})$.

To prove the second relation in (2.4.1), we use the same idea as above. We consider the ω-limit set

$$\omega(\gamma) = \bigcap_{\tau > 0} \overline{\cup\{u(t) : t \ge \tau\}}$$

which is a nonempty compact strictly invariant set. As above, it follows from the monotonicity of Φ and the invariance of $\omega(\gamma)$ that the Lyapunov function $\Phi(x)$ is a constant on $\omega(\gamma)$ and hence $\Phi(S_t u) = \Phi(u)$ for all $t > 0$ and $u \in \omega(\gamma)$. This implies that $\omega(\gamma) \subset \mathcal{N}$. As above, by the contradiction argument,

$$\text{dist}(u(t), \mathcal{N}) \leq \text{dist}(u(t), \omega(\gamma)) \to 0 \quad \text{as } t \to +\infty.$$

This completes the proof of Theorem 2.4.5. $\qquad\square$

Remark 2.4.6. It follows from the first equality in (2.4.1) that under the hypotheses of Theorem 2.4.5 the following relation is valid:

$$\sup\{\Phi(u) : u \in \mathfrak{A}\} \leq \sup\{\Phi(u) : u \in \mathcal{N}\}, \tag{2.4.4}$$

where $\Phi(u)$ is the corresponding Lyapunov function. If $\Phi(u)$ topologically dominates the metric of the phase space X, then the inequality in (2.4.4) can be used in order to provide an upper bound for the size of the attractor and an absorbing ball. This method can be applied to obtain uniform (with respect to the parameters of the problem) bounds for the attractor. We refer to Section 5.3 for an application of this idea for some class of second order in time models. $\qquad\blacksquare$

If the system (X, S_t) is gradient; i.e., if a strict Lyapunov function exists on the *whole* phase space, then the result of Theorem 2.4.5 can be improved (see, e.g., BABIN/VISHIK [9] or CHUESHOV [39]). More precisely, we can describe the long-time behavior of individual trajectories.

Theorem 2.4.7. *Assume that a gradient dynamical system* (X, S_t) *possesses a compact global attractor* $\mathfrak{A}$. *Then*

$$\lim_{t \to +\infty} \text{dist}_X(S_t x, \mathcal{N}) = 0 \quad \text{for any } x \in X; \tag{2.4.5}$$

that is, any trajectory stabilizes to the set $\mathcal{N}$ *of stationary points.*[9] *In particular, this means that the global minimal attractor* $\mathfrak{A}_{\min}$ *coincides with the set of the stationary points,* $\mathfrak{A}_{\min} = \mathcal{N}$.

Proof. For every $x \in X$ we consider the ω-limit set $\omega(x) = \cap_{\tau>0}\overline{\cup\{S_t x : t \geq \tau\}}$ and apply the same argument as in the end of the proof of Theorem 2.4.5. $\qquad\square$

Exercise 2.4.8. Show that relation (2.4.5) in the statement of Theorem 2.4.7 remains true if instead of the existence of a compact global attractor we assume that any semitrajectory of the system is Lagrange stable (see Definition 1.4.1). The assertion concerning global minimal attractors remains in force if we assume that the set $\mathcal{N}$ is bounded. $\qquad\blacksquare$

---

[9] This property is often referred to as *strong stability* of the set of equilibria.

Assume that $\mathcal{N} = \{z_1, \ldots, z_n\}$ is a finite set. In this case $\mathfrak{A} = \cup_{i=1}^{n} \mathcal{M}^u(z_i)$, where $\mathcal{M}^u(z_i)$ is the unstable manifold of the stationary point z_i. That is, $\mathcal{M}^u(z_i)$ consists of all $y \in X$ such that there exists a full trajectory $\gamma = \{u(t) : t \in \mathbb{R}\}$ with the properties $u(0) = y$ and $u(t) \to z_i$ as $t \to -\infty$.

Theorems 2.4.5 and 2.4.7 lead us to the following consequences.

Corollary 2.4.9. *Assume that a gradient dynamical system (X, S_t) possesses a compact global attractor $\mathfrak{A}$ and $\mathcal{N}$ is a finite set. Then*

(i) *The global attractor $\mathfrak{A}$ consists of full trajectories $\gamma = \{u(t) : t \in \mathbb{R}\}$ connecting pairs of stationary points: any $u \in \mathfrak{A}$ belongs to some full trajectory $\gamma \subset \mathfrak{A}$ and for any $\gamma \subset \mathfrak{A}$ there exists a pair $\{z, z^*\} \subset \mathcal{N}$ such that*

$$u(t) \to z \text{ as } t \to -\infty \text{ and } u(t) \to z^* \text{ as } t \to +\infty.$$

(ii) *For any $v \in X$ there exists a stationary point z such that $S_t v \to z$ as $t \to +\infty$.*

Remark 2.4.10. Assume that the hypotheses of Corollary 2.4.9 hold. Introduce m_0 distinct values $\Phi_1 < \Phi_2 < \cdots < \Phi_{m_0}$ of the set $\{\Phi(x) : x \in \mathcal{N}\}$ and let

$$\mathcal{N}^j = \left\{ x \in \mathcal{N} : \Phi(x) = \Phi_j \right\}, \quad j = 1, \ldots, m_0.$$

Then the sets $\mathcal{N}^1, \ldots, \mathcal{N}^{m_0}$ provide *Morse decomposition* of the attractor $\mathfrak{A}$. That is, (i) the subsets $\mathcal{N}^j$ are compact, invariant, and disjoint; and (ii) for any $x \in \mathfrak{A} \setminus \cup_j \mathcal{N}^j$ and every full trajectory $\gamma_x \subset \mathfrak{A}$ through x there exist $k > l$ such that $\alpha(\gamma_x) \in \mathcal{N}^k$ and $\omega(\gamma_x) \in \mathcal{N}^l$, where $\alpha(\gamma_x)$ and $\omega(\gamma_x)$ are the α- and ω-limit sets for γ_x (see (1.3.2)).

In the situation considered the set $\mathcal{N}^1$ is uniformly asymptotically stable (see Definition 2.3.19). Thus, $\mathcal{N}^1$ is a subattractor of the attractor $\mathfrak{A}$. We recall that by the definition (see BABIN [7]) any compact strictly invariant uniformly asymptotically stable subset of $\mathfrak{A}$ is called a *subattractor*. If the set $\mathcal{N}^1$ is not connected (e.g., it consists of isolated equilibria), then we can split $\mathcal{N}^1$ into several non-intersecting subattractors. This observation motivates (see BABIN [7]) the notion of a *fragmentation number* of the attractor $\mathfrak{A}$, which is defined as the maximal number of non-intersecting subattractors in $\mathfrak{A}$. This number characterizes the intrinsic complexity of the attractor. For further discussions we refer to BABIN [7] and the references therein.   ∎

The following example shows that the strictness of the corresponding Lyapunov function is important in the statements of Theorems 2.4.5 and 2.4.7.

Example 2.4.11 (Non-strict Lyapunov function). We consider the dynamical system $(\mathbb{R}^2, S_t)$ generated by the following equations:

$$\begin{cases} \dot{x}_1 = \mu x_1 - \alpha x_2 - x_1(x_1^2 + x_2^2), \\ \dot{x}_2 = \alpha x_1 + \mu x_2 - x_2(x_1^2 + x_2^2) \end{cases} \qquad (2.4.6)$$

where $\mu \in \mathbb{R}$ and $\alpha > 0$ are parameters. In the case $\alpha = 1$ this system was considered in Example 1.9.4 as a demonstration for the Andronov-Hopf bifurcation. It was shown that for $\mu \leq 0$ the system has a unique equilibrium $x_* = (0;0)$. If $\mu > 0$ and $\alpha > 0$ there is also the periodic orbit (the circle $C_{\sqrt{\mu}}$ with center at 0 and the radius $\sqrt{\mu}$). If in the latter case we take $\alpha = 0$, then the circle $C_{\sqrt{\mu}}$ consists of equilibria.

One can see that the function

$$V(x_1, x_2) = \frac{1}{2}(x_1^2 + x_2^2)^2 - \mu(x_1^2 + x_2^2)$$

satisfies the equation

$$\frac{dV(x_1, x_2)}{dt} = -2[(x_1^2 + x_2^2) - \mu]^2(x_1^2 + x_2^2) \leq 0$$

on a solution $S_t y_0 = (x_1(t); x_2(t))$. Thus, V is a Lyapunov function. Moreover, we observe the following picture:

- If $\mu \leq 0$ this function is *strict* (and the global attractor consists of a single (zero) equilibrium);
- If $\mu > 0$ and $\alpha = 0$, the function V is still *strict* (the circle $C_{\sqrt{\mu}}$ consists of equilibria) and the global attractor is the disc $D_{\sqrt{\mu}} = \{(x_1^2 + x_2^2) \leq \mu\}$, which can be seen as a collection of trajectories connecting the zero equilibrium and an equilibrium lying on $C_{\sqrt{\mu}}$.
- If $\mu > 0$ and $\alpha > 0$, the function V is *not strict* and the global attractor is the disc $D_{\sqrt{\mu}}$ which contains a nontrivial periodic orbit.

∎

The following exercise demonstrates the non-uniqueness of the Lyapunov function V in Example 2.4.11. A similar effect for local Lyapunov functions was observed in Exercises 1.6.5 and 1.7.5(D).

Exercise 2.4.12. Show that

$$W(x_1, x_2) = \frac{1}{3}(x_1^2 + x_2^2)^3 - \frac{\mu}{2}(x_1^2 + x_2^2)$$

is also a Lyapunov function for (2.4.6) which is strict when either $\mu \leq 0$ or $\alpha = 0$.

∎

Another example with non-strict Lyapunov function provides the Krasovskii system (see Exercises 1.8.22 and 2.1.10) under the condition that the damping coefficient $k(r)$ is non-negative and has a nonzero root.

To describe additional properties of global attractors for gradient systems, we introduce the following definition.

Definition 2.4.13. Let X be a Banach space. Assume that the evolution operator S_t of a dynamical system (X, S_t) is of class C^1; that is, $S_t u$ has a continuous Fréchet derivative[10] with respect to $u \in X$ for each $t > 0$. An equilibrium point z of dynamical system (X, S_t) is said to be *hyperbolic* if the Fréchet derivative $S' \equiv DS_1(z)$ of $S_t z$ at the moment $t = 1$ is a linear operator in X with the spectrum $\sigma(S')$ possessing the property

$$\sigma(S') \cap \{w \in \mathbb{C} : |w| = 1\} = \emptyset.$$

We also define the index ind (z) (of instability) of the equilibrium z as a dimension of the spectral subspace of the operator S' corresponding to the set $\sigma_+(S') \equiv \{z \in \sigma(S') : |z| > 1\}$.   ∎

The following assertion is proved in BABIN/VISHIK [9].

Theorem 2.4.14. *Assume that a gradient dynamical system (X, S_t) in a Banach space X with a strict Lyapunov function $\Phi(u)$ possesses the following properties.*

(i) It admits a compact global attractor $\mathfrak{A}$.
(ii) $S_t \in C^{1+\alpha}$ for some $\alpha > 0$ and there exists a vicinity $\mathcal{O} \supset \mathfrak{A}$ such that

$$\|DS_t(u) - DS_t(v)\|_{X \mapsto X} \leq C_T \|u - v\|_X^\alpha, \quad u, v \in \mathcal{O}, \ t \in [0, T].$$

(iii) $(t, u) \mapsto S_t u$ is continuous over $\mathbb{R}_+ \times \mathfrak{A}$.
(iv) The operators S_t are injective on $\mathfrak{A}$ for any $t > 0$ and S_t^{-1} are continuous on $\mathfrak{A}$.
(v) The Fréchet derivatives $DS_t(u)$ of $S_t u$ at any point $u \in \mathfrak{A}$ have zero kernel.
(vi) The set $\mathcal{N} = \{z_1, \ldots, z_n\}$ of equilibrium points is finite and every point $z_j \in \mathcal{N}$ is hyperbolic.

Let the indexation of equilibrium points be such that

$$\Phi(z_1) \leq \Phi(z_2) \leq \cdots \leq \Phi(z_n)$$

and $M_k = \cup_{j=1}^k \mathcal{M}^u(z_j)$, $M_0 = \emptyset$, where $\mathcal{M}^u(z_j)$ is the unstable manifold emanating from z_j. Assume that the function $t \mapsto \Phi(S_t u)$ is strictly decreasing for $u \notin \mathcal{N}$.
  Then $\mathfrak{A} = M_n$ and the following properties hold.

(i) $\mathcal{M}^u(z_i) \cap \mathcal{M}^u(z_j) = \emptyset$ when $i \neq j$.
(ii) M_k is a compact invariant set.
(iii) $\partial \mathcal{M}^u(z_i) \equiv \overline{\mathcal{M}^u(z_i)} \setminus \mathcal{M}^u(z_i)$ is an invariant set and $\partial \mathcal{M}^u(z_i) \subset M_{i-1}$.
(iv) For any compact set $K \subset \mathcal{M}^u(z_i) \setminus \{z_i\}$ we have

$$\lim_{t \to +\infty} \max\{\text{dist}_X(S_t k, M_{i-1}) : k \in K\} = 0.$$

---

[10]See Section A.5 in the Appendix for the definitions.

(v) *Every set $\mathcal{M}^u(z_i)$ is a C^1-manifold of finite dimension d_i, this manifold is diffeomorphic to $\mathbb{R}^{d_i}$, and the embedding $\mathcal{M}^u(z_i) \subset X$ is of class C^1 in a vicinity of any point $v \in \mathcal{M}^u(z_i)$. Moreover, $d_i = \mathrm{ind}(z_i)$.*

In many cases it is important to know how fast the trajectories starting from bounded sets converge to global attractors. The result stated below provides conditions sufficient for an exponential rate of stabilization to the attractor along with some additional properties of the attractor (see, e.g., BABIN/VISHIK [9], HALE [116] and also Theorems 4.7 and 4.8 in the survey RAUGEL [188]).

Theorem 2.4.15. *Let (X, S_t) be a dynamical system in a Banach space X. Assume that (i) an evolution operator S_t is C^1, (ii) the set $\mathcal{N}$ of equilibrium points is finite and all equilibria are hyperbolic, (iii) there exists a function Lyapunov $\Phi(x)$ on X such that $\Phi(S_t x) < \Phi(x)$ for all $x \in X$, $x \notin \mathcal{N}$ and for all $t > 0$, and (iv) there exists a compact global attractor $\mathfrak{A}$. Then*

- *For any $y \in X$ there exists $e \in \mathcal{N}$ such that*

$$\|S_t y - e\|_X \le C_y e^{-\omega t}, \quad t > 0.$$

Moreover,

$$\sup \{\mathrm{dist}\,(S_t y, \mathfrak{A}) \,:\, y \in B\} \le C_B e^{-\omega t}, \quad t > 0, \tag{2.4.7}$$

for any bounded set B in X. Here C_y, C_B, and ω are positive constants, and ω in (2.4.7) depends on the minimum, over $e \in \mathcal{N}$, of the distance of the spectrum of $D[S_1 e]$ to the unit circle in $\mathbb{C}$.
- *If we assume in addition that (i) S_1 is injective on the attractor and (ii) the linear map $D[S_1 y]$ is injective for every $y \in \mathfrak{A}$, then for each $e \in \mathcal{N}$ the unstable manifold $\mathcal{M}^u(e)$ is an embedded C^1-submanifold of X of finite dimension $\mathrm{ind}\,(e)$.*

We note that the proof of this result (see BABIN/VISHIK [9] or HALE [116]) relies on geometric consideration of the behavior of trajectories in a vicinity of equilibrium points. The critical assumption for this is that the evolution S_t is C^1 and that equilibria are finite and hyperbolic. The above assumptions allow us to reduce the problem of convergence in the vicinity of equilibria to a linear problem.

We also refer to CARVALHO/LANGA [25] and CARVALHO/LANGA/ROBINSON [26, Chapter 5] for some generalizations of the notion of a gradient system. These generalizations are related to the Morse decomposition of attractors and deal with families of isolated invariant sets rather than with collections of (isolated) equilibria.

Another important issue is persistence of the regular structure of a global attractor under perturbations. On this topic we mention the paper by BABIN/VISHIK [8], which presents some results on the persistence of the gradient structure (i.e., the existence of a strict Lyapunov function) for some classes of PDEs. Recently this question was discussed in great detail in ARAGÃO ET AL. [3] and CARVALHO/LANGA/ROBINSON [26, Chapter 5].

2.4.3 Criteria of existence of global attractors for gradient systems

In this section we prove several assertions on the existence of global attractors which do not assume any dissipativity properties of the system in explicit form.

We start with the following criterion for the existence of a global attractor for gradient systems (see, e.g., RAUGEL [188, Theorem 4.6]), which is useful in many applications.

Theorem 2.4.16. *Let (X, S_t) be an asymptotically smooth gradient system which has the property that for any bounded set $B \subset X$ there exists $\tau > 0$ such that $\gamma_\tau(B) \equiv \cup_{t \geq \tau} S_t B$ is bounded. If the set $\mathcal{N}$ of stationary points is bounded, then (X, S_t) has a compact global attractor $\mathfrak{A}$.*

Remark 2.4.17. By Theorem 2.4.5 the global attractor $\mathfrak{A}$ given by Theorem 2.4.16 coincides with the unstable set $\mathcal{M}_+(\mathcal{N})$ emanating from the set $\mathcal{N}$ of stationary points (see Definition 2.3.10), i.e., $\mathfrak{A} = \mathcal{M}_+(\mathcal{N})$.   ∎

Proof of Theorem 2.4.16. Let B be a bounded set in X and $B_\tau = \overline{\gamma_\tau(B)}$. We consider the restriction (B_τ, S_t) of the dynamical system (X, S_t) on the (forward invariant) set B_τ. Since B_τ is bounded, (B_τ, S_t) is a dissipative asymptotically smooth dynamical system. By Theorem 2.3.5 this system possesses a compact global attractor $\mathfrak{A}_B$. By Theorem 2.4.5 $\mathfrak{A}_B = \mathcal{M}_+(\mathcal{N}_B)$, where $\mathcal{N}_B = \mathcal{N} \cap B_\tau$. Under the condition $B \supset \mathcal{N}$ we have that $\mathcal{N} = S_t \mathcal{N} \subset S_t B$ for every $t > 0$. Thus $\mathcal{N} \subset B_\tau$. This implies that $\mathfrak{A}_B = \mathcal{M}_+(\mathcal{N})$ and thus the attractor $\mathfrak{A}_B$ is independent of B when $B \supset \mathcal{N}$. Since $\mathfrak{A}_{B_1} \subset \mathfrak{A}_{B_2}$ for $B_1 \subset B_2$, we have that $\mathfrak{A} := \mathcal{M}_+(\mathcal{N})$ attracts all bounded sets from X.   □

Using Theorem 2.4.16 we can obtain the following assertion (see Corollary 2.29 in CHUESHOV/LASIECKA [56]).

Theorem 2.4.18. *Assume that (X, S_t) is a gradient asymptotically smooth dynamical system. Assume its Lyapunov function $\Phi(x)$ is bounded from above on any bounded subset of X and the set $\Phi_R = \{x : \Phi(x) \leq R\}$ is bounded for every R. If the set $\mathcal{N}$ of stationary points of (X, S_t) is bounded, then (X, S_t) possesses a compact global attractor $\mathfrak{A} = \mathcal{M}^u(\mathcal{N})$.*

Proof. Due to Theorem 2.4.16, it is sufficient to show that for any bounded set $B \subset X$ the set $\gamma_+(B) \equiv \cup_{t \geq 0} S_t B$ is bounded. To see this, we note that $B \subset \Phi_R$ for some $R > 0$. Since Φ_R is invariant, we have that $\gamma_+(B) \subset \Phi_R$ and thus $\gamma_+(B)$ is bounded.   □

Exercise 2.4.19. Show that in the statement of Theorem 2.4.18 the condition concerning the boundedness of the set of stationary points can be changed to the requirement that (X, S_t) is point dissipative (see Definition 2.1.1).   ∎

Example 2.4.20 (Two-mode fluid-structure model). This model is the lowest Galerkin mode approximation of a system arising in the study of interaction of a fluid filling a bounded vessel with an elastic wall (see, e.g., CHUESHOV/RYZHKOVA [68] and the references therein). The equations appear as follows:

$$\frac{d}{dt}[z + \gamma\dot{y}] + \alpha z = f, \tag{2.4.8a}$$

$$\frac{d}{dt}[\gamma z + 2\dot{y}] - \lambda y + y^3 = h, \tag{2.4.8b}$$

where $\alpha > 0$, $|\gamma| \le 1$, and $\lambda, f, h \in \mathbb{R}$ are constants. We endow these equations with initial data

$$z(0) = z_0, \quad y(0) = y_0, \quad \dot{y}(0) = y_1. \tag{2.4.9}$$

One can see that problem (2.4.8) and (2.4.9) has a unique local solution for all initial data $(z_0; y_0; y_1) \in \mathbb{R}^3$. Using the multipliers $z - f/\alpha$ for the first equation and $\dot{y}$ for the second one, we obtain the following energy balance relation:

$$\frac{d}{dt}E(z(t) - f/\alpha, y(t), \dot{y}(t)) + \alpha[z(t) - f/\alpha]^2 = 0 \tag{2.4.10}$$

on the existence interval, where the energy functional E has the form

$$E(z, y, \dot{y}) = \frac{1}{2}z^2 + \gamma z\dot{y} + \dot{y}^2 + \frac{1}{4}y^4 - \frac{\lambda}{2}y^2 - hy.$$

The energy relation in (2.4.10) allows us to use the non-explosion criterion in Theorem A.1.2 and show that problem (2.4.8) and (2.4.9) generates a dynamical system in $\mathbb{R}^3$. Moreover, one can see that

$$V(z, y, \dot{y}) = E(z - f/\alpha, y, \dot{y})$$

is a strict global Lyapunov function for this system. Since the set

$$\{(z; y) \; : \; \alpha z = f, \; y^3 - \alpha y = h\}$$

of stationary solutions is finite, by Theorem 2.4.18 the system generated by (2.4.8) and (2.4.9) possesses a global attractor which coincides with the unstable set emanating from the set of equilibria.   ∎

If a system (X, S_t) is not gradient but possesses a Lyapunov function (which is not strict), we cannot guarantee that $\mathfrak{A} = \mathcal{M}^u(\mathcal{N})$. However, we can prove the following assertion (see also CHUESHOV [39, Theorem 6.2, Chapter 1] and CHUESHOV/LASIECKA [56, Theorem 2.30]).

Theorem 2.4.21. *Let (X, S_t) be an asymptotically smooth dynamical system in some complete metric space X. Assume that there exists a Lyapunov function $\Phi(x)$*

for (X, S_t) on X such that $\Phi(x)$ is bounded from above on any bounded subset of X and the set $\Phi_R = \{x : \Phi(x) \leq R\}$ is bounded for every R. Let $\mathcal{B}$ be the set of elements $x \in X$ such that there exists a full trajectory $\{u(t) : t \in \mathbb{R}\}$ with the properties $u(0) = x$ and $\Phi(u(t)) = \Phi(x)$ for all $t \in \mathbb{R}$. If $\mathcal{B}$ is bounded, then (X, S_t) possesses a compact global attractor and $\mathfrak{A} = \mathcal{M}^u(\mathcal{B})$.

Proof. We choose R_0 such that $\mathcal{B} \subset \Phi_{R_0}$. By Theorem 2.3.5 the dynamical system (Φ_R, S_t) possesses a compact global attractor $\mathfrak{A}_R$ for every R. Let $R \geq R_0$. In this case we have that $\mathcal{B} \subset \Phi_R$. By the same argument as in the proof of Theorem 2.4.5 we can show that for any full trajectory $\gamma = \{u(t) : t \in \mathbb{R}\}$ from the attractor $\mathfrak{A}_R$ we have that $\alpha(\gamma) \subset \mathcal{B}$ and thus $u(t) \to \mathcal{B}$ as $t \to -\infty$. This means that $\mathfrak{A}_R \subset \mathcal{M}^u(\mathcal{B})$. Since $\mathcal{B}$ is a bounded strictly invariant set, we have that $\mathcal{B} \subset \mathfrak{A}_R$. This implies that $\mathcal{M}^u(\mathcal{B}) \subset \mathfrak{A}_R$ and thus $\mathfrak{A}_R = \mathcal{M}^u(\mathcal{B})$ for all $R \geq R_0$. Therefore, $\mathfrak{A} := \mathcal{M}^u(\mathcal{B})$ is a global attractor for (X, S_t). $\square$

The following two exercises illustrate Theorem 2.4.21.

Exercise 2.4.22. Apply Theorem 2.4.21 to the model in Example 2.4.11 to describe the global attractor in the case when μ and α are positive.   ∎

Exercise 2.4.23. Apply Theorem 2.4.21 to describe the structure of the global attractor for the Krasovskii system (see Exercise 2.1.10).   ∎

Chapter 3
Finite-Dimensional Behavior and Quasi-Stability

This chapter deals mainly with the dimension theory of global attractors. We present some background and develop a relatively new approach which is based on some ideas due to O. Ladyzhenskaya (see LADYZHENSKAYA [142] and the literature cited there) and assumes minimal smoothness properties of evolutions. We also discuss a wide class of dynamical systems which admits what is called the stabilizability (or quasi-stability) estimate. The notion of quasi-stability originally arose in the study of some plate models with nonlinear critical damping. However, the extension developed in this chapter allows us to consider a wider class of second order models (see Chapter 5) and also to cover several classes of parabolic and delayed models (see Chapters 4 and 6). In addition to attractors, other long-time behavior objects such as exponential attractors and determining functional sets are considered from the point of view of quasi-stability in this chapter.

3.1 Dimension of global attractors

Finite dimensionality is an important property of global attractors that can be established for many dissipative dynamical systems. There are several approaches that provide effective estimates for the dimension of attractors of dissipative infinite-dimensional systems (see, e.g., BABIN/VISHIK [9], LADYZHENSKAYA [142], TEMAM [216]). Here we primarily focus on an approach that does not require smoothness of the evolutionary operator (as in BABIN/VISHIK [9], TEMAM [216]) and relies on the idea introduced by Ladyzhenskaya's theorem (see, e.g., LADYZHENSKAYA [142]) on the finite dimensionality of invariant sets. The development of the Ladyzhenskaya approach is based on some types of quasi-stability estimates (see CHUESHOV/LASIECKA [51, 56, 58] for a primary idea and also PRAŽÁK [187] for a similar method based on a squeezing property). This approach also covers the method suggested in MALLET-PARET [159] and MAÑÉ [161], which

© Springer International Publishing Switzerland 2015

I. Chueshov, *Dynamics of Quasi-Stable Dissipative Systems*, Universitext,
DOI 10.1007/978-3-319-22903-4_3

requires differentiability of the corresponding evolution operator. However, we wish to point out that the estimates of the dimension obtained in the framework of quasi-stability usually tend to be conservative. In the case of smooth dynamics the method based on control of contractions of finite-dimensional volumes leads to much sharper bounds for the attractor dimension. This volume contraction method was developed and applied by many authors (see, e.g., the discussion in the monographs BABIN/VISHIK [9], BOICHENKO/LEONOV/REITMANN [15], CHEPYZHOV/VISHIK [31], TEMAM [216]). Below, for the sake of self-containment, we will discuss this method following the presentation given in CHEPYZHOV/VISHIK [31] and TEMAM [216].

3.1.1 Fractal and Hausdorff dimensions

Fractal and Hausdorff dimensions are the most commonly used measures in the theory of infinite-dimensional dynamical systems. They can be defined as follows (see, e.g., FALCONER [95]).

Definition 3.1.1 (Fractal and Hausdorff dimensions). Let M be a compact set in a metric space X.

- The *fractal (box-counting) dimension* $\dim_f M$ of M is defined by

$$\dim_f M = \limsup_{\varepsilon \to 0} \frac{\ln n(M,\varepsilon)}{\ln(1/\varepsilon)} \ ,$$

  where $n(M,\varepsilon)$ is the minimal number of closed balls of radius ε which cover the set M.
- For positive d we define the (ball-based) *d-dimensional Hausdorff measure* by the formula

$$\mu(M,d) = \sup_{\varepsilon > 0} \mu(M,d,\varepsilon),$$

  where the value $+\infty$ is allowed, and

$$\mu(M,d,\varepsilon) = \inf \left\{ \sum_j (r_j)^d \ : \ M \subset \bigcup_i B(x_j, r_j), \ r_j \leq \varepsilon \right\}.$$

Here $B(x_j, r_j)$ is the ball in X with center x_j and radius r_j. The corresponding covering can be countable. One can show (check this!) that (i) if $\mu(M,d_*) < \infty$ for some $d_* > 0$, then $\mu(M,d) = 0$ for all $d > d_*$, and (ii) if $\mu(M,d_*) > 0$ for $d_* > 0$, then $\mu(M,d) = +\infty$ for all $d < d_*$. Thus, there exist $d_H \geq 0$ such that

$$\mu(M,d) = \begin{cases} +\infty, & \text{if } d < d_H; \\ 0, & \text{if } d > d_H. \end{cases}$$

The *Hausdorff dimension* $\dim_H M$ of M is defined as the value d_H separating two regions of values for μ, e.g., we can take $\dim_H M = \inf\{d : \mu(M,d) = 0\}$. ∎

Example 3.1.2. Let M be an interval of length l. It is clear that

$$\frac{l}{2\varepsilon} - 1 \le n(M,\varepsilon) \le \frac{l}{2\varepsilon} + 1.$$

Therefore,

$$\ln\frac{1}{\varepsilon} + \ln\frac{l - 2\varepsilon}{2} \le \ln n(M,\varepsilon) \le \ln\frac{1}{\varepsilon} + \frac{l + 2\varepsilon}{2}.$$

This implies that the fractal dimension $\dim_f M$ coincides with the value of the standard geometric dimension. ∎

The same effect demonstrates the following statement.

Proposition 3.1.3. *Let M be a bounded set with nonempty interior in $\mathbb{R}^d$. Then $\dim_f M = \dim_H M = d$.*

Proof. We consider the case of the fractal dimension only (for the Hausdorff case, see FALCONER [95], for instance). Since both dimensions are monotone with respect to set inclusion, it is sufficient to prove the result for a ball in $\mathbb{R}^d$. For this we use the following lemma on coverings.

Lemma 3.1.4. *Let $\mathbb{R}^d$ be equipped with Euclidean norm $|\cdot|$ and*

$$B_R = \{x \in \mathbb{R}^d : |x - y_*| \le R\}$$

be a ball in $\mathbb{R}^d$ with radius R. Then for any $\varepsilon > 0$ there exists a finite set $\{x_k : k = 1,\ldots n_\varepsilon\} \subset B_R$ such that

$$B_R \subset \bigcup_{k=1}^{n_\varepsilon}\{x \in \mathbb{R}^d : |x - x_k| \le \varepsilon\} \text{ and } n_\varepsilon \le \left(1 + \frac{2R}{\varepsilon}\right)^d.$$

Moreover, the maximal number of points $\{x_k\}$ inside the ball B_R possessing the property $|x_j - x_i| > \varepsilon$ for any $i \ne j$ admits the same bound $(1 + 2R/\varepsilon)^d$.

Proof. Because B_R is compact in $\mathbb{R}^d$, there exists a set $\{x_k : k = 1,\ldots,n_\varepsilon\} \subset B_R$ such that (i) for any $y \in B_R$ we can find x_i such that $|y - x_i| \le \varepsilon$, and (ii) $|x_j - x_i| > \varepsilon$ for any $i \ne j$. Thus, we need only prove the estimate for n_ε. Consider the balls

$$B_k = \{x \in \mathbb{R}^d : |x - x_k| < \varepsilon/2\}, \quad k = 1,\ldots,n_\varepsilon.$$

These balls possess the properties $B_k \cap B_j = \emptyset$ for $k \neq j,\, k, j = 1, \ldots n_\varepsilon$, and

$$B_k \subset \tilde{B} \equiv \left\{ x \in \mathbb{R}^d \ : \ |x - y_*| \leq R + \frac{\varepsilon}{2} \right\}, \quad k = 1, \ldots, n_\varepsilon.$$

Hence $n_\varepsilon \cdot \mathrm{Vol}\,(B_1) = \sum_{k=1}^{n_\varepsilon} \mathrm{Vol}(B_k) \leq \mathrm{Vol}(\tilde{B})$. This implies the estimate for n_ε.   □

Using this lemma one can see that

$$\left(\frac{R}{\varepsilon} \right)^d \leq n(B_R, \varepsilon) \leq \left(1 + \frac{2R}{\varepsilon} \right)^d.$$

This implies that $\dim_f B_R = d$ for every $R > 0$. Since we have $B_r \subset M \subset B_R$ for some $0 < r < R < \infty$, by monotonicity we obtain the conclusion of Proposition 3.1.3.   □

Example 3.1.5. Let M be the Cantor set obtained from the interval $[0, 1]$ by the sequential removal of the central thirds. First we remove all the points between $1/3$ and $2/3$. Then we remove the central thirds $(1/9, 2/9)$ and $(7/9, 8/9)$ of the two remaining intervals $[0, 1/3]$ and $[2/3, 1]$. After that we do the same with the central parts of the four remaining intervals, and so on. If we continue this process to infinity, we obtain the Cantor set M. Let us calculate its fractal dimension. First of all, we note that

$$M = \bigcap_{k=0}^{\infty} J_k,$$

where

$$
\begin{aligned}
J_0 &= [0, 1], \\
J_1 &= [0, 1/3] \cup [2/3, 1], \\
J_2 &= [0, 1/9] \cup [2/9, 1/3] \cup [2/3, 7/9] \cup [8/9, 1], \quad \text{and so on.}
\end{aligned}
$$

Each set J_k can be considered as a union of 2^k intervals of length 3^{-k}. Since $M \subset J_k$ for each k and the boundary points of J_k lie in M, the minimal number of intervals of length 3^{-k} covering the set M equals to 2^k. Therefore, one can show (see Exercise 3.1.11 below) that

$$\dim_f M = \lim_{k \to \infty} \frac{\ln 2^k}{\ln(2 \cdot 3^k)} = \frac{\ln 2}{\ln 3}.$$

Thus, the fractal dimension of the Cantor set is not an integer (if a set possesses this property, it is called a *fractal set*). One can also show that the Hausdorff dimension of this Cantor set has the same value; see, e.g., FALCONER [95].   ■

Some additional properties of the dimension are collected in the following exercises.

Exercise 3.1.6. Let M be a compact set in a complete metric space.

(A) Prove that $\dim_H M \leq \dim_f M$; i.e., the Hausdorff dimension does not exceed the fractal one.

(B) Verify that $\dim_f (M_1 \cup M_2) = \max\{\dim_f M_1, \dim_f M_2\}$ for the fractal dimension and

$$\dim_H \left(\cup_{j=1}^{\infty} M_j \right) = \max_j \dim_H M_j \quad \text{(Hausdorff dimension)}. \tag{3.1.1}$$

In particular, every *countable* set has Hausdorff dimension zero.

(C) Assume that $M_1 \times M_2$ is a direct product of two sets. Then for the fractal[1] dimensions we have

$$\dim_f (M_1 \times M_2) \leq \dim_f M_1 + \dim_f M_2. \tag{3.1.2}$$

(D) Let G be a Lipschitz mapping of one metric space into another. Then

$$\dim_f G(M) \leq \dim_f M.$$

Moreover, if G is Hölder, i.e., $\text{dist}(G(x), G(y)) \leq L[\text{dist}(x, y)]^{\alpha}$ for some $\alpha \leq 1$, then $\dim_f G(M) \leq \alpha^{-1} \dim_f M$. Check whether the same properties are valid for the Hausdorff dimension.

∎

Remark 3.1.7. In the case of the Hausdorff dimension an inequality like (3.1.2) is not true in general. We refer to Example 7.8 in FALCONER [95, p. 97] which shows that there exist sets E and F on $\mathbb{R}$ such that $\dim_H E = \dim_H F = 0$ and $\dim_H E \times F = 1$. In the case of the Hausdorff dimension we can only prove that

$$\dim_H E + \dim_H F \leq \dim_H (E \times F) \leq \dim_H E + \dim_f F$$

for any couple of sets E and F (see FALCONER [95, p. 94] for the proof for sets in $\mathbb{R}^d$).

∎

Exercise 3.1.8. Make the following calculations on the real line.

(A) Show that the fractal dimension coincides with the Hausdorff one in Example 3.1.2.

(B) Let $M = \{1/n\}_{n=1}^{\infty} \subset \mathbb{R}$. Show that $\dim_f M = 1/2$. Hint: $n < n(M, \varepsilon) < n + 1 + (2\varepsilon)^{-1}(n+1)^{-1}$ provided $[(n+1)(n+2)]^{-1} \leq 2\varepsilon < [n(n+1)]^{-1}$.

(C) Let $M = \{1/\ln n\}_{n=2}^{\infty} \subset \mathbb{R}$. Prove that $\dim_f M = 1$.

∎

---

[1] See Remark 3.1.7 for the case of the Hausdorff dimension.

The following facts can be found in ROBINSON [195] and BOICHENKO/LEONOV/ REITMANN [15].

Exercise 3.1.9. Let $\{e_n\}$ be an orthonormal basis in a Hilbert space X.

(A) Consider the set

$$M = \{0\} \cup \left\{ \frac{1}{\ln n} e_n : n = 1, 2, \ldots \right\}$$

Show that $\dim_f M = \infty$. What can we say about the Hausdorff dimension of this set? Hint: See BOICHENKO/LEONOV/REITMANN [15, p. 199] or ROBINSON [195, p. 352].

(B) Let

$$M_\alpha = \{0\} \cup \{n^{-\alpha} e_n : n = 1, 2, \ldots\} \quad \text{with } \alpha > 0. \tag{3.1.3}$$

Prove that $\dim_f M = \alpha^{-1}$. Hint: See the idea presented in ROBINSON [195, p. 329].

(C) For $s \geq 0$ we consider the Hilbert space X_s defined by the relation

$$X_s = \left\{ u = \sum_{k=1}^{\infty} c_k e_k : \|u\|_s^2 \equiv \sum_{k=1}^{\infty} k^{2s} |c_k|^2 < \infty \right\}.$$

Let M_α be given by (3.1.3). Show that (i) M_α is compact in X_s if and only if $s < \alpha$, and (ii) $\dim_f^{H_s} M_\alpha = [\alpha - s]^{-1}$ for all $0 \leq s < \alpha$, where $\dim_f^{H_s} M$ denotes the fractal dimension of a set M in the space H_s. Hint: The set M_α can be written in the form $M_\alpha = \{0\} \cup \{n^{-\alpha+s} e_n^s : n = 1, 2, \ldots\}$, where $\{e_n^s \equiv n^{-s} e_n\}$ is an orthonormal basis in H_s.

∎

Exercise 3.1.10. Show that a set M and its closure have the same fractal dimension. This is not true for the Hausdorff dimension of M. Hint: Take $M = \mathbb{Q} \cap [0, 1]$ (all rational numbers in $[0, 1]$) and show that

$$\dim_H\{\mathbb{Q} \cap [0, 1]\} = 0 \quad \text{and} \quad \dim_H\{[0, 1]\} = 1.$$

For the fractal dimension we have $\dim_f\{\mathbb{Q} \cap [0, 1]\} = \dim_f\{[0, 1]\} = 1$.   ∎

The following facts are useful in the dimension calculations.

Exercise 3.1.11. Let $N(M, \varepsilon)$ be the minimal number of closed sets of diameter 2ε that cover a compact set M. Prove the following statements.

(A) The fractal dimension $\dim_f M$ can be written in the form

$$\dim_f M = \limsup_{\varepsilon \to 0} \frac{\ln N(M, \varepsilon)}{\ln(1/\varepsilon)}. \tag{3.1.4}$$

(B) The dimension $\dim_f M$ can also be represented by the formula

$$\dim_f M = \limsup_{n \to 0} \frac{\ln N(M, \varepsilon_n)}{\ln(1/\varepsilon_n)} \tag{3.1.5}$$

for every monotone sequence $\varepsilon_n \to +0$ such that $\varepsilon_{n+1}/\varepsilon_n \geq \alpha > 0$ for some $\alpha > 0$.

∎

Remark 3.1.12. The facts presented in the exercises above show that the fractal dimension dominates the Hausdorff one. In contrast with the fractal dimension, the Hausdorff dimension is countably additive (in the sense of (3.1.1)). The examples in Exercises 3.1.8(B,C), 3.1.9(A), and 3.1.10 show that these dimensions do not coincide. The example in Exercise 3.1.9(A) even shows that the same set can have zero Hausdorff dimension and infinite fractal dimension. Moreover, as we can see from Exercise 3.1.9(C), the value of the dimension may depend on the topology chosen. We also refer to the paper SHUBOV [211], which provides an example of a set with finite fractal dimension in one space and infinite fractal dimension in another (smaller) space.

Below we mainly deal with the fractal dimension of attractors for the following reasons: (i) the fractal dimension is more convenient in calculations, and (ii) it estimates the Hausdorff dimension from above. We note that the importance of the notion of finite fractal dimension is also illustrated by the following property (see FOIAS/OLSON [102] or HUNT/KALOSHIN [125]): if M is a compact set in a Hilbert space X such that $\dim_f M < n/2$ for some $n \in \mathbb{N}$, then M can be placed in the graph of a Hölder continuous mapping which maps a compact subset of $\mathbb{R}^n$ onto M. We refer to FALCONER [95] for details and for other properties of Hausdorff and fractal (box-counting) dimension. We also mention the monograph ROBINSON [196], which discusses various aspects of dimension theory with applications to attractors of infinite-dimensional systems. ∎

We conclude this section with an assertion which shows that under some conditions even an uncountable union of finite-dimensional sets may have a finite fractal dimension (we use this fact in our constructions of fractal exponential attractors).

Proposition 3.1.13. *Let M be a compact set in a complete metric space X and $V(t)$ be a family of continuous mappings from M into X, $t \in [a, b]$. We assume that there exist $K, L > 0$ and $0 < \gamma \leq 1$ such that*

$$\mathrm{dist}_X(V(t)x, V(t)y) \leq L\,[\mathrm{dist}_X(x, y)]^\gamma, \quad x, y \in M, \ t \in [a, b], \qquad '$$

and

$$\mathrm{dist}_X(V(t_1)x, V(t_2)x) \leq K\,|t_1 - t_2|^\gamma, \quad x \in M, \ t_1, t_2 \in [a, b].$$

Then the set $M_V(a,b) = \bigcup_{t\in[a,b]} V(t)M$ has a finite fractal dimension in X,

$$\dim_f^X M_V(a,b) \leq \frac{1}{\gamma}\left[1 + \dim_f^X M\right].$$

Proof. Let $\{F_j\}$ be a minimal covering of M by its closed subsets with diameters less than 2ϵ. Then the family $\{V(t)F_j\}$ is a covering of $V(t)M$ with diameters less than $2^\gamma L\epsilon^\gamma$. Consider the sets

$$G_{kj} = V(t_k)F_j \quad \text{with } t_k = a + k\epsilon, \quad k = 0, 1, \ldots, n_\epsilon \leq \frac{b-a}{\epsilon}.$$

For every $y \in M_V(a,b)$ we can find k and j such that $y = V(t)x$ for some $t \in [t_k, t_{k+1}]$ and $x \in F_j$. In this case,

$$\text{dist}_X(y, G_{kj}) \leq \text{dist}_X(V(t)x, V(t_k)x) \leq K\,|t - t_k|^\gamma \leq K\,\epsilon^\gamma.$$

Therefore, the sets

$$\overline{\mathscr{O}}_{K\epsilon^\gamma}(G_{kj}) = \big\{w \in X \,:\, \text{dist}_X(w, G_{kj}) \leq K\,\epsilon^\gamma\big\}$$

give a covering for $M_V(a,b)$ with

$$\text{diam}\big(\overline{\mathscr{O}}_{K\epsilon^\gamma}(G_{kj})\big) \leq 2K\,\epsilon^\gamma + 2^\gamma L\,\epsilon^\gamma = (2K + 2^\gamma L)\,\epsilon^\gamma.$$

Thus,

$$N(M_V(a,b), (2K + 2^\gamma L)\,\epsilon^\gamma) \leq \frac{b-a+1}{\epsilon} N(M, \epsilon).$$

This implies the conclusion.   □

3.1.2   Criteria for finite dimension of invariant sets: Lipschitz case

In further considerations the following criterion (see CHUESHOV/LASIECKA [56]) turns out to be very useful (see also CHUESHOV/LASIECKA [48, 51] for related results). Its main advantage is that we do not involve any smoothness properties for evolutions except the Lipschitz continuity. The basic idea behind this criterion is some kind of splitting of evolution into stable and compact parts which refers to the difference of two trajectories. The compact part is described by means of compact seminorms. We recall (see, e.g., YOSIDA [229, Chapter 1]) that a real-valued function $n(x)$ defined on a linear space X is called a seminorm on X if

$$n(x + y) \leq n(x) + n(y) \quad \text{and} \quad n(\lambda x) = |\lambda|n(x) \quad \text{for all } x, y \in X, \ \lambda \in \mathbb{R}.$$

The difference between norms and seminorms is that a seminorm can be degenerate, i.e., $n(x) = 0$ does not imply $x = 0$. For instance, $n(x_1, x_2) = |x_1| + |x_2|$ is a norm on $\mathbb{R}^2$, and $n(x_1, x_2) = |x_1|$ is a seminorm.

Definition 3.1.14 (Compact seminorm). A seminorm $n(x)$ on a Banach space X is said to be compact if any bounded sequence $\{x_m\} \subset X$ contains a subsequence $\{x_{m_k}\}$ which is Cauchy with respect to n, i.e., $n(x_{m_k} - x_{m_l}) \to 0$ as $k, l \to \infty$. ∎

Our basic result in this section is the following theorem.

Theorem 3.1.15. *Let X be a Banach space and M be a bounded closed set in X. Assume that there exists a mapping $V : M \mapsto X$ such that*

(i) $M \subseteq VM$.
(ii) V is Lipschitz on M; that is, there exists $L > 0$ such that

$$\|Vv_1 - Vv_2\| \le L\|v_1 - v_2\|, \quad v_1, v_2 \in M. \tag{3.1.6}$$

(iii) *There exist compact seminorms $n_1(x)$ and $n_2(x)$ on X such that*

$$\|Vv_1 - Vv_2\| \le \eta\|v_1 - v_2\| + c_0 \cdot [n_1(v_1 - v_2) + n_2(Vv_1 - Vv_2)] \tag{3.1.7}$$

for any $v_1, v_2 \in M$, where $0 < \eta < 1$ and $c_0 > 0$ are constants.

Then M is a compact set in X of a finite fractal dimension. Moreover,

$$\dim_f M \le \left[\ln \frac{2}{1 + \eta}\right]^{-1} \cdot \ln m_0 \left(\frac{4c_0(1 + L^2)^{1/2}}{1 - \eta}\right), \tag{3.1.8}$$

where $m_0(R)$ is the maximal number of pairs (x_i, y_i) in $X \times X$ possessing the properties

$$\|x_i\|^2 + \|y_i\|^2 \le R^2, \quad n_1(x_i - x_j) + n_2(y_i - y_j) > 1, \quad i \ne j. \tag{3.1.9}$$

Exercise 3.1.16. Show that under the compactness hypothesis concerning the seminorms n_1 and n_2, the characteristic $m_0(R)$ defined in Theorem 3.1.15 is finite for every fixed $R > 0$. Hint: Apply the contradiction argument and use the compactness of the seminorms. ∎

Remark 3.1.17. We also note that if X is a separable Hilbert space and the seminorms n_1 and n_2 have the form $n_i(v) = \|P_i v\|$, $i = 1, 2$, where P_1 and P_2 are finite-dimensional orthoprojectors, then

$$\dim_f M \le (\dim P_1 + \dim P_2) \cdot \ln\left(1 + \frac{8(1 + L^2)^{1/2}\sqrt{2c_0}}{1 - \eta}\right) \cdot \left[\ln \frac{2}{1 + \eta}\right]^{-1}. \tag{3.1.10}$$

Indeed, in this case the set $\{(x_i, y_i)\}_{i=1}^{m_0(R)}$ satisfying (3.1.9) possesses the properties

$$\|P_1 x_i\|^2 + \|P_2 y_i\|^2 \le R^2, \quad \|P_1(x_i - x_j)\|^2 + \|P_2(y_i - y_j)\|^2 > 1/2, \quad i \ne j.$$

This means that the points $z_i = (P_1 x_i; P_2 y_i)$ belong to the ball of radius R in the space $Z = P_1 X \times P_2 X$ and possess the property $\|z_i - z_j\|_Z^2 1/\sqrt{2}$, $i \ne j$. We have that $\dim Z = \dim P_1 + \dim P_2 < \infty$. Thus, we can apply Lemma 3.1.4 to conclude that $m_0(R) \le (1 + 2\sqrt{2}R)^{\dim Z}$ in this case. This implies (3.1.10). ∎

Theorem 3.1.15 was derived in CHUESHOV/LASIECKA [56] as a consequence of some general dimension-type results established in the case of metric spaces. For Hilbert spaces the proof can be found in CHUESHOV/LASIECKA [51]; see also CHUESHOV/LASIECKA [58]. A similar approach based on the quasi-stability inequality (3.1.7) with compact norms instead of seminorms n_1 and n_2 was also discussed recently in FEIREISL/PRAŽÁK [100, Proposition 2.6]. Below we prove a more general version of Theorem 3.1.15 (see Theorem 3.1.21) which allows us to include several new models in the list of applications.

As a simple consequence of Theorem 3.1.15 we can obtain the following well-known assertion (see, e.g., LADYZHENSKAYA [141]).

Corollary 3.1.18 (Ladyzhenskaya's theorem). *Let M be a compact set in a Hilbert space H. Assume that V is a continuous mapping in H such that $V(M) \supseteq M$ and there exists a finite-dimensional projector P in H such that*

$$\|P(Vv_1 - Vv_2)\| \le l\|v_1 - v_2\|, \quad v_1, v_2 \in M, \tag{3.1.11}$$

and

$$\|(I - P)(Vv_1 - Vv_2)\| \le \delta\|v_1 - v_2\|, \quad v_1, v_2 \in M, \tag{3.1.12}$$

where $\delta < 1$. Then the fractal dimension $\dim_f M$ is finite and there exists a constant $c = c(\delta, l) > 0$ such that

$$\dim_f M \le c \cdot \dim P. \tag{3.1.13}$$

Proof. It follows from (3.1.11) and (3.1.12) that

$$\|Vv_1 - Vv_2\| \le \delta\|v_1 - v_2\| + \|P(Vv_1 - Vv_2)\|, \quad v_1, v_2 \in M.$$

Thus, we can apply Theorem 3.1.15 and also Remark 3.1.17 with $n_1 \equiv 0$ and $n_2(v) = \|Pv\|$. □

Exercise 3.1.19. Prove that M is a single point set when $l < 1 - \delta$ in (3.1.11) and (3.1.12). ∎

Roughly speaking, the assumptions (3.1.11) and (3.1.12) mean that the mapping V squeezes the set M along the space $(I-P)H$, although it does not stretch M too much along PH. The negative invariance of M gives us that $M \subseteq V^k M$ for all $k \in \mathbb{N}$. Thus, the set M must be initially *squeezed*. This property is expressed by the assertion on finite dimensionality of M. In a similar way we can interpret relation (3.1.7): V is a contraction up to modulo compact seminorm.

The following assertion demonstrates the role of another version of squeezing (which we call the Foias-Temam squeezing property; see the discussion in CON-STANTIN/FOIAS/TEMAM [79] and TEMAM [216]). We also refer to Chapter 4, where this property is discussed for several parabolic models.

Corollary 3.1.20 (Foias-Temam squeezing). *Let M be a compact set in a Banach space X and V be a Lipschitz continuous mapping in X such that $V(M) \supseteq M$. Assume that V satisfies a squeezing property on M in the following form: for some $\eta < 1$ and $\gamma > 0$ there exists a finite-dimensional projector P on X such that for every $v_1, v_2 \in M$ we have either*

$$\|(I - P)(Vv_1 - Vv_2)\| \le \gamma \|P(v_1 - v_2)\|,$$

or

$$\|Vv_1 - Vv_2\| \le \eta \|v_1 - v_2\|.$$

Then the fractal dimension $\dim_f M$ is finite.

Proof. Obviously under the conditions above,

$$\|Vv_1 - Vv_2\| \le \eta \|v_1 - v_2\| + (1 + \gamma)\|P(v_1 - v_2)\|, \quad v_1, v_2 \in M. \qquad (3.1.14)$$

Therefore, we can apply Theorem 3.1.15.   □

We also refer to CHUESHOV/LASIECKA [53] for an analysis of the dimension problem in the case of nonlinear relations of the type (3.1.7) and to CHUESHOV/LASIECKA [56] for statements in metric spaces and some other corollaries of Theorem 3.1.15.

We derive Theorem 3.1.15 from the following more general result which is central to this section.

Theorem 3.1.21. *Let X be a Banach space and M be a bounded closed set in X. Assume that there exists a mapping $V : M \mapsto X$ such that*

(i) $M \subseteq VM$;
(ii) There exist a Lipschitz mapping K from M into some Banach space Z and a compact seminorm $n_Z(x)$ on Z such that

$$\|Vv_1 - Vv_2\| \le \eta \|v_1 - v_2\| + n_Z(Kv_1 - Kv_2) \qquad (3.1.15)$$

for any $v_1, v_2 \in M$, where $0 < \eta < 1$ is a constant.

Then M is a compact set in X of a finite fractal dimension and

$$\dim_f M \le \left[\ln \frac{2}{1 + \eta}\right]^{-1} \cdot \ln m_Z \left(\frac{4L_K}{1 - \eta}\right), \tag{3.1.16}$$

where $L_K > 0$ is the Lipschitz constant for K:

$$\|Kv_1 - Kv_2\|_Z \le L_K \|v_1 - v_2\|, \quad v_1, v_2 \in M, \tag{3.1.17}$$

and $m_Z(R)$ is the maximal number of elements z_i in the ball $\{z \in Z : \|z_i\|_Z \le R\}$ possessing the property $n_Z(z_i - z_j) > 1$ when $i \ne j$.

In the proof of Theorem 3.1.21 we follow the line of the argument given in CHUESHOV/LASIECKA [51] and rely on the following lemma.

Lemma 3.1.22. *Assume that $V : M \mapsto X$ is a mapping such that (3.1.15) with some $\eta > 0$ holds. Then*

$$\alpha(VB) \le \eta \cdot \alpha(B) \text{ for any } B \subseteq M, \tag{3.1.18}$$

where $\alpha(B)$ is Kuratowski's α-measure of noncompactness of the set B (for the definition see Section 2.2.2). Thus, V is an α-contraction on M in the case when $\eta < 1$ (see Definition 2.2.15).

Proof. By the definition of $\alpha(B)$, for any $\varepsilon > 0$ there exist sets $F_1, \ldots, F_n$ such that

$$B = F_1 \cup \ldots \cup F_n, \quad \text{diam } F_i < \alpha(B) + \varepsilon.$$

Let $\mathcal{N} = \{x_i : i = 1, 2 \ldots m\} \subset B$ be a finite set such that for every $y \in B$ there is $i \in \{1, 2, \ldots, m\}$ with the property $n_Z(Ky - Kx_i) \le \varepsilon$. If there is no such set for some $\varepsilon > 0$, then there exists a sequence $\{z_n\} \subset B$ such that

$$n_Z(Kz_n - Kz_m) \ge \varepsilon \text{ for all } n \ne m. \tag{3.1.19}$$

The sequence $\{Kz_n\}$ contains a subsequence $\{Kz_{n_l}\}$ which is Cauchy with respect to n_Z, i.e., $n_Z(Kz_{n_l} - Kz_{n_m}) \to 0$ when $n = l, m \to \infty$. This is impossible due to (3.1.19). Thus, such a finite set $\mathcal{N}$ exists, and

$$B = \cup_{i=1}^m C_i, \quad C_i = \{y \in B : n_Z(Ky - Kx_i) \le \varepsilon\}, \quad x_i \in \mathcal{N}.$$

By exploiting the representations $B = \cup_{i,j}(C_i \cap F_j)$ and $VB = \cup_{i,j}(V(C_i \cap F_j))$, one can see from (3.1.15) that $\text{diam}(V(C_j \cap F_i)) \le \eta \cdot \alpha(B) + \varepsilon \cdot [2 + \eta]$. This implies (3.1.18). $\qquad\square$

Remark 3.1.23. We note that the α-contraction property of V given by the previous lemma is not sufficient for finite dimensionality of the set M. This can be shown by means of an example. Indeed, following CHUESHOV/LASIECKA [60] we suppose that

$$X = l_2 = \left\{ x = (x_1; x_2; \ldots) : \sum_{i=1}^{\infty} x_i^2 < \infty \right\}$$

and

$$M = \left\{ x = (x_1; x_2; \ldots) \in l_2 : |x_i| \leq i^{-2}, \ i = 1, 2, \ldots \right\}$$

We define a mapping V in X by the formula

$$[Vx]_i = f_i(x_i), \quad i = 1, 2, \ldots$$

where $f_i(s) = s$ for $|s| \leq i^{-2}$, $f_i(s) = i^{-2}$ for $s \geq i^{-2}$, and $f_i(s) = -i^{-2}$ for $s \leq -i^{-2}$. One can see that V is globally Lipschitz on X and $VX = M = VM$. Since M is a compact set, the mapping V is an α-contraction (with $\eta = 0$). On the other hand, it is clear that $\dim_f M = \infty$.

We also note that this example means that the statement of Theorem 2.8.1 in HALE [116] is not true without additional hypotheses concerning the mapping. ∎

Proof of Theorem 3.1.21. Lemma 3.1.22 implies that $\alpha(M) \leq \eta \cdot \alpha(M)$. Since $0 < \eta < 1$, this is possible only if $\alpha(M) = 0$. Thus, M is compact.

Assume that $\{F_i : i = 1, \ldots, N(M, \varepsilon)\}$ is the minimal covering of M by its closed subsets with a diameter equal to or less than 2ε with $0 < \varepsilon < 1$. Let $0 < \delta < 1$ and

$$\varrho(x, y) = n_Z(Kx - Ky), \quad x, y \in M. \tag{3.1.20}$$

Let $\{x_j^i : j = 1, \ldots, n_i\} \subset F_i$ be a maximal subset of F_i such that

$$\varrho(x_j^i, x_k^i) > \delta\varepsilon, \quad x_j^i, x_k^i \in F_i, \ j, k = 1, \ldots, n_i, \ j \neq k.$$

Obviously,

$$n_i \equiv m_\varrho(F_i, \delta\varepsilon) \leq \exp\left\{ \sigma_\varrho(M, \delta) \right\}, \tag{3.1.21}$$

where $m_\varrho(B, \varepsilon)$ is the maximal cardinality of a subset z_k in B such that $\varrho(z_k, z_l) > \varepsilon$ and

$$\sigma_\varrho(M, \delta) = \sup_{0 < \varepsilon < 1} \sup \left\{ \ln m_\varrho(F, \delta\varepsilon) : F \subset M, \ \mathrm{diam}\, F \leq 2\varepsilon \right\}. \tag{3.1.22}$$

The value $\sigma_\varrho(M, \delta)$ is finite and admits the estimate

$$\sigma_\varrho(M, \delta) \leq \ln m_Z \left(\frac{2L_K}{\delta} \right). \tag{3.1.23}$$

Indeed, let $B \subset M$. In the space Z we consider the set $\mathscr{B} = \{z = Kx : x \in B\}$. It is clear that

$$m_\varrho(B, \varepsilon) = \aleph \{z_i \in \mathscr{B} : n_Z(z_i - z_j) > \varepsilon, \ i \neq j\},$$

where $\aleph\{\ldots\}$ denotes the maximal number of elements with the given properties. Since by (3.1.17)

$$\mathrm{diam}\mathscr{B} = \sup_{x,y \in B} \|Kx - Ky\| \leq R \equiv L_K \mathrm{diam} B,$$

there exists $y_0 \in \mathscr{B}$ such that

$$\mathscr{B} \subset B_R(y_0) \equiv \{z \in Z : \|z - y_0\|_Z \leq R\}$$

Therefore, using the property $n_Z(\lambda z) = \lambda n_Z(z)$ for any $\lambda > 0$, we obtain that

$$m_\varrho(B, \varepsilon) \leq \aleph \{z_i \in B_R(y_0) : n_Z(z_i - z_j) > \varepsilon, \ i \neq j\}$$
$$= \aleph \{z_i \in B_R(0) : n_Z(z_i - z_j) > \varepsilon, \ i \neq j\}$$
$$= \aleph \{z_i \in B_{R/\varepsilon}(0) : n_Z(z_i - z_j) > 1, \ i \neq j\} = m_Z(R/\varepsilon).$$

This implies (3.1.23).

To continue with the proof, we note that

$$F_i \subset \bigcup_{j=1}^{n_i} B_j^i, \ B_j^i \equiv \{v \in F_i : \rho(v, x_j^i) \leq \delta \cdot \varepsilon\}.$$

Therefore,

$$VM \subset \bigcup_{i=1}^{N(M,\varepsilon)} \bigcup_{j=1}^{n_i} VB_j^i.$$

If $y_1, y_2 \in B_j^i$, then from (3.1.15) we have

$$\|Vy_1 - Vy_2\| \leq \eta \|y_1 - y_2\| + \rho(y_1, x_j^i) + \rho(y_2, x_j^i) \leq 2(\eta + \delta)\varepsilon.$$

Thus, $\mathrm{diam}\{VB_j^i\} \leq 2(\eta + \delta)\varepsilon$ for any $\varepsilon > 0$ and $0 < \delta < 1$. Therefore,

$$N(VM, (\eta + \delta)\varepsilon) \leq \exp\{\sigma_\varrho(M, \delta)\} \cdot N(M, \varepsilon). \tag{3.1.24}$$

For further use we emphasize that relation (3.1.24) remains true *without* the hypothesis $M \subseteq VM$.

If we choose $0 < \delta < 1 - \eta$, then from (3.1.24) under the assumption $M \subseteq VM$ we obtain that

$$\ln N \, (M, q\varepsilon) \leq \sigma_\varrho \, (M, \delta) + \ln N(M, \varepsilon)$$

for any $\varepsilon > 0$, where $q = \eta + \delta < 1$. Let $\varepsilon_n = q^n \varepsilon_0$ for some $\varepsilon_0 > 0$. It is clear that

$$\ln N \, (M, \varepsilon_n) \leq n\sigma_\varrho(M, \delta) + \ln N(M, \varepsilon_0), \quad n = 1, 2, \ldots. \tag{3.1.25}$$

Now for any $\varepsilon < \varepsilon_0$ we can find $n = n_\varepsilon$ and $\tilde{\varepsilon} \in [\varepsilon_1, \varepsilon_0)$ such that

$$\varepsilon_{n+1} \leq \varepsilon < \varepsilon_n, \quad \varepsilon = q^n \tilde{\varepsilon}. \tag{3.1.26}$$

Hence,

$$\ln N \, (M, \varepsilon) \leq n_\varepsilon \cdot \sigma_\varrho(M, \delta) + \ln N(M, \tilde{\varepsilon}) \leq n_\varepsilon \cdot \sigma_\varrho(M, \delta) + \ln N(M, \varepsilon_1).$$

Thus, by (3.1.4) we obtain that

$$\dim_f M \leq \sigma_\varrho(M, \delta) \cdot \limsup_{\varepsilon \to 0} \frac{n_\varepsilon}{\ln(1/\varepsilon)}.$$

It follows from (3.1.26) that

$$n_\varepsilon = \frac{\ln(\tilde{\varepsilon}/\varepsilon)}{\ln(1/q)} \leq \frac{\ln(\varepsilon_0/\varepsilon)}{\ln(1/q)}.$$

Therefore,

$$\dim_f M \leq \sigma_\varrho(M, \delta) \cdot \frac{1}{\ln(1/q)} \leq \ln m_Z \left(\frac{2L_K}{\delta} \right) \cdot \frac{1}{\ln(1/q)}.$$

If we take $\delta = (1 - \eta)/2$ and $q = (1 + \eta)/2$, we obtain (3.1.16). The proof of Theorem 3.1.21 is complete. $\qquad \square$

Remark 3.1.24. The analysis of the argument given in the proof of Theorem 3.1.21 shows that the statement of this theorem remains true if we assume that M is a compact set but the relation in (3.1.15) holds in a weaker (local) form:

(ii*)   There exist (a) a Lipschitz mapping K from M into some Banach space Z, (b) a compact seminorm $n_Z(x)$ on Z, and (c) $\varepsilon_0 > 0$ such that

$$\|Vv_1 - Vv_2\| \leq \eta\|v_1 - v_2\| + n_Z(Kv_1 - Kv_2) \tag{3.1.27}$$

for any $v_1, v_2 \in M$ possessing the property $\|v_1 - v_2\| \leq \varepsilon_0$, where $0 < \eta < 1$ is a constant.

We use this observation in Section 3.1.3 in order to derive from Theorem 3.1.21 the results on dimension for C^1 mappings given in MALLET-PARET [159] and MAÑÉ [161]. ∎

Proof of Theorem 3.1.15. We take the space $Z = X \times X$ endowed with the norm

$$\|z\|_Z = \left(\|x\|^2 + \|y\|^2\right)^{1/2}, \quad z = (x; y) \in Z,$$

and define the operator $K : M \mapsto Z$ by the formula $Kx = c_0(x; Vx)$. We also take $n_Z(z) = n_1(x) + n_2(y)$ for $z = (x; y)$. In this case the Lipschitz constant for K is $L_K = c_0(1 + L^2)^{1/2}$. We also have $m_z(R) = m_0(R)$. Thus, we can apply Theorem 3.1.21 to conclude the proof. □

Another consequence of Theorem 3.1.21 is the following assertion, which was proved in CHUESHOV/LASIECKA [56] by another method.

Corollary 3.1.25. *Let X be a Banach space and M be a bounded closed set in X. Assume that there exists a mapping $V : M \mapsto X$ such that (i) $M \subseteq VM$, and (ii) the mapping V admits the splitting*

$$V = S + K, \tag{3.1.28}$$

where S is Lipschitz and stable on M, i.e., there exists $0 < \eta < 1$ such that

$$\|Sv_1 - Sv_2\| \leq \eta \|v_1 - v_2\|, \quad v_1, v_2 \in M, \tag{3.1.29}$$

and K is a Lipschitz mapping from M into some Banach space $Y \subset X$, i.e.,

$$\|Kv_1 - Kv_2\|_Y \leq L_K \|v_1 - v_2\|, \quad v_1, v_2 \in M. \tag{3.1.30}$$

We assume that Y is compactly embedded in X.
   Then M is a compact set in X of a finite fractal dimension and

$$\dim_f M \leq \left[\ln \frac{2}{1 + \eta}\right]^{-1} \cdot \ln m_{Y,X}\left(\frac{4L_K}{1 - \eta}\right), \tag{3.1.31}$$

where $m_{Y,X}(R)$ is the maximal number of points x_i in the ball of radius R in Y possessing the properties $\|x_i - x_j\| > 1$, $i \neq j$.

Proof. It follows from (3.1.28) and (3.1.29) that

$$\|Vv_1 - Vv_2\| \leq \eta \|v_1 - v_2\| + \|J(Kv_1 - Kv_2)\|, \quad v_1, v_2 \in M,$$

where we have denoted by J the embedding operator Y into X. Thus, we can apply Theorem 3.1.21 with $Z = Y$ and $n_Z(z) = \|Jz\|$. Since J is compact, $n_Z(z)$ is a compact seminorm. □

If in the splitting (3.1.28) we have $S \equiv 0$, then we easily arrive at the following assertion which was proved in ZELIK [232, Theorem 4.1] (see also MÁLEK/PRAŽÁK [158, Lemma 1.3] and the recent monograph FEIREISL/PRAŽÁK [100, Theorem 2.4]). In these sources it was applied for some types of parabolic problems.

Corollary 3.1.26. *Let X and Y be Banach spaces such that Y is compactly embedded in X. Let M be a bounded closed set in X. Assume that $V : M \mapsto Y$ is a Lipschitz mapping from M into Y, i.e.,*

$$\|Vv_1 - Vv_2\|_Y \leq L\|v_1 - v_2\|_X, \quad v_1, v_2 \in M.$$

If $M \subset VM$, then M is a compact set in X and its fractal dimension (in X) admits the estimate

$$\dim_f M \leq \frac{\ln m_{Y,X}(4L)}{\ln 2},$$

where $m_{Y,X}(R)$ is the same as in Corollary 3.1.25.

Proof. We apply Corollary 3.1.25 with $S \equiv 0$ and $\eta = 0$. $\qquad\qquad\square$

3.1.3 Criteria for finite dimension of invariant sets: C^1 case

In this section we consider the case of smooth mappings. Our primary goal is to present the main idea of the volume contraction method (see CONSTANTIN/FOIAS [77], CONSTANTIN/FOIAS/TEMAM [79] and also the monographs BABIN/VISHIK [9], BOICHENKO/LEONOV/REITMANN [15], CHEPYZHOV/VISHIK [31], TEMAM [216]).

We start with several important statements which show how the results of the previous section can be applied in the smooth case. Namely, using Theorem 3.1.21 and also the observation made in Remark 3.1.24, we can give an alternative proof of some results established in MALLET-PARET [159] and MAÑÉ [161]. For this we first recall the following definition (see Section A.5 in the Appendix for more details concerning calculus in infinite-dimensional spaces).

Definition 3.1.27 (Fréchet derivative). Let $\mathscr{O}$ be an open set in a Banach space X. A mapping $V : \mathscr{O} \mapsto X$ is said to be *Fréchet differentiable* on $\mathscr{O}$ if for any $u \in \mathscr{O}$ there exists a bounded linear operator $V'(u)$ such that

$$\frac{\|V(v) - V(u) - V'(u)(v - u)\|}{\|v - u\|} \to 0 \quad \text{as} \quad \|v - u\| \to 0. \tag{3.1.32}$$

The operator $V'(u)$ is called the *(Fréchet) derivative* of V at the point $u \in \mathcal{O}$. The relation in (3.1.32) means that for every $u \in \mathcal{O}$ there exist $\delta > 0$ and a scalar function $\gamma(s)$ on $[0, \delta]$ such that $\gamma(s) \to 0$ as $s \to 0$ and

$$\|V(v) - V(u) - V'(u)(v - u)\| \leq \gamma(\|v - u\|)\|v - u\|.$$

If the number δ and the function $\gamma(s)$ do not depend on u, then the mapping V is said to be *uniformly (Fréchet) differentiable* on $\mathcal{O}$.   ∎

The following assertion generalizes Theorem 2.1 of MALLET-PARET [159] and provides a version of the result presented in MAÑÉ [161].

Theorem 3.1.28. *Let M be a compact set in a Banach space X. Assume that there exists an open set $\mathcal{O}$ such that $M \subset \mathcal{O} \subset X$. Suppose that $V : \mathcal{O} \mapsto X$ is Fréchet differentiable on $\mathcal{O}$ and $M \subseteq VM$. If the derivative $V'(u)$ is continuous with respect to u in the operator topology and there exists a finite-dimensional projector[2] P such that*

$$\|V'(u)(I - P)\| < 1 \quad \text{for every } u \in M, \tag{3.1.33}$$

then the set M has a finite fractal dimension.

Proof. Let $u \in M$. Since V is continuously differentiable on $\mathcal{O}$, we have that

$$V(v) - V(u) = \int_0^1 V'(\lambda v + (1 - \lambda)u)(v - u)d\lambda \tag{3.1.34}$$

for every $v \in X$ such that $\|u - v\| \leq \varepsilon < \text{dist}(M, X \setminus \mathcal{O})$. Using the continuity of V', the compactness of M, and the relation in (3.1.33), we can choose $\varepsilon > 0$ such that

$$\|V'(u + w)(I - P)\| \leq q < 1 \quad \text{and} \quad \|V'(u + w)\| \leq K, \quad \forall\, u \in M,\ \|w\| \leq \varepsilon.$$

This implies the property (ii*) in Remark 3.1.24 with $\eta = q$, $Z = X$, $n_Z(y) = K\|Py\|$. Thus, applying the observation made in Remark 3.1.24, we conclude the proof.   □

Instead of the existence of the continuous derivative with property (3.1.33) in Theorem 3.1.28, we can assume that V is uniformly quasi-differentiable[3] on M.

Definition 3.1.29 (Uniform quasi-differentiability). Let M be a set in a Banach space X and $V : M \mapsto X$ be a continuous mapping. This mapping is called *uniformly quasi-differentiable* in X on the set M if for any $u \in M$ there exists a bounded linear operator $L(u)$ on X such that

$$\|V(v) - V(u) - L(u)(v - u)\| \leq \gamma(\|v - u\|)\|v - u\| \tag{3.1.35}$$

---

[2]This means that P is a bounded operator on X such that $P^2 = P$ and $\dim P \equiv \dim PX < \infty$.

[3]For dimensionality considerations this notion was used by many authors. See, e.g., BABIN/VISHIK [9], CHEPYZHOV/VISHIK [31], TEMAM [216] and the references therein.

for all $u, v \in M$, where the scalar function $\gamma(s)$ does not depend on $u, v \in M$ and $\gamma(s) \to 0$ as $s \to 0$. The operator $L(u)$ is called the *quasi-derivative* of V at the point $u \in M$. ∎

We have the following version of Theorem 3.1.28.

Theorem 3.1.30. *Let M be a compact set in a Banach space X and* $V : M \mapsto X$ *be uniformly quasi-differentiable on M. Suppose that* $M \subseteq VM$ *and there exists a finite-dimensional projector P such that*

$$\|L(u)P\| < C \quad and \quad \|L(u)(I - P)\| \le q < 1 \quad for \ every \ u \in M. \tag{3.1.36}$$

Then the set M has a finite fractal dimension.

Proof. Let $u, v \in M$. It follows from (3.1.35) that

$$\|V(v) - V(u)\| \le \gamma(\|v - u\|)\|v - u\| + \|L(u)(I - P)(u - v)\| + \|L(u)P(u - v)\|$$

$$\le [q + \gamma(\|v - u\|)]\|v - u\| + C\|P(u - v)\|.$$

This implies the property (ii*) in Remark 3.1.24 for ε_0 chosen such that $\gamma(s) \le (1 - q)/2$ when $0 \le s \le \varepsilon_0$. In this case $\eta = (1 + q)/2$, $Z = X$, $n_Z(y) = C\|Py\|$. Thus, as in the previous case, we can conclude the proof. □

Obviously in both Theorems 3.1.28 and 3.1.30 we can take an arbitrary linear *compact* operator K instead of the projector P. Moreover, this observation can be improved. Namely, we can obtain the following result which basically was proved in MAÑÉ [161].

Theorem 3.1.31. *Let M be a compact set in a Banach space X and* $V : M \mapsto X$ *be uniformly quasi-differentiable on M. Assume that the quasi-derivative L(u) can be split into two parts*

$$L(u) = L^1(u) + L^2(u), \quad u \in M,$$

where

$$\sup_{u \in M} \|L^1(u)\| \equiv q < 1$$

and $L^2(u)$ *is a compact operator on X for each* $u \in M$. *We also assume that the function* $u \mapsto L^2(u)$ *is continuous in the operator norm. If* $M \subseteq VM$, *then* $\dim_f M$ *is finite.*

Proof. Let $u, v \in M$. As in the proof of Theorem 3.1.30, using (3.1.35) and also the splitting of $L(u)$, we obtain that

$$\|V(v) - V(u)\| \le [q + \gamma(\|v - u\|)]\|v - u\| + \|L^2(u)(u - v)\|.$$

We know that the family of operators $L^2(u)$ is continuous in the operator norm. Thus, the function $u \mapsto L^2(u)$ is uniformly continuous, and hence for every $\eta > 0$ there exists a finite δ-net $\{w_k\}_{k=1}^{N(\eta)}$ in M such that

$$\forall u \in M \; \exists k : \; \|L^2(u) - L^2(w_k)\| \leq \eta.$$

In this case,

$$\|L^2(u)(u-v)\| \leq \|[L^2(u) - L^2(w_k)](v-u)\| + \|L_2(w_k)(v-u)\|$$

$$\leq \eta \|v-u\| + \sum_{k=1}^{N(\eta)} \|L^2(w_k)(v-u)\|.$$

Therefore,

$$\|V(u) - V(u)\| \leq [q + \eta + \gamma(\|v-u\|)]\|v-u\| + \sum_{k=1}^{N(\eta)} \|L^2(w_k)(v-u)\|.$$

This allows us to apply Remark 3.1.24 and conclude the proof.   □

In all Theorems 3.1.28, 3.1.30, and 3.1.31, bounds for the dimension can be derived from relation (3.1.16) in the statement of Theorem 3.1.21. However, as was already mentioned, the bounds which follow from the results of Section 3.1.2 are rather conservative, and the volume contraction method makes it possible to improve these bounds significantly. This method requires both differentiability of evolution mapping V and the Hilbert structure of the phase space.

To introduce finite-dimensional volumes[4] and describe their properties we need the following definitions.

Let L be a linear bounded operator on a Hilbert space X. Following CONSTANTIN/FOIAS/TEMAM [79] and TEMAM [216] we introduce the numbers

$$\alpha_m(L) = \sup_{\substack{F \subset X \\ \dim F = m}} \inf_{\substack{u \in F \\ \|u\|=1}} \|Lu\|, \tag{3.1.37}$$

where F is a subspace in X, and we suppose that

$$\omega_m(L) = \alpha_1(L) \cdot \ldots \cdot \alpha_m(L). \tag{3.1.38}$$

The following assertion can be found in TEMAM [216, Chapter V].

---

[4]For the reader's convenience we mention that the linear algebra required is presented particularly clearly in CARVALHO/LANGA/ROBINSON [26].

Proposition 3.1.32. *Let L be a linear bounded operator on a Hilbert space X. Then the following minimax properties hold:*

$$\alpha_m(L) \equiv \sup_{\substack{F \subset X \\ \dim F = m}} \inf_{\substack{u \in F \\ \|u\| = 1}} \|Lu\| = \inf_{\substack{F \subset X \\ \dim F = m-1}} \sup_{\substack{u \perp F \\ \|u\| = 1}} \|Lu\|. \tag{3.1.39}$$

Moreover, $\{\alpha_m(L)\}$ *is a non-increasing sequence and*

$$\omega_m(L) = \sup_{\substack{\phi_1, \dots, \phi_m \in X \\ \|\phi_i\| = 1}} |L\phi_1 \wedge L\phi_2 \wedge \dots \wedge L\phi_m|, \tag{3.1.40}$$

where $|\phi_2 \wedge \dots \wedge \phi_m| \equiv \left[\det (\phi_i, \phi_j)_{i,j=1}^m\right]^{1/2}$ *is the m-dimensional volume of the parallelepiped spanned by* $\phi_1, \dots, \phi_m$.

Thus, the value $\omega_m(L)$ characterizes the behavior of m-dimensional volumes under the action of L. It is also clear that $\omega_1(L) = \alpha_1(L) = \|L\|_{X \mapsto X}$. Below we say that L *contracts* m-dimensional volumes if $\omega_m(L) < 1$. There is a simple characterization of volume contractive linear operators.

Proposition 3.1.33. *A linear bounded operator L contracts m-dimensional volumes for some m if and only if* $L = C + K$, *where C is a contraction, i.e.,* $\|C\|_{X \mapsto X} < 1$, *and K is a compact operator. Moreover, if* $\omega_m(L) < 1$, *then in the representation* $L = C + K$ *we can choose C to be a contraction and K a finite-dimensional operator such that* $\dim KX \leq m$.

Proof. Let $L = C + K$ with a contraction C and a compact operator K. Since $\{\alpha_n(L)\}$ is a non-increasing sequence, it is sufficient to show that $\alpha_m(L) < 1$ for some m. It follows from (3.1.39) that

$$\alpha_m(L) = \inf_{\substack{F \subset X \\ \dim F = m-1}} \sup_{\substack{u \perp F \\ \|u\| = 1}} \|Cu + Ku\| \leq \|C\|_{X \mapsto X} + \alpha_m(K).$$

Since K is a compact operator, we have that $\alpha_m(K) \to 0$ as $m \to \infty$ (see, e.g., TEMAM [216, Chapter V]). Thus, for some m we have that

$$\|C\|_{X \mapsto X} + \alpha_m(K) < 1.$$

Assume now that $\omega_m(L) < 1$ for some m. Since $\{\alpha_n(L)\}$ is a non-increasing sequence, we have that $\alpha_m(L) < 1$. It follows from results in TEMAM [216, Chapter V, Section 1.3] that the space X can be split into the direct sum

$$X = X_0 + X_0^\perp,$$

where X_0 is spanned by an orthogonal family $\{e_j\}_{j=1}^{m-1}$ of eigenvectors of the operator $(L^*L)^{1/2}$. Moreover, we have that

$$\|L\phi\|_X \le \alpha_m(L)\|\phi\|_X, \quad \phi \in X_0^\perp.$$

This means that the operator $R = (L^*L)^{1/2}$ is a sum of a contraction and a finite-dimensional projector. Using the polar representation of a bounded operator (see, e.g., DUNFORD/SCHWARTZ [89, Chapter 12, Section 7]), we can conclude that the operator L has the same structure.   □

To state the main theorem of the volume contraction method, we need to extend the characteristic $\omega_m(L)$ defined for integers m on all non-negative reals by the following interpolation formulas:

$$\omega_0(L) = 1, \quad \omega_d(L) = \omega_m(L)^{1-s}\omega_{m+1}(L)^s = \omega_m(L)\alpha_{m+1}(L)^s,$$

for all $d = m + s$, $m \in \mathbb{Z}_+$, $0 \le s < 1$. One can show that the function $d \mapsto \omega_d(L)$ is non-increasing (see, e.g., TEMAM [216, Chapter 5]).

The proof of the following result can be found in TEMAM [216]; see also the references therein.

Theorem 3.1.34 (Basic Hausdorff dimension bound). *Let M be a compact set in a Hilbert space X and $V : M \mapsto X$ be uniformly quasi-differentiable on M. Assume that $VM = M$ and the quasi-derivative $L(u)$ possesses the properties*

$$\sup_{u \in M} \|L(u)\|_{X \mapsto X} < +\infty$$

and

$$\omega_d \equiv \sup_{u \in M} \omega_d(L(u)) < 1 \quad \text{for some } d \in \mathbb{R}_+. \tag{3.1.41}$$

Then the Hausdorff dimension $\dim_H M$ of M is finite and $\dim_H M \le d$.

In contrast with the argument given in Theorem 3.1.21, which is based on covering results for finite-dimensional balls (see Lemma 3.1.4), the proof of Theorem 3.1.34 involves more refined covering lemmas concerning ellipsoids. This is the main reason why it is possible to obtain the relation (3.1.41) for the dimension, leading to substantial improvement of the dimension bounds.

By Proposition 3.1.33 the requirement in (3.1.41) implies that the quasi-derivative $L(u)$ for each $u \in M$ can be split as

$$L(u) = C(u) + K(u), \tag{3.1.42}$$

where $C(u)$ is a contraction and K is a finite-dimensional operator such that

$$\sup_{u \in M} \|C(u)\|_{X \mapsto X} \le [\omega_d]^{1/d} < 1, \quad \dim KX \le d + 1.$$

However, in order to apply the quasi-stability method via Theorem 3.1.31 here, we need to assume that in the splitting (3.1.42) the operators are continuous with respect to u in the operator norm. In this sense the fact on the finite dimension cannot be derived from Theorem 3.1.21.

A result similar to Theorem 3.1.34 is also valid for the fractal dimension. Namely, we have the following assertion.

Theorem 3.1.35 (Basic fractal dimension bound). *Let V be uniformly quasi-differentiable on a compact set M in a Hilbert space X and $M \subseteq VM$. Assume that*

$$\omega_j = \sup_{u \in M} \omega_j(L(u)) < k_j < \infty, \quad j = 1, 2, \dots, n,$$

and for $d = n + s$ with $0 \le s < 1$ we have

$$\omega_d \equiv \sup_{u \in M} \omega_d(L(u)) \le k_d < 1.$$

Then the dimension $\dim_f M$ of M is finite and admits the estimate

$$\dim_f M \le d \max_{0 \le j \le n} \left(\frac{\log k_j}{\log(1/k_d)} + \frac{j}{d} \right).$$

For the proof we refer to CONSTANTIN/FOIAS/TEMAM [79]; see also CHEP-YZHOV/VISHIK [31]. Another version of the corresponding statement concerning fractal dimension can be found in TEMAM [216]. It is also possible to show (see CHEPYZHOV/ILYIN [29]) that $\dim_f M$ has the same bound as $\dim_H M$ in Theorem 3.1.34 under the additional hypotheses that $VM = M$ and the quasi-differential $L(u)$ is continuous with respect to u in the operator norm.

To control volume contractions and optimize bounds for dimension it is convenient to use the uniform Lyapunov numbers introduced in CONSTANTIN/FOIAS [77] and CONSTANTIN/FOIAS/TEMAM [79].

We first note that $V^p = V \circ \dots \circ V$ is also uniformly quasi-differentiable with the quasi-derivative

$$L_p(u) = L(V^{p-1}u)L(V^{p-2}u)\dots L(Vu)L(u).$$

Thus, we can define the numbers

$$\bar{\omega}_j(p) = \sup_{u \in M} \omega_j(L^p(u)).$$

One can show (see TEMAM [216, Chapter V]) that $\bar{\omega}_j(p)$ is subexponential with respect to p, i.e.,

$$\bar{\omega}_j(p + q) \le \bar{\omega}_j(p)\bar{\omega}_j(q).$$

This implies that the values

$$\Pi_j \equiv \lim_{p \to \infty} \left[\bar{\omega}_j(p)\right]^{1/p} = \inf_{p \in \mathbb{N}} \left[\bar{\omega}_j(p)\right]^{1/p} \tag{3.1.43}$$

exist. With this notation Theorems 3.1.34 and 3.1.35 lead to the following assertion.

Corollary 3.1.36. *Let $V : M \mapsto X$ be uniformly quasi-differentiable on a compact set M in a Hilbert space X. Assume that $VM = M$ and the quasi-derivative $L(u)$ possesses the property*

$$\sup_{u \in M} \|L(u)\|_{X \mapsto X} < +\infty.$$

If $\Pi_d < 1$ for some $d > 0$, then

$$\dim_H M \le d \quad and \quad \dim_f M \le d \max_{0 \le j \le n} \left(\frac{\log \Pi_j}{\log(1/\Pi_d)} + \frac{j}{d} \right),$$

where n is the integer part of d.

The statement of Corollary 3.1.36 can be rewritten in another form. To do this, following CONSTANTIN/FOIAS [77] and CONSTANTIN/FOIAS/TEMAM [79] we introduce the notation

$$\Lambda_1 = \Pi_1, \quad \Lambda_m = \Pi_m/\Pi_{m-1}, \quad m \ge 2.$$

We obviously have that

$$\Lambda_m \equiv \lim_{p \to \infty} \left(\frac{\bar{\omega}_m(p)}{\bar{\omega}_{m-1}(p)} \right)^{1/p}.$$

The numbers Λ_m are called *uniform Lyapunov numbers* on M for the mapping V, and

$$\mu_m = \log \Lambda_m \text{ are } \textit{global Lyapunov exponents}, \quad m \ge 1.$$

The following result can be found in TEMAM [216].

Theorem 3.1.37. *Let $V : M \mapsto X$ be uniformly quasi-differentiable on a compact set M in a Hilbert space X. Assume that $VM = M$ and the quasi-derivative $L(u)$ possesses the property*

$$\sup_{u \in M} \|L(u)\|_{X \mapsto X} < +\infty.$$

If $\mu_1 + \ldots + \mu_{n+1} < 0$ *for some* $n > 0$, *then*

$$\mu_{n+1} < 0, \quad \frac{\mu_1 + \ldots + \mu_n}{|\mu_{n+1}|} < 1,$$

and

$$\dim_H M \le \dim_{\text{Lyap}} M \equiv n_0 + \frac{\mu_1 + \ldots + \mu_{n_0}}{|\mu_{n_0+1}|}, \tag{3.1.44}$$

where n_0 *is the minimal integer such that*

$$\mu_1 + \ldots + \mu_{n_0} \ge 0 \quad \text{and} \quad \mu_1 + \ldots + \mu_{n_0+1} < 0.$$

The right-hand side $\dim_{\text{Lyap}} M$ in (3.1.44) is called the *Lyapunov dimension* of the set M, and relation (3.1.44), which means that the Lyapunov dimension dominates the Hausdorff dimension, is known as the *Kaplan-Yorke formula*. See the references in CONSTANTIN/FOIAS [77] and CONSTANTIN/FOIAS/TEMAM [79] or in CHEPYZHOV/VISHIK [31] and TEMAM [216]. We also mention that for the fractal dimension the following bound:

$$\dim_f M \le (n_0 + 1) \max_{1 \le j \le n_0} \left[1 + \frac{\mu_1 + \ldots + \mu_j}{|\mu_1 + \ldots + \mu_{n_0+1}|} \right] \tag{3.1.45}$$

is valid (see, e.g., TEMAM [216] and CHEPYZHOV/VISHIK [31]). Under some conditions of a different nature concerning the mapping V and its (quasi-) derivative, one can show that the fractal dimension admits the same bound as the Hausdorff one in the Kaplan-Yorke formula (3.1.44). See the discussions in CHEPYZHOV/VISHIK [31] and also in CHEPYZHOV/ILYIN [29].

3.2 Exponential attractors for discrete systems

The dimension theorems discussed in the previous section pertain to negatively or strictly invariant sets M ($M \subseteq V(M)$). As for positively invariant sets, the finite dimensionality is not guaranteed. However, one can show that the latter sets are attracted by finite-dimensional compacts at an exponential rate. For instance, the method presented in the proof of Theorem 3.1.21 allows us to obtain the following assertion, which is a version of the result proved in CHUESHOV/LASIECKA [56] for metric spaces.

Theorem 3.2.1. *Let* $V : M \mapsto M$ *be a mapping defined on a closed bounded set M of a Banach space X. Assume that there exist a Lipschitz mapping K from M into some Banach space Z and a compact seminorm $n_Z(x)$ on Z such that*

$$\|Vv_1 - Vv_2\| \le \eta\|v_1 - v_2\| + n_Z(Kv_1 - Kv_2) \tag{3.2.1}$$

for any $v_1, v_2 \in M$, where $0 < \eta < 1$ is a constant. Then for any $\theta \in (\eta, 1)$ there exists a positively invariant compact set $A_\theta \subset M$ of finite fractal dimension satisfying

$$\sup\left\{\mathrm{dist}(V^k u, A_\theta) : u \in M\right\} \le r\theta^k, \quad k = 1, 2, \ldots, \tag{3.2.2}$$

for some constant $r > 0$. Moreover,

$$\dim_f A_\theta \le \ln m_Z\left(\frac{2L_K}{\theta - \eta}\right) \cdot \left[\ln \frac{1}{\theta}\right]^{-1}, \tag{3.2.3}$$

where, as in Theorem 3.1.21, L_K is the Lipschitz constant for K (see (3.1.17)) and $m_Z(R)$ is the maximal number of elements z_i in the ball $\{z \in Z : \|z_i\|_Z \le R\}$ possessing the property $n_Z(z_i - z_j) > 1$ when $i \ne j$.

Note that the condition (3.2.2) means an exponential rate of attraction.

Proof. It follows from Lemma 3.1.22 that V is an α-contraction. Therefore, by Exercise 2.3.8 the set $M_0 = \cap_{n \ge 1} V^n M$ is a compact global attractor for the discrete dynamical system (M, V^k). By Theorem 3.1.21, $\dim_f M_0 < \infty$. We construct a set A_θ as an extension of M_0.

Since V is an α-contraction on M, due to invariance we can assume that $\alpha(M) \le 2$ and thus $N(M, 1) < \infty$. Here and below $N(B, \varepsilon)$ denotes the cardinality of the minimal covering of B by its closed subsets of diameter equal to or less than 2ε.

It follows from (3.1.24) that

$$N(VM, q\varepsilon) \le \exp\left\{\sigma_\varrho(M; \delta)\right\} \cdot N(M, \varepsilon)$$

for any $\varepsilon > 0$, where $q = \eta + \delta$ and $\sigma_\varrho(M, \delta)$ is given by (3.1.22) and admits the estimate (3.1.23). Taking $V^{n-1}M$ instead of M in the previous formula, we obtain

$$N(V^n M, q\varepsilon) \le \exp\left\{\sigma_\varrho(V^{n-1}M, \delta)\right\} \cdot N(V^{n-1}M, \varepsilon), \quad n = 1, 2, \ldots$$

Since $V^{n-1}M \subset M$, we have that

$$\sigma_\varrho(V^{n-1}M, \delta) \le \sigma_\varrho(M, \delta), \quad n = 1, 2, \ldots$$

Thus, we have

$$N(V^n M, q\varepsilon) \le \exp\left\{\sigma_\varrho(M, \delta)\right\} \cdot N(V^{n-1}M, \varepsilon), \quad n = 1, 2, \ldots,$$

and hence

$$N\left(V^{n}M, \varepsilon_{n}\right) \leq \exp\left\{n\sigma_{\varrho}(M, \delta)\right\} \cdot N(M, \varepsilon_{0}), \quad n = 1, 2, \ldots, \tag{3.2.4}$$

where we choose $\varepsilon_{n} = q^{n}\varepsilon_{0}$ with some $1/2 \leq \varepsilon_{0} \leq 1$.

In the construction of inertial sets we rely on some ideas presented in EDEN ET AL. [92]. For this we need the following assertion.

Lemma 3.2.2. *Assume that* $\theta > \eta$. *Then there exists a collection of finite sets* $\{E_{m}\}_{m=0}^{\infty}$ *possessing the properties:*

(i) $E_{m} \subset V^{m}M$ *for every* $m = 0, 1, \ldots,$ *and*

$$V^{m}M \subset \bigcup_{v \in E_{m}} \left(V^{m}M \cap B_{2\theta^{m}}(v)\right), \quad m = 0, 1, \ldots, \tag{3.2.5}$$

where $B_{\rho}(v) = \{w \in X : \|w - v\| \leq \rho\}$ *is a ball with the center at* v.

(ii) *There exists a constant* $N_{0} > 0$ *such that for every* $m \geq 0$ *we have*

$$\operatorname{Card} E_{m} \leq N(V^{m}M, \theta^{m}) \leq N_{0} \exp\left(m \cdot \sigma_{\varrho}(M, \theta - \eta)\right), \tag{3.2.6}$$

where $\sigma_{\varrho}(M, \delta)$ *is given by* (3.1.22) *and admits estimate* (3.1.23).

Proof. Let $E_{m} = \{a_{i}^{m} : i = 1, \ldots, N_{m}\}$ be a maximal set in $V^{m}M$ possessing the property $\|a_{i}^{m} - a_{j}^{m}\| > 2\theta^{m}$, $i \neq j$. Then it is clear that (3.2.5) holds. To establish relation (3.2.6) we note that the inequality

$$N_{m} \equiv \operatorname{Card} E_{m} \leq N(V^{m}M, \theta^{m}), \quad m = 0, 1, \ldots, \tag{3.2.7}$$

follows from the fact that two different elements from E_{m} cannot belong to the same set of diameter $2\theta^{m}$. By (3.2.4) this implies (3.2.6). $\qquad\square$

Completion of the proof of Theorem 3.2.1. We prove that the set

$$A_{\theta} = M_{0} \cup \{V^{k}E_{m} : k, m = 0, 1, 2, \ldots\}$$

satisfies the conclusion of the theorem.

It is easy to see that A_{θ} is a compact, positively invariant set. By (3.2.5),

$$\operatorname{dist}(V^{m}y, A_{\theta}) \leq \operatorname{dist}(V^{m}y, E_{m}) \leq 2\theta^{m}, \quad m = 0, 1, 2, \ldots,$$

for every $y \in M$. This implies (3.2.2).

To estimate the fractal dimension of A_{θ}, we use the idea presented in the monograph CHUESHOV/LASIECKA [56]. We first note that

$$A_{\theta} \subset V^{n}M \cup \{V^{k}E_{m} : k + m \leq n - 1, k, m \geq 0\}$$

for every $n \geq 1$. Therefore,

$$N(A_\theta, \varepsilon) \leq N(V^n M, \varepsilon) + \sum_{m=0}^{n-1} (n - m) \operatorname{Card} E_m$$

for every $n \geq 1$ and $\varepsilon > 0$. Consequently, choosing $\varepsilon = \theta^n$ from Lemma 3.2.2, we obtain

$$N(A_\theta, \theta^n) \leq N_0 \exp\left(n \cdot \sigma_\varrho(M, \theta - \eta)\right) \left\{1 + \sum_{j=1}^{\infty} j \exp\left(-j \cdot \sigma_\varrho(M, \theta - \eta)\right)\right\}$$

$$= N_0 \exp\left(n \cdot \sigma_\varrho(M, \theta - \eta)\right) \left\{1 + \frac{\exp\left(\sigma_\varrho(M, \theta - \eta)\right)}{\left(\exp\left(\sigma_\varrho(M, \theta - \eta)\right) - 1\right)^2}\right\}$$

for every $n \geq 1$. As in the proof of Theorem 3.1.21, we take $0 < \varepsilon < 1$ and choose $n = n_\varepsilon$ such that $\theta^n \leq \varepsilon < \theta^{n-1}$. Thus,

$$\ln N(A_\theta, \varepsilon) \leq \ln N(A_\theta, \theta^{n_\varepsilon}) \leq n_\varepsilon \sigma_\varrho(M, \theta - \eta) + C(K, L, \theta, \eta).$$

Because $n_\varepsilon \leq 1 + \ln(1/\varepsilon) \left[\ln(1/\theta)\right]^{-1}$, this implies (3.2.3). $\square$

As an application of Theorem 3.2.1 we obtain the following assertion (which was also established in CHUEHSOV/LASIECKA [56] and FEIREISL/PRAŽÁK [100] by other methods).

Theorem 3.2.3. *Let $V : M \mapsto M$ be a mapping defined on a closed bounded set M of a Banach space X. Assume that the Lipschitz condition for V in (3.1.6) holds, and that there exist compact seminorms n_1 and n_2 on H such that*

$$\|V v_1 - V v_2\| \leq \eta \|v_1 - v_2\| + c_0 \cdot [n_1(v_1 - v_2) + n_2(V v_1 - V v_2)] \qquad (3.2.8)$$

holds for any $v_1, v_2 \in M$, where $0 < \eta < 1$ and $c_0 > 0$ are constants. Then for any $\theta \in (\eta, 1)$ there exists a positively invariant compact set $A_\theta \subset M$ of finite fractal dimension satisfying (3.2.2). Moreover,

$$\dim_f A_\theta \leq \ln m_0 \left(\frac{2 c_0 (1 + L^2)^{1/2}}{\theta - \eta}\right) \cdot \left[\ln \frac{1}{\theta}\right]^{-1}, \qquad (3.2.9)$$

where, as in Theorem 3.1.15, $m_0(R)$ is the maximal number of pairs (x_i, y_i) in $X \times X$ possessing the properties

$$\|x_i\|^2 + \|y_i\|^2 \leq R^2, \quad n_1(x_i - x_j) + n_2(y_i - y_j) > 1, \quad i \neq j.$$

If X is a Hilbert space and $n_1(x) = \|P_1 x\|$ and $n_2(x) = \|P_2 x\|$, where P_1 and P_2 are finite-dimensional projectors in X, then

$$\dim_f A_\theta \leq \ln\left(1 + \frac{4\sqrt{2}(1 + L^2)^{1/2} c_0}{\theta - \eta}\right)\left[\ln \frac{1}{\theta}\right]^{-1} (\dim P_1 + \dim P_2). \quad (3.2.10)$$

Proof. We have the same choice of Z, K and n_Z as in the proof of Theorem 3.1.15. Thus, in estimate (3.2.3) we need to set $L_K = c_0(1 + L^2)^{1/2}$. In the case when the seminorms n_1 and n_2 are generated by orthogonal projectors, we use the observation made in Remark 3.1.17. □

As a consequence of Theorem 3.2.3 we can easily derive the following assertion, which is compatible with the construction presented in EDEN ET AL. [92].

Corollary 3.2.4. *Let $V : M \mapsto M$ be a mapping defined on a closed bounded set M of a Banach space X. Assume that the Lipschitz condition for V in (3.1.6) holds and V possesses on M the Foias-Temam squeezing property: for some $\eta < 1$ and $\gamma > 0$ there exists a finite-dimensional projector P on X such that for every $v_1, v_2 \in M$ we have either*

$$\|(I - P)(Vv_1 - Vv_2)\| \leq \gamma \|P(v_1 - v_2)\|,$$

or

$$\|Vv_1 - Vv_2\| \leq \eta \|v_1 - v_2\|.$$

Then for any $\theta \in (\eta, 1)$ there exists a forward invariant compact finite-dimensional set A_θ satisfying (3.2.2).

Proof. As in Corollary 3.1.20 we have relation (3.1.14), which allows us to apply Theorem 3.2.3. □

Another consequence of Theorem 3.2.1 is the following assertion.

Theorem 3.2.5. *Let $V : M \mapsto M$ be a mapping defined on a closed bounded set M of a Banach space X. Assume that V admits the splitting $V = S + K$ such that relations (3.1.29) and (3.1.30) are in force. Then there exists a forward invariant compact finite-dimensional set A_θ such that (3.2.2) holds for some $0 < \theta < 1$ and $r > 0$.*

The proof of Theorem 3.2.5 follows from Theorem 3.2.1 and uses the same observation as in Corollary 3.1.25. We also note that in the case when (3.1.29) holds with $\eta < 1/2$ the statement of Theorem 3.2.5 is well known from the paper EFENDIEV/MIRANVILLE/ZELIK [91].

Under the hypotheses of Theorem 3.2.1 the discrete dynamical system (M, V^k) possesses a compact global attractor M_0. This attractor uniformly attracts all the trajectories of the system (M, V^k), and by Theorem 3.1.21 $\dim_f M_0 < \infty$.

Unfortunately, in general the rate of convergence to the attractor cannot be estimated. This rate may be very slow. However, Theorem 3.2.1 attests that the global attractor is contained in a finite-dimensional positively invariant set which attracts M uniformly and exponentially fast. Thus, the dynamics of the system becomes finite-dimensional exponentially fast independent of the speed of convergence to the global attractor. Moreover, the reduction principle (see Theorem 2.3.22) is applicable in this case. Thus, finite-dimensional, positively invariant, exponentially attracting sets can be useful to describe the qualitative behavior of infinite-dimensional systems. These sets are frequently called *inertial sets* or *fractal exponential attractors* (see EDEN ET AL. [92] and also Section 3.4.1 below). In some cases they turn out to be surfaces in the phase space. For details we refer to EDEN ET AL. [92] and to the references therein; see also the survey MIRANVILLE/ZELIK [166].

3.3 Determining functionals

In many applications it is important to search for minimal (or close to minimal) sets of natural parameters of the problem that uniquely determine the long-time behavior of the system. This question was first discussed in FOIAS/PRODI [105] and LADYZHENSKAYA [140] for the 2D Navier-Stokes equations. Later on, other equations and models were considered (see, e.g., CONSTANTIN/DOERING/TITI [76], FOIAS ET AL. [104], FOIAS/TEMAM [107], FOIAS/TITI [108], LADYZHEN-SKAYA [141], SERMANGE/TEMAM [205] and the references quoted therein). The concepts of determining nodes (FOIAS ET AL. [104], FOIAS/TEMAM [107], SER-MANGE/TEMAM [205]) and determining local volume averages (FOIAS/TITI [108], JONES/TITI [126, 127]) were also introduced. The general concept of determining functionals in framework interpolation theory was introduced as well (see COCK-BURN/JONES/TITI [73, 74]). These functionals can be interpreted as some kinds of measurements of the state of the system. For further details we refer to the survey CHUESHOV [38] and to the references quoted therein (see also CHUESHOV [39, Chapter 5]). Recently the theory of determining functionals was applied in the study of the (discrete) data assimilation problem, which originated from weather prediction (see, e.g., HAYDEN/OLSON/TITI [122] and also CHUESHOV [44]).

3.3.1 Main concepts

The following definition is based on the property established in FOIAS/PRODI [105] for the Fourier modes of solutions to the 2D Navier-Stokes system with periodic boundary conditions.

Definition 3.3.1. Let (X, S_t) be a dynamical system in some Banach space X. Assume that there exists a complete linear topological space V which is continuously embedded into X (the case $V = X$ is allowed) and for any $x \in X$ there exists a moment t_* such that $S_t x \in V$ for all $t \geq t_*$. Let $\mathscr{L} = \{l_j : j = 1, \ldots, N\}$ be a set of linear continuous functionals on V and let two semitrajectories $\{u(t) = S_t u : t > 0\}$ and $\{v(t) = S_t v : t > 0\}$ be given. Then $\mathscr{L}$ is said to be *a set of (asymptotically) determining functionals* on V for those trajectories of the system (X, S_t) if the condition

$$\lim_{t \to \infty} |l_j(u(t)) - l_j(v(t))| = 0 \quad \text{for } j = 1, \ldots, N \tag{3.3.1}$$

implies that

$$\lim_{t \to \infty} \| u(t) - v(t) \|_X = 0. \tag{3.3.2}$$

If the implication above is true for *any* two semitrajectories $\{u(t) = S_t u : t > 0\}$ and $\{v(t) = S_t v : t > 0\}$, we call $\mathscr{L}$ determining on V for the system (X, S_t). ∎

Remark 3.3.2. **1.** The property in (3.3.1) can be written as

$$\lim_{t \to +\infty} n_{\mathscr{L}}(u(t) - v(t)) = 0 \quad \text{with } n_{\mathscr{L}}(u) = \max_{j=1,\ldots,N} |l_j(u)|.$$

It is clear that $n_{\mathscr{L}}$ is a seminorm on $\mathscr{L}$. This observation allows us to introduce the notion of a *determining seminorm*: a continuous seminorm n on V is said to be (asymptotically) determining for (X, S_t) if the property $n(u(t) - v(t)) \to 0$ as $t \to \infty$ implies (3.3.2).

2. We note that sometimes it is convenient to use other types of convergence in (3.3.1) and (3.3.2). For instance, instead of (3.3.1), in CHUESHOV [38, Definition 1.1] (see also CHUESHOV [39, Chapter 5]) the following weaker property is assumed:

$$\lim_{t \to \infty} \int_t^{t+1} |l_j(u(\tau)) - l_j(v(\tau))|^2 d\tau = 0 \quad \text{for } j = 1, \ldots, N. \tag{3.3.3}$$

We can also consider the convergence in (3.3.1) along some sequence $\{t_n\}$ tending to infinity. This can be interpreted as a measurement of the state of the system made from time to time (for instance, every two hours in weather prediction).

3. Other approaches and definitions characterizing determining functionals are possible and have been used in the literature (see, e.g., LADYZHENSKAYA [140] and COCKBURN/JONES/TITI [73, 74] and the references therein). For instance, there is a definition (see CHUESHOV [38] or CHUESHOV [39, Chap. 5] and also Remarks 3.3.14 and 3.4.13(2) below) based on an extension to a general dynamical system of the property of finite-dimensional projections proved in LADYZHENSKAYA [140] for trajectories lying in the global attractor of the 2D Navier-Stokes equations.

∎

For characterization of a set of determining functionals it is convenient to use the following concept of completeness defect which has been suggested in CHUESHOV [37, 38] in the case of pairs of embedded Banach spaces. To consider a certain class of quasi-stable systems (see Section 3.4 below), we introduce this notion in the case of a single space equipped with an additional seminorm.

Definition 3.3.3. Let V be a Banach space and μ be a seminorm on V. The *completeness defect* of a set $\mathscr{L}$ of linear functionals on V with respect to μ is the value

$$\epsilon_{\mathscr{L}}(V, \mu) = \sup \{\mu(w) : w \in V, l(w) = 0, l \in \mathscr{L}, \|w\|_V \le 1\} . \tag{3.3.4}$$

In the case when V is continuously and densely embedded into another Banach space X (with the property $\| \cdot \|_X \le c\| \cdot \|_V$), the value

$$\epsilon_{\mathscr{L}}(V, X) \equiv \epsilon_{\mathscr{L}}(V, \| \cdot \|_X)$$

$$= \sup\{\|w\|_X : w \in V, l(w) = 0, l \in \mathscr{L}, \|w\|_V \le 1\} \tag{3.3.5}$$

is said to be the completeness defect of a set $\mathscr{L}$ of linear functionals on V with respect to X (see CHUESHOV [37, 38]).   ∎

We note that finite dimensionality of the Span$\mathscr{L}$ of the set $\mathscr{L}$ is not assumed at this point. It is also obvious that $\epsilon_{\mathscr{L}_1}(V, \mu) \ge \epsilon_{\mathscr{L}_2}(V, \mu)$ provided Span$\mathscr{L}_1 \subset$ Span$\mathscr{L}_2$. In addition, $\epsilon_{\mathscr{L}}(V, \mu) = 0$ if and only if $\mu(w) = 0$ for every element

$$w \in \mathscr{L}^{\perp} = \{w \in V : l(w) = 0, \forall l \in \mathscr{L}\}.$$

Thus, if μ is norm, then the relation $\epsilon_{\mathscr{L}}(V, \mu) = 0$ is equivalent to the statement that the class of functionals $\mathscr{L}$ is complete in V; that is, the following uniqueness condition holds: $l(w) = 0$ for all $l \in \mathscr{L}$ implies $w = 0$.

The basic properties of completeness defect which we use in the subsequent considerations are described in the following assertions (see CHUESHOV [38] and [39, Chapter 5] for the case of pairs of spaces).

Proposition 3.3.4. *Let* $\epsilon_{\mathscr{L}} = \epsilon_{\mathscr{L}}(V, \mu)$ *be the completeness defect of a finite set* $\mathscr{L} = \{l_j : j = 1, \ldots, N\}$ *of linear functionals on* V *with respect to some seminorm* μ. *Then there exists a positive constant* $C_{\mathscr{L}}$ *such that*

$$\mu(w) \le C_{\mathscr{L}} \cdot \max\{|l_j(w)| : j = 1, \ldots, N\} + \epsilon_{\mathscr{L}} \cdot \|w\|_V \quad \text{for any } w \in V. \tag{3.3.6}$$

Proof. Obviously we can assume that $\{l_j\}$ are linearly independent functionals. This allows us to construct a biorthogonal system $\{e_j : j = 1, \ldots, N\} \subset V$ for $\mathscr{L}$ (i.e., we have that $l_j(v_i) = 0$ if $j \ne i$ and $l_j(e_j) = 1$). In this case for any $w \in V$ the element $v = w - \sum_{i=1}^{N} l_i(w)e_i$ possesses the properties $l_j(v) = 0$ for $j = 1, \ldots, N$. By the definition in (3.3.4) we have that $\mu(v) \le \epsilon_{\mathscr{L}}\|v\|_V$. Therefore, from the representation for v we obtain (3.3.6).   □

The following assertion gives a condition under which the completeness defect is small.

Proposition 3.3.5. *Let V be a separable Hilbert space and μ be a weakly continuous seminorm on V, i.e., $\mu(x_n) \to 0$ when $x_n \to x$ weakly in V as $n \to \infty$. Then*

(1) *For any $\varepsilon > 0$ there exist a constant K_ε and finite-dimensional orthoprojectors P^ε in V such that*

$$\mu(v) \le \varepsilon \|v\| + K_\varepsilon \|P^\varepsilon v\|, \quad v \in V. \tag{3.3.7}$$

(2) *For any $\varepsilon > 0$ there exists a a finite family of functionals $\mathscr{L} = \{l_j : j = 1, \dots, N\}$ such that $\epsilon_{\mathscr{L}}(V, \mu) \le \varepsilon$.*

Proof. To establish the first statement we follow the line of argument given in CHUESHOV/LASIECKA [58, Chapter 7]. Assume that (3.3.7) is not true. Then there exist $\varepsilon_0 > 0$ and a sequence of orthoprojectors $\{P_m\}$ such that $P_m \to I$ strongly in V and

$$\mu(v_m) \ge \varepsilon_0 + c_m \|P_m v_m\|, \quad m = 1, 2, \dots, \tag{3.3.8}$$

for some sequence $\{v_m\} \subset V$ with the property $\|v_m\| = 1$, where $c_m \to \infty$ as $m \to \infty$. We can also assume that $v_m \to v$ weakly in V for some $v \in V$. It follows from (3.3.8) that $\|P_m v_m\| \to 0$ as $m \to \infty$.

Because

$$P_m v = P_m(v - v_m) + P_m v_m \to 0 \quad \text{weakly in } V,$$

we conclude that $v = 0$. Since μ is weakly continuous, this implies that $\mu(v_m) \to 0$ as $m \to \infty$, which contradicts (3.3.8). Thus (3.3.7) holds.

To prove the second part we take a basis $\{e_k\}$ in the finite-dimensional space $P^\varepsilon V$ and consider the functionals $L_j(v) = (v, e_j)$, $j = 1, \dots, N \equiv \dim P^\varepsilon$. The property $\epsilon_{\mathscr{L}}(V, \mu) \le \varepsilon$ follows from (3.3.7). □

Exercise 3.3.6. Show that any weakly continuous seminorm on a reflexive Banach space is compact in the sense of Definition 3.1.14. We do not know whether the inverse statement is valid. ∎

Exercise 3.3.7. Let $V \subset W \subset X$ be Banach spaces such that all the embeddings are continuous and dense. Assume that the (interpolation) inequality

$$\|u\|_W \le a_\theta \|u\|_X^\theta \|u\|_V^{1-\theta}, \quad u \in V,$$

is valid with some constants $a_\theta > 0$ and $0 < \theta < 1$. Then for any set $\mathscr{L}$ of the linear functionals on W the following estimate holds:

$$\left[a_\theta^{-1} \epsilon_{\mathscr{L}}(V, W) \right]^{1/\theta} \le \epsilon_{\mathscr{L}}(V, X) \le \left[a_\theta \epsilon_{\mathscr{L}}(W, X) \right]^{1/(1-\theta)}. \tag{3.3.9}$$

Hint: See [38] or [39, Chapter 5]. ∎

We also note that the completeness defect is closely related with some concepts of the approximation theory (see CHUESHOV [38] or [39, Chapter 5] for details). Below we use the so-called interpolation operators which are related with the set of functionals given. To describe their properties we need the following notion.

Definition 3.3.8. Let $V \subset H$ be separable Hilbert spaces and R be a linear operator from V into H. As in AUBIN [5], the value

$$e_V^H(R) = \sup\{\|u - Ru\|_H \; : \; \|u\|_V \le 1\} \equiv \|I - R\|_{V \mapsto H}$$

is said to be the *global approximation error* in H arising in the approximation of elements $v \in V$ by elements Rv. Here and below $\| \cdot \|_{V \mapsto H}$ denotes the operator norm for linear mappings from V into H.   ∎

The following assertion (see CHUESHOV [38, 39] for the proof) shows that the completeness defect provides us with a bound from below for the best possible global approximation error.

Theorem 3.3.9. *Let V and H be separable Hilbert spaces such that V is compactly and densely embedded into H. Let $\mathscr{L}$ be a finite set of linear functionals on V. Then we have the following relations:*

$$\epsilon_{\mathscr{L}}(V,H) = \min\{e_V^H(R) \; : \; R \in \mathscr{R}_{\mathscr{L}}\},$$

where $\mathscr{R}_{\mathscr{L}}$ is the family of linear bounded operators $R : V \mapsto H$ and such that $Rv = 0$ for all $v \in \mathscr{L}^{\perp} = \{v \in V \; : \; l(v) = 0, \; l \in \mathscr{L}\}$. Moreover, we have that

$$\epsilon_{\mathscr{L}}(V,H) = e_V^H(I - Q_{\mathscr{L}}) = \sup\{\|Q_{\mathscr{L}}u\|_H \; : \; \|u\|_V \le 1\}, \tag{3.3.10}$$

where $Q_{\mathscr{L}}$ is the orthoprojector in V onto $\mathscr{L}^{\perp}$.

One can show (see CHUESHOV [39]) that any operator $R \in \mathscr{R}_{\mathscr{L}}$ has the form

$$Rv = \sum_{j=1}^{N} l_j(v)\psi_j, \quad \forall \, v \in V, \tag{3.3.11}$$

where $\{\psi_j\}$ is an arbitrary finite set of elements from V. This is why $\mathscr{R}_{\mathscr{L}}$ is called the set of interpolation operators corresponding to the set $\mathscr{L}$. An operator $R \in \mathscr{R}_{\mathscr{L}}$ is called a *Lagrange* interpolation operator, if it has form (3.3.11) with $\{\psi_j\}$ such that $l_k(\psi_j) = \delta_{kj}$. In the case of Lagrange operators we have that $R^2 = R$, i.e., R is a projector.

We also note that the operator $Q_{\mathscr{L}}$ in (3.3.10) has the following structure:

$$Q_{\mathscr{L}} = I - P_{\mathscr{L}} \quad \text{with} \quad P_{\mathscr{L}}v = \sum_{j=1}^{N}(\xi_j, v)_V \xi_j, \quad \forall \, v \in V,$$

where $\{\xi_j\}$ is the orthonormal basis in the orthogonal supplement $\mathscr{M}_{\mathscr{L}}$ to the annihilator $\mathscr{L}^{\perp}$ in V. We call $P_{\mathscr{L}}$ the *optimal* interpolation operator corresponding to the set $\mathscr{L}$.

Example 3.3.10 (Modes). Let A be a positive operator with a discrete spectrum in a separable Hilbert space H with domain $\mathscr{D}(A)$; that is, there exists the orthonormal basis $\{e_k\}$ in H such that

$$Ae_k = \omega_k e_k, \quad 0 < \omega_1 \le \omega_2 \le \cdots, \quad \lim_{k\to\infty} \omega_k = \infty. \tag{3.3.12}$$

Let $\{H_s\}_{s\in\mathbb{R}}$ be the scale of spaces generated by A; that is, $H_s = \mathscr{D}(A^s)$ if $s \ge 0$ and H_s is the completion of H with respect to the norm $\|A^s \cdot \|$ when $s < 0$. Denote by $\mathscr{L}$ the set of functionals $\mathscr{L} = \{l_j(u) = (u, e_j)_H : j = 1, 2, \ldots, N\}$. A simple calculation (see CHUESHOV [38] or [39, Chapter 5]) shows that $\epsilon_{\mathscr{L}}(H_s, H_{\sigma}) = \omega_{N+1}^{\sigma-s}$ for every $s > \sigma$. ∎

The following two families of functionals on the Sobolev spaces[5] $H^s(\Omega)$ are important from the point of view of applications.

Let Ω be either a smooth domain or a parallelepiped in $\mathbb{R}^n$. Assume that Ω is divided into subdomains $\{\Omega_j : j = 1, 2, \ldots, N\}$ such that

$$\overline{\Omega} = \bigcup\{\overline{\Omega}_j : j = 1, 2, \ldots, N\}, \quad \Omega_j \bigcap \Omega_i = \emptyset, \ j \ne i,$$

where the bar denotes the closure of a set.

Example 3.3.11 (Generalized local volume averages). Assume that $\lambda_j(x)$ is a function from $L_{\infty}(\Omega_j)$ such that

$$\operatorname{supp} \lambda_j \subset\subset \overline{\Omega}_j, \quad \int_{\Omega_j} \lambda_j(x)dx = 1$$

and Ω_j is a star-like domain with respect to the support $\operatorname{supp} \lambda_j$.[6] We define the set $\mathscr{L}$ of generalized local volume averages as a family of functionals of the form

$$\mathscr{L} = \left\{ l_j(u) = \int_{\Omega_j} \lambda_j(x)u(x) \, dx, \ j = 1, 2, \ldots, N \right\}.$$

---

[5]We refer to ADAMS [1] or LIONS/MAGENES [152] for definitions and basic facts from the theory of these spaces.

[6]This means that for every $x \in \Omega_j$ there is $y \in \operatorname{supp} \lambda_j$ such that the interval $\{\lambda x + (1 - \lambda)y : 0 \le \lambda \le 1\}$ lies in Ω_j.

It follows from CHUESHOV [38, Theorem 3.1] that there exist constants c_1 and c_2 depending on s, σ, and Ω such that for $\epsilon_{\mathscr{L}}(s, \sigma) \equiv \epsilon_{\mathscr{L}}(H^s(\Omega), H^{\sigma}(\Omega))$ the estimate

$$c_1 \cdot \left[\max_j d_j \right]^{s-\sigma} \leq \epsilon_{\mathscr{L}}(s, \sigma) \leq c_2 \cdot \left[\max_j d_j \right]^{s-\sigma} \tag{3.3.13}$$

holds for every $0 \leq \sigma \leq s$, where $d_j = \operatorname{diam} \Omega_j \equiv \sup\{|x - y| \, : \, x, y \in \Omega_j\}$.   ∎

Example 3.3.12 (Nodes). Let the domain Ω be divided into subdomains $\{\Omega_j\}$ as described above. We choose the point x_j in each subdomain Ω_j and define the set of functionals on $H^m(\Omega)$, $m = [n/2] + 1$ (we call them nodes):

$$\mathscr{L} = \{l_j(u) = u(x_j) \, : \, x_j \in \Omega_j, \, j = 1, 2, \ldots, N\}.$$

By Theorem 3.2 from CHUESHOV [38] estimate (3.3.13) remains true for $s \geq m$ and $0 \leq \sigma \leq s$.   ∎

Below we present some results that involve the completeness defect to characterize sets of determining functionals.

3.3.2 A result on the existence of determining functionals

It is clear that to establish the existence of a finite number of determining functionals we need to control the difference of two trajectories of the system. For different classes this can be done in different ways depending on the properties of the system. However, in all cases some basic stability (and quasi-stability) type calculations are present in all approaches (we refer to CHUESHOV [38] and [39, Chapter 5] for a survey). As an illustration we provide one particular result in which the spaces V and X are the same.

Theorem 3.3.13. *Let (X, S_t) be a dynamical system in a Banach space X. Assume that for two semitrajectories $u(t)$ and $v(t)$ there exist a seminorm μ and a function $\psi(t) \in L_1^{loc}(\mathbb{R}_+)$ such that*

$$\Gamma_{\psi}^+ \equiv \limsup_{t \to \infty} \int_t^{t+1} |\psi(\tau)| \, d\tau < \infty$$

and

$$\|u(t) - v(t)\|^2 + \int_s^t \psi(\tau) \cdot \|u(\tau) - v(\tau)\|^2 \, d\tau$$

$$\leq \|u(s) - v(s)\|^2 + \int_s^t \mu(u(\tau) - v(\tau))^2 \, d\tau \tag{3.3.14}$$

holds for all $t \geq s \geq 0$. Then μ is a determining[7] seminorm for this couple of solutions provided

$$\gamma_\psi^+ \equiv \liminf_{t \to \infty} \frac{1}{a} \int_t^{t+a} \psi(\tau)\,d\tau > 0 \ \text{for some}\ a > 0.$$

Let $\mathscr{L} = \{l_j : j = 1, \dots, N\}$ be a family of linear continuous functionals on X. If we assume in addition that $\varepsilon_{\mathscr{L}}^2(X, \mu) < \gamma_\psi^+$, then $\mathscr{L}$ is determining for the semitrajectories $u(t)$ and $v(t)$.

Proof. For the proof we use the line of argument given in CHUESHOV [38].
  Let $h(t) = \|u(t) - v(t)\|^2$. Then (3.3.14) yields

$$h(t) + \int_s^t \tilde{\psi}(\tau)h(\tau)\,d\tau \leq h(s) + \int_s^t g(\tau)\,d\tau \tag{3.3.15}$$

for $t \geq s \geq 0$, where either

$$\tilde{\psi}(\tau) = \psi(\tau) \ \text{and}\ g(\tau) = \mu(u(\tau) - v(\tau))^2,$$

in the first case, or else

$$\tilde{\psi}(\tau) = \psi(\tau) - (1 + \delta)\varepsilon_{\mathscr{L}}^2(X, \mu) \ \text{and}\ g(\tau) = C_{\mathscr{L}}^\delta \max_j |l_j(u(\tau) - v(\tau))|^2$$

with arbitrary $\delta > 0$ in the second case (we also use Proposition 3.3.4).
  Solving the inequality[8] in (3.3.15), we obtain

$$h(t) \leq h(s) \exp\left\{-\int_s^t \tilde{\psi}(\sigma)\,d\sigma\right\} + \int_s^t g(\tau)\exp\left\{-\int_\tau^t \tilde{\psi}(\sigma)d\sigma\right\} d\tau \tag{3.3.16}$$

when $t \geq s$. One can see that there exists $\tau_* > 0$ such that

$$\int_\tau^{\tau+a} \tilde{\psi}(\sigma)\,d\sigma \geq a\beta_\psi > 0 \ \text{for all}\ t > \tau \geq \tau_*,$$

where $\beta_\psi > 0$ is any number less than γ_ψ^+ in the first case and $\gamma_\psi^+ - \varepsilon_{\mathscr{L}}^2(X, \mu)$ in the second case. This implies that

$$\int_\tau^t \tilde{\psi}(\sigma)d\sigma \geq \beta_\psi(t - \tau) - C \ \text{for all}\ t > \tau \geq \tau_*,$$

---

[7] See the definition in Remark 3.3.2.
[8] See Section A.2 in the Appendix.

where C depends on a, Γ_ψ^+, and γ_ψ^+. Therefore, (3.3.16) yields

$$\limsup_{t\to\infty} h(t) \le \limsup_{t\to\infty} \int_{\tau_*}^t g(\tau) \exp\left\{-\int_\tau^t \tilde\psi(\sigma)d\sigma\right\} d\tau.$$

Since $g(t) \to 0$ as $t \to \infty$, this relation via l'Hôpital's rule allows us to complete the proof of the theorem. □

Remark 3.3.14. Under the conditions of Theorem 3.3.13 the set $\mathscr{L}$ is also determining in the following Ladyzhenskaya type (see LADYZHENSKAYA [140]) sense: for any two full trajectories $u(t)$ and $v(t)$ defined on the whole time axis and such that

$$\sup\{\|u(t)\| + \|v(t)\| \; : \; -\infty < t < \infty\} \le R < \infty$$

the condition

$$\exists\, t_* \in \mathbb{R}: \; l_j(u(t)) = l_j(v(t)) \text{ for almost all } t < t_* \text{ and } j = 1,\dots,N, \qquad (3.3.17)$$

implies that $u(t) \equiv v(t)$ for all $t \in \mathbb{R}$ provided

$$\gamma_\psi^- \equiv \liminf_{t\to-\infty} \frac{1}{a}\int_t^{t+a} \psi(\tau)\, d\tau > \varepsilon_\mathscr{L}^2(X,\mu) \text{ for some } a > 0.$$

Indeed, it follows from (3.3.16) that

$$h(t) \le h(s) \exp\left\{-\int_s^t \tilde\psi(\sigma)\, d\sigma\right\} \text{ for all } -\infty < s \le t < t_*.$$

Thus, in the limit $s \to -\infty$ we obtain that $u(t) \equiv v(t)$ for all $t < t_*$. Therefore,

$$u(t+\tau) = S_\tau u(t) = S_\tau v(t) = v(t+\tau), \quad \forall\, t < t_*, \tau \ge 0.$$

This implies the conclusion.

We note that the property above appeared for the first time in LADYZHEN-SKAYA [140] (see also LADYZHENSKAYA [141, 142]) for the case of modes (see Example 3.3.10) with $t_* = +\infty$ as a statement that finite-dimensional projections of full trajectories from the attractor of the 2D Navier-Stoke equations uniquely determinate these trajectories. ■

3.4 Quasi-stable systems

We complete this chapter with a section which collects some general facts based on unifying specific criteria that lead to the existence and desired properties of attractors such as their finite dimension and the existence of exponential attractors.

We single out a class of "quasi-stable" systems that enjoy some kind of stabilizability inequalities written in some general form. Although these inequalities are often difficult to establish, once proved they provide a string of consequences that describe various properties of attractors. For the first time this kind of stability attracted attention in the paper of CHUESHOV/LASIECKA [51, Theorem 3.11], devoted to dynamics of second order in time evolution equations. Later on the quasi-stability method was developed in CHUESHOV/LASIECKA [56, 58] in order to cover wave/plate-type models with nonlinear (CHUESHOV/LASIECKA [52, 56, 58]) and thermal (CHUESHOV/LASIECKA [57, 58]) damping. In that form the method covers a large variety of hyperbolic-type models; see the discussion in Remark 7.9.3 of CHUESHOV/LASIECKA [58]. Here we extend the notion of quasi-stability to include models with "parabolic"-type behavior. Then we specify two subclasses of quasi-stable systems. Both are motivated by different evolution models and demonstrate additional properties of dynamics such as smoothness of attractors and existence of finite families of determining functionals. The first subclass is designed mainly to cover some variety of semilinear parabolic-type problems. The second one corresponds to models generated by the second order in time evolution equations.

3.4.1 General concept of a quasi-stable system

We start with quasi-stability inequality at fixed time. This (unified) notion was motivated by several classes of PDE models, both parabolic and hyperbolic. Moreover, the idea behind this notion can be applied in many other cases (see, e.g., CHUESHOV/LASIECKA [56, 59]). Systems with delay/memory terms[9] can also be included in this framework (see, e.g., CHUESHOV/REZOUNENKO [66, 67], FASTOVSKA [98, 99], POTOMKIN [186], RYZHKOVA [201] and also the proof of Theorem 9.3.5 in CHUESHOV/LASIECKA [58]). The same idea was recently applied in CHUESHOV/KOLBASIN [49] (see also Chapter 5) for the analysis of long-time dynamics in a degenerate hyperbolic-type model. The quantum Zakharov system (see CHUESHOV [43] and the references therein) and several classes of fluid-structure interaction models (see, e.g., CHUESHOV [45, 46] and CHUESHOV/RYZHKOVA [68, 69]) also demonstrate some applications of quasi-stability idea.

Definition 3.4.1. Let (X, S_t) be a dynamical system in some Banach space X. This system is said to be *quasi-stable* on a set $\mathscr{B} \subset X$ (at time t_*) if there exist (a) time $t_* > 0$, (b) a Banach space Z, (c) a globally Lipschitz mapping $K : \mathscr{B} \mapsto Z$, and (d) a compact seminorm[10] $n_Z(\cdot)$ on the space Z, such that

---

[9]For more details in the case of delay models we refer to Chapter 6.

[10]See Definition 3.1.14.

$$\|S_{t_*}y_1 - S_{t_*}y_2\|_X \leq q \cdot \|y_1 - y_2\|_X + n_Z(Ky_1 - Ky_2) \tag{3.4.1}$$

for every $y_1, y_2 \in \mathscr{B}$ with $0 \leq q < 1$. We emphasize that the space Z, the operator K, the seminorm n_Z, and the time moment t_* may depend on $\mathscr{B}$.   ∎

The definition of *quasi-stability* is rather natural from the point of view of long-time behavior. It pertains to decomposition of the flow into exponentially stable and compact parts (see (3.4.1)). This represents some sort of analogy with the "splitting" method (BABIN/VISHIK [9] and TEMAM [216]); however, the decomposition refers to the difference of two trajectories, rather than a single trajectory. We mention that in the degenerate case when $n_Z \equiv 0$ the relation in (3.4.1) transforms into the following one:

$$\|S_{t_*}y_1 - S_{t_*}y_2\|_X \leq q \cdot \|y_1 - y_2\|_X \quad \text{for every } y_1, y_2 \in \mathscr{B}. \tag{3.4.2}$$

Thus, S_{t_*} is a contraction on the closure $\overline{\mathscr{B}}$ of $\mathscr{B}$. If we assume that $\mathscr{B}$ is forward invariant, then there is a unique fixed point $\tilde{y}$ for S_{t_*} in $\overline{\mathscr{B}}$. The invariance of $\overline{\mathscr{B}}$ implies that $S_t\tilde{y}$ is also a fixed point for every $t > 0$. Thus, by the uniqueness we have that $S_t\tilde{y} = \tilde{y}$ for all $t > 0$, i.e., $\tilde{y}$ is a unique equilibrium in $\overline{\mathscr{B}}$. Moreover, it follows from (3.4.2) that this equilibrium is exponentially stable in $\overline{\mathscr{B}}$, i.e.,

$$\|S_t y - \tilde{y}\|_X \leq Ce^{-\alpha t} \sup_{\tau \in [0, t_*]} \|S_\tau y - \tilde{y}\|_X \quad \text{for every } y \in \overline{\mathscr{B}}$$

with some $\alpha > 0$. This observation explains why the property in (3.4.1) is called quasi-stability. We also refer to Remark 3.4.16 for a discussion of a quasi-stability notion in the case when a model possesses some structural properties with a particular form of the operator K.

In the following exercise we point out an important special case of the quasi-stability introduced in Definition 3.4.1.

Exercise 3.4.2. Let Y be a Banach space compactly embedded in X. Instead of (3.4.1), assume that S_{t_*} is globally Lipschitz from $\mathscr{B}$ into Y. Show that (X, S_t) is quasi-stable on $\mathscr{B}$ at time t_*. Hint: See the argument in Corollaries 3.1.25 and 3.1.26.   ∎

In what follows our first task is to show that quasi-stable systems enjoy many nice properties that include the existence of global finite-dimensional attractors and fractal exponential attractors. Then we switch on particular forms of quasi-stability.

We first show that the fixed time quasi-stability property implies asymptotic compactness.

Proposition 3.4.3 (Asymptotic smoothness). *Let a dynamical system (X, S_t) be quasi-stable on every bounded forward invariant set $\mathscr{B}$ in X. Then (X, S_t) is asymptotically smooth.*

Proof. For every forward invariant set $\mathscr{B}$ there exists $t_* = t_*(\mathscr{B}) > 0$ such that (3.4.1) holds. Therefore, we can apply Theorem 2.2.21 with $g(s) = qs$, $T = t_*$, and $\varrho(y_1, y_2) = n_Z(Ky_1 - Ky_2)$ to obtain the result.   □

Corollary 3.4.4 (Global attractor). *Let a system (X, S_t) be dissipative and satisfy the hypothesis of Proposition 3.4.3. Then this system possesses a compact global attractor.*

Proof. Since by Proposition 3.4.3 the system (X, S_t) is asymptotically smooth, the result follows from Theorem 2.3.5.   □

The following assertion shows that quasi-stability implies the finite dimensionality of a global attractor.

Theorem 3.4.5 (Finite-dimensional attractor). *Assume that a system (X, S_t) possesses a compact global attractor $\mathfrak{A}$ and is quasi-stable on $\mathfrak{A}$ at some point $t_* > 0$ (see Definition 3.4.1). Then the attractor $\mathfrak{A}$ has a finite fractal dimension $\dim_f \mathfrak{A}$ in X. Moreover, we have the estimate*

$$\dim_f \mathfrak{A} \leq \left[\ln \frac{2}{1+q} \right]^{-1} \cdot \ln m_Z \left(\frac{4L_K}{1-q} \right), \tag{3.4.3}$$

where $L_K > 0$ is the Lipschitz constant for K (see (3.1.17)) and $m_Z(R)$ is the maximal number of elements z_i in the ball $\{z \in Z : \|z_i\|_Z \leq R\}$ possessing the property $n_Z(z_i - z_j) > 1$ when $i \neq j$.

Proof. We apply Theorem 3.1.21 with $V = S_{t_*}$.   □

For quasi-stable systems we have several results pertaining to (generalized) fractal exponential attractors. We start with the following definition (see EDEN ET AL. [92]).

Definition 3.4.6. A compact set $A_{\exp} \subset X$ is said to be *inertial* (or a *fractal exponential attractor*) of the dynamical system (X, S_t) if $A_{\exp}$ is a positively invariant set of finite fractal dimension and for every bounded set $D \subset X$ there exist positive constants t_D, C_D, and γ_D such that

$$d_X\{S_t D \,|\, A_{\exp}\} \equiv \sup_{x \in D} \text{dist}_X(S_t x, A_{\exp}) \leq C_D \cdot e^{-\gamma_D(t-t_D)}, \quad t \geq t_D.$$

If the exponential attractor has finite fractal dimension in some extended space $\tilde{X} \supset X$, we frequently call this exponentially attracting set a *generalized* fractal exponential attractor.   ■

For more details concerning fractal exponential attractors we refer to EDEN ET AL. [92] and also to the recent survey of MIRANVILLE/ZELIK [166]. We only mention that (i) a global attractor can be non-exponential (see Exercise 2.3.29), and (ii) an exponential global attractor is not unique, but contains the global attractor.

We note that the standard technical tool (see, e.g., EDEN ET AL. [92] and a discussion in MIRANVILLE/ZELIK [166]) in the construction of fractal exponential attractors is the squeezing property in the Foias-Temam sense (see a discussion and the references in CONSTANTIN/FOIAS/TEMAM [79] and TEMAM [216]). This property says (see, e.g., the statement of Corollary 3.2.4), roughly speaking, that either the higher modes are dominated by the lower ones or that the semiflow is contracted exponentially. We also refer to the survey of MIRANVILLE/ZELIK [166] for some generalization of this method. Instead our approach is based on the quasi-stability property, which says that the semiflow is asymptotically contracted up to a homogeneous compact additive term.

Theorem 3.4.7 (Fractal exponential attractor). *Assume that a dynamical system* (X, S_t) *is dissipative and quasi-stable (in the sense of Definition 3.4.1) on some bounded absorbing set* $\mathscr{B}$ *at some moment* $t_* > 0$. *We also assume that*

$$\|S_t y_1 - S_t y_2\|_X \leq C_{\mathscr{B}} \cdot \|y_1 - y_2\|_X \quad \text{for every } y_1, y_2 \in \mathscr{B} \text{ and } t \in [0, t_*] \quad (3.4.4)$$

and there exists a space $\tilde{X} \supseteq X$ *such that* $t \mapsto S_t y$ *is Hölder continuous in* $\tilde{X}$ *for every* $y \in \mathscr{B}$ *in the sense that there exist* $0 < \gamma \leq 1$ *and* $C_{\mathscr{B}, t_*} > 0$ *such that*

$$\|S_{t_1} y - S_{t_2} y\|_{\tilde{X}} \leq C_{\mathscr{B}} |t_1 - t_2|^{\gamma}, \quad t_1, t_2 \in [0, t_*], \ y \in \mathscr{B}. \quad (3.4.5)$$

Then the dynamical system (X, S_t) *possesses a (generalized) fractal exponential attractor whose dimension is finite in the space* $\tilde{X}$.

Proof. We can assume that $\mathscr{B}$ is forward invariant. In this case $V := S_{t_*}$ maps $\mathscr{B}$ into itself, and we can apply Theorem 3.2.1. This theorem implies that the mapping V possesses a fractal exponential attractor; that is, there exists a compact set $\mathscr{A} \subset \mathscr{B}$ and a number $0 < \eta < 1$ such that $\dim_f^X \mathscr{A} < \infty$, $V\mathscr{A} \subset \mathscr{A}$, and

$$\sup \left\{ \text{dist}_X(V^k U, \mathscr{A}) \ : \ U \in \mathscr{B} \right\} \leq C\eta^k, \quad k = 1, 2, \dots, \quad (3.4.6)$$

for some constant $C > 0$. One can also see that

$$A_{\exp} = \cup \{ S_t \mathscr{A} \ : \ t \in [0, t_*] \}$$

is a compact forward invariant set with respect to S_t; that is, $S_t A_{\exp} \subset A_{\exp}$. Moreover, it follows from (3.4.5) and from Proposition 3.1.13 that

$$\dim_f^{\tilde{X}} A_{\exp} \leq \gamma^{-1} \left[1 + \dim_f^X \mathscr{A} \right] < \infty.$$

We also have from (3.4.6) and (3.4.4) that

$$\sup \left\{ \text{dist}_X(S_t y, A_{\exp}) \ : \ y \in \mathscr{B} \right\} \leq Ce^{-\delta t}, \quad t \geq 0,$$

for some $\delta > 0$. Thus, $A_{\exp}$ is a (generalized) fractal exponential attractor. $\qquad\square$

We do not know whether the finiteness of the fractal dimension $\dim_f^{\tilde{X}} A_{\exp}$ holds true without the Hölder continuity property (3.4.5) imposed in some vicinity of $A_{\exp}$. This is because $A_{\exp}$ is a *uncountable* union of (finite-dimensional) sets $S_t \mathscr{A}$. We also emphasize that fractal dimension depends on the topology; see Remark 3.1.12.

The following assertion is a version of the theorem proved in CHUESHOV [39, Chapter 1] (see also FABRIE ET AL. [94] for a similar approach to the construction of exponential attractors).

Theorem 3.4.8 (Exponential attractors via transitivity). *Assume that a dynamical system (X, S_t) on a separable Banach space X possesses the following properties:*

- *There exist a positively invariant compact set F and positive constants C and γ such that*

$$\sup \{\mathrm{dist}_X(S_t x, F) \ : \ x \in D\} \le C \cdot e^{-\gamma(t - t_D)}$$

  for every bounded set $D \subset X$ and for $t \ge t_D$.
- *There exist a neighborhood $\mathcal{O}$ of the compact F and numbers Δ_1 and α_1 such that*

$$\|S_t x_1 - S_t x_2\| \le \Delta_1 e^{\alpha_1 t} \|x_1 - x_2\|$$

  provided that $S_t x_i$ belongs to the closure $\overline{\mathcal{O}}$ of $\mathcal{O}$ for all $t \ge 0$.
- *The mapping $t \mapsto S_t x$ is uniformly Hölder continuous on F; that is, there exist constants $C_F(T) > 0$ and $\eta \in (0, 1]$ such that*

$$\|S_{t_1} x - S_{t_2} x\| \le C_F(T) |t_1 - t_2|^\eta, \quad t_1, t_2 \in [0, T], \ x \in F.$$

- *The system (X, S_t) is quasi-stable at some time $t_* > 0$ on F.*

Then there exists a fractal exponential attractor $A_{\exp}$ for (X, S_t) whose dimension is finite in X.

Proof. We consider the restriction (F, S_t) of the system (X, S_t) on the compact invariant set F. As in the proof of Theorem 3.4.7, we can conclude that there exists a compact forward invariant set $A_{\exp}$ with finite dimension $\dim_f^X A_{\exp}$ in X such that

$$\sup \{\mathrm{dist}_X(S_t y, A_{\exp}) \ : \ y \in F\} \le C e^{-\nu t}, \quad t \ge 0,$$

for some $\nu > 0$. Therefore, we can apply Lemma 2.3.27 on reduction in the vicinity $\mathcal{O}$ to obtain the conclusion. $\qquad\square$

We emphasize that the Hölder time continuity property in Theorem 3.4.8 is imposed on some exponentially attracting set F only. In applications related to nonlinear evolutionary PDEs this set F can be some space of smooth functions.[11] So we deal with dynamics in a smoother space and there is a chance to estimate time derivatives of solutions in norms which are stronger than in the case of energy-type solutions. For parabolic problems this effect is demonstrated in Chapters 4 and 6. We also refer FABRIE ET AL. [94] for a nontrivial application of this idea in the case of wave dynamics.

We also note that after the basic monograph of EDEN ET AL. [92] exponential attractors were studied by many authors for a large variety of PDE systems, and the theory was refined in several directions (see the survey in MIRANVILLE/ZELIK [166] and the references therein). Moreover, as was mentioned in MIRANVILLE/ZELIK [166], there is a common opinion that exponential attractors exist for all equations of mathematical physics for which it is possible to prove the existence of a finite-dimensional compact global attractor.

Now we split our considerations into two special cases demonstrating additional features of dynamics.

3.4.2 Quasi-stable systems: special case

Here we consider the quasi-stability inequality with a special choice of the space Z, the seminorm n_Z, and the operator K. This choice was motivated by several classes of parabolic PDE problems (see Chapters 4 and 6 below). In the case considered we can also show the existence of finite families of determining functionals.

Assumption 3.4.9. Let (X, S_t) be a dynamical system in some Banach space X and $\mathscr{B} \subset X$. Assume that there exist (a) compact seminorms $n_1(\cdot)$ and $n_2(\cdot)$ on the space X, and (b) numbers $a_*, t_* > 0$ and $0 \le q < 1$ such that

$$\|S_t y_1 - S_t y_2\|_X \le a_* \cdot \|y_1 - y_2\|_X \quad \text{for every } y_1, y_2 \in \mathscr{B} \text{ and } t \in [0, t_*] \qquad (3.4.7)$$

and

$$\|S_{t_*} y_1 - S_{t_*} y_2\|_X \le q \cdot \|y_1 - y_2\|_X + n_1(y_1 - y_2) + n_2(S_{t_*} y_1 - S_{t_*} y_2) \qquad (3.4.8)$$

for every $y_1, y_2 \in \mathscr{B}$.

Proposition 3.4.10. *Under Assumption 3.4.9 the system (X, S_t) is quasi-stable on $\mathscr{B} \subset X$.*

Proof. We take $Z = X \times X$, $n_z(x, y) = n_1(x) + n_2(y)$, and define $K : X \mapsto Z$ by the relation $Kx = (x; S_{t_*} x)$.   □

---

[11]This is definitely true for parabolic models because they possess smoothening properties. See Chapter 4.

This simple observation allows us to apply the results presented in the previous section to establish the following assertion.

Theorem 3.4.11 (Global and exponential attractor). *Assume that a dynamical system (X, S_t) is dissipative and satisfies Assumption 3.4.9 on some bounded absorbing set $\mathscr{B}$. Then this system possesses a compact global attractor of finite fractal dimension $\dim_f A$ in X satisfying the estimate*

$$\dim_f A \leq \left[\ln \frac{2}{1+q}\right]^{-1} \cdot \ln m_0 \left(\frac{4(1+a_*^2)^{1/2}}{1-q}\right), \tag{3.4.9}$$

where $m_0(R)$ is the maximal number[12] of pairs (x_i, y_i) in $X \times X$ possessing the properties

$$\|x_i\|^2 + \|y_i\|^2 \leq R^2, \quad n_1(x_i - x_j) + n_2(y_i - y_j) > 1, \quad i \neq j.$$

If in addition we assume that there exists a space $\tilde{X} \supseteq X$ such that $t \mapsto S_t y$ is Hölder continuous in $\tilde{X}$ for every $y \in \mathscr{B}$; that is, there exist $0 < \gamma \leq 1$ and $C_{\mathscr{B},T} > 0$ such that

$$\|S_{t_1} y - S_{t_2} y\|_{\tilde{X}} \leq C_{\mathscr{B},T} |t_1 - t_2|^\gamma, \quad t_1, t_2 \in [0, T], \ y \in \mathscr{B}, \tag{3.4.10}$$

then the dynamical system (X, S_t) possesses a (generalized) fractal exponential attractor whose dimension is finite in the space $\tilde{X}$.

Proof. To prove the existence of a global attractor with estimate (3.4.9) for its dimension we apply Corollary 3.4.4 and Theorem 3.4.5 with the same choice of Z, n_Z, and K as in Proposition 3.4.10. The existence of a fractal exponential attractor follows from Theorem 3.4.7. □

Using the structure of the quasi-stability inequality in (3.4.8), we can also establish the following assertion on determining functionals.

Theorem 3.4.12 (Determining functionals). *Assume that a system (X, S_t) is dissipative and satisfies Assumption 3.4.9 on some bounded absorbing set $\mathscr{B}$. Let $\mathscr{L} = \{l_j : j = 1, \ldots, N\}$ be a set of linearly independent functionals on X. Assume that*

$$\epsilon_{\mathscr{L}}(n_1) + \epsilon_{\mathscr{L}}(n_2) < 1 - q, \tag{3.4.11}$$

where $\epsilon_{\mathscr{L}}(n_j) \equiv \epsilon_{\mathscr{L}}(X, n_j)$ is the completeness defect of the family $\mathscr{L}$ with respect to the seminorm n_j (see Definition 3.3.3), the constant $q < 1$, and the seminorms n_i are the same as in relation (3.4.7). Then $\mathscr{L}$ is the set of asymptotically determining functionals, i.e., the relation

---

[12]This number is finite for every R. See Exercise 3.1.16 and Remark 3.1.17.

$$\lim_{t \to \infty} l_j(S_t y_1 - S_t y_2) = 0, \quad j = 1, 2, \ldots, N, \tag{3.4.12}$$

implies that $\lim_{t \to \infty} \|S_t y_1 - S_t y_2\|_X = 0$.

Proof. By Proposition 3.3.4 we have that

$$n_i(v) \le \epsilon_{\mathscr{L}}(n_i) \|v\|_X + C_{\mathscr{L}} \max_{j=1,\ldots,N} |l_j(v)|, \quad \forall v \in X. \tag{3.4.13}$$

Let $V = S_{t_*}$. We can assume that $\mathscr{B}$ is forward invariant with respect to V. Then using (3.4.13), from (3.4.8) we obtain that

$$\|V y_1 - V y_2\|_X \le \eta \cdot \|y_1 - y_2\|_X + \mathscr{N}(y_1 - y_2) + \mathscr{N}(V y_1 - V y_2) \tag{3.4.14}$$

for every $y_1, y_2 \in \mathscr{B}$, where

$$\eta \equiv [q + \epsilon_{\mathscr{L}}(n_1)][(1 - \epsilon_{\mathscr{L}}(n_2)]^{-1} < 1 \quad \text{and} \quad \mathscr{N}(v) = C_{\mathscr{L}} \max_{j=1,\ldots,N} |l_j(v)|$$

for some positive constant $C_{\mathscr{L}}$. After iteration of (3.4.14) we obtain

$$\|V^m y_1 - V^m y_2\|_X \le \eta^m \cdot \|y_1 - y_2\|_X + \sum_{k=1}^{m} \eta^{k-1} \mathscr{N}(V^{m-k} y_1 - V^{m-k} y_2)$$

$$+ \sum_{k=0}^{m-1} \eta^k \mathscr{N}(V^{m-k} y_1 - V^{m-k} y_2), \quad m = 1, 2, \ldots \tag{3.4.15}$$

By (3.4.12) we have that

$$\mathscr{N}(V^n y_1 - V^n y_2) \to 0 \text{ as } n \to +\infty.$$

Therefore, one can see that

$$\lim_{m \to \infty} \left\{ \sum_{k=1}^{m} \eta^{k-1} \mathscr{N}(V^{m-k} y_1 - V^{m-k} y_2) + \sum_{k=0}^{m-1} \eta^k \mathscr{N}(V^{m-k} y_1 - V^{m-k} y_2) \right\} = 0.$$

Hence (3.4.15) yields

$$\lim_{m \to \infty} \|V^m y_1 - V^m y_2\|_X = 0,$$

where $V^m = S_{mt_*}$. Using (3.4.7) we can conclude now that $\lim_{t \to \infty} \|S_t y_1 - S_t y_2\|_X = 0$ and complete the proof. $\square$

Remark 3.4.13. **1.** As one can see from the argument given in the proof of Theorem 3.4.12, the conclusion follows if instead of (3.4.12) we assume that

$$\lim_{n \to \infty} l_j(S_{t_n} y_1 - S_{t_n} y_2) = 0, \quad j = 1, 2, \ldots, N,$$

where $t_n = nt_*$, i.e., it is sufficient to know that the corresponding measurements become close to each other along some sequence of time moments. Moreover, if the system (X, S_t) is a point quasi-stable for some range of time moments, say $t_* \in [\alpha, \beta]$ with $\alpha > 0$, we can take any sequence $\{t_n\}$ satisfying the inequality $\alpha \le t_{n+1} - t_n \le \beta$ in the relation above. We refer to Theorem 1.3 in Chapter 5 of CHUESHOV [39] for a discussion of a similar result.

2. Under the hypotheses of Theorem 3.4.12 we can prove that the set $\mathscr{L}$ of functionals is also determining in the Ladyzhenskaya sense: for any two full trajectories $\gamma_j = \{u_j(t) : t \in \mathbb{R}\}$ which belong to $\mathscr{B}$ the property

$$\exists t_* \in \mathbb{R} : \quad l(u_1(t)) = l(u_2(t)) \quad \text{for every } t < t_*, \ l \in \mathscr{L},$$

implies that $u_1(t) \equiv u_2(t)$ for all $t \in \mathbb{R}$. Indeed, in this case instead of (3.4.15) we can write

$$\|u_1(t) - u_2(t)\|_X \le \eta^m \|u_1(t - mt_*) - u_2(t - mt_*)\|_X \le C_{\mathscr{B}} \eta^m$$

for every $m \in \mathbb{Z}_+$ and $t < t_*$. As in Remark 3.3.14, in the limit $m \to \infty$ this implies the conclusion.

■

3.4.3 Asymptotically quasi-stable systems

Now we discuss properties of quasi-stable systems whose phase space admits an additional structure. Our main motivation is related to nonlinear PDEs of second order in time possibly interacting with parabolic equations. The results presented were established earlier in CHUESHOV/LASIECKA [58, Section 7.9] by another method.

We assume the following structure of the model.

Assumption 3.4.14 (Structure). Let X, Y, and Θ be reflexive Banach spaces; X is compactly embedded in Y. We endow the space $H = X \times Y \times \Theta$ with the norm

$$\|y\|_H^2 = \|u_0\|_X^2 + \|u_1\|_Y^2 + \|\theta\|_Z^2, \quad y = (u_0; u_1; \theta_0).$$

The trivial case $\Theta = \{0\}$ is allowed. We assume that (H, S_t) is a dynamical system in $H = X \times Y \times \Theta$ with the evolution operator of the form

$$S_t y = (u(t); u_t(t); \theta(t)), \quad y = (u_0; u_1; \theta_0) \in H, \tag{3.4.16}$$

where the functions $u(t)$ and $\theta(t)$ possess the properties

$$u \in C(\mathbb{R}_+, X) \cap C^1(\mathbb{R}_+, Y), \quad \theta \in C(\mathbb{R}_+, \Theta). \tag{3.4.17}$$

The structure of the phase space H and the evolution operator S_t in Assumption 3.4.14 is motivated by the study of some systems generated by equations of the second order in time in $X \times Y$ possibly interacting with first order evolutions in the space Θ. This type of interaction arises in modeling of thermoelastic plates (see, e.g., CHUESHOV/LASIECKA [57] and [58, Chapters 5 and 11]). We also refer to Section A.3 in the Appendix concerning the functional spaces in (3.4.17) and also L_p-classes which we use below.

Definition 3.4.15 (Asymptotic quasi-stability). A dynamical system of the form (3.4.16) is said to be asymptotically quasi-stable on a set $\mathscr{B} \subset H$ if there exist a compact seminorm $\mu_X(\cdot)$ on the space X and non-negative scalar functions $a(t)$, $b(t)$, and $c(t)$ on $\mathbb{R}_+$ such that (i) $a(t)$ and $c(t)$ are locally bounded on $[0, \infty)$, (ii) $b(t) \in L_1(\mathbb{R}_+)$ possesses the property $\lim_{t \to \infty} b(t) = 0$, and (iii) for every $y_1, y_2 \in \mathscr{B}$ and $t > 0$ the following relations:

$$\|S_t y_1 - S_t y_2\|_H^2 \le a(t) \cdot \|y_1 - y_2\|_H^2 \tag{3.4.18}$$

and

$$\|S_t y_1 - S_t y_2\|_H^2 \le b(t) \cdot \|y_1 - y_2\|_H^2 + c(t) \cdot \sup_{0 \le s \le t} \left[\mu_X(u^1(s) - u^2(s)) \right]^2 \tag{3.4.19}$$

hold. Here we denote $S_t y_i = (u^i(t); u_t^i(t); \theta^i(t))$, $i = 1, 2$. ∎

Remark 3.4.16. Relation (3.4.19), in the context of long-time dynamics, was introduced in CHUESHOV/LASIECKA [51] (see also ELLER/CHUESHOV/LASIECKA [48] and the discussion in CHUESHOV/LASIECKA [56]). Roughly speaking, it means asymptotic stability modulo compact terms. The inequality in (3.4.19) was called a *stabilizability estimate*. To obtain such an estimate proves fairly technical (in critical problems) and requires rather subtle PDE tools to prove it. Illustrations of the method are given later in Chapters 5 and 6 for some abstract models (see also Chapters 9 and 11 in CHUESHOV/LASIECKA [58] for a variety of von Karman models and BUCCI/CHUESHOV/LASIECKA [19, 20], CHUESHOV ET AL. [41, 48, 51, 54–56, 61, 62, 68], NABOKA [168] for similar considerations for other models). ∎

Proposition 3.4.17 (Quasi-stability). *Let structural Assumption 3.4.14 be in force. Assume that the dynamical system (H, S_t) is asymptotically quasi-stable on some set $\mathscr{B}$ in H. Then this system is quasi-stable on the set $\mathscr{B}$ at every time $T > 0$ such that $b(T) < 1$.*

Proof. We first show that the seminorm μ possesses the following property:

$$\forall \, \varepsilon > 0 \; \exists \, C_\varepsilon > 0 : \quad \mu_X(u) \le \varepsilon \|u\|_X + C_\varepsilon \|u\|_Y \quad \text{for any} \quad u \in X. \tag{3.4.20}$$

To see this we introduce the Banach space

$$\tilde{X} = \text{Closure} \left\{ v \in X : \|v\|_{\tilde{X}} \equiv \mu_X(v) + \|v\|_Y < \infty \right\}. \tag{3.4.21}$$

One can see that X is compactly embedded in the Banach space $\tilde{X}$. To prove (3.4.20) we need to show that

$$\forall \varepsilon > 0 \; \exists C_\varepsilon > 0 : \quad \|u\|_{\tilde{X}} \le \varepsilon \|u\|_X + C_\varepsilon \|u\|_Y \quad \text{for any} \quad u \in X. \tag{3.4.22}$$

This can be done by the following argument (due to LIONS [151]). If (3.4.22) is not true, then there exist $\varepsilon_0 > 0$ and a sequence $\{u_n\} \subset X$ such that

$$\|u_n\|_{\tilde{X}} \ge \varepsilon_0 \|u_n\|_X + n\|u_n\|_Y \quad \text{for every } n = 1, 2, \ldots.$$

Thus, for $v_n = u_n \|u_n\|_{\tilde{X}}^{-1}$ we have that

$$\varepsilon_0 \|v_n\|_X + n\|v_n\|_Y \le 1 \quad \text{and} \quad \|v_n\|_{\tilde{X}} = 1 \quad \text{for } n = 1, 2, \ldots.$$

Therefore, the sequence $\{v_n\}$ is relatively compact in $\tilde{X}$ and $\|v_n\|_Y \to 0$ as $n \to \infty$. This implies that $\{v_n\}$ contains a subsequence which converges to 0 in $\tilde{X}$. This contradicts the fact $\|v_n\|_{\tilde{X}} = 1$ for all n. Thus, the relation in (3.4.20) is proved. Using this relation and also (3.4.18), we obtain

$$\sup_{0 \le s \le T} [\mu_X(u_1(s) - u_2(s))]^2 \le \varepsilon \|y_1 - y_2\|_H^2 + C(\varepsilon, T) \sup_{0 \le s \le T} \|u_1(s) - u_2(s)\|_Y^2,$$

where we use the notation $y_i = (u_0^i; u_1^i; \theta_0^i)$, $i = 1, 2$. This allows us to rewrite (3.4.19) at the moment T in the following form:

$$\|S_T y_1 - S_T y_2\|_H^2 \le [b(T) + \varepsilon] \cdot \|y_1 - y_2\|_H^2 + C_{\varepsilon, T} \sup_{0 \le s \le T} \|u^1(s) - u^2(s)\|_Y^2 \tag{3.4.23}$$

for every $\varepsilon > 0$ with $S_t y_i = (u^i(t); u_t^i(t); \theta^i(t))$, $i = 1, 2$.

Now we take $Z = W_1(0, T)$, where

$$W_1(0, T) = \left\{ z \in L_2(0, T; X) : \|z\|_{W_1(0,T)}^2 \equiv \int_0^T \left(\|z(t)\|_X^2 + \|z_t(t)\|_Y^2 \right) dt < \infty \right\} \tag{3.4.24}$$

and choose T and $\varepsilon > 0$ such that $b(T) + \varepsilon < q < 1$ in (3.4.23). We define operator $K : H \mapsto Z$ by the relation

$$(Ky)(t) = P_X S_t y, \quad t \in [0, T], \quad \text{where} \quad y = (u_0; u_1; \theta_0) \in H.$$

Here P_X is the projection in H on the first component; i.e., in the case when $S_t y = (u(t); u_t(t); \theta(t))$ we have that $P_X S_t y = u(t)$. It follows from (3.4.18) that K is globally Lipschitz.

By the Aubin-Dubinskii-Lions theorem (see Theorem A.3.7 in the Appendix) the space $W_1(0, T)$ is compactly embedded into $C(0, T; Y)$. This implies that the seminorm $n_Z(z) = C_{\varepsilon, T} \sup_{0 \le s \le T} \|z(s)\|_Y$ is compact on Z. Thus, the evolution operator S_t given by (3.4.16) satisfies the requirements of Definition 3.4.1 for every point $t_* = T$ with $b(T) < 1$. This completes the proof of the proposition.   □

In the same way as in the previous section we can derive from the results of Section 3.4.1 the following assertion.

Theorem 3.4.18 (Global and exponential attractor). *Let Assumption* 3.4.14 *be valid. Assume that the system* (H, S_t) *is dissipative and asymptotically quasi-stable on a bounded forward invariant absorbing set* $\mathscr{B}$ *in* H. *Then the system* (H, S_t) *possesses a compact global attractor A. This attractor A has a finite fractal dimension* $\dim_f^H A$.

Assume in addition that there exists a space $\tilde{H} \supseteq H$ *such that* $t \mapsto S_t y$ *is Hölder continuous in* $\tilde{H}$ *for every* $y \in \mathscr{B}$; *that is, there exist* $0 < \gamma \le 1$ *and* $C_{\mathscr{B}, T} > 0$ *such that*

$$\|S_{t_1} y - S_{t_2} y\|_{\tilde{H}} \le C_{\mathscr{B}, T} |t_1 - t_2|^\gamma, \quad t_1, t_2 \in [0, T], \ y \in \mathscr{B}. \tag{3.4.25}$$

Then the dynamical system (H, S_t) *possesses a (generalized) fractal exponential attractor whose dimension is finite in the space* $\tilde{H}$.

We note that this theorem was proved in CHUESHOV/LASIECKA [56, 58] by the method of "short" trajectories initially suggested in some form in MÁLEK/NEČAS [157] and MÁLEK/PRAŽÁK [158]; see also FEIREISL/PRAŽÁK [100]. We also note that we can use the relation in (3.4.3) in Theorem 3.4.5 to provide some bounds for the dimension $\dim_f^H A$. Similar estimates on the abstract level can be found in CHUESHOV/LASIECKA [56, 58] or [60] for different situations.

The asymptotic quasi-stability allows us to obtain additional regularity of trajectories lying on the global attractor. The theorem below provides regularity for time derivatives. The needed "space" regularity usually follows from the analysis of the respective PDE via elliptic theory (see the corresponding results in CHUESHOV/LASIECKA [58]).

Theorem 3.4.19 (Regularity). *Let Assumption* 3.4.14 *be valid. Assume that the dynamical system* (H, S_t) *possesses a compact global attractor* $\mathfrak{A}$ *and is asymptotically quasi-stable on the attractor* $\mathfrak{A}$. *Moreover, we assume that* (3.4.19) *holds with the function* $c(t)$ *possessing the property* $c_\infty = \sup_{t \in \mathbb{R}_+} c(t) < \infty$. *Then any full trajectory* $\{(u(t); u_t(t); \theta(t)) : t \in \mathbb{R}\}$ *that belongs to the global attractor enjoys the following regularity properties:*

$$u_t \in L_\infty(\mathbb{R}; X) \cap C(\mathbb{R}; Y), \quad u_{tt} \in L_\infty(\mathbb{R}; Y), \quad \theta_t \in L_\infty(\mathbb{R}; Z) \tag{3.4.26}$$

Moreover, there exists $R > 0$ such that

$$\|u_t(t)\|_X^2 + \|u_{tt}(t)\|_Y^2 + \|\theta_t(t)\|_Z^2 \le R^2, \quad t \in \mathbb{R}, \tag{3.4.27}$$

where R depends on the constant c_∞, on the seminorm μ_X in Definition 3.4.15, and also on the embedding properties of X into Y.

Proof. It follows from (3.4.19) that for any two full trajectories

$$\gamma = \{U(t) \equiv (u(t); u_t(t); \theta(t)) : t \in \mathbb{R}\},$$
$$\gamma^* = \{U^*(t) \equiv (u^*(t); u_t^*(t); \theta^*(t)) : t \in \mathbb{R}\}$$

from the global attractor we have that

$$\|Z(t)\|_H^2 \le b(t-s)\|Z(s)\|_H^2 + c(t-s) \sup_{s \le \tau \le t} [\mu_X(z(\tau))]^2 \tag{3.4.28}$$

for all $s \le t$, $s, t \in \mathbb{R}$, where $Z(t) = U^*(t) - U(t)$ and $z(t) = u^*(t) - u(t)$. In the limit $s \to -\infty$ relation (3.4.28) gives us that

$$\|Z(t)\|_H^2 \le c_\infty \sup_{-\infty \le \tau \le t} [\mu_X(z(\tau))]^2$$

for every $t \in \mathbb{R}$ and for every couple of trajectories γ and γ^*. Using relation (3.4.20) we can conclude that

$$\sup_{-\infty \le \tau \le t} \|Z(\tau)\|_H^2 \le C \sup_{-\infty \le \tau \le t} \|z(\tau)\|_Y^2, \tag{3.4.29}$$

for every $t \in \mathbb{R}$ and for every couple of trajectories γ and γ^* from the attractor.

Now we fix a trajectory γ and for $0 < |h| < 1$ we consider the shifted trajectory $\gamma^* \equiv \gamma_h = \{y(t+h) : t \in \mathbb{R}\}$. Applying (3.4.29) for this pair of trajectories and using the fact that all terms in (3.4.29) are quadratic with respect to Z, we obtain that

$$\sup_{-\infty \le \tau \le t} \{\|u^h(\tau)\|_X^2 + \|u_t^h(\tau)\|_Y^2 + \|\theta_t^h(\tau)\|_Z^2\} \le C \sup_{-\infty \le \tau \le t} \|u^h(\tau)\|_Y^2, \tag{3.4.30}$$

where $u^h(t) = h^{-1} \cdot [u(t+h) - u(t)]$ and $\theta^h(t) = h^{-1} \cdot [\theta(t+h) - \theta(t)]$. On the attractor we obviously have that

$$\|u^h(t)\|_Y \le \frac{1}{h} \cdot \int_0^h \|u_t(\tau+t)\|_Y d\tau \le C, \quad t \in \mathbb{R},$$

with uniformity in h. Therefore, (3.4.30) implies that

$$\|u^h(t)\|_X^2 + \|u_t^h(t)\|_Y^2 + \|\theta_t^h(t)\|_Z^2 \le C, \quad t \in \mathbb{R}.$$

Passing with the limit on h this yields relations (3.4.26) and (3.4.27).   □

Another consequence of the quasi-stability estimate is the following assertion borrowed from CHUESHOV/LASIECKA [58] and pertaining to determining functionals (see Theorem 3.4.12 and also Section 3.3 for a general discussion of the theory of determining functionals).

Theorem 3.4.20 (Determining functionals). *Let Assumption* 3.4.14 *be valid. Assume that the dynamical system* (H, S_t) *is dissipative and asymptotically quasi-stable on some bounded absorbing set* $\mathscr{B}$. *Let* $\mathscr{L} = \{l_j : j = 1, \ldots, N\}$ *be a set of linearly independent functionals on* X *and* $\epsilon_{\mathscr{L}}(\mu_X)$ *be its completeness defect with respect to the seminorm* μ_X *(see Definition* 3.3.3*). If there exists* $\tau > 0$ *such that*

$$\eta_\tau \equiv b(\tau) + \epsilon_{\mathscr{L}}^2(\mu_X)c(\tau) \cdot \sup_{s \in [0,\tau]} a(s) < 1, \tag{3.4.31}$$

then the relation

$$\lim_{t \to \infty} l_j(u^1(s) - u^2(s)) = 0, \quad j = 1, 2, \ldots, N, \tag{3.4.32}$$

implies that $\lim_{t \to \infty} \|S_t y_1 - S_t y_2\|_H = 0$. *Here* $S_t y_i = (u^i(t); u_t^i(t); \theta^i(t))$, $i = 1, 2$.

Proof. We first note that the convergence in (3.4.32) is equivalent to the convergence

$$\Delta_{\mathscr{L}}(t) \equiv \sup_{s \in [t,t+\tau]} \max_j |l_j(u^1(s) - u^2(s))| = 0, \quad t \to \infty, \tag{3.4.33}$$

for every fixed $\tau > 0$. Assume now that

$$S_t y_i = (u^i(t); u_t^i(t); \theta^i(t)) \in \mathscr{B} \text{ for } t \ge t_0, \quad i = 1, 2.$$

Then from (3.4.19) we have that

$$\|S_{t+\tau} y_1 - S_{t+\tau} y_2\|_H^2 \le b(\tau)\|S_t y_1 - S_t y_2\|_H^2 + c(\tau) \sup_{t \le s \le t+\tau} \left[\mu_X(u^1(s) - u^2(s))\right]^2, \tag{3.4.34}$$

for any $t \ge t_0$. By Proposition 3.3.4 we have that

$$\mu_X(v) \le \epsilon_{\mathscr{L}}(\mu_X)\|v\|_X + C_{\mathscr{L}} \max_{j=1,\ldots,N} |l_j(v)|, \quad \forall v \in X. \tag{3.4.35}$$

From (3.4.35) we have

$$[\mu_X(v)]^2 \le (1+\delta)\epsilon_{\mathscr{L}}^2(\mu_X)\|v\|_X^2 + C_{\mathscr{L},\circ} \max_{j=1,\dots,N} |l_j(v)|^2, \quad \forall v \in X,$$

for each $\delta > 0$. By (3.4.18) this implies that

$$\sup_{t \le s \le t+\tau} \left[\mu_X(u^1(s) - u^2(s))\right]^2$$

$$\le \left[(1+\delta)\epsilon_{\mathscr{L}}^2(\mu_X) \sup_{s\in[0,\tau]} a(s)\right] \|S_t y_1 - S_t y_2\|_H^2 + C_{\mathscr{L},\delta}\Delta_{\mathscr{L}}^2(t).$$

Consequently, (3.4.34) yields

$$\|S_{t+\tau}y_1 - S_{t+\tau}y_2\|_H^2 \le \eta\|S_t y_1 - S_t y_2\|_H^2 + C_{\mathscr{L},\delta}\Delta_{\mathscr{L}}^2(t),$$

where $\eta = (1+\delta)\epsilon_{\mathscr{L}}^2(\mu_X)c(\tau)\cdot\sup_{s\in[0,\tau]} a(s) + b(\tau)$. Under condition (3.4.31) we can choose $\delta > 0$ such that $\eta < 1$ and find that

$$\|S_{t_0+n\tau}y_1 - S_{t_0+n\tau}y_2\|_H^2 \le \eta^n \cdot \|S_{t_0}y_1 - S_{t_0}y_2\|_H^2 + C\sum_{m=0}^{n-1} \eta^{n-m-1}\Delta_{\mathscr{L}}^2(t_0 + m\tau).$$

It is easy to see now that $\lim_{n\to\infty} \|S_{t_0+n\tau}y_1 - S_{t_0+n\tau}y_2\|_H^2 = 0$ under conditions (3.4.31) and (3.4.33). Application of (3.4.18) completes the proof. $\quad\square$

Chapter 4
Abstract Parabolic Problems

The main goal of this chapter is to show how the general methods developed in the previous chapters can be applied in the study of properties of qualitative dynamics for a class of abstract evolution equations of the form

$$u_t + Au = B(u), \quad u\big|_{t=0} = u_0 \in H,$$

in a Hilbert space H, where A is a positive self-adjoint operator and $B(u)$ is a nonlinear mapping on H. This kind of model covers many classes of parabolic-type equations including heat and reaction-diffusion models and also some hydrodynamical problems. The first several sections are based on the substantially revised content of CHUESHOV [39, Chapter 2] and present an approach involving the notion of a mild solution. The final (hydrodynamical) sections are new. They use another concept of a solution. We deal here with what are called weak (or variational) solutions. In many cases weak and mild solutions lead to the same functions. However, well-posedness statements for each of them require different tools.

Our considerations are concentrated around well-posedness and long-time dynamics issues. In the latter case we apply the idea of the quasi-stability method. For the considered class of models this method is based on the Ladyzhenskaya squeezing property.

Many results presented in this chapter are known from other sources. Nevertheless, we include them in order to demonstrate the advantages of the developed technology in the most transparent way.

4.1 Positive operators with discrete spectrum

In this section we consider properties of the linear part of the problem above. The class of linear operators we deal with arises in many PDEs on bounded domains.

© Springer International Publishing Switzerland 2015
I. Chueshov, *Dynamics of Quasi-Stable Dissipative Systems*, Universitext,
DOI 10.1007/978-3-319-22903-4_4

Definition 4.1.1. A linear self-adjoint positive densely defined operator A with domain $\mathscr{D}(A)$ on a separable Hilbert space H is said to be an operator with discrete spectrum if there exists an orthonormal basis $\{e_k\}$ in H consisting of the eigenvectors of the operator A:

$$Ae_k = \lambda_k e_k, \quad 0 < \lambda_1 \leq \lambda_2 \leq \cdots, \quad \lim_{k\to\infty} \lambda_k = \infty.$$

∎

In this case for every element $u \in H$ we have the relations

$$u = \sum_{k=1}^{\infty} (u, e_k)e_k \quad \text{and} \quad \|u\|^2 = \sum_{k=1}^{\infty} |(u, e_k)|^2,$$

where $\|\cdot\|$ and $(\cdot, \cdot)$ are the norm and scalar product in H. Moreover,

$$u \in \mathscr{D}(A) \quad \text{if and only if} \quad \sum_{k=1}^{\infty} \lambda_k^2 |(u, e_k)|^2 < \infty,$$

and

$$Au = \sum_{k=1}^{\infty} \lambda_k (u, e_k)e_k \quad \text{for} \ u \in \mathscr{D}(A).$$

This structure of the operator A allows us to define an operator $f(A)$ for a wide class of functions $f(s)$ defined on the positive semi-axis $\mathbb{R}_+$. We can take

$$\mathscr{D}(f(A)) = \left\{ h = \sum_{k=1}^{\infty} c_k e_k \in H : \sum_{k=1}^{\infty} c_k^2 |f(\lambda_k)|^2 < \infty \right\}$$

as a domain and assume that

$$f(A)h = \sum_{k=1}^{\infty} c_k f(\lambda_k)e_k, \quad h = \sum_{k=1}^{\infty} c_k e_k \in \mathscr{D}(f(A)).$$

In particular, one can define operators A^α for every $\alpha \in \mathbb{R}$. This allows us to introduce the space H_s (with $s > 0$) as the domain $\mathscr{D}(A^s)$ equipped with the graph norm $\|\cdot\|_s = \|A^s \cdot \|$. If s is negative, we define H_s as the completion of H with respect to the norm $\|\cdot\|_s = \|A^{-|s|} \cdot \|$. We note that $H_0 = H$ and the spaces $H_{-\sigma}$ with $\sigma > 0$ can be identified with the space of formal series $f \sim \sum_{k=1}^{\infty} c_k e_k$ such that

$$\|f\|_{-\sigma}^2 \equiv \sum_{k=1}^{\infty} c_k^2 \lambda_k^{-2\sigma} < \infty.$$

Thus, we obtain the *scale* of spaces H_s with $s \in \mathbb{R}$, which possesses several important properties. Some of them are collected in the following exercise.

Exercise 4.1.2. Prove the following statements.

(A) Each H_s is a separable Hilbert space with the inner product given by $(u, v)_s = (A^s u, A^s v)$.

(B) $H_s \subset H_\sigma$ for every $s > \sigma$, Moreover, H_s is dense in H_σ.

(C) The operator A can be extended on the whole scale $\{H_s\}$ as a bounded linear operator from H_s into H_{s-1}.

(D) Similarly, the operator A^{-1} exists and is bounded from H_s into H_{s+1} for every $s \in \mathbb{R}$.

(E) Prove the following interpolation inequality:

$$\|A^\theta u\| \leq \|Au\|^\theta \|u\|^{1-\theta}, \quad \forall u \in \mathscr{D}(A), \ \theta \in [0, 1]. \tag{4.1.1}$$

Moreover, if for some $\alpha, \beta, \mu \in \mathbb{R}$ we have that $\mu = \theta\alpha + (1 - \theta)\beta$ with $\theta \in [0, 1]$, then

$$\|A^\mu u\| \leq \|A^\alpha u\|^\theta \|A^\beta u\|^{1-\theta}, \quad \forall u \in \cap_n \mathscr{D}(A^n). \tag{4.1.2}$$

Hint: Write the norm on the left-hand side of (4.1.1) in coordinates and apply the Hölder inequality:

$$\left|\sum x_i y_i\right| \leq \left(\sum |x_i|^p\right)^{1/p} \left(\sum |y_i|^q\right)^{1/q}, \quad \frac{1}{p} + \frac{1}{q} = 1, \ 1 < p, q + \infty,$$

with $p = \theta^{-1}$ and $q = (1 - \theta)^{-1}$. Then derive (4.1.2) from (4.1.1).

(F) Using the result in (4.1.1) and the inequality

$$|xy| \leq \frac{1}{p}|x|^p + \frac{1}{q}|y|^q, \quad \frac{1}{p} + \frac{1}{q} = 1, \ 1 < p, q + \infty,$$

show that

$$\|A^\theta u\| \leq \varepsilon \|Au\| + \left(\frac{\theta}{\varepsilon}\right)^{\theta/(1-\theta)} (1 - \theta)\|u\|, \quad \forall u \in \mathscr{D}(A), \ \varepsilon > 0, \ \theta \in (0, 1).$$

∎

The following exercise means that H_{-s} is dual for H_s with respect to the inner product in H.

Exercise 4.1.3. Let $f \in H_\sigma$ for $\sigma > 0$. Show that the linear functional $F(u) = (f, u)$ can be continuously extended from the space H to $H_{-\sigma}$ and $|(f, g)| \leq \|f\|_\sigma \|g\|_{-\sigma}$ for any $f \in H_\sigma$ and $g \in H_{-\sigma}$. Moreover, for every $\sigma \in \mathbb{R}$ any continuous linear functional $F(u)$ on H_σ has the form $F(u) = (u, g)$, where $g \in H_{-\sigma}$. Thus, $H_{-\sigma}$ is the space of continuous linear functionals on H_σ for every $\sigma \in \mathbb{R}$. ∎

Remark 4.1.4. A similar scale of spaces H_s can be constructed based on a general positive operator A. In this case the properties presented in Exercises 4.1.2 and 4.1.3 remain true. We can even consider more general operators for the generation of the continuous scales of spaces (see, e.g., ENGEL/NAGEL [93] or HENRY [123]).   ∎

The following fact involves the discrete spectrum property.

Exercise 4.1.5. Show that the relation

$$P_N u = \sum_{k=1}^{N} (u, e_k) e_k, \quad u \in H_\sigma, \quad \sigma \in \mathbb{R}, \tag{4.1.3}$$

defines an orthoprojector onto $\mathrm{Span}\{e_k : k = 1, 2, \ldots, N\}$ in each space H_σ. Moreover, $P_N u \in \cap_{s>0} H_s$ and $\|P_N u - u\|_\sigma \to 0$ as $N \to \infty$. Thus, we can approximate any element from the scale by its smooth projections.   ∎

The following compactness statement is important in our further considerations.

Proposition 4.1.6. *Let $s > \sigma$. Then the space H_s is compactly embedded into H_σ. This means that every sequence bounded in H_s is relatively compact in H_σ.*

Proof. It is well known that every bounded set in a separable Hilbert space is weakly relatively compact. Thus, it contains a weakly convergent sequence (see, e.g., DUNFORD/SCHWARTZ [88, Chapter 5]). Therefore, it is sufficient to prove that any sequence weakly tending to zero in H_s converges to zero with respect to the norm of the space H_σ.

Let $\{f_n\}$ be a sequence weakly convergent to zero in H_s. This means that

$$\lim_{n \to \infty} (f_n, g)_s = 0, \quad \forall g \in H_s, \quad \text{and} \quad c_f \equiv \sup_n \|f_n\|_s < \infty. \tag{4.1.4}$$

For every $N \geq 2$ we obviously have

$$\|f_n\|_\sigma^2 = \sum_{k=1}^{\infty} \lambda_k^{2\sigma} |(f_n, e_k)|^2 \leq \sum_{k=1}^{N-1} \lambda_k^{2\sigma} |(f_n, e_k)|^2 + \frac{1}{\lambda_N^{2(s-\sigma)}} \sum_{k=N}^{\infty} \lambda_k^{2s} |(f_n, e_k)|^2$$

$$\leq \sum_{k=1}^{N-1} \lambda_k^{2\sigma} |(f_n, e_k)|^2 + \frac{1}{\lambda_N^{2(s-\sigma)}} \|f_n\|_s^2$$

Therefore, it follows from (4.1.4) that

$$\limsup_{n \to \infty} \|f_n\|_\sigma \leq \frac{c_f}{\lambda_N^{s-\sigma}}, \quad N = 2, 3 \ldots$$

Passing to the limit $N \to \infty$ yields that $\lim_{n \to \infty} \|f_n\|_\sigma^2 = 0$. This completes the proof of Proposition 4.1.6.   □

Exercise 4.1.7. The resolvent $R(\lambda, A) = (A - \lambda)^{-1}$, $\lambda \neq \lambda_k$, is a compact operator in each space H_σ. ∎

Now we define an exponential function of A by the formula

$$e^{-tA} u = \sum_{k=1}^{\infty} e^{-\lambda_k t}(u, e_k)e_k \quad \text{for} \quad u = \sum_{k=1}^{\infty}(u, e_k)e_k. \tag{4.1.5}$$

Exercise 4.1.8. Show that $S_t = e^{-tA}$ is a strongly continuous contraction semigroup on each space H_s whose generator is $-A$. This means (see, e.g., PAZY [181]) that

(i) the linear operators S_t are contractive on H_s for each $t \geq 0$, i.e., $\|S_t\|_{H_s \mapsto H_s} \leq 1$;

(ii) S_t satisfies the semigroup property

$$S_0 = I, \quad S_{t+\tau} = S_t S_\tau \quad \text{for all} \ \ t, \tau \in \mathbb{R}_+;$$

(iii) any semitrajectory $t \mapsto S_t x$ is strongly continuous from $\mathbb{R}_+$ into H_s for every $x \in H_s$; and

(iv) the function $t \mapsto S_t x$ is strongly differentiable in H_s for every $x \in H_{1+s}$ and $\frac{d}{dt} S_t x \big|_{t=0} = -Ax$. The latter property is equivalent to the relation

$$\lim_{t \to +0} \left\| \frac{1}{t}(S_t - I)x + Ax \right\|_s = 0, \quad \forall x \in H_{s+1}.$$

For $t > 0$ we also have that $e^{-tA} H_\sigma \subset \cap_{s>0} H_s$ for every $\sigma \in \mathbb{R}$ and e^{-tA} is a compact operator on each H_σ. ∎

The following assertion describes the decay rates of e^{-tA}.

Proposition 4.1.9. *Let* $Q_N = I - P_N$, *where* P_N *is the projector defined in* (4.1.3). *We also suppose that* $Q_0 = I$. *Then for every* $\alpha \geq 0$ *the following inequality:*

$$\|A^\alpha Q_N e^{-tA} u\| \leq \left[\left(\frac{\alpha}{t} \right)^\alpha + \lambda_{N+1}^\alpha \right] e^{-t\lambda_{N+1}} \|Q_N u\|, \quad u \in H, \ N = 0, 1, \ldots, \tag{4.1.6}$$

holds. In the case $\alpha = 0$ *we suppose that* $0^0 = 0$.

Proof. One can see that

$$\|A^\alpha Q_N e^{-tA} u\|^2 = \sum_{k=N+1}^{\infty} \lambda_k^{2\alpha} e^{-2\lambda_k t} |(u, e_k)|^2 \leq \sup_{k \geq N+1} \left[\lambda_k^{2\alpha} e^{-2\lambda_k t} \right] \sum_{k=N+1}^{\infty} |(u, e_k)|^2.$$

This yields that

$$\|A^\alpha Q_N e^{-tA} u\| \leq \frac{1}{t^\alpha} \sup_{\mu \geq t\lambda_{N+1}} \left[\mu^\alpha e^{-\mu} \right] \|Q_N u\|.$$

Since the maximum of the function $\mu^\alpha e^{-\mu}$ on $\mathbb{R}_+$ is $\alpha^\alpha e^{-\alpha}$ and is attained when $\mu = \alpha$, we can see that

$$
\sup_{\mu \geq t\lambda_{N+1}} [\mu^\alpha e^{-\mu}] = \begin{cases} (t\lambda_{N+1})^\alpha e^{-\lambda_{N+1}t}, & \text{if } t\lambda_{N+1} \geq \alpha; \\ \alpha^\alpha e^{-\alpha}, & \text{if } t\lambda_{N+1} < \alpha, \end{cases}
$$

$$
\leq [\alpha^\alpha + (t\lambda_{N+1})^\alpha] e^{-\lambda_{N+1}t}.
$$

This implies (4.1.6).   □

The estimate in (4.1.6) with $N = 0$ implies uniform stability of e^{-tA} in each space H_s with the estimate

$$
\|A^\alpha e^{-tA}\| \leq \left[\left(\frac{\alpha}{t}\right)^\alpha + \lambda_1^\alpha \right] e^{-t\lambda_1}, \quad t > 0, \ \forall \alpha > 0. \tag{4.1.7}
$$

Here and below we denote the operator norm in the space H by $\|\cdot\|$ (if there is no confusion) or $\|\cdot\|_{H \mapsto H}$.

We also have that

$$
\|A^{-\beta}(1 - e^{-tA})\| \leq \sup_{\lambda > 0} \frac{1 - e^{-t\lambda}}{\lambda^\beta} \leq \sup_{\lambda > 0} \frac{\min\{1, t\lambda\}}{\lambda^\beta}
$$

$$
= t^\beta \sup_{\lambda > 0} \frac{\min\{1, \lambda\}}{\lambda^\beta} = t^\beta, \quad t > 0, \ \forall \beta \in [0, 1]. \tag{4.1.8}
$$

This means that the semigroup e^{-tA} is strongly continuous in H and uniformly continuous (even Hölder) as a mapping from H into $H_{-\beta}$ for $\beta > 0$.

In some calculations (on finite time intervals near zero) it is convenient to use another version of (4.1.7), which we show in the following exercise.

Exercise 4.1.10. Show that

$$
\|A^\alpha e^{-tA}\| \leq \left(\frac{\alpha}{et}\right)^\alpha, \quad t > 0, \tag{4.1.9}
$$

where $\alpha \geq 0$ is arbitrary (with the rule that $0^0 = 1$).   ∎

Below we also use the following inequality.

Exercise 4.1.11. Let $0 < \alpha < 1$. Using Proposition 4.1.9, show that

$$
\int_0^t \|Q_N A^\alpha e^{-(t-\tau)A} f\| d\tau \leq \frac{1 + \varkappa_\alpha}{\lambda_{N+1}^{1-\alpha}} \|f\|, \quad t > 0, \tag{4.1.10}
$$

where

$$\varkappa_\alpha = \alpha^\alpha \int_0^\infty \xi^{-\alpha} e^{-\xi} \, d\tau. \tag{4.1.11}$$

Hint: After applying (4.1.6) change the integration variable τ into $\xi = (t - \tau)\lambda_{N+1}$.

■

4.2 Well-posedness of semilinear parabolic equations

As we already mentioned, our main point of interest in this chapter is an abstract parabolic problem of the form

$$u_t + Au = B(u), \quad u\big|_{t=0} = u_0 \in H, \tag{4.2.1}$$

Our basic requirements concerning this model are listed below.

Assumption 4.2.1 (Basic hypotheses). We assume that H is a separable Hilbert space and

(A) A is a positive self-adjoint operator with discrete spectrum on H (see Definition 4.1.1);

(B) B is a (nonlinear) locally Lipschitz mapping from $H_\alpha = \mathscr{D}(A^\alpha)$ into H with some $0 \le \alpha < 1$; i.e., we assume that for every $\rho > 0$ there exists L_ρ such that

$$\|B(u_1) - B(u_2)\| \le L_\rho \|A^\alpha(u_1 - u_2)\|, \quad u_i \in H_\alpha, \quad \|A^\alpha u_i\| \le \rho, \quad i = 1, 2. \tag{4.2.2}$$

The requirements in Assumption 4.2.1 are used to show that equation (4.2.1) generates a dynamical system with several important compactness properties. When we are interested in solvability and/or uniqueness only, the hypotheses in Assumption 4.2.1 can be relaxed. For instance, using the result presented in Chapter 6 we can avoid the Lipschitz condition (4.2.2) in the local existence part. Moreover, we can also show that the uniqueness is a *generic* property for non-Lipschitz (but continuous) nonlinearities (see Section A.6 in the Appendix).

We already know that under Assumption 4.2.1(A) the operator $-A$ generates a strongly continuous compact semigroup e^{-tA} of contractions (see Exercise 4.1.8 in Section 4.1). This allows us, similar to HENRY [123], to introduce the following definition.

Definition 4.2.2 (Mild solution). A function[1] $u \in C([0, T); H_\alpha)$ is said to be a *mild solution* to (4.2.1) on an interval $[0, T)$ if

---

[1] We refer to Section A.3 in the Appendix for the description of the basic vector-valued functional spaces which we use below.

$$u(t) = e^{-tA}u_0 + \int_0^t e^{-(t-\tau)A}B(u(\tau))\,d\tau, \quad t \in [0, T), \tag{4.2.3}$$

and similarly for the closed interval $[0, T]$.   ∎

4.2.1 Basic well-posedness theorem

Our main result in this section is the following theorem.

Theorem 4.2.3 (Well-posedness). *Let Assumption* 4.2.1 *be in force. Then*

- *Local solutions: For each element $u_0 \in H_\alpha$ we can find $t_{max} \leq \infty$ such that there is a unique mild solution $u(t)$ to problem (4.2.1) defined on $[0, t_{max})$.*
- *We have $\lim_{t \to t_{max}} \| u(t) \|_\alpha = \infty$, provided $t_{max} < \infty$.*
- *Any two mild solutions $u_1(t)$ and $u_2(t)$ with initial data u_{10} and u_{20} on the joint interval $[0, T]$ of existence admit the estimate*

$$\| u_1(t) - u_2(t) \|_\alpha \leq C_T(R) \| u_{10} - u_{20} \|_\alpha, \quad t \in [0, T], \tag{4.2.4}$$

under the condition $\sup_{[0,T]} \|u_i(t)\|_\alpha \leq R$, $i = 1, 2$.
- **Global solutions:** *Let $u(t)$ be a mild solution with some initial data u_0. Assume that there exist $T > 0$ and $C_T(u_0)$ such that*

$$\sup_{[0,T^*)} \|u(t)\|_\alpha \leq C_T(u_0) \tag{4.2.5}$$

for every interval $[0, T^*) \subset [0, T]$ of the existence of the solution $u(t)$. Then the solution $u(t)$ exists on $[0, T]$. Moreover, if for any $T > 0$ there exists $C_T(u_0)$ such that (4.2.5) holds, then the solution $u(t)$ can be extended on $[0, +\infty)$; that is, $t_{max} = \infty$.
- *If $B(u)$ is globally Lipschitz, i.e., (4.2.2) is satisfied with $L_\rho \equiv L$ independent of ρ, then for every $u_0 \in H_\alpha$ there exists a unique mild solution to (4.2.1) for every interval $[0, T]$. In this case (4.2.4) can be written in the form*

$$\| u_1(t) - u_2(t) \|_\alpha \leq C e^{\omega t} \| u_{10} - u_{20} \|_\alpha, \quad t > 0, \tag{4.2.6}$$

for every pair of mild solutions $u_1(t)$ and $u_2(t)$, where $C > 0$ and $\omega \geq 0$ are constants.

In the case $\alpha = 0$ the statement of Theorem 4.2.3 can be easily derived from the results presented in PAZY [181, p. 184]. The case $\alpha > 0$ requires some modifications.

Proof.   *Step 1: B is globally Lipschitz.* In this case we apply the standard fixed point argument in a slightly modified form in comparison with PAZY [181]. Here we briefly describe this argument.

Let $W_T = C([0, T], H_\alpha)$ endowed with the corresponding sup-norm. For a given $u_0 \in H_\alpha$ we define a mapping $\mathscr{B}$ on W_T by the formula

$$\mathscr{B}[u](t) = e^{-At}u_0 + \int_0^t e^{-(t-\tau)A} B(u(\tau)) \, d\tau, \quad t \in [0, T].$$

Using the relation (see Exercise 4.1.10)

$$\|A^\alpha e^{-tA}\| \leq \left(\frac{\alpha}{et}\right)^\alpha, \quad t > 0, \quad \alpha > 0, \tag{4.2.7}$$

where we suppose that $0^0 = 1$, and also the estimate (see (4.1.8))

$$\|A^{-\beta}(1 - e^{-tA})\| \leq t^\beta, \quad t > 0, \quad \forall \beta \in [0, 1], \tag{4.2.8}$$

one can see that $\mathscr{B}[u] \in W_T$ and thus $\mathscr{B} : W_T \mapsto W_T$. Moreover, using (4.2.7) we have that

$$\|\mathscr{B}[u](t) - \mathscr{B}[v](t)\|_\alpha \leq L \left(\frac{\alpha}{e}\right)^\alpha T^{1-\alpha}(1 - \alpha)^{-1} \|u - v\|_{W_T}, \quad t \in [0, T],$$

and thus

$$\|\mathscr{B}[u] - \mathscr{B}[v]\|_{W_T} \leq q(T) \|u - v\|_{W_T} \text{ with } q(T) \equiv L \left(\frac{\alpha}{e}\right)^\alpha T^{1-\alpha}(1 - \alpha)^{-1}.$$

Taking T such that $q(T) < 1$, we obtain the existence and uniqueness of a fixed point for $\mathscr{B}$. This implies the local existence result with the time interval *independent* of the initial data. This allows us to use the standard step-by-step procedure to construct a global solution.

Step 2: Locally Lipschitz case. We define a truncating operator π_R on H_α by the formula

$$\pi_R(x) = \begin{cases} x, & \text{if } \|x\|_\alpha \leq R; \\ R\,x\,\|x\|_\alpha^{-1}, & \text{if } \|x\|_\alpha > R, \end{cases} \tag{4.2.9}$$

for $R > 0$ and introduce a (truncated) function $B_R(u)$ by the relation

$$B_R(u) = B(\pi_R(u)), \quad u \in H_\alpha.$$

One can see that $B_R(u) = B(u)$ for $\|u\|_\alpha \leq R$. According to Lemma A.4.1 from the Appendix applied in the Hilbert space H_α, the mapping B_R is globally Lipschitz for every R and the Lipschitz constant is L_R. Thus, we can guarantee that for any $u_0 \in H_\alpha$ the equation

$$u_t + Au = B_R(u) \tag{4.2.10}$$

has a unique mild solution u on $\mathbb{R}_+$. By (4.2.3) and (4.2.7) we have that

$$\|u(t)\|_\alpha \leq \|u_0\|_\alpha + c_\alpha T^{1-\alpha} \left[\|B(0)\| + L_R \max_{[0,T]} \|u(\tau))\|_\alpha \right], \quad t \in [0, T].$$

Thus,

$$\|u\|_{W_T} \leq \left[\|u_0\|_\alpha + c_\alpha T^{1-\alpha} \|B(0)\| \right] (1 - c_\alpha L_R T^{1-\alpha})^{-1}, \tag{4.2.11}$$

provided $c_\alpha L_R T^{1-\alpha} < 1$. Let $u_0 \in H_\alpha$ be arbitrary. We choose R such that $\|u_0\|_\alpha < R$ and assume for now that u is a solution to (4.2.10). Due to (4.2.11) we can choose T^* such that $\|u(t)\| \leq R$ for $t \leq T_*$. Consequently, $B_R(u(t)) = B(u(t))$ for $t \leq T^*$, and this means that $u(t)$ solves equation (4.2.1) on that interval.

Step 3: Lipschitz continuity. To prove the Lipschitz properties in (4.2.4) and (4.2.6), we note that (4.2.3) and also (4.2.2) and (4.2.7) imply that

$$\|u_1(t) - u_2(t))\|_\alpha \leq \|u_{10} - u_{20}\|_\alpha + C_{R,\alpha} \int_0^t \frac{1}{(t-\tau)^\alpha} \|u_1(\tau)) - u_2(\tau)\|_\alpha \, d\tau.$$

Therefore, the Henry-Gronwall lemma (see Lemma A.2.3 in the Appendix) yields

$$\|u_1(t) - u_2(t))\|_\alpha \leq 2e^{a(R,\alpha)t} \|u_{10} - u_{20}\|_\alpha, \quad \forall t \in [0, T],$$

where $a(R, \alpha)$ is a positive constant. This implies (4.2.4) and also (4.2.6), because in the latter case $C_{R,\alpha}$ and hence $a(R, \alpha)$ do not depend on R.

Step 4: Maximal existence interval. It is clear from the argument in Step 2 that if u is a solution to (4.2.3) on $[0, t^*]$, it can be extended to a solution on the interval $[0, t^* + \delta]$ for some $\delta > 0$. For this we use the same method as outlined above with the initial value t^* and with a larger R. Of course, δ depends on upper bounds for $\|u(t^*)\|_\alpha$.

Let $[0, t_{\max})$ be the maximal interval of existence of the solution. If $t_{\max} < \infty$, then $\lim_{t \nearrow t_{\max}} \|u(t)\| = \infty$. Otherwise there exists a sequence $t_n \nearrow t_{\max}$ such that $\|u(t_n)\|_\alpha \leq C$. This would allow us to extend u as a solution to (4.2.3) to an interval $[0, t_n + \delta]$ with $\delta > 0$ independent of n. Hence u can be extended beyond $t_{\max}$, which contradicts the construction of $t_{\max}$.

To prove the global existence part we need only note that (4.2.5) allows us to extend a solution inside the interval $[0, T]$ with time steps independent of $T^* < T$. This completes the proof of Theorem 4.2.3.   □

As we see from Theorem 4.2.3, to guarantee the global existence and uniqueness of mild solutions it is sufficient to assume that B is *globally* Lipschitz. The following exercise shows that this hypothesis can be slightly relaxed by assuming that B is *locally* Lipschitz but linearly bounded.

Exercise 4.2.4. Let Assumption 4.2.1 be in force. Assume that $B(u)$ is linearly bounded, i.e.,

$$\|B(u)\| \leq C_1 + C_2\|A^\alpha u\| \text{ for all } u \in H_\alpha.$$

Then problem (4.2.1) has a unique mild solution for every interval $[0, T]$ and for any initial data. Hint: Use the non-explosion criterion in (4.2.5) and also the Henry-Gronwall lemma (see Lemma A.2.3 in the Appendix). ∎

Remark 4.2.5. Theorem 4.2.3 can also be applied in the situation when, instead of (4.2.2), we assume that B is a locally Lipschitz mapping from $H_{s+\alpha}$ into H_s with some $s \in \mathbb{R}$ and $0 \leq \alpha < 1$ and for every $\rho > 0$ there exists L_ρ such that

$$\|B(u_1)-B(u_2)\|_s \leq L_\rho\|u_1-u_2\|_{s+\alpha}, \ u_i \in H_\alpha, \ \|u_i\|_{s+\alpha} \leq \rho, \ i = 1, 2. \quad (4.2.12)$$

Indeed, in this case we can consider the problem in the space $\hat{H} = H_s$. The semi-group e^{-At} possesses the same properties in each space H_s, and thus Theorem 4.2.3 gives a result on well-posedness of problem (4.2.1) in the space $C([0, T], H_{s+\alpha})$.

We also note that if (4.2.12) holds for some $\alpha = \alpha_*$, then the same relation is true for all $\alpha \in [\alpha_*, 1)$. Thus, using Theorem 4.2.3 we can conclude that a solution from $C([0, T], H_{s+\alpha_*})$ belongs to $C([0, T], H_{s+\alpha})$ for every $\alpha \in [\alpha_*, 1)$ provided $u_0 \in H_\alpha$. Below we will see that this conclusion remains true even for $u_0 \in H_{\alpha_*}$ but on the intervals $[\delta, T]$ with $\delta > 0$.

Finally we note that we can avoid the requirements of the discrete spectrum for A by assuming that A is a positive self-adjoint operator only. ∎

Exercise 4.2.6. Let $u(t)$ be a mild solution. Using the integral relation in (4.2.3) and the smoothening estimate in (4.2.7), show that $u(t) \in C((0, T], H_\beta)$ for every $\beta \in [\alpha, 1)$ and

$$\|u(t)\|_\beta \leq C_{\alpha,\beta}\left[\frac{1}{t^{\beta-\alpha}}\|u_0\|_\alpha + m_R\right], \quad t \in (0, T],$$

for every $t \in (0, T]$ and some constant m_R under the condition that $\|u(t)\|_\alpha \leq R$. ∎

Exercise 4.2.7. Using the same idea as in Exercise 4.2.6, show that the Lipschitz property in (4.2.4) can be improved in the following way: for every $\beta \in [\alpha, 1)$ and every pair $u_1(t)$ $u_2(t)$ of mild solutions on $[0, T]$ such that $\|u_j(t)\|_\alpha \leq R$ for all $t \in [0, T]$, we have the estimate

$$\|u_1(t) - u_2(t)\|_\beta \leq C_{\alpha,\beta}\left[a_R(T) + \frac{1}{t^{\beta-\alpha}}\right]\|u_{10} - u_{20}\|_\alpha, \quad t \in (0, T],$$

with positive constants $C_{\alpha,\beta}$ and $a_R(T)$. ∎

Exercise 4.2.8. Using (4.2.7) show that every mild solution $u(t)$ to (4.2.1) satisfies the following inequality:

$$\|u(t_1) - u(t_2)\|_\beta \le \left\|\left(1 - e^{-|t_1 - t_2|A}\right)u_0\right\|_\beta + C_{\alpha,\beta}m_R|t_1 - t_2|^{1-\beta}, \quad t_1, t_2 \in (0, T],$$

for every $\beta \in [0, 1)$ under the condition that $\|u(t)\|_\alpha \le R$ for every $t \in (0, T]$. In particular, by (4.2.8) this implies that $u(t)$ is Hölder continuous in each space H_θ with $\theta < \alpha$ and in H_α if $u_0 \in H_\gamma$ with $\gamma > \alpha$.   ∎

Proposition 4.2.9. *Let Assumption 4.2.1 be in force. Then any mild solution $u(t)$ is also weak; i.e., it satisfies the relation*

$$(u(t), v) = (u_0, v) - \int_0^t (u(\tau), Av)\, d\tau + \int_0^t (B(u(\tau)), v)\, d\tau, \quad v \in \mathscr{D}(A), \qquad (4.2.13)$$

for t from every existence interval $[0, T]$. Moreover, we have that

- *$u \in L_2(0, T; H_{1/2})$,*
- *$u(t)$ is absolutely continuous in $H_{-1/2}$,*
- *$u(t)$ satisfies (4.2.1) as an equality in $H_{-1/2}$ for almost $t \in [0, T]$,*
- *the following energy balance relation holds:*

$$\frac{1}{2}\|u(t)\|^2 + \int_0^t \|u(\tau)\|_{1/2}^2 d\tau = \frac{1}{2}\|u_0\|^2 + \int_0^t (B(u(\tau)), u(\tau))\, d\tau. \qquad (4.2.14)$$

Proof. Let P_N be the orthoprojector onto Span $\{e_1, \ldots, e_N\}$. Then $u_N = P_N u(t)$ satisfies the integral equation

$$u_N(t) = e^{-tA}P_N u_0 + \int_0^t e^{-(t-\tau)A}f_N(\tau)\, d\tau \quad \text{with } f_N(\tau) = P_N B(u(\tau))$$

in the finite-dimensional space $P_N H$. Thus, the standard finite-dimensional calculations after the limit transition lead to the desired results.   □

Exercise 4.2.10. Show that under the conditions of Proposition 4.2.9 any mild solution possesses the property $u_t \in C((0, T); H_{-\delta})$ for every $\delta > 0$. Hint: Use equation (4.2.1) and apply the result of Exercise 4.2.6.   ∎

4.2.2   A case study: Henry-Miklavčič model

To demonstrate different issues related to the well-posedness of mild solutions, we consider an abstract version of the model considered in HENRY [123] (see also CHUESHOV [39, Chapter 2]).

Let A satisfy Assumption 4.2.1. Consider the problem

$$u_t + Au = -\mu \|A^\alpha u\|^m u, \quad u\big|_{t=0} = u_0, \tag{4.2.15}$$

where $0 \leq \alpha < 1$, $m \geq 1$, and $\mu \in \mathbb{R}$. We note that a one-dimensional version of this model was considered in Exercise 1.7.19.

Exercise 4.2.11. Make sure that problem (4.2.15) satisfies the Lipschitz condition in (4.2.2) with $L_\rho = |\mu| \lambda_1^{-\alpha} (1 + m) \rho^m$.   ∎

The model in (4.2.15) is exactly solvable. Indeed, any mild solution $u(t)$ can be seen as a solution to the linear equation

$$u_t + Au + f(t)u = 0, \quad \text{with } f(t) = \mu \|A^\alpha u(t)\|^m. \tag{4.2.16}$$

Thus, we can write

$$u(t) = \exp\left\{ -tA - \int_0^t f(\tau)\, d\tau \right\} u_0,$$

which implies that

$$\|A^\alpha e^{-tA} u_0\|^m = \|A^\alpha u(t)\|^m \exp\left\{ \mu m \int_0^t \|A^\alpha u(\tau)\|^m\, d\tau \right\}$$

$$= \frac{1}{m\mu} \frac{d}{dt} \exp\left\{ \mu m \int_0^t \|A^\alpha u(\tau)\|^m\, d\tau \right\}.$$

Therefore, after integration we obtain

$$\exp\left\{ -\mu \int_0^t \|A^\alpha u(\tau)\|^m\, d\tau \right\} = \frac{1}{\left[1 + \mu m \int_0^t \|A^\alpha e^{-\tau A} u_0\|^m\, d\tau \right]^{1/m}}$$

and thus arrive at the following explicit formula for a mild solution:

$$u(t) = \frac{\exp\{-At\}\, u_0}{\left[1 + \mu m \int_0^t \|A^\alpha e^{-\tau A} u_0\|^m\, d\tau \right]^{1/m}} \tag{4.2.17}$$

This formula allows us to state explosion criteria for solutions in the system. We recall that explosion of a local solution $u(t)$ means that $u(t)$ is defined on some interval $[0, T)$ and blows up at the end of the interval, i.e., $\|u(t)\|_\alpha \to +\infty$ as $t \to T$ with $t < T$.

- **Case** $\mu \geq 0$: There is no explosion; a mild solution exists globally for every initial data $u_0 \in H_\alpha$.
- **Case** $\mu < 0$: In this case a solution $u(t)$ with initial data u_0 blows up if and only if

$$\int_0^{+\infty} \|A^\alpha e^{-\tau A} u_0\|^m \, d\tau > \frac{1}{|\mu| m}. \qquad (4.2.18)$$

In particular, this means that solutions with small initial data exist globally and solutions with large u_0 blow up. This statement is confirmed by the following exercises.

Exercise 4.2.12. Let $\mu < 0$. Show that solutions to (4.2.15) exist globally under the condition $\|u_0\|_\alpha \leq (\lambda_1 |\mu|^{-1})^{1/m}$. Hint: We know that $\|e^{-tA} u_0\|_\alpha \leq e^{-\lambda_1 t} \|u_0\|_\alpha$. ∎

Exercise 4.2.13. Let $\mu < 0$. Show that for every $v_0 \in H_\alpha$ there exists a positive number $\eta_* = \eta(v_0, \alpha, \mu, m)$ such that the solution to (4.2.15) with $u_0 = \eta v_0$ exists globally when $\eta \leq \eta_*$ and this is not true when $\eta > \eta_*$. Calculate this critical barrier η_*. ∎

Exercise 4.2.14. Show that in the case $m = 2$ the explosion criterion in (4.2.18) can be written as $\|A^{\alpha-1/2} u_0\| > 1/\sqrt{|\mu|}$. ∎

We note that the formula in (4.2.17) can also be used to construct solutions in the case when $u_0 \notin H_\alpha$. Indeed, let $u_0 \in H_\beta$ for some $0 \leq \beta < \alpha$ such that $(\alpha-\beta)m < 1$. In this case,

$$\int_0^t \|A^\alpha e^{-\tau A} u_0\|^m \, d\tau \leq C_{\alpha,\beta} \|u_0\|_\beta^m \int_0^t \left(\frac{1}{\tau}\right)^{(\alpha-\beta)m} d\tau = C_{\alpha,\beta} \|u_0\|_\beta^m \frac{t^{1-(\alpha-\beta)m}}{1-(\alpha-\beta)m}.$$

Thus, for every $\mu \in \mathbb{R}$ there exists $T_\mu \leq +\infty$ such that the function $u(t)$ given by (4.2.17) possesses the property

$$u \in C([0, T_\mu), H_\beta) \cap C^\infty((0, T_\mu), H_s), \quad \forall s \in \mathbb{R}.$$

The behavior of this function near zero is described in the following exercise.

Exercise 4.2.15. Show that $\limsup_{t \to +0} \{t^{\alpha-\beta} \|A^\alpha u(t)\|\} < +\infty$. ∎

Thus, one can show that the integral in (4.2.3) for this u exists, and thus $u(t)$ satisfies integral equation (4.2.3) with $B(u) = -\mu \|A^\alpha u\|^m u$ on the interval $[0, T_\mu)$. Hence we have a mild solution for the data $u_0 \notin H_\alpha$. Moreover, $u(t)$ satisfies (4.2.15) in the classical sense.

In the case $\mu < 0$ we can observe the non-uniqueness effect when the initial data do not belong to H_α. Indeed, following an idea presented in HENRY [123], it is easy to show that the function

$$u(t) = \exp\left\{\int_t^1 f(\tau)\,d\tau\right\} v(t) \quad \text{with} \quad v(t) = e^{-At}v_0$$

solves the differential equation in (4.2.15) on the interval $(0, 1)$. Here $f(t)$ is the same as in (4.2.16). Thus, by the same calculations as used above, one can see that any solution $v(t)$ to the equation $v_t + Av = 0$ generates a solution to (4.2.15) on the interval $(0, 1)$ by the formula

$$u(t) = \frac{v(t)}{\left[1 + |\mu|m \int_t^1 \|A^\alpha v(\tau)\|^m \, d\tau\right]^{1/m}}. \tag{4.2.19}$$

We take $v(t) = e^{-tA}v_0$, where $v_0 \in H$. In this case, $u \in C^\infty((0, 1], H_s)$ for all $s \in \mathbb{R}$. Assume that

$$\liminf_{t \to +0}\left\{t^\beta \|A^\alpha e^{-tA}v_0\|\right\} > 0 \quad \text{for some } \beta > 1/m. \tag{4.2.20}$$

Then we have

$$\liminf_{t \to +0}\left\{t^{m\beta-1}\int_t^1 \|A^\alpha e^{-\tau A}v_0\|^m \, d\tau\right\} > 0.$$

This implies $u(t) \to 0$ in H as $t \to +0$. Thus, the function u given by (4.2.19) possesses the property

$$u \in \mathcal{L} \equiv C([0, 1], H) \cap C^\infty((0, 1), \mathcal{D}(A))$$

and solves (4.2.15) with the zero initial data. Hence, under the condition (4.2.20), we have non-uniqueness of mild solutions to (4.2.15) in the class $\mathcal{L}$.

We note that the requirement in (4.2.20) can be satisfied if, for instance, we assume that the spectrum of the operator A has the form $\lambda_k = k^2$, $k = 1, 2, \ldots$ (this corresponds to the spectrum of the 1D Laplace operator). Indeed, let $\rho > 1/2$; then

$$v_0 = \sum_{k=1}^{\infty} \frac{1}{k^\rho} e_k \in H.$$

In this case with $4\alpha - 2\rho > -1$ we have that

$$\|A^\alpha e^{-tA}v_0\|^2 = \sum_{k=1}^{\infty} k^{4\alpha-2\rho}e^{-k^2t} \geq c_0\int_1^\infty \xi^{4\alpha-2\rho}e^{-\xi^2t}\,d\xi$$

$$= c_0 \left(\frac{1}{\sqrt{t}} \right)^{4\alpha - 2\rho + 1} \int_{\sqrt{t}}^{\infty} s^{4\alpha - 2\rho} e^{-s^2} \, d\xi \geq c_1 \left(\frac{1}{t} \right)^{2\alpha - \rho + 1/2}.$$

Thus, we obtain (4.2.20) with $\beta = \alpha - (\rho - 1/2)/2$. Thus, in the case when $\alpha m > 1$ we can choose $\rho > 1/2$ such that (4.2.20) holds.

4.2.3   Galerkin method

Well-posedness issues for the problem in (4.2.1) can also be studied with the help of the Galerkin method.

Let P_N be the orthoprojector in H onto Span$\{e_1, \ldots, e_N\}$. A Galerkin approximate solution to (4.2.1) of order N with respect to the basis $\{e_k\}$ is defined as a continuously differentiable function

$$u^N(t) = \sum_{k=1}^{N} g_k(t) e_k,$$

with values in the finite-dimensional space $P_N H$ which satisfies the equations

$$u_t^N + A u^N = P_N B(u^N), \quad u^N \big|_{t=0} = P_N u_0 \in H. \tag{4.2.21}$$

In the exercises below we assume that Assumption 4.2.1 is in force.

Exercise 4.2.16. Show that (4.2.21) can be rewritten as a system of N ordinary differential equations for the functions $g_k(t)$. Using Theorem A.1.2 make sure that for each N the problem in (4.2.21) is locally well-posed.   ∎

Exercise 4.2.17. Show that problem (4.2.21) is equivalent to the problem of finding a continuous function u^N with values in $P_N H$ satisfying the integral equation

$$u^N(t) = e^{-tA} P_N u_0 + \int_0^t e^{-(t-\tau)A} P_N B(u^N(\tau)) \, d\tau. \tag{4.2.22}$$

∎

Exercise 4.2.18. Let $u_0 \in H_\alpha$. Using the same method as in Theorem 4.2.3, prove the local solvability well-posedness of problem (4.2.22) on a segment $[0, T]$, where the parameter T can be chosen to be independent of N. Moreover, the following uniform estimate is valid:

$$\sup_N \max_{t \in [0,T]} \| u^N(t) \|_\alpha \leq R \quad \text{for some } R > 0. \tag{4.2.23}$$

The parameter R depends on $\|u_0\|_\alpha$.   ∎

Theorem 4.2.19. *Let Assumption 4.2.1 be in force and $u_0 \in H_\alpha$. Assume that a sequence of approximate solutions u^N admits estimate (4.2.23) on some interval $[0, T]$. Then there exists a mild solution $u(t)$ to problem (4.2.1) on the segment $[0, T]$ and*

$$\max_{t \in [0,T]} \|u(t) - u^N(t)\|_\alpha \leq C \left(\|(1 - P_N)u_0\| + \frac{1}{\lambda_{N+1}^{1-\alpha}} \right), \tag{4.2.24}$$

where $C = C(T, R, \alpha) > 0$ is a constant independent of N.

Proof. It follows from (4.2.21) that

$$u^N(t) = e^{-(t-s)A} u^N(s) + \int_s^t e^{-(t-\tau)A} P_N B(u^N(\tau)) \, d\tau, \quad N = 1, 2, \ldots.$$

Therefore, using Proposition 4.1.9 and also estimate (4.2.7), we can see that for $N < M$:

$$\|u^M(t) - u^N(t)\|_\alpha \leq \|u^M(s) - u^N(s)\|_\alpha$$
$$+ \int_s^t \left[\left(\frac{\alpha}{t-\tau} \right)^\alpha + \lambda_{N+1}^\alpha \right] e^{-(t-\tau)\lambda_{N+1}} \|B(u^N(\tau))\| \, d\tau$$
$$+ \int_s^t \left(\frac{\alpha}{e(t-\tau)} \right)^\alpha \|B(u^N(\tau)) - B(u^M(\tau))\| \, d\tau$$

for all $0 \leq s < t \leq T$. This implies that

$$\|u^M(t) - u^N(t)\|_\alpha \leq \|u^M(s) - u^N(s)\|_\alpha + M(R)J_N(t, s)$$
$$+ L_R \left(\frac{\alpha}{e} \right)^\alpha \frac{|t - s|^{1-\alpha}}{1 - \alpha} \max_{\tau \in [s,t]} \|u^N(\tau) - u^M(\tau)\|_\alpha \tag{4.2.25}$$

for all $0 \leq s < t \leq T$, where $M(R) = \|B(0)\| + L_R R$ and

$$J_N(t, s) = \int_s^t \left[\left(\frac{\alpha}{t-\tau} \right)^\alpha + \lambda_{N+1}^\alpha \right] e^{-(t-\tau)\lambda_{N+1}} \, d\tau, \quad t > s. \tag{4.2.26}$$

Introducing the new variable $\xi = (t - \tau)\lambda_{M+1}$, one can see that

$$J_N(t, s) \leq J_N(t, -\infty) = \lambda_{N+1}^{-1+\alpha} \left[1 + \alpha^\alpha \int_0^\infty \xi^{-\alpha} e^{-\xi} \, d\xi \right] \equiv \lambda_{M+1}^{-1+\alpha} [1 + \varkappa_\alpha]. \tag{4.2.27}$$

Thus, it follows from (4.2.25) that there exists $T_* = T_*(R, \alpha) \leq T$ such that

$$\max_{t \in [s, s+T_*]} \|u^M(t) - u^N(t)\|_\alpha \leq 2\|u^M(s) - u^N(s)\|_\alpha + C(R, \alpha)\lambda_{N+1}^{-1+\alpha}$$

for all $s \in [0, T)$ such that $s + T_* \leq T$. Hence the step-by-step procedure leads to the relation

$$\max_{t \in [0,T]} \|u^M(t) - u^N(t)\|_\alpha \leq C \left(\|(P_M - P_N)u_0\| + \lambda_{N+1}^{-(1-\alpha)} \right)$$

for all $M > N$ and for any interval $[0, T]$ for which (4.2.23) holds. Thus, there exists a function $u \in C(0, T; H_\alpha)$ such that (4.2.24) holds. It is easy to see that this u is a mild solution to (4.2.1).   □

4.2.4 Nonlinearity with the main potential part

Now we consider a case in which we can guarantee the global solvability of problem (4.2.1). We impose the following hypothesis.

Assumption 4.2.20. Assumption 4.2.1 holds with $\alpha = 1/2$, i.e.,

 (i) A is a positive self-adjoint operator with discrete spectrum on H (see Definition 4.1.1);
 (ii) B is a locally Lipschitz mapping from $H_{1/2} = \mathscr{D}(A^{1/2})$ into H such that for every $\rho > 0$ there exists L_ρ such that

$$\|B(u_1) - B(u_2)\| \leq L_\rho \|u_1 - u_2\|_{1/2}, \quad u_i \in H_{1/2}, \quad \|u_i\|_{1/2} \leq \rho, \quad i = 1, 2. \tag{4.2.28}$$

Assume in addition that

$$B(u) = -B_0(u) + B_1(u),$$

where $B_1 : H_{1/2} \mapsto H$ is continuous and linearly bounded, i.e.,

$$\|B_1(u)\| \leq c_1 + c_2 \|u\|_{1/2}, \quad u \in H_{1/2}, \tag{4.2.29}$$

and $B_0 : H_{1/2} \mapsto H$ is a potential operator on the space $H_{1/2}$. This means (see Section A.5 in the Appendix) that there exists a Frechét differentiable functional $\Pi(u)$ on $H_{1/2}$ such that $B_0(u) = \Pi'(u)$, i.e.,

$$\lim_{\|v\|_{1/2} \to 0} \frac{1}{\|v\|_{1/2}} \left[\Pi(u + v) - \Pi(u) - (B_0(u), v) \right] = 0. \tag{4.2.30}$$

Remark 4.2.21. We note that under Assumption 4.2.20 it follows from Proposition A.5.2 in the Appendix (see also (A.5.4)) that for every $u \in C^1([a, b], H_{1/2})$ the scalar function $t \mapsto \Pi(u(t))$ is a C^1 function on $[a, b]$ and

$$\frac{d}{dt}\Pi(u(t)) = (B_0(u(t)), u_t(t)), \quad t \in [a, b]. \tag{4.2.31}$$

Moreover, the relation in (A.5.5) in this case gives us that

$$\Pi(u + v) - \Pi(u) = \int_0^1 (B_0(u + \lambda v), v)\, d\lambda \quad \text{for every } u, v \in H_{1/2}.$$

This formula with $u = 0$ yields

$$\sup\{|\Pi(v)| \; : \; \|v\|_{1/2} \le R\} < \infty \quad \text{for every } R > 0$$

under the condition that $\sup\{\|B_0(v)\| \; : \; \|v\|_{1/2} \le R\} < \infty$ for every $R > 0$. ∎

Theorem 4.2.22. *Let Assumption 4.2.20 be in force. Assume that the functional $\Pi(u)$ possesses the following property: there exist $\beta < 1/2$ and $\gamma \ge 0$ such that*

$$\beta\|A^{1/2}u\|^2 + \Pi(u) + \gamma \ge 0, \quad \forall u \in H_{1/2}. \tag{4.2.32}$$

Then for every $u_0 \in H_{1/2}$ problem (4.2.1) has a unique mild solution $u(t)$ in the space $C(\mathbb{R}_+; H_{1/2})$. Moreover, there exists the time derivative u_t which belongs to $L_2(0, T; H)$ for every $T > 0$, and the following balance relation:

$$E(u(t)) + \int_0^t \|u_t(\tau)\|^2 d\tau = E(u_0) + \int_0^t (B_1(u(\tau)), u_t(\tau))d\tau \tag{4.2.33}$$

holds for every $t > 0$. Here

$$E(u) = \frac{1}{2}\|A^{1/2}u\|^2 + \Pi(u), \quad u \in H_{1/2}.$$

Proof. To obtain the result we use Theorem 4.2.19 on approximations.
Let

$$u^N(t) = \sum_{k=1}^{N} g_k(t)e_k$$

be the Galerkin approximation of order N. Then $u^N(t)$ solves problem (4.2.21); i.e., we have that

$$u_t^N + Au^N = P_N B(u^N), \quad u^N\big|_{t=0} = P_N u_0 \in H.$$

To apply Theorem 4.2.19 we need to check (4.2.23). To do this we multiply the equation above by u_t^N. This yields

$$\|u_t^N\|^2 + \frac{d}{dt}\left[\frac{1}{2}\|A^{1/2}u^N\|^2 + \Pi(u^N)\right] = (B_1(u^N), u_t^N)$$

$$\leq \frac{1}{2}\|u_t^N\|^2 + c_1^2 + c_2^2\|A^{1/2}u^N\|^2,$$

where c_1 and c_2 are the constants from (4.2.29). By (4.2.32),

$$W(u^N) := \frac{1}{2}\|A^{1/2}u^N\|^2 + \Pi(u^N) + \gamma \geq \left(\frac{1}{2} - \beta\right)\|A^{1/2}u^N\|^2.$$

It is also easy to see that

$$\frac{1}{2}\|u_t^N\|^2 + \frac{d}{dt}W(u^N) \leq a_1 W(u^N) + a_2$$

on any interval $[0, T]$ with some $a_i > 0$. This relation allows us to obtain the estimate

$$\|A^{1/2}u^N(t)\|^2 + \int_0^t \|u_t^N(\tau)\|^2 d\tau = C_T\big(1 + E(P_N u_0)\big) \leq C_{R,T} \qquad (4.2.34)$$

for every initial data $u_0 \in H_{1/2}$ such that $\|u_0\|_{1/2} \leq R$. This allows us to apply Theorem 4.2.19 and show the existence of a mild solution. Moreover, using (4.2.34) we have that $\{u_t^N\}$ is a weakly compact sequence in $L_2(0, T; H)$ for every $T > 0$. Therefore, the convergence $u^N \to u$ given by Theorem 4.2.19 implies that the generalized time derivative u_t exists and belongs to $L_2(0, T; H)$.

To prove (4.2.33), we note that from (4.2.13) in Proposition 4.2.9 it follows that $w(t) = P_N u(t)$ solves the following finite-dimensional equation:

$$w_t + Aw = f_N(t) \equiv -P_N B_0(u(t)) + P_N B_1(u(t)).$$

Thus, using the multiplier w_t we obtain

$$E(P_N u(t)) + \int_0^t \|P_N u_t(\tau)\|^2 d\tau = E(P_N u_0) + \int_0^t (B_1(u(\tau)), P_N u_t(\tau))d\tau$$

$$+ \int_0^t (B_0(P_N u(\tau)) - B_0(u(\tau)), P_N u_t(\tau))d\tau.$$

After the limit transition $N \to \infty$, we obtain (4.2.33).   □

Remark 4.2.23. By Proposition 4.2.9 the solution $u(t)$ given by Theorem 4.2.22 is also weak. Moreover, it follows from (4.2.33) that $u_t(t) \in H$ for almost all t, and the equation in (4.2.1) is satisfied as an equality in H. In particular, this allows us to see that the mild solution possesses the properties

$$u \in L_2(0, T; \mathscr{D}(A)), \quad u_t \in L_2(0, T; H), \quad \forall T > 0.$$

Following PAZY [181] it is natural to call this solution strong.

The method which leads to the a priori estimate (4.2.34) and the balance relation in (4.2.33) is very close to the approach developed in Chapter 5 for second order in time models. The multiplier u_t is crucial in both cases. We refer to Section 5.1.3 for a further discussion of this issue.   ∎

4.2.5 Semilinear heat (reaction-diffusion) equation

One of the standard applications of the results presented above is a semilinear heat equation.

In a bounded domain $\Omega \subset \mathbb{R}^d$ we consider the following problem:

$$u_t(x, t) - \Delta u(x, t) + f(u(x, t), \nabla u(x, t)) = 0 \qquad (4.2.35)$$

endowed with boundary and initial conditions

$$u\big|_{\partial\Omega} = 0, \quad u\big|_{t=0} = u_0. \qquad (4.2.36)$$

Here $f(u, \xi)$ is a function on $\mathbb{R}^{1+d}$ which is specified below.

We consider (4.2.35) in the space $H = L_2(\Omega)$ and suppose $A = -\Delta$ on the domain

$$\mathscr{D}(A) = H^2(\Omega) \cap H_0^1(\Omega) \equiv \{u \in L_2(\Omega) : \partial_{x_i x_j} u \in L_2(\Omega), \ u\big|_{\partial\Omega} = 0\},$$

where we use the notation $H^s(\Omega)$ for the Sobolev space of order s and $H_0^s(\Omega)$ denotes the closure of $C_0^\infty(\Omega)$ in $H^s(\Omega)$. It is well known that $\mathscr{D}(A^{1/2}) = H_0^1(\Omega)$. The nonlinear mapping B is defined by the relation

$$[B(u)](x) = f(u(x), \nabla u(x)), \quad u \in H_0^1(\Omega).$$

There are two important cases in which B satisfies (4.2.2) with $\alpha = 1/2$:

- $f : \mathbb{R}^{1+d} \mapsto \mathbb{R}$ is globally Lipschitz. In this case (4.2.2) holds with L_ρ independent of ρ.
- $f : \mathbb{R}^{1+d} \mapsto \mathbb{R}$ possesses a polynomially bounded main part, i.e.,

$$f(u, \xi) = -f_0(u) + f_1(u, \xi), \qquad (4.2.37)$$

where $f_1 : \mathbb{R}^{1+d} \mapsto \mathbb{R}$ is globally Lipschitz and $f_0 : \mathbb{R}^1 \mapsto \mathbb{R}$ satisfies the inequality

$$|f_0(u) - f_0(v)| \leq C(1 + |u|^r + |v|^r)|u - v|, \qquad (4.2.38)$$

where $r \in [0, +\infty)$ when $d \leq 2$ and $r \leq 2(d-2)^{-1}$ for $d \geq 3$. To prove (4.2.2) with $\alpha = 1/2$ for this case, we use the following embedding (see ADAMS [1]):

$$H^1(\Omega) \subset L_p(\Omega) \quad \text{if} \quad \begin{cases} d \leq 2, \ \forall p < \infty; \\ d > 2, \ p \leq 2d(d-2)^{-1}. \end{cases}$$

Under the conditions above, $B_0(u) = f_0(u(x))$ has a potential $\Pi(u)$ which is given by

$$\Pi(u) = \int_\Omega \left[\int_0^{u(x)} f_0(\xi) \, d\xi \right] dx, \quad u \in H_0^1(\Omega).$$

The condition in (4.2.32) is satisfied when

$$\liminf_{|s| \to \infty} \frac{f_0(s)}{s} > -\lambda_1, \tag{4.2.39}$$

where λ_1 is a first eigenvalue of the operator $-\Delta$ with the Dirichlet boundary conditions.

The conditions above can be relaxed in several directions. For instance, for $d = 3$ we know from ADAMS [1] that $H^s(\Omega) \subset L_\infty(\Omega)$ when $s > 3/2$. Thus, if $f_0 \in C^1(\mathbb{R})$, then

$$\|f_0(u_1) - f_0(u_2)\|^2_{L^2(\Omega)}$$

$$\leq \int_\Omega \left[\int_0^1 |f_0'(\xi u_1(\xi) + (1-\xi)u_2(x))| \, d\xi \right]^2 |u_1(x) - u_2(x)|^2 dx$$

$$\leq \sup \left\{ |f'(s)| \ : \ |s| \leq c_0 \left(\|u_1\|_{H^s(\Omega)} + \|u_2\|_{H^s(\Omega)} \right) \right\} \|u_1 - u_2\|^2_{L^2(\Omega)}$$

for every $s > 3/2$. This observation makes it possible to show that the corresponding operator B is locally Lipschitz from $\mathscr{D}(A^\alpha)$ into H for every $3/4 < \alpha < 1$. Thus, we can apply Theorem 4.2.3 to prove the local well-posedness of the problem in (4.2.35) and (4.2.36) without any growth requirements like (4.2.38). For more details and other examples we refer to HENRY [123] and CHUESHOV [39, Chapter 2].

In particular, we can consider a vector version of the model considered in (4.2.35) by assuming that

$$u = (u^1; \ldots; u^m), \quad f(u) = (f^1(u); \ldots; f^m(u)).$$

This situation describes a reaction-diffusion process of m reagents with concentrations $u^1, u^2, \ldots, u^m$, whose diffusion coefficients can also be different. Thus (see, e.g., HENRY [123]) we obtain the system

$$u_t^i - \nu_i \Delta u + f^i(u^1, \ldots, u^m) = 0, \quad i = 1, 2, \ldots, m. \tag{4.2.40}$$

In this case the Neumann boundary condition

$$\frac{\partial u^i}{\partial n} = 0, \quad \text{on } \partial\Omega, \quad i = 1, 2, \ldots, m, \quad (n \text{ is outward normal to } \partial\Omega), \tag{4.2.41}$$

is more natural than the Dirichlet condition in (4.2.36), because the former states the absence of flows of reagents crossing $\partial\Omega$.

We mention that the dynamics in system (4.2.40) and (4.2.41) can present rather complicated behaviors. For instance, every bifurcation picture discussed in Section 1.9 can be detected in some reaction-diffusion system of the form (4.2.40) and (4.2.41). The point is that the subspace $\mathscr{L}_0$ of the constant (independent of x) vectors is invariant with respect to the dynamics governed by (4.2.40) and (4.2.41). The restriction of (4.2.40) and (4.2.41) on the subspace $\mathscr{L}_0$ is an ODE of the form $\dot{u} + f(u) = 0$. We also refer to the survey of POLÁČEK [185] and the references therein for general results on the realization of ODE dynamics by several classes of parabolic PDEs.

4.3 Long-time dynamics in semilinear parabolic equations

In this section we continue our studies of the qualitative properties of parabolic systems which are modeled by equation (4.2.1). We concentrate on long-time dynamics and attractors. To proceed we assume that problem (4.2.1) has a unique mild solution on $\mathbb{R}_+$.[2] By Theorem 4.2.3 this solution is continuous both in time and initial data. Thus, we can define a dynamical system (H_α, S_t) on H_α with an evolution operator S_t given by the formula

$$S_t u_0 = u(t), \quad \text{where } u(t) \text{ solves } (4.2.1).$$

Our goal in this section is to apply the method developed in Chapters 2 and 3 to describe the asymptotic properties of the dynamical system (H_α, S_t).

4.3.1 Dissipativity and compactness

We start with the following assertion, which shows that dissipativity implies compactness.

---

[2] We refer to Theorem 4.2.22, which provides sufficient conditions for this global well-posedness.

Proposition 4.3.1. *Assume that the problem (4.2.1) satisfies Assumption 4.2.1 and generates a dynamical system* (H_α, S_t) *in* H_α*. Then* (H_α, S_t) *is conditionally compact (see Definition 2.2.1). If this system is dissipative, then it is compact (see Definition 2.2.1). Moreover, if the ball* $B_0 = \{u \in H_\alpha : \|u\|_\alpha \le R_0\}$ *is absorbing, then for every* $\theta \in [0, 1)$ *there exists* $R_\theta > 0$ *such that the set* $B_\theta = \{u \in H_\theta : \|u\|_\theta \le R_\theta\}$ *is also absorbing for* (H_α, S_t)*. The latter property with* $\theta \in (\alpha, 1)$ *means that the system has a compact absorbing set.*

Proof. First we prove that dissipativity implies compactness. Due to the compactness statement in Proposition 4.1.6, it is sufficient to show that the ball B_θ with some radius $R_\theta > 0$ is absorbing when $\theta \in (\alpha, 1)$.

It follows from (4.2.3) and (4.1.9) that $u(t) = S_t u_0$ satisfies the relation

$$\|u(t)\|_\theta \le \left(\frac{\theta - \alpha}{e(t - s)}\right)^{\theta - \alpha} \|u(s)\|_\alpha + \int_s^t \left(\frac{\theta}{e(t - \tau)}\right)^\theta \|B(u(\tau))\| \, d\tau \qquad (4.3.1)$$

for all $t > s$. Let D be a bounded set in H_α and $u_0 \in D$. We have $u(t) = S_t u_0 \in B_0$ for $t \ge t_D$. Thus,

$$\|u(t)\|_\theta \le \left(\frac{\theta - \alpha}{e(t - s)}\right)^{\theta - \alpha} R_0 + L(R_0, B) \left(\frac{\theta}{e}\right)^\theta \frac{|t - s|^{1 - \theta}}{1 - \theta} \qquad (4.3.2)$$

for all $t > s \ge t_D$, where $L(R_0, B) = \sup\{\|B(u)\| : u \in B_0\}$. Thus taking $t = s + 1$ we obtain that the set $B_\theta = \{u \in H_\theta : \|u\|_\theta \le R_\theta\}$ is absorbing with the radius

$$R_\theta = \left(\frac{\theta - \alpha}{e}\right)^{\theta - \alpha} R_0 + L(R_0, B) \left(\frac{\theta}{e}\right)^\theta \frac{1}{1 - \theta}.$$

To prove the conditional compactness of (H_α, S_t), we note that for every bounded forward invariant set D we have that $\|B(u(\tau))\| \le C_D$ for all $\tau \ge 0$. Therefore, the conclusion follows from (4.3.1) with $s = 0$ in the same manner as above. This concludes the proof. $\square$

The following theorem gives some conditions under which equation (4.2.1) generates a dissipative dynamical system.

Theorem 4.3.2. *Assume that Assumption 4.2.20 is in force; that is,*

(i) A *is a positive self-adjoint operator with discrete spectrum on* H*;*
(ii) B *is a locally Lipschitz mapping from* $H_{1/2} = \mathscr{D}(A^{1/2})$ *into* H *such that for every* $\rho > 0$ *there exists* L_ρ *such that (4.2.28) holds;*
(iii) $B(u) = -B_0(u) + B_1(u)$*, where* $B_1 : H_{1/2} \mapsto H$ *is continuous and linearly bounded, i.e., (4.2.29) holds, and* $B_0 \equiv \Pi' : H_{1/2} \mapsto H$ *is a potential operator on the space* $H_{1/2}$*.*

Assume in addition that

(iv) *there exist $\beta < 1/2$ and $\gamma \geq 0$ such that (4.2.32) holds, i.e.,*

$$\beta \|A^{1/2}u\|^2 + \Pi(u) + \gamma \geq 0, \quad \forall u \in H_{1/2}; \tag{4.3.3}$$

(v) *there exist $\mu < 1$ and $\delta > 0$ such that*

$$(B_0(u), u) + \mu \|A^{1/2}u\|^2 + \delta \geq 0, \quad \forall u \in H_{1/2}; \tag{4.3.4}$$

(vi) *for every $\varkappa > 0$ there exists $C_\varkappa$ such that*

$$\|B_1(u)\| \leq C_\varkappa + \varkappa \|A^{1/2}u\| \quad \text{for all } u \in H_{1/2}. \tag{4.3.5}$$

Then problem (4.2.1) generates a dissipative dynamical system $(H_{1/2}, S_t)$ in $H_{1/2}$.

Proof. By Theorem 4.2.22 problem (4.2.1) possesses a unique global solution $u(t)$ and generates a dynamical system in $H_{1/2}$. Using the balance relation in (4.2.14), we obtain that

$$\frac{1}{2}\frac{d}{dt}\|u\|^2 + \|A^{1/2}u\|^2 + (B_0(u), u) = (B_1(u), u) \leq \eta \|A^{1/2}u\|^2 + C_\eta, \tag{4.3.6}$$

for every $\eta > 0$ and for almost all $t \geq 0$. The balance relation in (4.2.33) yields

$$\|u_t\|^2 + \frac{d}{dt}\left[\frac{1}{2}\|A^{1/2}u\|^2 + \Pi(u)\right] = (B_1(u), u_t) \leq \frac{1}{2}\|u_t\|^2 + \eta \|A^{1/2}u\|^2 + C_\eta \tag{4.3.7}$$

for almost all $t \geq 0$. On the space $H_{1/2}$ we define the functional

$$\mathscr{W}(u) := \frac{1}{2}\|u\|^2 + \frac{1}{2}\|A^{1/2}u\|^2 + \Pi(u) + \gamma.$$

Since by Remark 4.2.21 $\Pi(u)$ is bounded on every bounded set, we conclude from (4.3.3) that

$$a_1 \|A^{1/2}u\|^2 \leq \mathscr{W}(u) \leq \phi(\|A^{1/2}u\|), \tag{4.3.8}$$

where $a_1 > 0$ and $\phi(r) = \gamma + a_2 r^2 + \sup\{|\Pi(u)| : \|u\|_{1/2} \leq r\}$. It also follows from (4.3.6), (4.3.7), and (4.3.4) that

$$\frac{d}{dt}\mathscr{W}(u(t)) + \nu \|A^{1/2}u(t)\|^2 \leq b, \quad t > 0, \tag{4.3.9}$$

for some $v, b > 0$. The relations in (4.3.8) and (4.3.9) allow us to apply Theorem 2.1.2 (see also Exercise 2.1.3) and conclude that $(H_{1/2}, S_t)$ is a dissipative system.   □

Exercise 4.3.3. Show that (4.3.5) is valid when B_1 has sublinear growth; i.e., there exists $\theta \in [0, 1)$ such that $\|B_1(u)\| \le c_1 + c_2 \|A^{1/2}u\|^\theta$ for all $u \in H_{1/2}$.   ∎

Exercise 4.3.4. Show that the conditions of Theorem 4.3.2 are valid for the heat equation in (4.2.35) when f has the form (4.2.37), where (i) f_0 satisfies (4.2.38) and (4.2.39); (ii) f_1 is globally Lipschitz and has sublinear growth

$$|f_1(u, \xi)| \le c_1 + c_2(|u| + |\xi|)^\theta \quad \text{with } \theta \in [0, 1).$$
                                                                                                                      ∎

Corollary 4.3.5. *Under the hypotheses of Theorem* 4.3.2, *the system* $(H_{1/2}, S_t)$ *generated by* (4.2.1) *is compact. Thus, it possesses a compact global attractor.*

Proof. By Theorem 4.3.2 the system is dissipative. Hence by Proposition 4.3.1 the system is compact. Thus, the existence of a compact global attractor follows from Theorem 2.3.5.   □

As we will see below, the global attractor given by Corollary 4.3.5 is finite-dimensional. We derive this fact from quasi-stability properties of the system considered, which we establish in Section 4.3.3 below.

4.3.2 Stationary solutions

Under the hypotheses of Theorem 4.3.2 we can also establish the existence of stationary solutions (which definitely belong to the attractor).

Proposition 4.3.6. *Let* $B(u) = -B_0(u) + B_1(u)$ *and let Assumption* 4.2.20 *be in force. Assume in addition that relations* (4.3.4) *and* (4.3.5) *hold. Then the set*

$$\mathscr{N} = \{u \in H_{1/2} : Au + B_0(u) = B_1(u)\} \tag{4.3.10}$$

of stationary solutions is a nonempty bounded set in H_1 *(hence it is compact in* $H_{1/2}$).

We note that the hypotheses in Proposition 4.3.6 are not optimal and can be relaxed substantially (see, e.g., LIONS [151] or SHOWALTER [210]). Our motivation behind Proposition 4.3.6 is to describe equilibria in the case when the system $(H_{1/2}, S_t)$ possesses a global attractor.

Proof. We consider a sequence $\{u^N\}$ of Galerkin approximations which are defined as elements in $P_N H$ satisfying the equation

$$Au^N + P_N B_0(u^N) = P_N B_1(u^N), \quad N = 1, 2, \ldots \tag{4.3.11}$$

To show the existence of these solutions, we apply the following "acute angle" lemma (see, e.g., SHOWALTER [210, Proposition 2.1, p. 37]).

Lemma 4.3.7. *Let $f : B_R := \{x \in R^d : |x| \le R\} \mapsto R^d$ be a continuous mapping such that $(f(x), x)_{R^d} \ge 0$ for every $x \in \partial B_R$. Then f has a zero in the ball B_R.*

Exercise 4.3.8. State and prove Lemma 4.3.7 in the 1D case ($d = 1$).   ∎

To check the conditions of Lemma 4.3.7, we note that (4.3.4) and (4.3.5) with an appropriate $\varkappa > 0$ imply that

$$(Au^N + P_N B_0(u^N) - P_N B_1(u^N), u^N) \ge \frac{1}{2}(1 - \mu)\|A^{1/2}u^N\|^2 - c(\delta, \mu).$$

By Lemma 4.3.7 this implies that for every N there exists at least one solution to (4.3.11) satisfying the estimate

$$\|A^{1/2}u^N\|^2 \le R \equiv 2(1 - \mu)^{-1}c(\delta, \mu)$$

This yields that

$$\|Au^N\|^2 \le \|B_0(u^N)\| + \|B_1(u^N)\| \le C_R.$$

Thus, the set $\{u^N\}$ is bounded in H_1 and hence compact in $H_{1/2}$. This allows us to make the limit transition (along a subsequence) and prove the existence of stationary solutions. It is clear that the set $\mathcal{N}$ of all stationary solutions is bounded in H_1.   □

As we know from Chapter 2, except for equilibria the global attractor may also contain periodic orbits. In this relation we note that, similar to finite-dimensional Theorem 1.8.8 on periodic orbits, we can give a low bound for possible periods of orbits of the system generated by (4.2.1) (see ROBINSON [196, p. 156]). However, it seems that the bound in ROBINSON [196] is not sharp.

4.3.3 Squeezing and quasi-stability

Two facts presented in this section allow us to involve the powerful tools developed in Chapter 3 for studies of parabolic models. The first fact is the following Ladyzhenskaya squeezing property (see LADYZHENSKAYA [142] and the references therein) of the evolution operator S_t. This is an important component of many qualitative considerations for parabolic PDEs. It can be treated as a strong form of quasi-stability.

Proposition 4.3.9 (Ladyzhenskaya squeezing property). *Let Assumption 4.2.1 be in force. Assume that (4.2.1) generates a dynamical system (H_α, S_t). We denote $Q_N = I - P_N$, where P_N is the orthoprojector onto Span $\{e_1, \ldots, e_N\}$ in H. Then for every $0 < q < 1$, $0 < a \le b < +\infty$, and $R > 0$ there exists $N_* = N(a, b, R, q)$ such that*

$$\|Q_N[S_t u - S_t u_*]\|_\alpha \leq q \|u - u_*\|_\alpha, \quad \forall t \in [a, b], \quad \forall N \geq N_*, \tag{4.3.12}$$

for any u and u_* from the set

$$\mathscr{D} = \{u \in H_\alpha : \|S_t u\|_\alpha \leq R \text{ for all } t \in [0, b]\}.$$

Proof. Let $u(t) = S_t u$ and $u_*(t) = S_t u_*$. Using Proposition 4.1.9, one can see that

$$\|Q_N(u(t) - u_*(t))\|_\alpha \leq e^{-t\lambda_{N+1}} \|Q_N(u(0) - u_*(0))\|_\alpha$$

$$+ L_R \int_0^t \left[\left(\frac{\alpha}{t - \tau}\right)^\alpha + \lambda_{N+1}^\alpha\right] e^{-(t-\tau)\lambda_{N+1}} \|u(\tau)) - u_*(\tau)\|_\alpha \, d\tau \tag{4.3.13}$$

for all $t \in [0, b]$. It follows from (4.2.4) (see also (4.2.6) in the case of globally Lipschitz B) that there exists a constant $C(R, b) > 0$ such that

$$\|u(t) - u_*(t)\|_\alpha \leq C(R, b) \|u(0) - u_*(0)\|_\alpha, \quad t \in [0, b], \ u(0), u_*(0) \in \mathscr{D}.$$

Therefore, (4.3.13) yields

$$\|Q_N(u(t) - u_*(t))\|_\alpha \leq \left[e^{-t\lambda_{N+1}} + L_R C(R, b) J_N(t, 0)\right] \|u(0) - u_*(0)\|_\alpha, \tag{4.3.14}$$

where $J_N(t, 0)$ is defined in (4.2.26). Thus, using (4.2.27) we obtain that

$$\|Q_N(S_t u - S_t u_*)\|_\alpha \leq \left[e^{-t\lambda_{N+1}} + C_*(R, b)\lambda_{N+1}^{-1+\alpha}\right] \|u - u_*\|_\alpha \tag{4.3.15}$$

for all $t \in [0, b]$ and $u, u_* \in \mathscr{D}$. This inequality implies that for every $q < 1$, $0 < a \leq b < +\infty$, and $R > 0$ we can choose N_* such that (4.3.12) holds. $\square$

Exercise 4.3.10. If $B(u)$ is a globally Lipschitz function, then (4.3.12) is valid for every $u, u_* \in H_\alpha$ with N_* independent of R. ∎

Proposition 4.3.11 (Quasi-stability). *Let the hypotheses of Proposition 4.3.9 be in force. Then for every $q < 1$, $0 < a \leq b < +\infty$, and forward invariant bounded set $\mathscr{B}$ there exists $N = N(a, b, q, \mathscr{B})$ such that*

$$\|S_t u - S_t u_*\|_\alpha \leq q \|u - u_*\|_\alpha + \|P_N[S_t u - S_t u_*]\|_\alpha, \quad \forall t \in [a, b], \tag{4.3.16}$$

for all $u, u_* \in \mathscr{B}$. This means that the system (H_α, S_t) satisfies Assumption 3.4.9 with arbitrary $t^* > 0$ and implies that (H_α, S_t) is quasi-stable on $\mathscr{B}$ at every time t_*.

Proof. We have that

$$\|S_t u - S_t u_*\|_\alpha \leq \|Q_N[S_t u - S_t u_*]\|_\alpha + \|P_N[S_t u - S_t u_*]\|_\alpha.$$

Thus (4.3.16) follows from (4.3.12).

It is clear that $\|P_N[\cdot]\|_\alpha$ is a compact seminorm on H_α. Thus (3.4.8) in Assumption 3.4.9 holds with $X = H_\alpha$, $n_1 \equiv 0$, and $n_2 = \|P_N[\cdot]\|_\alpha$. The Lipschitz property in (3.4.7) follows from (4.2.4). Therefore, we can apply Proposition 3.4.10. □

Exercise 4.3.12. Show that the relation in (4.3.16) can be written in the following uniform form:

$$\|S_t u - S_t u_*\|_\alpha \leq q\|S_r u - S_r u_*\|_\alpha + \|P_N[S_t u - S_t u_*]\|_\alpha, \quad u, u_* \in \mathscr{B}, \qquad (4.3.17)$$

for all $r \geq 0$ and $t \in [a + r, b + r]$ with q and N independent of r. ■

Remark 4.3.13. It follows from Exercise 4.2.7 that for every forward invariant bounded set $\mathscr{B}$ in H_α the evolution operator S_t is a Lipschitz mapping from $\mathscr{B}$ into H_β with $\beta \in [\alpha, 1)$. More precisely, for every $\beta \in [\alpha, 1)$ and $0 < a \leq b < +\infty$ and forward invariant bounded set $\mathscr{B}$ there exists a constant $C = C_{\mathscr{B}}(\alpha, \beta; a, b)$ such that

$$\|S_t u - S_t u_*\|_\beta \leq C_{\mathscr{B}}(\alpha, \beta; a, b)\|u - u_*\|_\alpha, \quad \forall t \in [a, b], \quad \forall u, u_* \in \mathscr{B}. \qquad (4.3.18)$$

Since H_β is compactly embedded into H_α for $\beta > \alpha$, the inequality in (4.3.18) allows us to establish *another form* of quasi-stability of the system (H_α, S_t) using the same idea as in Corollaries 3.1.25 and 3.1.26. ■

Remark 4.3.14 (Foias-Temam squeezing property). The Ladyzhenskaya squeezing inequality (4.3.12) implies that for every $\gamma > 0$ the following property holds: for each $t \in [a, b]$ and $u, u_* \in \mathscr{D}$ we have *either*

$$\|Q_N(S_t u - S_t u_*)\|_\alpha \leq \gamma\|P_N[S_t u - S_t u_*]\|_\alpha, \qquad (4.3.19)$$

or else

$$\|S_t u - S_t u_*\|_\alpha \leq q\left(1 + \frac{1}{\gamma}\right)\|u - u_*\|_\alpha. \qquad (4.3.20)$$

Indeed, if (4.3.19) is not true for some t and u, u_*, then

$$\|P_N(S_t u - S_t u_*)\|_\alpha < \frac{1}{\gamma}\|Q_N[S_t u - S_t u_*]\|_\alpha.$$

Thus,

$$\|S_t u - S_t u_*\|_\alpha \leq \|P_N(S_t u - S_t u_*)\|_\alpha + \|Q_N[S_t u - S_t u_*]\|_\alpha$$

$$\leq \left(1 + \frac{1}{\gamma}\right)\|Q_N[S_t u - S_t u_*]\|_\alpha.$$

Therefore, (4.3.12) implies (4.3.20). In the case when $q(1 + \gamma^{-1}) < 1$, the property above is a dynamical form of the Foias-Temam squeezing property restricted to the interval $[a, b]$; see CONSTANTIN/FOIAS/TEMAM [79] or TEMAM [216].   ∎

The results of Propositions 4.3.9 and 4.3.11 allow us to make several important conclusions concerning the long-time dynamics of (H_α, S_t).

4.3.4   Global and exponential attractors

The following assertion is a consequence of Theorem 3.4.11.

Theorem 4.3.15 (Global and exponential attractor). *Let Assumption 4.2.1 be in force. Assume that (4.2.1) generates a dissipative dynamical system (H_α, S_t). Then this system possesses a compact global attractor $\mathfrak{A}$ of finite fractal dimension $\dim_f \mathfrak{A}$ in H_α. This attractor is a bounded set in H_1, and for any full trajectory $\{u(t) : t \in \mathbb{R}\}$ from the attractor we have that $u(t)$ is an absolutely continuous function with values in H_β for every $\beta < 1$ and*

$$\sup_{t \in \mathbb{R}} \left\{ \|u_t(t)\|_\beta + \|Au(t) - B(u(t))\|_\beta \right\} \leq C. \tag{4.3.21}$$

Moreover, the system (H_α, S_t) possesses a fractal exponential attractor $\mathfrak{A}_{exp}$ (whose dimension is finite in the phase space H_α).

We note that the bounds for the dimensions of $\mathfrak{A}$ and $\mathfrak{A}_{exp}$ can be derived from Theorems 3.4.11 and 3.2.3.

Proof. By Proposition 4.3.11 the system satisfies Assumption 3.4.9 on every bounded forward invariant set. Thus, we can apply Theorem 3.4.11 to conclude the existence of finite-dimensional global attractor $\mathfrak{A}$.

To prove the claimed smoothness of the attractor, we first note that, by Proposition 4.3.1, for each $\beta < 1$ there is an absorbing set which is bounded in H_β. Thus, the attractor is bounded in H_β for every $\alpha \leq \beta < 1$. Next, we can restrict the system (H_α, S_t) on the space H_β with $\alpha < \beta < 1$ and apply Proposition 4.3.11 on the set $\mathfrak{A}$. It follows from (4.3.17) that

$$\|u(t + h) - u(t)\|_\beta \leq q\|u(t + h - 1) - u(t - 1)\|_\beta + \lambda_N \|u(t + h) - u(t)\|_{-1+\beta}$$

for every $t \in \mathbb{R}$ and for any full trajectory $\{u(t) : t \in \mathbb{R}\}$ from the attractor. Since on the attractor

$$\|u(t + h) - u(t)\|_{-1+\beta} \leq \int_t^{t+h} \left[\|u(\tau)\|_\beta + \|B(u(\tau))\| \right] d\tau \leq C_\mathfrak{A}|h|,$$

we obtain that

$$(1 - q) \sup_{t \in \mathbb{R}} \|u(t + h) - u(t)\|_\beta \leq C_{\mathfrak{A}} \lambda_N |h|.$$

After the limit transition $h \to 0$ this yields (4.3.21), which, in particular, means that $\mathfrak{A}$ is a bounded subset of H_1.

To prove the existence of the fractal exponential attractors $\mathfrak{A}_{exp}$, we use the second part of Theorem 3.4.11. For this we need to check the Hölder continuity property (3.4.10) on some absorbing set for (H_α, S_t) with $\tilde{X} = H_\alpha$. To do this, we note that by Proposition 4.3.1 there exists an absorbing set B_θ which is bounded in H_θ with some $\theta > \alpha$. Therefore, Exercise 4.2.8 yields the desired Hölder continuity. $\quad\square$

Corollary 4.3.16. *Let the hypotheses of Theorem 4.3.2 be in force. Then the system $(H_{1/2}, S_t)$ generated by (4.2.1) possesses a finite-dimensional compact global attractor $\mathfrak{A}$ which is a bounded set in H_1. There also exists a fractal exponential attractor $\mathfrak{A}_{exp}$ (with finite dimension in $H_{1/2}$).*

Proof. We apply Theorem 4.3.2 to show that the system $(H_{1/2}, S_t)$ is dissipative. Thus, the result follows from Theorem 4.3.15. $\quad\square$

4.3.5 Gradient structure and rates of stabilization to equilibria

In this section we consider the situation presented in Corollary 4.3.16 under additional conditions which guarantee a gradient structure of the system.

Theorem 4.3.17. *Let the hypotheses of Theorem 4.3.2 be in force and $B_1(u) \equiv 0$. Then the system $(H_{1/2}, S_t)$ generated by (4.2.1) is gradient, and thus the global attractor given by Corollary 4.3.16 has a regular structure, i.e., $\mathfrak{A} = \mathcal{M}^u(\mathcal{N})$, where $\mathcal{M}^u(\mathcal{N})$ denotes the unstable manifold emanating from the set*

$$\mathcal{N} = \{u \in H_\alpha : Au + B_0(u) = 0\}$$

of stationary points (see Definition 2.3.10). The global attractor $\mathfrak{A}$ consists of full trajectories $\gamma = \{u(t) : t \in \mathbb{R}\}$ such that

$$\lim_{t \to -\infty} \text{dist}_X(u(t), \mathcal{N}) = 0 \quad \text{and} \quad \lim_{t \to +\infty} \text{dist}_X(u(t), \mathcal{N}) = 0. \tag{4.3.22}$$

For any $u \in H_{1/2}$ we have

$$\lim_{t \to +\infty} \text{dist}_X(S_t u, \mathcal{N}) = 0, \tag{4.3.23}$$

that is, any trajectory stabilizes to the set $\mathcal{N}$ of stationary points. In particular, if the set $\mathcal{N}$ is finite, then for every $u \in H_{1/2}$ there exists $\phi \in \mathcal{N}$ such that

$$\lim_{t \to +\infty} \|S_t u - \phi\|_\alpha = 0. \tag{4.3.24}$$

Proof. Due to the balance relation (4.2.33) with $B_1 \equiv 0$, the functional

$$E(u) = \frac{1}{2}\|A^{1/2}u\|^2 + \Pi(u),$$

is a strict Lyapunov function for $(H_{1/2}, S_t)$. Thus, the structure of the attractor follows from Theorem 2.4.5. The convergence property in (4.3.24) follows from Theorem 2.4.7.   □

Under additional conditions one can show that the convergence to an equilibrium in (4.3.24) takes place at an exponential rate.

Theorem 4.3.18 (Rate of stabilization). *In addition to the hypotheses of Theorem 4.3.17, we assume that the set $\mathcal{N}$ of stationary points is finite and $B_0(u)$ is Fréchet differentiable on $H_{1/2}$ at every stationary point. This means[3] that for every $\phi \in \mathcal{N}$ there exists a bounded linear operator $B_0'(\phi) : H_{1/2} \mapsto H$ such that*

$$\lim_{\|v\|_{1/2} \to 0} \frac{\|B_0(\phi + v) - B_0(\phi) - B_0'(\phi)v\|}{\|v\|_{1/2}} = 0.$$

Let all equilibria be hyperbolic in the sense that the equation $Au + B_0'(\phi)u = 0$ has only a trivial solution for each $\phi \in \mathcal{N}$. Then for any $u \in H_{1/2}$ there exist an equilibrium $\phi \in \mathcal{N}$ and constants $\gamma > 0, C > 0$ such that

$$\|S_t u - \phi\|_{1/2} \le Ce^{-\gamma t} \text{ for all } t > 0. \tag{4.3.25}$$

We note that this type of stabilization theorem is well known in the literature for different classes of gradient systems, and several approaches to the question of stabilization rates are available (see, e.g., BABIN/VISHIK [9] and also CHUESHOV/LASIECKA [51, 56, 58] and the references therein). The approach presented in BABIN/VISHIK [9] relies on the analysis of linearized dynamics near each equilibrium and requires the hyperbolicity condition in a dynamical form (we need this condition in a weaker form). Here we use the method developed for second order in time evolution equations in CHUESHOV/LASIECKA [51, 56] (see also the discussion in Chapter 5).

Proof. Let $u \in H_{1/2}$ and $u(t) = S_t u$. Then by Theorem 4.3.17 there exists an equilibrium $\phi \in \mathcal{N}$ satisfying (4.3.24). Thus, we need only to prove that $S_t u$ tends to ϕ with the stated rate. We can assume that $\sup_{t \ge 0} \|S_t u\|_{1/2} \le R$ for some $R > 0$.

The function $z(t) = u(t) - \phi$ solves the following equation:

---

[3] For details see Definition A.5.1 in the Appendix.

$$z_t(t) + Az(t) + B_0(\phi + z(t)) - B_0(\phi) = 0, \quad t > 0, \tag{4.3.26}$$

in both the mild and the weak sense. Let

$$\mathscr{E}(t) = \frac{1}{2}\|z(t)\|_{1/2}^2 + \Phi(t),$$

where

$$\Phi(t) = \Pi(\phi + z(t)) - \Pi(\phi) - (B_0(\phi), z) \equiv \int_0^1 (B_0(\phi + \lambda z) - B_0(\phi), z)d\lambda.$$

In the same way as in Theorem 4.2.22 one can see that

$$\mathscr{E}(t) + \int_0^t \|z_t(\tau)\|^2 d\tau = \mathscr{E}(0). \tag{4.3.27}$$

In particular, $\mathscr{E}(t)$ is non-increasing. Moreover, since $z \to 0$ in $H_{1/2}$ as $t \to +\infty$, we have that $\mathscr{E}(t) \to 0$ when $t \to +\infty$. Thus $\mathscr{E}(t) \geq 0$ for all $t \geq 0$. Since

$$|\Phi(t)| \leq C_R\|z(t)\|_{1/2}\|z(t)\| \leq \varepsilon\|z(t)\|_{1/2}^2 + C_R\varepsilon^{-1}\|z(t)\|^2, \quad \forall \varepsilon > 0,$$

we have that

$$\frac{1}{4}\|z(t)\|_{1/2}^2 - C_R\|z(t)\|^2 \leq \mathscr{E}(t) \leq \|z(t)\|_{1/2}^2 + C_R\|z(t)\|^2, \quad \forall \varepsilon > 0, \tag{4.3.28}$$

Multiplying (4.3.26) by z one can show that

$$\|z(t)\|^2 + \int_0^t \|z(\tau)\|_{1/2}^2 d\tau \leq \|z(0)\|^2 + C_R \int_0^t \|z(\tau)\|^2 d\tau.$$

Since $\mathscr{E}(t)$ is monotone, we have that

$$T\mathscr{E}(T) \leq \int_0^T \mathscr{E}(t)dt \leq \int_0^T \|z(t)\|_{1/2}^2 dt + C_R \int_0^T \|z(t)\|^2 dt.$$

Thus, we obtain that

$$T\mathscr{E}(T) + \int_0^T \|z(t)\|_{1/2}^2 dt \leq C_{R,T} \max_{[0,T]} \|z(t)\|^2 \quad \text{for every } T \geq 0. \tag{4.3.29}$$

Now we prove the following lemma.

Lemma 4.3.19. *Let $z(t)$ be a mild solution to (4.3.26) such that*

$$\int_{T-1}^{T} \|z(t)\|_{1/2}^2 dt \leq \delta \quad and \quad \sup_{t \in \mathbb{R}_+} \|z(t)\|_{1/2} \leq \varrho \tag{4.3.30}$$

with some $\delta, \varrho > 0$ and $T > 1$. Then there exists $\delta_0 > 0$ such that

$$\max_{[0,T]} \|z(t)\|^2 \leq C \int_0^T \|z_t(t)\|^2 dt \tag{4.3.31}$$

for every $0 < \delta \leq \delta_0$, where the constant C may depend on δ, ϱ, and T.

Proof. Assume that (4.3.31) is not true. Then for some $\delta > 0$ small enough there exists a sequence of solutions $\{z^n(t)\}$ satisfying (4.3.30) and such that

$$\lim_{n \to \infty} \left\{ \max_{[0,T]} \|z^n(t)\|^2 \left[\int_0^T \|z_t^n(t)\|^2 dt \right]^{-1} \right\} = \infty. \tag{4.3.32}$$

By (4.3.30) $\max_{[0,T]} \|z^n(t)\|^2 \leq C_\varrho$ for all n. Thus (4.3.32) implies that

$$\lim_{n \to \infty} \int_0^T \|z_t^n(t)\|^2 dt = 0. \tag{4.3.33}$$

Therefore,

$$\max_{[0,T]} \|z^n(t) - z^n(0)\| \leq \sqrt{T} \left[\int_0^T \|z_t^n(t)\|^2 dt \right]^{1/2} \to 0, \quad n \to \infty.$$

Thus, due to (4.3.30) we can assume that there exists $z^* \in H_{1/2}$ such that

$$z^n(\cdot) \to z^* \quad *\text{-weakly in } L_\infty(0, T; H_{1/2}). \tag{4.3.34}$$

It follows from (4.3.33), (4.3.26), and (4.3.30) that

$$\int_0^T \|Az^n(t)\|^2 dt \leq C \left[\int_0^T \|z_t^n(t)\|^2 dt + \int_0^T \|z^n(t)\|_{1/2}^2 dt \right] \leq C_T(\varrho)$$

for every n. Thus, we can see that $u^* = \phi + z^* \in H_1$ solves the problem $Au + B_0(u) = 0$. From (4.3.30) we have that $\|A^{1/2}(u^* - \phi)\|^2 \leq \delta$. If we choose $\delta_0 > 0$ such that $\|A^{1/2}(\phi_1 - \phi_2)\|^2 > 2\delta_0$ for every couple ϕ_1 and ϕ_2 of stationary solutions (we can do it because the set $\mathcal{N}$ is finite), then we can conclude that $u^* = \phi$ provided $\delta \leq \delta_0$. Thus we have $z^* = 0$ in (4.3.34). Moreover, applying Theorem A.3.7 on compactness, we can conclude that

$$\max_{[0,T]} \|z^n(t)\|^2 + \int_0^T \|z^n(t)\|_{1-\eta}^2 dt + \int_0^T \|z_t^n(t)\|^2 dt \to 0, \quad n \to \infty, \quad \forall \, \eta > 0.$$

$$(4.3.35)$$

Now we normalize the sequence z^n by defining

$$\hat{z}^n \equiv c_n^{-1} z^n \quad \text{with} \quad c_n = \max_{[0,T]} \|z^n(t)\|.$$

By (4.3.35), $c_n \to 0$ as $n \to \infty$. The relation in (4.3.32) yields

$$\int_0^T \|\hat{z}_t^n(t)\|^2 dt \to 0 \quad \text{as} \quad n \to \infty.$$

$$(4.3.36)$$

It follows from (4.3.27) and (4.3.28) that

$$\frac{1}{4} \|A^{1/2} z^n(t)\|^2 \le \mathscr{E}_n(t) + C_R \|z^n(t)\|^2 \le \mathscr{E}_n(T) + \int_0^T \|z_t^n(\tau)\|^2 d\tau + C_R \|z^n(t)\|^2,$$

for all $t \in [0, T]$. Hence using (4.3.29) we obtain that

$$\frac{1}{4} \sup_{t \in [0,T]} \|A^{1/2} z^n(t)\|^2 \le \mathscr{E}_n(T) + \int_0^T \|z_t^n(\tau)\|^2 d\tau + C_R \max_{[0,T]} \|z^n(t)\|^2,$$

$$\le C_{R,T} \max_{[0,T]} \|z^n(t)\|^2, \quad T > 1.$$

Therefore, (4.3.35) implies that

$$\sup_{t \in [0,T]} \|A^{1/2} z^n(t)\| \to 0, \quad n \to \infty,$$

and also the following uniform estimate:

$$\sup_{t \in [0,T]} \|A^{1/2} \hat{z}^n(t)\|^2 \le C, \quad n = 1, 2, \ldots.$$

Using (4.3.26) we can conclude that

$$\|A\hat{z}^n(t)\|^2 \le 2\|\hat{z}_t^n(t)\|^2 + \frac{2}{c_n^2} \|B(\phi + z^n(t)) - B(\phi)\|^2$$

$$\le 2\|\hat{z}_t^n(t)\|^2 + C_{R,T} \|A^{1/2} \hat{z}^n(t)\|^2.$$

This implies that

$$\int_0^T \|A\hat{z}^n(t)\|^2 dt \le C_{R,T}, \quad n = 1, 2, \ldots.$$

Thus, relying on (4.3.36) we can assume that there exists $\hat{z}^* \in H_1$ such that

$$\hat{z}^n(\cdot) \to \hat{z}^* \quad *\text{-weakly in } L_2(0, T; H_1). \tag{4.3.37}$$

Moreover, by the Aubin-Dubinskii-Lions theorem (see Theorem A.3.7 in the Appendix),

$$\int_0^T \|\hat{z}^n(t) - \hat{z}^*\|_{1-\eta}^2 dt \to 0, \quad n \to \infty, \quad \forall \eta > 0. \tag{4.3.38}$$

The function $\hat{z}^n$ satisfies the equation

$$\hat{z}_t^n + A\hat{z}^n + \frac{1}{c_n}[B_0(\phi + z^n) - B_0(\phi)] = 0. \tag{4.3.39}$$

It follows from (4.3.38) that

$$\frac{1}{c_n}[B_0(\phi + z^n) - B_0(\phi)] \to B_0'(\phi)\hat{z}^* \quad \text{strongly in } L_2(0, T; H).$$

Therefore, after the limit transition in (4.3.39) we have that $\hat{z}^*$ satisfies

$$A\hat{z}^* + B_0'(\phi)\hat{z}^* = 0$$

and, by hyperbolicity of ϕ, we conclude that $\hat{z}^* = 0$. Thus (4.3.38) with $\hat{z}^* = 0$ and (4.3.36) imply that

$$\max_{[0,T]} \|\hat{z}^n\| \to 0 \quad \text{as } n \to \infty,$$

which is impossible. □

Completion of the proof of Theorem 4.3.18. By (4.3.24) we can choose $T_0 > 0$ such that (4.3.30) holds with $\delta \le \delta_0$ and $T > T_0$. Lemma 4.3.19, inequality (4.3.29), and the energy relation in (4.3.27) imply that

$$T\mathscr{E}(T) \le C_{R,T} \int_0^T \|z_t(t)\|^2 dt \le C_{R,T}[\mathscr{E}(0) - \mathscr{E}(T)]. \tag{4.3.40}$$

Therefore, $\mathscr{E}(T) \le \gamma_R\mathscr{E}(0)$ for some $0 < \gamma_R < 1$. This yields that $\mathscr{E}(mT) \le \gamma_R^m\mathscr{E}(0)$ for $m = 1, 2, \ldots$. We also have that

$$\|z(mT)\|_{1/2}^2 \le 4\mathscr{E}(mT) + C_R \max_{[mT,(m+1)T]} \|z^n(t)\|^2 \le C_R\mathscr{E}(mT), \quad m = 1, 2, \ldots$$

Thus, $\|z(mT)\|_{1/2}^2 \le C_R\gamma_R^m$ for $m = 1, 2, \ldots$. This implies (4.3.25) and completes the proof. □

4.3.6 Determining functionals

We now present several results on determining functionals. For a general discussion of the theory of determining functionals we refer to Section 3.3 and to the references therein. As was already mentioned in that section, finite families of determining functionals provide an important tool for parameterizing the long-time dynamics.

We start with a result based on quasi-stability which specifies the method suggested in Theorem 3.4.12 for the case of abstract parabolic problems.

Theorem 4.3.20 (Determining functionals). *Let Assumption* 4.2.1 *be in force. Assume that* (4.2.1) *generates a dissipative system* (H_α, S_t).

- **Modes:** *There exists* $N_* > 0$ *such that the modes* $\{e_k : k = 1, 2, \ldots, N\}$ *are determining for every* $N \geq N_*$; *i.e., for every* $y_1, y_2 \in H_\alpha$ *the property*

$$\lim_{t \to \infty} (S_t y_1 - S_t y_2, e_j) = 0, \quad j = 1, 2, \ldots, N, \qquad (4.3.41)$$

implies

$$\lim_{t \to \infty} \|S_t y_1 - S_t y_2\|_\alpha = 0. \qquad (4.3.42)$$

Here $\{e_k\}_{k=1}^\infty$ *is the eigenbasis of the operator* A.
- **General functionals:** *Let* $\mathcal{L} = \{l_j : j = 1, \ldots, N\}$ *be a set of linearly independent functionals on* H_α *and* $\epsilon_{\mathcal{L}}(H_\alpha, H)$ *be the corresponding completeness defect (see Definition 3.3.3). Then there exists* $\epsilon_* > 0$ *such that under the condition* $\epsilon_{\mathcal{L}}(H_\alpha, H) \leq \epsilon_*$ *the set* $\mathcal{L}$ *is determining; i.e., the relation*

$$\lim_{t \to \infty} l_j(S_t y_1 - S_t y_2) = 0, \quad j = 1, 2, \ldots, N, \qquad (4.3.43)$$

implies (4.3.42).

Proof. In the case of modes, the statement is a direct consequence of (4.3.16) and the argument given in the proof of Theorem 3.4.12. In the case of general functionals, we note that (4.3.16) implies the quasi-stability in (3.4.8) with $X = H_\alpha$, $n_1 \equiv 0$, and $n_2(y) = \lambda_N^\alpha \|y\|$. In the case considered, $\epsilon_{\mathcal{L}}(n_2) = \lambda_N^\alpha \epsilon_{\mathcal{L}}(H_\alpha, H)$. $\qquad \square$

Another kind of result on determining functionals is based on the following observation, which involves functionals defined on a certain model space. This structure allows us to cover several situations which are important from an applications point of view (see below).

As above, we consider an abstract parabolic equation defined on a separable Hilbert space H of the form

$$u_t + Au = B(u), \quad t > 0, \quad u|_{t=0} = u_0. \qquad (4.3.44)$$

Here A is a positive operator with discrete spectrum; see Definition 4.1.1.

Theorem 4.3.21. *Let Assumption 4.2.1 be in force with $\alpha = 1/2$. Assume that problem (4.3.44) is uniquely solvable within the class $\mathscr{W} = C(\mathbb{R}_+; H_{1/2})$ and it is point dissipative; that is, there exists $R > 0$ such that*

$$\|A^{1/2}u(t)\| \le R, \quad \text{when } t \ge t_0(u_0), \text{ for every } u(t) \in \mathscr{W}. \tag{4.3.45}$$

Let Z_0 and Z_1 denote Hilbert spaces with the norms $\|\cdot\|_{Z_0}$ and $\|\cdot\|_{Z_1}$ such that $Z_1 \subset Z_0$. Assume that there exists a linear operator $J : V = H_{1/2} \mapsto Z_1$ such that $\|Ju\|_{Z_1} \le K \cdot \|u\|_{1/2}$, and the following monotonicity-type inequality:

$$\frac{1}{2}\|A^{1/2}(u_1 - u_2)\|^2 - (B(u_1) - B(u_2), u_1 - u_2) \ge -C(R)\|J(u_1 - u_2)\|_{Z_0}^2 \tag{4.3.46}$$

is fulfilled for all $u_j \in H_{1/2}$ possessing the properties $\|A^{1/2}u_j\| \le R$, where $R > 0$ is the constant from (4.3.45) and $C(R)$ is a positive number. Let $\mathscr{L} = \{l_i : i = 1, \ldots, N\}$ be a set of linear continuous functionals on Z_1. Then the set

$$J^*\mathscr{L} = \{l_i^J(w) = l_i(Jw) : l_i \in \mathscr{L}\}$$

is determining for problem (4.3.44) provided $\epsilon_{\mathscr{L}}(Z_1, Z_0) < (2C(R)K^2)^{-1/2}$.

Proof. For the proof we use the line of argument given in CHUESHOV [38]. One can see that

$$\frac{1}{2}\frac{d}{dt}\|w(t)\|^2 + (Aw(t), w(t)) = (B(u_1(t)) - B(u_2(t)), w(t))$$

for the difference $w(t) = u_1(t) - u_2(t)$ of two solutions $u_1(t)$ and $u_2(t)$ from $\mathscr{W}$. Hence the condition in (4.3.46) gives

$$\frac{d}{dt}\|w(t)\|^2 + \|A^{1/2}w(t)\|^2 \le 2C(R)\|Jw(t)\|_{Z_0}^2$$

for t large enough. Proposition 3.3.4 implies that

$$\|Jw\|_{Z_0} \le K\epsilon_{\mathscr{L}}(Z_1, Z_0)\|A^{1/2}w\| + C_{\mathscr{L}} \cdot \max_i |l_i(Jw)|.$$

Therefore,

$$\frac{d}{dt}\|w(t)\|^2 + \beta_{\mathscr{L}} \cdot \|A^{1/2}w(t)\|^2 \le C_{\mathscr{L},\eta} \cdot \max_i |l_i(Jw(t))|^2$$

for t large enough, where

$$\beta_{\mathscr{L}} = 1 - 2(1 + \eta)\epsilon_{\mathscr{L}}^2(Z_1, Z_0)C(R)K^2$$

with arbitrary $\eta > 0$. Obviously we can choose $\eta > 0$ such that $\beta_{\mathscr{L}} > 0$. Applying Gronwall's lemma, we obtain

$$\|w(t)\|^2 \le \|w(s)\|^2 \exp\{-\lambda_1 \beta_{\mathscr{L}}(t-s)\}$$
$$+ C_\eta \int_s^t \exp\{-\lambda_1 \beta_{\mathscr{L}}(t-\tau)\} \mathscr{N}_{\mathscr{L}}[w(\tau)] d\tau \qquad (4.3.47)$$

for $t \ge s$ with $s > 0$ large enough, where $\mathscr{N}_{\mathscr{L}}[w] = \max_i |l_i(Jw)|^2$. This implies that

$$\limsup_{t \to \infty} \|w(t)\|^2 \le C_\delta \limsup_{t \to \infty} \int_s^t \exp\{-\beta_{\mathscr{L}}(t-\tau)\} \mathscr{N}_{\mathscr{L}}[w(\tau)] d\tau.$$

Thus, l'Hôpital's rule yields $\|w(t)\| \to 0$ as $t \to \infty$.

By interpolation (see Exercise 4.1.2(E)), using (4.3.45) we have that $\|w(t)\|_s \to 0$ as $t \to \infty$ for every $s < 1/2$. To obtain the result in $H_{1/2}$ we need to use the fact of the smoothening of the trajectories (see Exercises 4.2.6 and 4.2.7). This completes the proof. □

Exercise 4.3.22. Show that in the case when the condition (4.3.46) is met with the constant C independent of R, we can get rid of the dissipativity condition (4.3.45) in Theorem 4.3.21. ∎

Remark 4.3.23. Under the conditions of Theorem 4.3.21, the set $\mathscr{L}$ is also determining in the Ladyzhenskaya sense (see Remark 3.3.2(3) and also Remark 3.3.14): for any two solutions $u_1(t)$ and $u_2(t)$ to problem (4.3.44) defined on the whole time axis and such that

$$\sup\{\|A^{1/2}u_i(t)\| : -\infty < t < \infty\} \le R, \; i = 1, 2,$$

the condition

$$\exists t_* \in \mathbb{R} : \; l_j(u_1(t)) = l_j(u_2(t)) \text{ for almost all } t < t_* \text{ and } j = 1, \ldots, N$$

implies that $u_1(t) \equiv u_2(t)$ for all $t \in \mathbb{R}$. The proof of this assertion follows from relation (4.3.47), which gives us

$$\|w(t)\|^2 \le \|w(s)\|^2 \exp\{-\beta_{\mathscr{L}}(t-s)\} \text{ for all } -\infty < s < t < t_*.$$

As in Remark 3.3.14, in the limit $s \to -\infty$ this implies the conclusion. Similar determining properties can also be established in the quasi-stability framework of Theorem 4.3.20; see Remark 3.4.13(2). ∎

We note that the standard situation covered by the framework of Theorem 4.3.21 is the case of boundary determining functionals, that is, the case when $Z_1 \subset Z_0$ are Sobolev-type spaces on the boundary of nonlinear parabolic PDEs in a bounded domain (see CHUESHOV [38] and [39, Section 5.8] for details). However, we can also apply this theorem in the following situation.

Corollary 4.3.24. *Let Assumption 4.2.1 with $\alpha = 1/2$ be in force. Assume that problem (4.3.44) is uniquely solvable within the class $\mathcal{W} = C(\mathbb{R}_+; H_{1/2})$ and it is point dissipative;[4] i.e., it satisfies (4.3.45). Let $\mathcal{L} = \{l_j : j = 1, \ldots, N\}$ be a set of linear continuous functionals on the space $H_{1/2}$. Then $\mathcal{L}$ is a set of asymptotically determining functionals for problem (4.3.44) provided the completeness defect $\epsilon_{\mathcal{L}}(H_{1/2}, H)$ fulfills the inequality*

$$\varepsilon_{\mathcal{L}} \equiv \epsilon_{\mathcal{L}}(H_{1/2}, H) < L_R^{-1}, \tag{4.3.48}$$

where L_R and R are the same as in (4.2.2) and (4.3.45).

Proof. We show that the condition in (4.3.46) of Theorem 4.3.21 holds with $Z_0 = H$, $Z_1 = H_{1/2}$, and $J = Id$. Indeed, it is clear that

$$\frac{1}{2}\|A^{1/2}(u_1 - u_2)\|^2 - (B(u_1) - B(u_2), u_1 - u_2)$$

$$\geq \frac{1}{2}\|A^{1/2}(u_1 - u_2)\|^2 - L_R\|A^{1/2}(u_1 - u_2)\| \cdot \|u_1 - u_2\| \geq -\frac{L_R^2}{2}\|u_1 - u_2\|^2.$$

This implies (4.3.46) with $C(R) = L_R^2/2$ and concludes the proof.   □

For other applications and generalizations of the idea presented in Theorem 4.3.21 and Corollary 4.3.24, we refer to CHUESHOV [38] and [39, Chapter 5].

4.3.7   Discrete data assimilation

The data assimilation problem is the question of how to incorporate available observation data into computational schemes to improve the quality of the predictions of the future evolution of the corresponding dynamical system. This problem has a long history and has been studied by many authors at different levels (see, e.g., the monographs of LAHOZ ET AL (EDS) [82], KALNAY [129] and the references therein).

We consider the case when observations of the system are made in some sequence $\{t_n\}$ of moments of time and use the same formulation of the data assimilation problem as in HAYDEN/OLSON/TITI [122].

Our main goal is to demonstrate the role of the Ladyzhenskaya squeezing property (see Proposition 4.3.9) in the solving of data assimilation problems. As in

---

[4]As above (see Exercise 4.3.22), we can avoid this condition in the case when *(4.2.2)* holds with L_ρ independent of ρ.

HAYDEN/OLSON/TITI [122], we also involve the notion of determining modes or, more generally, determining functionals. However, our method is different from the approach developed in HAYDEN/OLSON/TITI [122] for the 2D Navier-Stokes equations.

We assume that Assumption 4.2.1 is in force and problem (4.2.1) generates a dynamical system (H_α, S_t). Let $\mathscr{L} = \{l_j : j = 1, \ldots, N\}$ be a finite family of functionals on H_α (each functional l_j can be interpreted as a single observational measurement). Let $R_{\mathscr{L}}$ be some Lagrange interpolation operator related with $\mathscr{L}$, i.e.,

$$R_{\mathscr{L}} v = \sum_{j=1}^{N} l_j(v) \psi_j, \quad \forall v \in H_\alpha, \tag{4.3.49}$$

where $\{\psi_j\}$ is a finite set of elements from H_α such that $l_k(\psi_j) = \delta_{kj}$ (in this case $R_{\mathscr{L}}^2 = R_{\mathscr{L}}$, i.e., $R_{\mathscr{L}}$ is a projector). We refer to Section 3.3 for further details concerning interpolation operators.

Definition 4.3.25 (Data assimilation). For a given solution $U(t) = S_t U_0$ to (4.2.1) with initial data U_0, we consider the sequence $\{r_{\mathscr{L}}^n \equiv R_{\mathscr{L}} U(t_n)\}$ which represents the (joint) observational measurements of the solution $U(t)$ at a sequence $\{t_n\}$ of times. The sequence $\{r_{\mathscr{L}}^n\}$ is called a sequence of *observation values*. Now we can construct *prognostic values* at time t_n for $U(t)$ by the formula

$$u_n = (1 - R_{\mathscr{L}}) S_{t_n - t_{n-1}} u_{n-1} + r_{\mathscr{L}}^n, \quad n = 1, 2, \ldots, \tag{4.3.50}$$

where u_0 is an (unknown) vector, which, according to HAYDEN/OLSON/TITI [122], corresponds to an initial guess U_0 of the reference solution. We can also define the prognostic (piecewise continuous) trajectory as

$$u(t) = S_{t-t_n} u_n \text{ for } t \in [t_n, t_{n+1}), \quad n = 0, 1, 2, \ldots. \tag{4.3.51}$$

We say that the prognosis is *asymptotically reliable* at a sequence of times t_n if

$$\|U(t_n) - u_n\|_\alpha \to 0 \text{ as } n \to +\infty.$$

■

Our goal is to find conditions on $R_{\mathscr{L}}$ and t_n which guarantee that the prognosis based on a finite number of single observations $\mathscr{L} = \{l_j : 1 \le j \le N\}$ is asymptotically reliable.

We assume that $0 < a \le t_{n+1} - t_n \le b < +\infty$ for some positive a and b.

The following assertion is our main result concerning problem (4.2.1).

Theorem 4.3.26. *Let Assumption 4.2.1 be in force with a globally Lipschitz B (i.e., L_ρ in (4.2.2) is independent of ρ). Assume that $\mathscr{L}$ is a finite family of functionals* on H_α and there is a Lagrange interpolation operator $R_{\mathscr{L}}$ of the form (4.3.49) with $\psi_j \in H_\beta$ for some $\beta \in (\alpha, 1)$ possessing the property

$$\|1 - R_{\mathscr{L}}\|_{H_\beta \mapsto H_\beta} \leq c_0 \qquad (4.3.52)$$

with the constant c_0 independent of $\mathscr{L}$. Let $U(t) = S_t U_0$ with $U_0 \in H_\beta$. Then there exists $\epsilon_* > 0$ such that under the condition[5] $\epsilon(H_\beta, H_\alpha) \leq \epsilon_*$ the prognosis in (4.3.50) is asymptotically reliable for every $u_0 \in H_\beta$.

In the case of the modes described in Example 3.3.10 and based on the eigen-elements of A there exists $N_* = N_*(\alpha)$ such that the prognosis (4.3.50) is asymptotically reliable for every $\beta \in [\alpha, 1)$ with $R_{\mathscr{L}} = P_{\mathscr{L}}$, where $P_{\mathscr{L}}$ is given by

$$P_{\mathscr{L}} v = \sum_{j=1}^{N} (e_j, v) e_j, \quad \forall v \in H_\alpha, \qquad (4.3.53)$$

with some $N \geq N_*$, where $\{e_k\}$ is the eigenbasis of A.

Proof. We note that in the globally Lipschitz case problem (4.2.1) generates a dynamical system (H_β, S_t) for each $\beta \in [\alpha, 1)$, see Theorem 4.2.3 and Remark 4.2.5. In each space H_β we obviously have that

$$U(t_n) - u_n = (1 - R_{\mathscr{L}})[S_{t_n - t_{n-1}} U(t_{n-1}) - S_{t_n - t_{n-1}} u_{n-1}].$$

In the case of modes we have that $I - R_{\mathscr{L}} = Q_N$, where Q_N is the orthoprojector onto $\overline{\text{Span}}\{e_k, k \geq N + 1\}$ in H. Therefore, using the squeezing property given in Proposition 4.3.9 (see also Exercise 4.3.10), we can choose N_* such that

$$\|U(t_n) - u_n\|_\beta \leq q\|U(t_{n-1}) - u_{n-1}\|_\beta \quad \text{with } q < 1.$$

This implies that

$$\|U(t_n) - u_n\|_\beta \to 0 \quad \text{as } n \to +\infty, \quad \forall \beta \in [\alpha, 1),$$

with exponential speed. Therefore, the statement of the theorem is valid in the case of modes.

In the general case, Proposition 4.3.9 (see also Exercise 4.3.10 and Proposition 4.3.11) implies that

$$\|S_{\Delta_n} U(t_{n-1}) - S_{\Delta_n} u_{n-1}\|_\beta$$

$$\leq q_N \|U(t_{n-1}) - u_{n-1}\|_\beta + \lambda_N^{\beta - \alpha} \|S_{\Delta_n} U(t_{n-1}) - S_{\Delta_n} u_{n-1}\|_\alpha \qquad (4.3.54)$$

---

[5] $\epsilon(H_\beta, H_\alpha)$ denotes the corresponding completeness defect; see Definition 3.3.3.

with $\Delta_n = t_n - t_{n-1}$, where $q_N < 1$ can be chosen as small as we need at the expense of N. By the Lipschitz property in (4.2.6) of the evolution operator S_t, we have that

$$\|S_{\Delta_n} U(t_{n-1}) - S_{\Delta_n} u_{n-1}\|_\alpha \leq Ce^{\omega b}\|U(t_{n-1}) - u_{n-1}\|_\alpha.$$

Since $l_j(U(t_{n-1})) = l_j(u_{n-1})$, this gives

$$\|S_{\Delta_n} U(t_{n-1}) - S_{\Delta_n} u_{n-1}\|_\alpha \leq \epsilon(H_\beta, H_\alpha)Ce^{\omega b}\|U(t_{n-1}) - u_{n-1}\|_\beta.$$

Thus (4.3.54) yields

$$\|U(t_n) - u_n\|_\beta \leq \tilde{q}\|U(t_{n-1}) - u_{n-1}\|_\beta,$$

where

$$\tilde{q} = \|I - R_{\mathscr{L}}\|_{H_\beta \mapsto H_\beta}\left[q_N + \lambda_N^{\beta-\alpha}\epsilon(H_\beta, H_\alpha)Ce^{\omega b}\right].$$

Hence we can choose N and $\epsilon(H_\beta, H_\alpha)$ such that $\tilde{q} < 1$. Therefore, the prognosis is asymptotically reliable with exponential speed.   □

We conclude our considerations with the following remark.

Remark 4.3.27.   **1.** As an example of set $\mathscr{L}$ functionals $\{l_j\}$ satisfying (4.3.52), we can consider *generalized modes* which are defined by the formulas:

$$l_j(u) = (Ke_j, u),\quad j = 1,\ldots,N,$$

where $\{e_j\}$ is the eigenbasis of the operator A and K is a linear invertible self-adjoint operator in H such that K and K^{-1} map H_β into itself. In this case the operator $R_{\mathscr{L}}$ has the form (4.3.49) with $\psi_j = K^{-1}e_j$.

2. Under the conditions of Theorem 4.3.26, we also have that

$$\lim_{t \to +\infty} \|U(t) - u(t)\|_\beta = 0$$

for the prognostic trajectory given by (4.3.51). Thus, the prognosis is also reliable in the sense used in HAYDEN/OLSON/TITI [122].

3. The most restrictive hypothesis of Theorem 4.3.26 is the global Lipschitz property for B. To overcome this restriction for general models we need to establish additional dissipativity-type properties for every sequence of prognostic values. This requires additional structural properties of the corresponding system. We demonstrate this approach below in the case of two-dimensional hydrodynamical systems.   ∎

4.4 Abstract model for 2D hydrodynamical systems

Now we consider another class of parabolic-type models. Namely, we deal with a certain abstract situation which covers a wide variety of models arising in two-dimensional hydrodynamics. The main advantage of the scheme developed here is that we do not use specific functional spaces of particular hydrodynamical models; we rely on general properties of abstract operators and spaces. We note that a similar approach to hydrodynamical problems was realized earlier by TEMAM [215, 216].

In contrast with the previous sections, our further considerations are based on the notion of a weak (variational) solution, demonstrating another approach to dynamics in parabolic models.

4.4.1 Abstract hypotheses and motivation

As above, let H denote a separable Hilbert space with the norm $\|\cdot\|$ and the inner product $(.,.)$. Assume that A is a self-adjoint positive linear operator on H. We set $V = H_{1/2} = \mathscr{D}(A^{1/2})$ and denote $\|v\|_V = \|A^{1/2}v\|$ for $v \in V$. Let $V' = H_{-1/2}$ be the dual of V (with respect to $(.,.)$). Thus, we have the triple $V \subset H \subset V'$. The duality between $u \in V$ and $v \in V'$ is denoted by the same symbol as the inner product in H.

The goal is to study the following abstract model in H:

$$u_t(t) + Au(t) + B\big(u(t), u(t)\big) + Ku(t) = f, \quad u\big|_{t=0} = u_0, \qquad (4.4.1)$$

where $f \in V'$ is given and $B : V \times V \mapsto V'$ and $K : H \mapsto H$ are continuous mappings satisfying the following hypotheses (for the basic motivation we refer to Example 4.4.4 below).

Assumption 4.4.1. • The operator $A = A^* > 0$ has a discrete spectrum (see Section 4.1).
- $B : V \times V \to V'$ is a bilinear continuous mapping.
- The trilinear form $b(u_1, u_2, u_3) = (B(u_1, u_2), u_3)$ possesses the following skew-symmetric property

$$(B(u_1, u_2), u_3) = -(B(u_1, u_3), u_2) \quad \text{for } u_i \in V, i = 1, 2, 3. \qquad (4.4.2)$$

- There exists a Banach (interpolation) space $\mathscr{H}$ possessing the properties

 (i) $V \subset \mathscr{H} \subset H$;
 (ii) there exists a constant $a_0 > 0$ such that

$$\|v\|_{\mathscr{H}}^2 \le a_0 \|v\| \, \|v\|_V \quad \text{for any } v \in V; \qquad (4.4.3)$$

(iii) there exists a constant $C > 0$ such that

$$|(B(u_1, u_2), u_3)| \leq C \|u_1\|_{\mathscr{H}} \|u_2\|_V \|u_3\|_{\mathscr{H}}, \quad \text{for } u_i \in V, \ i = 1, 2, 3. \tag{4.4.4}$$

1. $K : H \mapsto H$ is globally Lipschitz.

Remark 4.4.2. **(1)** The relation in (4.4.3) holds true in the case when $\mathscr{H} = H_{1/4}$. However, sometimes it can be another type of space. See the examples below.
(2) The relation in (4.4.4) means that B maps $\mathscr{H} \times V$ into $\mathscr{H}' \subset V'$ and also $\mathscr{H} \times \mathscr{H}$ into V' continuously. In particular, we have that

$$\|B(u_1, u_2)\|_{V'} \equiv \|A^{-1/2}B(u_1, u_2)\| \leq C \|u_1\|_{\mathscr{H}} \|u_2\|_{\mathscr{H}} \quad \text{for } u_1, u_2 \in \mathscr{H}. \tag{4.4.5}$$

(3) Since the operator A has a discrete spectrum, by Proposition 4.1.6 the space V is compactly embedded into H. This property implies that V is also compactly embedded into $\mathscr{H}$. Indeed, if $\{u_n\}$ is a sequence in V which is weakly convergent to zero in V, then due to the compactness of the embedding $V \subset H$ we have that

$$\lim_{n \to \infty} \|u_n\| = 0 \quad \text{and} \quad \sup_n \|u_n\|_V < \infty.$$

Therefore (4.4.3) yields that $\|u_n\|_{\mathscr{H}} \to 0$ as $N \to \infty$. This means that the embedding $V \subset \mathscr{H}$ is also compact. Due to (4.4.5) this implies that the mapping

$$(u_1; u_2) \mapsto B(u_1, u_2)$$

is *weakly* continuous from $V \times V$ into V' equipped with the strong topology. This observation is important for 2D hydrodynamical PDE models in bounded domains.

∎

Exercise 4.4.3. Using relations (4.4.2), (4.4.3), and (4.4.4), show that for every $\eta > 0$ there exists $C_\eta > 0$ such that

$$|(B(u_1, u_2), u_3)| \leq \eta \|u_3\|_V^2 + C_\eta \|u_1\|_{\mathscr{H}}^2 \|u_2\|_{\mathscr{H}}^2, \quad u_i \in V, \ i = 1, 2, 3, \tag{4.4.6}$$

$$|(B(u_1), u_2)| \leq \eta \|u_1\|_V^2 + C_\eta \|u_1\|^2 \|u_2\|_{\mathscr{H}}^4, \quad u_1, u_2 \in V, \tag{4.4.7}$$

and also

$$|(B(u_1) - B(u_2), u_1 - u_2)| = |(B(u_1 - u_2), u_2)| \tag{4.4.8}$$

$$\leq \eta \|u_1 - u_2\|_V^2 + C_\eta \|u_1 - u_2\|^2 \|u_2\|_{\mathscr{H}}^4, \quad u_1, u_2 \in V.$$

In the last two relations we use the notation $B(u) = B(u, u)$. ∎

The main motivation for the conditions in Assumption 4.4.1 is that it covers a wide class of 2D hydrodynamical models. In this section we mention only one of them; for others we refer to Section 4.6.

Example 4.4.4 (2D Navier-Stokes equation). Let D be a bounded simply connected domain of $\mathbb{R}^2$. We consider the Navier-Stokes equation with the Dirichlet (no-slip) boundary conditions:

$$\partial_t u - \nu \Delta u + u \nabla u + \nabla p = f, \quad \text{div}\, u = 0 \ \text{in}\ D, \qquad u = 0 \quad \text{on}\ \partial D, \qquad (4.4.9)$$

where $u = (u^1(x, t); u^2(x, t))$ is the velocity of a fluid, $p(x, t)$ is the pressure, $x = (x_1; x_2) \in D$, $\nu > 0$ is the kinematic viscosity, and $f(x)$ represents external forces. We also use the notation

$$u \nabla u = \left(\sum_{i=1,2} u_i \partial_i \right) u \ \text{and}\ \text{div}\, u = \sum_{i=1,2} \partial_i u_i.$$

Let n denote the outward normal to ∂D and let

$$H = \{ f \in \left[L^2(D) \right]^2 : \text{div} f = 0 \ \text{in}\ D \ \text{and}\ (f, n) = 0 \ \text{on}\ \partial D \}$$

be endowed with the usual L^2 scalar product. Projecting on the space H of divergence-free vector fields, problem (4.4.9) can be written in the form (4.4.1) (with $K \equiv 0$) in the space H (see, e.g., [215]), where A is the Stokes operator generated by the bilinear form

$$a(u_1, u_2) = \nu \sum_{j=1}^{2} \int_D \nabla u_1^j \cdot \nabla u_2^j \, dx, \qquad (4.4.10)$$

with $u_1, u_2 \in V = \left[H_0^1(D) \right]^2 \cap H$. Here $H^1(D)$ is the first order Sobolev space on D; see [1]. The map $B \equiv \tilde{B} : V \times V \to V'$ is defined by

$$(\tilde{B}(u_1, u_2), u_3) = \int_D [u_1(x) \nabla u_2(x)] \, u_3(x) dx \equiv \sum_{i,j=1}^{2} \int_D u_1^j \, \partial_j u_2^i \, u_3^i \, dx \qquad (4.4.11)$$

for $u_i \in V$. Using integration by parts, and Schwarz's and Young's inequalities, one checks that this map $\tilde{B}$ satisfies Assumption 4.4.1 with $\mathcal{H} = \left[L^4(D) \right]^2 \cap H$. The inequality in (4.4.3) is the well-known Ladyzhenskaya inequality (see, e.g., CONSTANTIN/FOIAS [78] or TEMAM [215]).

We can also include in (4.4.9) a Coriolis-type force by changing f into $f - Ku$, where $K(u^1, u^2) = c_0(-u^2, u^1)$, for some constant c_0. In this case we get (4.4.1) with $K \neq 0$.   ∎

4.4.2 Well-posedness

In this section we prove that the problem in (4.4.1) is well-posed in the class of weak (variational) solutions. To do this, we use the Galerkin method in combination with compactness[6] properties discussed in Remark 4.4.2(3). Note also that we cannot apply the result from Section 4.2. The point is that, from Assumption 4.4.1, the nonlinearity is locally Lipschitz from $\mathscr{H} \supset V = H_{1/2}$ into $V' = H_{-1/2}$. Thus, in general, B *does not* map $H_{s+\alpha}$ into H_s for some $s \in \mathbb{R}$ and $\alpha \in [0, 1)$. Hence we cannot guarantee the main requirement of Section 4.2 concerning the nonlinearity (cf. Assumption 4.2.1 and Remark 4.2.5).

We start with the notion of a weak solution for the model in (4.4.1).

Definition 4.4.5. A function[7] $u \in L_2(0, T; V)$ is said to be a weak solution for (4.4.1) on an interval $[0, T]$ if the following relation is satisfied:

$$-\int_0^T (u(t), v_t(t))dt + \int_0^T [(Au(t), v(t)) + (B(u(t)) + K(u(t)), v(t))]\, dt$$

$$= (u_0, v(0)) + \int_0^T (f, v(t))dt \qquad (4.4.12)$$

for any function v from the class

$$W_T = \{v \in L_2(0, T; V) : v_t \in L_2(0, T; H), \ v(T) = 0\}.$$

Here and below we use the notation $B(u) = B(u, u)$.

The following exercise provides another form of relation (4.4.12).

Exercise 4.4.6. Using a test function of the form

$$v(t) = \left[\int_t^T \phi(\tau)d\tau\right] \cdot v \quad \text{with } \phi \in C([0, T]), \ v \in V,$$

show that any weak solution satisfies the relation

$$(u(t), v) = (u_0, v) - \int_0^t [(Au(\tau), v) + (B(u(\tau)) + K(u(\tau)) - f, v)]\, d\tau \qquad (4.4.13)$$

for every $v \in V$ and for almost all t from the interval $[0, T]$.   ∎

Our main result in this section is the following theorem.

---

[6] In the Appendix we present another approach (based on the monotonicity idea presented in MENALDI/SRITHARAN [165]) which does not involve discreteness of the spectrum of the operator A and compactness of embeddings.

[7] See the Appendix for the basic definitions and properties of measurable vector functions.

Theorem 4.4.7 (Well-posedness). *Let Assumption* 4.4.1 *be in force. Then for any* $f \in V'$ *and* $u_0 \in H$, *problem* (4.4.1) *has a unique weak solution u on any interval* $[0, T]$. *This solution (i) belongs to* $C(0, T; H) \cap L_2(0, T; V)$, *(ii) satisfies the following energy balance relation:*

$$\frac{1}{2}\|u(t)\|^2 + \int_0^t \|u(\tau)\|_V^2 d\tau = \frac{1}{2}\|u(0)\|^2 + \int_0^t (f - K(u(\tau)), u(\tau))d\tau, \quad t > 0,$$
$$(4.4.14)$$

(iii) possesses the properties

$$\|u(t)\|^2 + \int_0^t \|u(\tau)\|_V^2 d\tau + \int_0^t \|u(\tau)\|_{\mathcal{H}}^4 d\tau \leq C_T(\|f\|_{V'}, \rho), \quad t \in [0, T], \quad (4.4.15)$$

for every $\|u_0\| \leq \rho$, *and*

$$\|u^1(t) - u^2(t)\|^2 + \int_0^t \||u^1(\tau) - u^2(\tau)\|_V^2 d\tau$$

$$\leq c_0\|u^1(0) - u^2(0)\|^2 \exp\left\{c_1 t + c_2 \int_0^t \|u^1(\tau)\|_{\mathcal{H}}^4 d\tau\right\} \quad (4.4.16)$$

for every pair of solutions $u^1(t)$ *and* $u^2(t)$, *where* c_i *are absolute constants.*

Proof. We use the Galerkin method. Let $\{\varphi_k\}_{k \geq 1}$ be an orthonormal basis[8] of the Hilbert space H such that $\varphi_k \in \mathscr{D}(A)$. We denote by $u^N(t) = \sum_{k=1}^{N} g_k(t)\varphi_k$ a function satisfying the relations

$$(u_t^N + Au^N + G(u^N), \varphi_k) = 0, \ k = 1, \ldots, N, \quad \text{and} \quad u^N\big|_{t=0} = P_N u_0 \in H, \quad (4.4.17)$$

where P_N is the orthogonal projector in H on $H_N = \text{Span}\{\varphi_1, \cdots, \varphi_N\}$ and $G : V \to V'$ is defined by

$$G(u) = B(u, u) + K(u) - f, \quad \forall u \in V.$$

One can see that Galerkin approximate solutions u^N exist at least locally. Moreover, multiplying the first relation (4.4.17) by $g_k(t)$ after summation we obtain the balance relation in (4.4.14) with u^N instead of u on every existence interval $[0, T]$. This leads to the following a priori estimate:

$$\|u^N(t)\|^2 + \int_0^t \|u^N(\tau)\|_V^2 d\tau \leq C_T(f, u_0), \quad t \in [0, T],$$

which, due to (4.4.3), yields

---

[8] We can use the eigenbasis $\{e_k\}$ of A. However, this is not necessary.

$$\|u^N(t)\|^2 + \int_0^t \|u^N(\tau)\|_V^2 d\tau + \int_0^t \|u^N(\tau)\|_{\mathscr{H}}^4 d\tau \le C_T(f, u_0), \quad t \in [0, T]. \quad (4.4.18)$$

From (4.4.17) using (4.4.5) we obtain

$$\|u_t^N(\tau)\|_{V'} \equiv \|A^{-1/2} u_t^N(\tau)\| \le c_0 \|u^N(\tau)\|_V + c_1 \|u^N(\tau)\|_{\mathscr{H}}^2 + c_2.$$

Thus, we arrive at the following additional a priori estimate:

$$\int_0^T \|u_t^N(\tau)\|_{V'}^2 d\tau \le C_T(f, u_0). \quad (4.4.19)$$

The estimates in (4.4.18) and (4.4.19) imply the global existence of approximate solutions and also the existence of the function

$$u(t) \in L_\infty(0, T; H) \cap L_2(0, T; V) \cap L_4(0, T; \mathscr{H}) \quad \text{with } u_t \in L_2(0, T; V') \quad (4.4.20)$$

such that along a subsequence we have

(i) $u^N \to u$ weakly in $L^2(0, T; V)$ and in $L^4(0, T; \mathscr{H})$, *-weakly in $L_\infty(0, T; H)$;
(ii) $u_t^N \to u_t$ weakly in $L^2(0, T; V')$.

By Proposition A.3.3 from the Appendix, $u(t) \in C([0, T]; H)$. Moreover, since the embedding $V \subset \mathscr{H}$ is compact, using the Aubin-Dubinskii-Lions theorem (see Section A.3.3 in the Appendix) we conclude that

$$\max_{[0,T]} \|u^N(\tau) - u(\tau)\|_{V'}^2 + \int_0^T \|u^N(\tau) - u(\tau)\|_{\mathscr{H}}^2 d\tau \to 0, \quad N \to \infty.$$

Using (4.4.5) we also have that

$$\int_0^T \|B(u^N(\tau)) - B(u(\tau))\|_{\mathscr{H}}^{4/3} d\tau \le C \int_0^T \|u^N(\tau)\|_{\mathscr{H}}^{4/3} \|u^N(\tau) - u(\tau)\|_{\mathscr{H}}^{4/3} d\tau$$

$$\le C \left[\int_0^T \|u^N(\tau)\|_{\mathscr{H}}^4 d\tau \right]^{1/3} \left[\int_0^T \|u^N(\tau) - u(\tau)\|_{\mathscr{H}}^2 d\tau \right]^{2/3} \to 0,$$

as $N \to \infty$. All these convergences stated above allow us to make the limit transition $N \to \infty$ in the integral form of (4.4.17) and show that $u(t)$ is a weak solution to problem (4.4.1) satisfying (4.4.15).

It follows from (4.4.20) that u satisfies (4.4.1) as an equality in V'. This implies the balance relation in (4.4.14) and also allows us to prove (4.4.16). Indeed, the function $u(t) = u^1(t) - u^2(t)$ satisfies the equation

$$u_t + Au + B(u^1(t), u^1(t)) - B(u^2(t), u^2(t)) + K(u^1(t)) - K(u^2(t)) = 0.$$

Therefore, multiplying[9] the equation by $u(t)$ and relying on (4.4.8) with $\eta = 1/2$, one can see that

$$\frac{d}{dt}\|u(t)\|^2 + \|u(t)\|_V^2 \le 2\big(R_1 + C_{1/2}\|u'(t)\|_{\mathscr{H}}^4\big)\|u(t)\|^2.$$

Thus, the standard argument via Gronwall's lemma yields (4.4.16).

Uniqueness of solutions to (4.4.1) follows from (4.4.16).   □

4.4.3   Generation of C^1 dynamical system

By Theorem 4.4.7 problem (4.4.1) generates a dynamical system (H, S_t) with evolution operator defined by weak solutions according to the formula

$$S_t u_0 = u(t), \quad \text{where } u(t) \text{ solves (4.4.1)}.$$

Our goal in this section is to show that the semiflow S_t is C^1 with respect to initial data (the corresponding definitions can be found in Section A.5 in the Appendix).

Theorem 4.4.8. *Let Assumption 4.4.1 be in force. Assume for simplicity that K is a linear operator. Then the semiflow S_t generated by (4.4.1) is C^1 in the sense that $u \mapsto S_t u$ has a Fréchet derivative[10] $S_t'[u]$ for each $u \in H$ which depends continuously on u in the operator norm. This derivative $S_t'[u]$ is a linear bounded mapping on H and can be calculated by the formula $S_t'[u]w_0 = w(t)$, where $w(t)$ is a weak (variational) solution to the following linear nonautonomous problem:*

$$w_t + Aw + B\big(w, u(t)\big) + B\big(u(t), w\big) + Kw = 0, \quad w\big|_{t=0} = w_0, \qquad (4.4.21)$$

with $u(t) = S_t u$.

The main ingredient in the proof of this theorem is the following assertion.

Lemma 4.4.9. *Let the hypotheses of Theorem 4.4.8 be in force. Then for every $u(t) \in C(0, T; H) \cap L_2(0, T; V)$ and $w_0 \in H$, problem (4.4.21) has a unique weak solution w on the interval $[0, T]$. This solution possesses the property*

$$w(t) \in C(0, T; H) \cap L_2(0, T; V)$$

and satisfies the following estimate:

---

[9]This multiplication can be justified via Proposition A.3.3.

[10]See details in the Appendix, Section A.5.

$$\|w(t)\|^2 + \int_0^t \|w(\tau)\|_V^2 d\tau \le \|w(0)\|^2 \exp\left\{2t\,\|K\|_{H \mapsto H} + c \int_0^t \|u(\tau)\|_V^2 d\tau\right\}$$

$$(4.4.22)$$

for all $t \in [0, T]$. If $w(t)$ and $\bar{w}(t)$ are solutions to (4.4.21) with different $u(t)$ and $\bar{u}(t)$, then

$$\|w(t) - \bar{w}(t)\|^2 + \int_0^t \|w(\tau) - \bar{w}(\tau)\|_V^2 d\tau$$

$$\le c_1 e(t, u, \bar{u})\|w_0\|^2 \left[\int_0^t \|u(\tau) - \bar{u}(\tau)\|_{\mathscr{H}}^4 d\tau\right]^{1/2}, \qquad (4.4.23)$$

where

$$e(t, u, \bar{u}) \equiv \exp\left\{4t\,\|K\|_{H \mapsto H} + c_2 \int_0^t \left[\|u(\tau)\|_V^2 + \|\bar{u}(\tau)\|_V^2\right] d\tau\right\}.$$

Proof. To obtain the result we can use the Galerkin method, and for this we need some a priori estimates. We obtain them at the formal level. They can be justified in the standard way (see, e.g., LIONS [151]). Multiplying (4.4.21) by $w(t)$ in H yields

$$\frac{1}{2}\frac{d}{dt}\|w\|^2 + \|w\|_V^2 + (B(w, u(t)), w) + (Kw, w) = 0.$$

Using (4.4.4) and the interpolation inequality in (4.4.3), one can see that

$$\frac{d}{dt}\|w\|^2 + \|w\|_V^2 \le 2\left[\|K\|_{H \mapsto H} + c_1\|u(t)\|_V^2\right]\|w\|^2.$$

This implies (4.4.22) and provides an appropriate a priori estimate.

To establish (4.4.23) we note that $z = w - \bar{w}$ satisfies the equation

$$z_t + Az + B(z, u(t)) + B(u(t), z) + B(\bar{w}, u(t) - \bar{u}(t)) + B(u(t) - \bar{u}(t), \bar{w}) + Kz = 0.$$

Using (4.4.4) we obtain

$$\frac{d}{dt}\|z\|^2 + \|z\|_V^2 \le 2\left[\|K\|_{H \mapsto H} + c_1\|u(t)\|_V^2\right]\|z\|^2 + c_2\|\bar{w}\|_{\mathscr{H}}^2\|u(t) - \bar{u}(t)\|_{\mathscr{H}}^2.$$

Therefore, using Gronwall's lemma and also the estimate in (4.4.22) for $\bar{w}$, we can obtain (4.4.23). □

To conclude the proof of Theorem 4.4.8, we follow the standard scheme (see, e.g., BABIN/VISHIK [9]). Let $w(t) \equiv w(t; u_0, w_0)$ solve (4.4.21) with $u(t) = S_t u_0$. Then the function

$$z(t) = S_t[u_0 + w_0] - S_t[u_0] - w(t)$$

satisfies the equation

$$z_t + Az + F(t) = 0 \tag{4.4.24}$$

with

$$F(t) = B(\bar{u}, \bar{u}) - B(u, u) - B(u, w) - B(w, u) + Kz,$$

where $\bar{u}(t) = S_t[u_0 + w_0]$ and $u(t) = S_t[u_0]$.

One can see that $F(t)$ can be written in the form

$$F = B(z, \bar{u}) + B(u, z) + B(w, \bar{u} - u) + Kz.$$

Therefore, using the skew symmetry of B, we have that

$$(F, z) = (B(z, \bar{u}), z) + (B(w, w), z) + (Kz, z).$$

Hence by (4.4.3) and (4.4.4) we obtain that

$$|(F, z)| \leq \frac{1}{2} \left[\|z\|_V^2 + c_0 \left(1 + \|\bar{u}\|_V^2\right) \|z\|^2 + c_1 \|w\|_{\mathscr{H}}^4 \right]$$

for some constants $c_0, c_1 > 0$. Therefore, multiplying (4.4.24) by z we obtain

$$\frac{d}{dt} \|z\|^2 + \|z\|_V^2 \leq c_0 \left(1 + \|\bar{u}\|_V^2\right) \|z\|^2 + c_1 \|w\|_{\mathscr{H}}^4.$$

Thus, Gronwall's lemma yields

$$\|z(t)\|^2 + \int_0^t \|z(\tau)\|_V^2 d\tau \leq c_1 \exp\left\{ c_0 \int_0^t \left(1 + \|\bar{u}\|_V^2\right) d\tau \right\} \int_0^t \|w\|_{\mathscr{H}}^4 d\tau.$$

Therefore, by the estimate in (4.4.15) and Lemma 4.4.9 we obtain that

$$\|S_t[u_0 + w_0] - S_t[u_0] - w(t; u_0, w_0)\|^2 \leq C_R(T) \|w_0\|^4$$

for every $t \in [0, T]$ and $\|u_0 + w_0\|, \|w_0\| \leq R$. Thus $w(t; u_0, w_0) = S_t'[u_0]w_0$. This implies the conclusion of Theorem 4.4.8.

4.5   Long-time dynamics in 2D hydrodynamical systems

In this section we study asymptotic properties of the system (H, S_t) generated by problem (4.4.1). We follow the schemes and rely on the results presented in Chapters 2 and 3.

4.5.1   Dissipativity

In the following assertion we collect several preliminary dissipativity properties of the system.

Proposition 4.5.1 (Primitive dissipativity estimates). *Let Assumption* 4.4.1 *be in force. Assume that K is skew-symmetric; i.e.,* $(Ku, u) = 0$ *for all* $u \in H$. *Let* λ_1 *be the minimal eigenvalue of the operator A. Then for any solution* $u(t)$ *to* (4.4.1) *we have*

$$\|u(t)\|^2 \le \|u_0\|^2 e^{-\lambda_1 t} + \frac{\|f\|_{V'}^2}{\lambda_1}\left(1 - e^{-\lambda_1 t}\right), \quad \forall\, t \ge 0. \tag{4.5.1}$$

Moreover,

$$\|u(t+1)\|^2 + \int_t^{t+1} \|u(\tau)\|_V^2 d\tau \le \|u_0\|^2 e^{-\lambda_1 t} + \|f\|_{V'}^2\left(1 + \frac{1}{\lambda_1}\right), \quad \forall\, t \ge 0. \tag{4.5.2}$$

and also

$$\int_t^{t+1} \|u(\tau)\|_{\mathscr{H}}^4 d\tau \le a_0^2 \left[\|u_0\|^2 e^{-\lambda_1 t} + \|f\|_{V'}^2\left(1 + \frac{1}{\lambda_1}\right)\right]^2, \quad \forall\, t \ge 0. \tag{4.5.3}$$

Proof. The standard multiplication by u gives that

$$\frac{1}{2}\frac{d}{dt}\|u\|^2 + (Au, u) = (f, u). \tag{4.5.4}$$

Since $|(f, u)| \le (\|f\|_{V'}^2 + \|u\|_V^2)/2$, this implies that

$$\frac{d}{dt}\|u(t)\|^2 + \lambda_1\|u(t)\|^2 \le \|f\|_{V'}^2.$$

Applying a Gronwall-type argument we have (4.5.1). Similarly, after integration from (4.5.4), we have that

$$\|u(t+1)\|^2 + \int_t^{t+1} \|u(\tau)\|_V^2 d\tau \le \|u(t)\|^2 + \|f\|_{V'}^2.$$

Thus, (4.5.2) follows from (4.5.1).

To prove (4.5.3) we use two previous estimates and also the interpolation inequality in (4.4.3).   □

Remark 4.5.2 (Absorbing properties). Proposition 4.5.1 implies the following absorbing properties: for every $\rho > 0$ there exists a time t_ρ such that we have

$$\|u(t+1)\|^2 + \int_t^{t+1} \|u(\tau)\|_V^2 d\tau \leq R_f^2 \equiv 1 + \|f\|_{V'}^2 \left(1 + \frac{1}{\lambda_1}\right), \quad \forall\, t \geq t_\rho, \quad (4.5.5)$$

and

$$\int_t^{t+1} \|u(\tau)\|_{\mathscr{H}}^4 d\tau \leq a_0^2 R_f^4, \quad \forall\, t \geq t_\rho, \quad (4.5.6)$$

for all u_0 with the property that $\|u_0\| \leq \rho$. In particular, the system (H, S_t) generated by (4.4.1) is dissipative; see Definition 2.1.1.   ■

4.5.2 Determining functionals

The dissipativity properties established above make it possible to obtain the existence of a finite number of (asymptotically) determining functionals (for the definitions we refer to Section 3.3), even without any compactness assumptions.

Theorem 4.5.3. *Assume that the hypotheses of Proposition 4.5.1 hold. Let u_1 and u_2 be two solutions to problem (4.4.1) with different initial data. Let $\mathscr{L} = \{l_i : i = 1, \ldots, N\}$ be a set of the linear continuous functionals on V such that*

$$\lim_{t \to \infty} l_j(u_1(t) - u_2(t)) = 0, \quad j = 1, 2, \ldots, N. \quad (4.5.7)$$

Assume that one of the solutions (say, u_1) possesses the property

$$\limsup_{t \to +\infty} \int_t^{t+1} \|u_1(\tau)\|_{\mathscr{H}}^4 d\tau \leq R_*^4. \quad (4.5.8)$$

Then the set $\mathscr{L}$ is asymptotically determining for these two solutions in the sense that $\|u_1(t) - u_2(t)\| \to 0$ as $t \to \infty$, provided the completeness defect $\epsilon_{\mathscr{L}} \equiv \epsilon_{\mathscr{L}}(V, H)$ satisfies the condition

$$\varepsilon_{\mathscr{L}} < \varepsilon_* \equiv \left(2R_1 + 2C_{1/2} R_*^4\right)^{-1/2}, \quad (4.5.9)$$

where R_1 is the Lipschitz constant of K and $C_{1/2}$ is the constant in (4.4.7) with $\eta = 1/2$.

Proof. Let $u = u_1 - u_2$. It follows from (4.4.8) with $\eta = 1/2$ that

$$\frac{d}{dt}\|u\|^2 + \|u\|_V^2 \leq 2\big(R_1 + C_{1/2}\|u_1\|_{\mathscr{H}}^4\big)\|u\|^2, \tag{4.5.10}$$

where R_1 and $C_{1/2}$ are the same as in the statement.

By Proposition 3.3.4,

$$\|A^{1/2}u\|^2 \geq \frac{1}{(1+\delta)\epsilon_{\mathscr{L}}^2}\|u\|^2 - C_{\mathscr{L},\delta}\mathscr{N}_{\mathscr{L}}^2(u), \quad \mathscr{N}_{\mathscr{L}}(u) = \max_j |l_j(u)|,$$

for every $\delta > 0$. Thus, we obtain

$$\frac{d}{dt}\|u\|^2 + \left[\frac{1}{(1+\delta)\epsilon_{\mathscr{L}}^2} - 2\big(R_1 + C_{1/2}\|u_1(t)\|_{\mathscr{H}}^4\big)\right]\|u\|^2 \leq C_{\mathscr{L},\delta}\mathscr{N}_{\mathscr{L}}^2(u(t)). \tag{4.5.11}$$

This allows us to complete the proof. Indeed, it follows from (4.5.11) that

$$\|u(t)\|^2 + \int_s^t \psi(\tau)\cdot\|u(\tau)\|^2\,d\tau \leq \|u(s)\|^2 + C_{\mathscr{L},\delta}\int_s^t \mathscr{N}_{\mathscr{L}}^2(u(\tau))\,d\tau \tag{4.5.12}$$

holds for all $t \geq s \geq 0$, where

$$\psi(t) = \frac{1}{(1+\delta)\epsilon_{\mathscr{L}}^2} - 2\big(R_1 + C_{1/2}\|u_1(t)\|_{\mathscr{H}}^4\big).$$

Under the condition (4.5.9) we have that

$$\psi_\infty \equiv \liminf_{t\to\infty}\int_t^{t+1}\psi(\tau)\,d\tau \geq \frac{1}{(1+\delta)\epsilon_{\mathscr{L}}^2} - 2\big(R_1 + C_{1/2}R_*^4\big) > 0.$$

for some $\delta > 0$. Thus, we can apply Theorem 3.3.13 to obtain the conclusion.　□

4.5.3 Compactness of the semiflow and a global attractor

The proof of the existence of global attractors requires some compactness properties of semiflow, and for this we need to impose additional requirements concerning the system. These conditions are motivated by the hydrodynamical systems considered in Section 4.6.

Proposition 4.5.4. *In addition to basic Assumption* 4.4.1, *we assume that*

$$H_{1/4} \subset \mathscr{H} \quad and \quad \|u\|_{\mathscr{H}} \leq C\|u\|_{1/4}, \quad u \in H_{1/4}. \tag{4.5.13}$$

Let $f \in H_{-1/4}$. *Then* $H_{1/4}$ *is invariant with respect to the semiflow* S_t. *Moreover, for every* $T > 0$ *and* $R > 0$ *there exists a constant* $C(T, R)$ *such that*

$$\|A^{1/4}S_t u_0\|^2 \leq \frac{1}{\sqrt{t}}\, C(T,R) \ \ for \ all \ \ t \in (0,T], \qquad (4.5.14)$$

provided $\|u_0\| \leq R$.

Remark 4.5.5. The condition in (4.5.13) is true in the case of the 2D Navier-Stokes equations (see Example 4.4.4). The point is that in this case we have

$$\mathscr{D}(A^{1/4}) \subset \left[L^4(D)\right]^2 \cap H = \mathscr{H}. \qquad (4.5.15)$$

This fact relies on space interpolation theory and certain embedding theorems (see, e.g., TRIEBEL [220]). Some self-contained details concerning (4.5.15) can be found in CONSTANTIN/FOIAS [78]. ■

Proof. The following argument can be justified on the Galerkin approximations.
  Multiplying (4.4.1) by $A^{1/2}u$ we obtain

$$\frac{1}{2}\frac{d}{dt}\|A^{1/4}u\|^2 + \|A^{3/4}u\|^2 + (B(u), A^{1/2}u) + (K(u) - f, A^{1/2}u) = 0.$$

By (4.4.4) and (4.5.13) we have

$$|(B(u), A^{1/2}u)| \leq C\|u\|_{\mathscr{H}} \|u\|_V \|A^{1/2}u\|_{\mathscr{H}} \leq C\,\|A^{3/4}u\| \|u\|_{\mathscr{H}} \|u\|_V.$$

By interpolation, $\|u\|_V \leq \|A^{1/4}u\|^{1/2}\|A^{3/4}u\|^{1/2}$. Thus,

$$|(B(u), A^{1/2}u)| \leq \frac{1}{4}\|A^{3/4}u\|^2 + C\|A^{1/4}u\|^2 \|u\|_{\mathscr{H}}^4.$$

We also have that

$$|(K(u) - f, A^{1/2}u)| \leq \frac{1}{4}\|A^{3/4}u\|^2 + C_1\|A^{1/4}u\|^2 + C_2(1 + \|f\|_{-1/4}^2).$$

Consequently,

$$\frac{d}{dt}\|A^{1/4}u\|^2 + \|A^{3/4}u\|^2 \leq c_1\left[1 + \|u\|_{\mathscr{H}}^4\right]\|A^{1/4}u\|^2 + c_2(1 + \|f\|_{-1/4}^2). \qquad (4.5.16)$$

Introducing

$$\Psi(t) = \|A^{1/4}u(t)\|^2 + \int_0^t \|A^{3/4}u(\tau)\|^2\, d\tau,$$

we can rewrite (4.5.16) as

$$\frac{d}{dt}\Psi(t) \leq c_1 \left[1 + \|u(t)\|_{\mathscr{H}}^4\right]\Psi(t) + c_2(1 + \|f\|_{-1/4}^2).$$

Thus, Gronwall's lemma yields

$$\|A^{1/4}u(t)\|^2 + \int_0^t \|A^{3/4}u(\tau)\|^2\,d\tau$$

$$\leq \left[\|A^{1/4}u_0\|^2 + c_0(1 + \|f\|_{-1/4}^2)\right]\exp\left\{c_1\left[t + \int_0^t \|u\|_{\mathscr{H}}^4 d\tau\right]\right\}.$$

This implies that $H_{1/4}$ is forward invariant with respect to S_t.

To obtain the smoothening property in (4.5.14), we multiply (4.5.16) by t. This allows us to show that the function $\psi(t) := t\|A^{1/4}u(t)\|^2$ satisfies the relation

$$\frac{d}{dt}\psi(t) \leq c_1\left[1 + \|u\|_{\mathscr{H}}^4\right]\psi(t) + \|A^{1/4}u(t)\|^2 + c_2 t(1 + \|f\|_{-1/4}^2).$$

Applying Gronwall's lemma we obtain that

$$t\|A^{1/4}u(t)\|^2 \leq h(t)\exp\left\{c_1\left[t + \int_0^t \|u\|_{\mathscr{H}}^4 d\tau\right]\right\},$$

where

$$h(t) = c_2 t^2(1 + \|f\|_{-1/4}^2) + \int_0^t \|A^{1/4}u(\tau)\|^2\,d\tau$$

$$\leq c_2 t^2(1 + \|f\|_{-1/4}^2) + \sqrt{t}\max_{[0,T]}\|u(\tau)\|\left[\int_0^T \|A^{1/2}u(\tau)\|^2\,d\tau\right]^{1/2}.$$

Here above we have used the Hölder inequality and also the interpolation relation $\|A^{1/4}u\|^2 \leq \|u\|\|A^{1/2}u\|$. Thus, by (4.4.15) we obtain that

$$h(t) \leq C(T,f,R)\sqrt{t}, \quad t \in (0,T].$$

This implies (4.5.14). □

Another possibility to get the smoothening property is to impose additional hypotheses concerning the nonlinearity B (again, the condition below is motivated by the 2D Navier-Stokes system; see, e.g., TEMAM [215]).

Proposition 4.5.6. *In addition to basic Assumption 4.4.1, we assume that $f \in H$ and $B : V \times \mathscr{D}(A) \mapsto H$ and also that there exists a constant $C > 0$ such that*

$$\|B(u_1,u_2)\| \leq C\|u_1\|_{\mathscr{H}}\|u_2\|_V^{1/2}\|Au_2\|^{1/2}, \quad \text{for } u_1 \in V, \; u_2 \in \mathscr{D}(A). \quad (4.5.17)$$

Then $H_{1/2}$ is invariant with respect to semiflow S_t and for every $T > 0$ and $R > 0$
there exists a constant $C(T, R)$ such that

$$\|A^{1/2}S_t u_0\|^2 \le \frac{1}{t} C(T, R) \text{ for all } t \in (0, T], \tag{4.5.18}$$

provided $\|u_0\| \le R$.

Proof. As in Proposition 4.5.4, our argument is formal. For justification we can use
the Galerkin approximations.

Multiplying (4.4.1) by Au and using (4.5.17), we obtain that

$$\frac{d}{dt}\|A^{1/2}u\|^2 + \|Au\|^2 \le c_1 \left[1 + \|u\|_{\mathcal{H}}^4\right]\|A^{1/2}u\|^2 + c_2(1 + \|f\|^2). \tag{4.5.19}$$

Therefore, the same argument as in Proposition 4.5.4 makes it possible to show that
$H_{1/2}$ is forward invariant with respect to S_t.

Multiplying (4.5.19) by t, we obtain that the function $\phi(t) := t\|A^{1/2}u(t)\|^2$
satisfies the relation

$$\frac{d}{dt}\phi(t) \le c_1 \left[1 + \|u\|_{\mathcal{H}}^4\right]\phi(t) + \|A^{1/2}u(t)\|^2 + c_2 t(1 + \|f\|^2).$$

Gronwall's lemma yields

$$\|A^{1/2}u(t)\|^2 \le t^{-1}g(t)\exp\left\{c_1\left[t + \int_0^t \|u(\tau)\|_{\mathcal{H}}^4 d\tau\right]\right\},$$

where

$$g(t) = c_2 t^2(1 + \|f\|^2) + \int_0^t \|A^{1/2}u(\tau)\|^2 d\tau.$$

Relation (4.4.15) yields $g(t) \le C(T, f, R)$ for $t \in (0, T]$. Thus (4.5.18) follows. $\square$

Now we are in position to formulate hypotheses which guarantee the existence of a
compact global attractor for the system considered.

Assumption 4.5.7. In addition to the requirements in Assumption 4.4.1, we assume
that

- $f \in H$ and $(K(u), u) = 0$ for every $u \in V$;
- one of the assumptions in (4.5.13) or in (4.5.17) holds.

Under this assumption we can apply either Proposition 4.5.4 or Proposition 4.5.6
and show that there exists R_* such that the set

$$B = \{u \in H : \|A^{1/4}u(t)\| \le R_*\}$$

is an absorbing set for the system (H, S_t). Since A has a discrete spectrum and, by Proposition 4.1.6, $H_{1/4}$ is compactly embedded into H, the set B is compact. Thus, the system (H, S_t) is compact and from Theorem 2.3.5 we have the following assertion.

Theorem 4.5.8 (Global attractor). *Let Assumption 4.5.7 be in force. Then the system (H, S_t) generated by (4.4.1) has a compact global attractor. This attractor is a bounded set in $H_{1/4}$ (in $H_{1/2}$ when (4.5.17) holds).*

4.5.4 Squeezing, quasi-stability, and finite-dimensional attractors

In this section we establish the Ladyzhenskaya squeezing property and also quasi-stability for 2D hydrodynamical-type systems. As was already mentioned in Section 4.3.3, the squeezing property demonstrates a strong form of quasi-stability. In principle this property allows us to study the long-time dynamics of the system directly (as was done in LADYZHENSKAYA [142], for instance). However, the quasi-stability method incorporates this class of models in a more general unified framework and provides useful tools for the further studies.

Proposition 4.5.9 (Ladyzhenskaya squeezing property). *Let Assumption 4.5.7 be in force and $Q_N = I - P_N$, where P_N is the orthoprojector onto Span $\{e_1, \ldots, e_N\}$ in H, where $\{e_k\}$ is the eigenbasis of the operator A. Then for every $0 < q < 1$, $0 < a \le b < +\infty$, and $R > 0$ there exists $N_* = N(a, b, R, q)$ such that*

$$\|Q_N[S_t u - S_t u_*]\| \le q\|u - u_*\|, \quad \forall t \in [a, b], \quad \forall N \ge N_*, \tag{4.5.20}$$

for any u and u_* from the set $\mathscr{D}$, where

$$\mathscr{D} = \{u \in H_{1/4} : \|S_t u\|_{1/4} \le R \text{ for all } t \in [0, b]\}$$

in the case when (4.5.13) holds and

$$\mathscr{D} = \{u \in H_{1/2} : \|S_t u\|_{1/2} \le R \text{ for all } t \in [0, b]\}$$

when (4.5.17) is valid.

Proof. Let $u(t) = S_t u$ and $u_*(t) = S_t u_*$. Then $w(t) = u(t) - u_*(t)$ satisfies the equation

$$w_t + Aw + B(w, u) + B(u_*, w) + K(u) - K(u_*) = 0. \tag{4.5.21}$$

Using (4.4.4) and (4.4.3) we have that

$$|(B(w,u),Q_Nw)| \leq C\|w\|_{\mathcal{H}}\|u\|_{\mathcal{H}}\|Q_Nw\|_V \leq \varepsilon\|Q_Nw\|_V^2 + C_\varepsilon\|w\|_{\mathcal{H}}^2\|u\|_{\mathcal{H}}^2$$
$$\leq \varepsilon(\|Q_Nw\|_V^2 + \|w\|_V^2) + C_\varepsilon\|w\|^2\|u\|_{\mathcal{H}}^4$$

for every $\varepsilon > 0$, and, similarly,

$$|(B(u_*,w),Q_Nw)| \leq \varepsilon(\|Q_Nw\|_V^2 + \|w\|_V^2) + C_\varepsilon\|w\|^2\|u_*\|_{\mathcal{H}}^4.$$

Therefore, multiplying (4.5.21) by Q_Nw we obtain that

$$\frac{d}{dt}\|Q_Nw\|^2 + \|A^{1/2}Q_Nw\|^2 \leq \varepsilon\|w\|_V^2 + C_\varepsilon\|w\|^2(\|u\|_{\mathcal{H}}^4 + \|u_*\|_{\mathcal{H}}^4) + C\|w\|^2, \tag{4.5.22}$$

for every $\varepsilon > 0$. Thus, for $u, u_* \in \mathcal{D}$ we have that

$$\frac{d}{dt}\|Q_Nw\|^2 + \lambda_{N+1}\|Q_Nw\|^2 \leq \varepsilon\|w\|_V^2 + C_\varepsilon(R)\|w\|^2 \tag{4.5.23}$$

for every $\varepsilon > 0$. Therefore,

$$\|Q_Nw(t)\|^2 \leq \|Q_Nw(0)\|^2 e^{-\lambda_{N+1}t}$$
$$+ \varepsilon\int_0^t \|w(\tau)\|_V^2 e^{-\lambda_{N+1}(t-\tau)}\,d\tau + C_\varepsilon(R)\int_0^t \|w(\tau)\|^2 e^{-\lambda_{N+1}(t-\tau)}\,d\tau$$
$$\leq \|w(0)\|^2 e^{-\lambda_{N+1}t} + \varepsilon\int_0^t \|w(\tau)\|_V^2\,d\tau + \frac{C_\varepsilon(R)}{\lambda_{N+1}}\max_{\tau\in[0,t]}\|w(\tau)\|^2.$$

Using (4.4.16) we obtain that

$$\|Q_Nw(t)\|^2 \leq \left[e^{-\lambda_{N+1}t} + \left(\varepsilon + \frac{C_\varepsilon(R)}{\lambda_{N+1}}\right)e^{c(R)t}\right]\|w(0)\|^2$$

for every $\varepsilon > 0$. This implies (4.5.20).   □

In the same way as in Section 4.3.3 we can prove quasi-stability.

Proposition 4.5.10 (Quasi-stability). *Let Assumption 4.5.7 be in force. Then for every $0 < q < 1$, $0 < a \leq b < +\infty$, and a forward invariant set $\mathscr{B}$ which is bounded in $H_{1/4}$ (in $H_{1/2}$ if (4.5.17) holds), there exists $N = N(a,b,q,\mathscr{B})$ such that*

$$\|S_tu - S_tu_*\| \leq q\|S_ru - S_ru_*\| + \|P_N[S_tu - S_tu_*]\|, \quad \forall\, t \in [a+r, b+r], \tag{4.5.24}$$

for all $u, u_* \in \mathscr{B}$ and $r \geq 0$. This means that the system (H, S_t) is quasi-stable on $\mathscr{B}$ at every time t_*. Moreover, (H, S_t) satisfies Assumption 3.4.9 with arbitrary $t^* > 0$.

Proof. Exactly as in the proof of Proposition 4.3.11 (see also Exercise 4.3.12), we can see that (4.5.24) follows from (4.5.20) and Assumption 3.4.9 holds with $X = H$, $n_1 \equiv 0$, and $n_2 = \|P_N[\cdot]\|$. □

Similar to the techniques in Section 4.3.4, we can derive from Propositions 4.5.9 and 4.5.10 several conclusions concerning global and exponential attractors of (H, S_t).

The following assertion is based on Theorem 3.4.11.

Theorem 4.5.11 (Global and exponential attractor). *Let Assumption 4.5.7 be in force. Then the global attractor $\mathfrak{A}$ of the system (H, S_t) generated by (4.4.1) is a bounded set in $H_{1/4}$ (in $H_{1/2}$ if (4.5.17) holds) and possesses the properties:*

- *$\mathfrak{A}$ has finite fractal dimension $\dim_f \mathfrak{A}$ in H.*
- *For any full trajectory $\{u(t) : t \in \mathbb{R}\}$ from the attractor, $u(t)$ is an absolutely continuous function with values in H and*

$$\sup_{t \in \mathbb{R}} \left\{ \|u_t(t)\| + \|A^{1/2}u(t)\| + \|Au(t) + B(u(t))\| \right\} \leq C. \tag{4.5.25}$$

Moreover, the system (H, S_t) possesses a fractal exponential attractor $\mathfrak{A}_{exp}$ (whose dimension is finite in the phase space H).

The bounds for the dimensions of $\mathfrak{A}$ and $\mathfrak{A}_{exp}$ can be derived from Theorems 3.4.11 and 3.2.3.

Proof. The existence of a global attractor $\mathfrak{A}$ and the basic smoothness was proved in Theorem 4.5.8.

By Proposition 4.5.10 the system satisfies Assumption 3.4.9 on every forward invariant set which is bounded in $H_{1/4}$ (or in $H_{1/4}$). Thus, we can apply Theorem 3.4.11 to conclude finite dimensionality of the global attractor.

To prove the claimed smoothness of the attractor, we note that it follows from (4.5.24) that

$$\|u(t + h) - u(t)\| \leq q\|u(t + h - 1) - u(t - 1)\| + \|P_N[u(t + h) - u(t)]\|$$

for every $t \in \mathbb{R}$ and for any full trajectory $\{u(t) : t \in \mathbb{R}\}$ from the attractor. One can see that

$$\|P_N[u(t + h) - u(t)]\| \leq \int_t^{t+h} \|P_N[B(u(\tau)) + K(u(t)) - f]\| \, d\tau \leq C_{\mathfrak{A}}^N |h|.$$

Substituting this in the previous formula, we obtain that

$$(1 - q) \sup_{t \in \mathbb{R}} \|u(t + h) - u(t)\| \leq C_{\mathfrak{A}}^N |h|.$$

After the limit transition $h \to 0$ this yields (4.5.25).

To prove the existence of the fractal exponential attractors $\mathfrak{A}_{exp}$, we use the second part of Theorem 3.4.11. For this we need to check the Hölder continuity property (3.4.10) on some forward invariant absorbing set for (H, S_t). On an absorbing set in $H_{1/4}$ (the case when (4.5.13) is valid) or in $H_{1/2}$ (if (4.5.17) holds) we definitely have that $\|u_t(t)\|_{-3/4}$ is bounded. Thus, interpolation gives that $u(t)$ is a Hölder function in H.   □

Remark 4.5.12 (Dimension via volume contraction method). Since the semiflow S_t is C^1 in the case considered with *linear* operator K, we can use the volume contraction method presented in Section 3.1.3. This method is based on a Liouville-type formula for the limiting volume contraction coefficients Π_j given by (3.1.43) and constructed with the help of the evolution $V = S_1$. Namely (see, e.g., one of the monographs BABIN/VISHIK [9], CHEPYZHOV/VISHIK [31], ROBINSON [195, 196], TEMAM [216]), using the structure of the derivative $S_t'[u]$ for $S_t u$ given in Theorem 4.4.8, one can show that

$$\Pi_j \le \exp\left\{ \limsup_{t\to\infty} \sup_{u_0\in\mathfrak{A}} \frac{1}{t} \int_0^t \mathrm{Tr}_j L(u_0, \tau)d\tau \right\},$$

where the operator $L(u_0, t)$ is given by

$$L(u_0, t)w = -Aw - B(u(t), w) - B(w, u(t)) - Kw, \quad w \in \mathscr{D}(A),$$

with $u(t) = S_t u_0$. The m-dimensional trace of the linear operator L is defined by the relation

$$\mathrm{Tr}_m L = \sup \left\{ \mathrm{Tr}\,(LQ) \; : \; \begin{array}{l} Q \text{ is orthoprojector in } H, \\ QH \subset \mathscr{D}(A), \quad \dim Q = m \end{array} \right\}.$$

Let

$$q_j = \limsup_{t\to\infty} \sup_{u_0\in\mathfrak{A}} \frac{1}{t} \int_0^t \mathrm{Tr}_j L(u_0, \tau)d\tau.$$

It follows from the results in Section 3.1.3 that if j_0 is the smallest number such that $q_{j_0} < 0$ and $q_{j_0-1} \ge 0$, then

$$\dim_H \mathfrak{A} \le j_0.$$

In the concrete 2D hydrodynamical models this formula usually leads to the best possible estimate for the Hausdorff dimension (see, e.g., BABIN/VISHIK [9] and TEMAM [216] in the case of the 2D Navier-Stokes equations). However, at the abstract level considered, we can perform the following calculations only.

It follows from the basic Assumption 4.4.1 and the properties of the trace operation that

$$\text{Tr}_j \left(L(u_0, \tau) \right) \leq -\frac{1}{2} \sum_{k=1}^{j} \lambda_k + j \left[c_1 + c_2 \|u(\tau)\|_V^2 \right]$$

for some constants c_1 and c_2. Therefore, using (4.5.2) one can see that

$$q_j \leq -\frac{1}{2} \sum_{k=1}^{j} \lambda_k + j \left[c_1 + c_2 \|f\|_{V'}^2 \left(1 + \frac{1}{\lambda_1} \right) \right].$$

Since $\lambda_k \to +\infty$ as $k \to \infty$, we can conclude that there exists j_0 such that $q_{j_0} < 0$. To obtain a more explicit form of bounds for j_0, we need additional information about the behavior of the spectrum λ_k when $k \to \infty$ and also more sophisticated estimates for the trace of linearization of the nonlinear term (see the corresponding calculations BABIN/VISHIK in [9], CHEPYZHOV/VISHIK [31], TEMAM [216]). Thus, as we already mentioned, the optimal estimates for dimension require more detailed information and involve model-dependent techniques. ∎

4.5.5 Data assimilation

The main goal in this section[11] is to demonstrate some additional properties of dynamics which directly follow from the Ladyzhenskaya squeezing property. We deal with a (hydrodynamical) data assimilation problem which is important from the point of view of weather prediction (see, e.g., KALNAY [129] and also HAYDEN/OLSON/TITI [122] where the 2D Navier-Stokes equations are considered).

As in Section 4.3.7, we consider the discrete data assimilation problem in the sense due to HAYDEN/OLSON/TITI [122]. We suppose that Assumption 4.5.7 is valid with the compactness condition (4.5.13). Instead of (4.5.13) we can also assume (4.5.17); however, the argument is different in the latter case at the final stage.

Let $\mathscr{L} = \{l_j : j = 1, \ldots, N\}$ be a finite family of functionals on $H_{1/2}$ and $R_{\mathscr{L}}$ be a Lagrange interpolation operator related to $\mathscr{L}$ of the form

$$R_{\mathscr{L}} v = \sum_{j=1}^{N} l_j(v) \psi_j, \quad \forall \, v \in H, \tag{4.5.26}$$

where $\{\psi_j\}$ is a finite set of elements from $H_{1/2}$ such that $l_k(\psi_j) = \delta_{kj}$.

For a given solution $U(t) = S_t U_0$ to (4.4.1) with initial data U_0, we consider the sequence $\{r_{\mathscr{L}}^n \equiv R_{\mathscr{L}} U(t_n)\}$ of observation values and define prognostic values at time t_n for $U(t)$ by the formula

---

[11]This section can be omitted at the first reading.

$$u_n = (1 - R_{\mathscr{L}})S_{t_n - t_{n-1}} u_{n-1} + r^n_{\mathscr{L}}, \quad n = 1, 2, \ldots, \tag{4.5.27}$$

where u_0 is an initial guess of the reference solution U_0 (see [122]). We recall (see Definition 4.3.25) that the prognosis is *asymptotically reliable* at a sequence of times t_n if

$$\|U(t_n) - u_n\| \to 0 \quad \text{as} \quad n \to +\infty.$$

We are looking for conditions on $R_{\mathscr{L}}$ and t_n which guarantee that the prognosis based on observations $\mathscr{L} = \{l_j : 1 \leq j \leq N\}$ is asymptotically reliable. To obtain results we follow the same line of argument as in CHUESHOV [44]. We also note that the approach here is different from the method applied in Section 4.3.7 in the case of *globally* Lipschitz nonlinearities.

We assume that $0 < a \leq t_{n+1} - t_n \leq b < +\infty$ for some positive a and b.

The following assertion gives us a dissipativity property for prognostic values, which is important for our application of the Ladyzhenskaya squeezing property.

Lemma 4.5.13. *Let Assumption 4.5.7 be valid. Assume that $\|U(t)\|_{1/4} \leq \varkappa$ for all $t \geq t_0$ and*

$$\|R_{\mathscr{L}}\|_{H \mapsto H} \leq c_0 \quad \text{and} \quad \|1 - R_{\mathscr{L}}\|_{H \mapsto H} \leq c_1 \text{ with } c_1 < e^{\lambda_1 a/2}.$$

Then there exists $n_* > 0$ such that

$$\|u_n\| \leq 1 + \varrho_* \quad \text{for all } n \geq n_*, \tag{4.5.28}$$

where

$$\varrho_* = \left(c_1 \|f\|_{V'} \lambda_1^{-1/2} + c_0 \lambda_1^{-1/4} \varkappa\right)\left(1 - c_1 e^{-\lambda_1 a/2}\right)^{-1}.$$

If we assume in addition that $\|1 - R_{\mathscr{L}}\|_{H_{1/4} \mapsto H_{1/4}} \leq c_2$, then there exists $C(a, b, \varrho_*)$ such that

$$\|u_n\|_{1/4} \leq \varrho \equiv c_2 C(a, b, \varrho_*) + (1 + c_2)\varkappa \quad \text{for all } n \geq m_* \equiv 1 + n_*. \tag{4.5.29}$$

Proof. One can see from Proposition 4.5.1 that

$$\|u_n\| \leq c_1 e^{-\lambda_1 a/2} \|u_{n-1}\| + c_1 \|f\|_{V'} \lambda_1^{-1/2} + c_0 \lambda_1^{-1/4} \varkappa, \quad n = 1, 2, \ldots$$

This implies that

$$\|u_n\| \leq q_*^n \|u_0\| + \varrho_*, \quad n = 1, 2, \ldots$$

where $q_* = c_1 e^{-\lambda_1 a/2}$. This yields (4.5.28).

To prove (4.5.29) we note that

$$\|u_n\|_{1/4} \le c_2 \|S_{t_n - t_{n-1}} u_{n-1}\|_{1/4} + (1 + c_2)\varkappa, \quad n = 1, 2, \ldots$$

Hence (4.5.29) follows from (4.5.28) and (4.5.14) or (4.5.18). □

In the case of (spectral) modes we have the following assertion.

Corollary 4.5.14 (Modes). *We take the functionals $l_k(v) = (v, e_k)$ and thus the interpolation operator $R_{\mathscr{L}}$ in (4.5.27) is given by (4.3.53), i.e.,*

$$R_{\mathscr{L}} v = \sum_{j=1}^{N} (e_j, v) e_j, \quad \forall\, v \in H, \tag{4.5.30}$$

with some N. Here $\{e_k\}$ is the eigenbasis of A. Then there exist positive constants $C(\|f\|_V, \varkappa, a, b)$ and m_* such that

$$\|u_n\|_{1/4} \le \varrho \equiv C(\|f\|_{V'}, \varkappa, a, b) \quad \text{for all } n \ge m_*.$$

Proof. In this case $c_0 = c_1 = c_2 = 1$. □

Now we are in position to obtain the main results. We start with the case of modes.

Theorem 4.5.15 (Modes). *Let Assumption 4.5.7 be valid with the compactness condition (4.5.13). Then there exists N_* such that the prognosis (4.5.27) is asymptotically reliable with $R_{\mathscr{L}}$ given by (4.5.30) with arbitrary $N \ge N_*$.*

Proof. We obviously have that

$$U(t_n) - u_n = (1 - R_{\mathscr{L}})[S_{t_n - t_{n-1}} U(t_{n-1}) - S_{t_n - t_{n-1}} u_{n-1}], \quad \forall\, n \ge 1. \tag{4.5.31}$$

In the case of modes we have that $I - R_{\mathscr{L}} = Q_N$. Therefore, using Corollary 4.5.14, by Proposition 4.5.9 we can choose N_* such that

$$\|U(t_n) - u_n\| = q\|U(t_{n-1}) - u_{n-1}\|, \quad n \ge m_*,$$

with $q < 1$. This implies that

$$\|U(t_n) - u_n\| \to 0 \quad \text{as } n \to +\infty$$

with exponential speed. Thus, the prognosis is asymptotically reliable. □

The next result deals with general functionals.

Theorem 4.5.16. *Let Assumption 4.5.7 be in force with the compactness condition (4.5.13). Assume that $\mathscr{L}$ is a finite family of functionals on H and there is a Lagrange interpolation operator $R_{\mathscr{L}}$ possessing the properties:*

$$\|1 - R_{\mathscr{L}}\|_{H \mapsto H} \leq c_1 \quad and \quad \|1 - R_{\mathscr{L}}\|_{H_{1/4} \mapsto H_{1/4}} \leq c_2$$

with the constants c_1 and c_2 independent of $\mathscr{L}$ such that $c_1 < e^{\lambda_1 a/2}$. Then there exists $\epsilon_* > 0$ such that under the condition $\epsilon_{\mathscr{L}}(H, H_{-\delta}) \leq \epsilon_*$ for some $0 < \delta < 1/4$ the prognosis in (4.5.27) is asymptotically reliable for every $u_0 \in H$. Here $\epsilon_{\mathscr{L}}(H, H_{-\delta})$ denotes the corresponding completeness defect for the set $\mathscr{L}$.

Proof. Proposition 4.5.10 and Lemma 4.5.13 imply that

$$\|S_{\Delta_n} U(t_{n-1}) - S_{\Delta_n} u_{n-1}\|$$
$$\leq q_N \|U(t_{n-1}) - u_{n-1}\| + \lambda_N^\delta \|P_N[S_{\Delta_n} U(t_{n-1}) - S_{\Delta_n} u_{n-1}]\|_{-\delta} \qquad (4.5.32)$$

for $n \geq m_*$ with $\Delta_n = t_n - t_{n-1}$, where $q_N < 1$ can be chosen as small as we need at the expense of N. Now we apply the following lemma (its proof will be given later).

Lemma 4.5.17. *Let* $u^1(t) = S_t u^1$ *and* $u^2(t) = S_t u^2$ *be such that* $\|u^i(t)\|_{1/4} \leq \varkappa$ *for all* $t \in [0, T]$ *and* $0 < \delta \leq 1/4$. *Then*

$$\|A^{-\delta}[S_t u^1 - S_t u^2]\| \leq C_{T,\varkappa} \|A^{-\delta}[u^1 - u^2]\| \quad for \ t \in [0, T].$$

This lemma implies that

$$\|S_{\Delta_n} U(t_{n-1}) - S_{\Delta_n} u_{n-1}\|_{-\delta} \leq C(b, \varrho) \|U(t_{n-1}) - u_{n-1}\|_{-\delta}, \quad n \geq m_*.$$

Since $l_j(U(t_{n-1})) = l_j(u_{n-1})$, this gives

$$\|S_{\Delta_n} U(t_{n-1}) - S_{\Delta_n} u_{n-1}\|_{-\delta} \leq \epsilon_{\mathscr{L}}(H, H_{-\delta}) C(b, \varrho) \|U(t_{n-1}) - u_{n-1}\|, \quad n \geq m_*.$$

Thus (4.5.31) and (4.5.32) yield

$$\|U(t_n) - u_n\| \leq \tilde{q} \|U(t_{n-1}) - u_{n-1}\|$$

for $n \geq m_*$, where

$$\tilde{q} = \|I - R_{\mathscr{L}}\|_{H \mapsto H} \left[q_N + \lambda_N^\delta \epsilon_{\mathscr{L}}(H, H_{-\delta}) C(b, \varrho) \right].$$

We can choose N and $\epsilon_{\mathscr{L}}(H, H_{-\delta})$ such that $\tilde{q} < 1$. Therefore, the prognosis is asymptotically reliable with exponential speed.   □

Remark 4.5.18. The number of functionals which provides an asymptotically reliable prognosis according to Theorems 4.5.15 and 4.5.16 is finite. However, the estimates for this number which follow from the statement of the theorem are not optimal and not constructive. The derivation of optimal bounds for the number requires more careful analysis of constants related to dissipativity and squeezing properties of individual trajectories. We refer to HAYDEN/OLSON/TITI [122] for a more constructive approach based on the multipliers technique and developed in the case of the 2D Navier-Stokes equations for the reference solution from the global attractor.   ∎

Proof of Lemma 4.5.17. The function $u(t) = u^1(t) - u^2(t)$ satisfies the equation

$$u_t + Au + B\big(u^1(t), u^1(t)\big) - B\big(u^2(t), u^2(t)\big) + K(u^1(t)) - K(u^2(t)) = 0.$$

This allows us to multiply the equation by $A^{-2\delta}u(t)$ and to show that

$$\frac{1}{2}\frac{d}{dt}\|A^{-\delta}u(t)\|^2 + \|A^{1/2-\delta}u(t)\|^2$$
$$\leq |B(u, u^2), A^{-2\delta}u)| + |B\big(u^1, u\big), A^{-2\delta}u)| + C\|u\|\|A^{-2\delta}u\|^2.$$

It follows from (4.4.4) that

$$|B(u, u^2), A^{-2\delta}u)| + |B\big(u^1, u\big), A^{-2\delta}u)| \leq C(\|u^1\|_{\mathscr{H}} + \|u^2\|_{\mathscr{H}})\|u\|_{\mathscr{H}}\|A^{1/2-2\delta}u\|.$$

Using the fact that $\|u\|_{\mathscr{H}} \leq C\|u\|_{1/4}$ and the interpolation we conclude that

$$\|u\|_{\mathscr{H}}\|A^{1/2-2\delta}u\| \leq C\|A^{-\delta}u\|^{1/2}\|A^{1/2-\delta}u\|^{3/2}.$$

This implies

$$\frac{d}{dt}\|A^{-\delta}u(t)\|^2 \leq C(1 + \|u^1\|_{\mathscr{H}}^4 + \|u^2\|_{\mathscr{H}}^4)\|A^{-\delta}u\|^2,$$

which allows us to make the conclusion via Gronwall's lemma.

4.6 Hydrodynamical applications

In this section we discuss several important hydrodynamical models for which we can apply the theory we have developed.

4.6.1 *2D magnetohydrodynamic equations*

We consider magnetohydrodynamic (MHD) equations for a viscous incompressible resistive fluid in a 2D bounded domain D, which have the form (see, e.g., MOREAU [167]):

$$\partial_t u - \nu_1 \Delta u + u \nabla u = -\nabla \left(p + \frac{s}{2} |b|^2 \right) + s b \nabla b + f, \tag{4.6.1}$$

$$\partial_t b - \nu_2 \Delta b + u \nabla b = b \nabla u + g, \tag{4.6.2}$$

$$\text{div } u = 0, \quad \text{div } b = 0, \tag{4.6.3}$$

where $u = (u^1(x,t); u^2(x,t))$ and $b = (b^1(x,t); b^2(x,t))$ denote velocity and magnetic fields, $p(x,t)$ is a scalar pressure, and $x = (x_1; x_2) \in D$. We recall that the term $v \nabla w$ for different 2D fields v and w has the form

$$v \nabla w = \left(\sum_{i=1,2} v_i \partial_i \right) w \quad \text{for} \quad v = (v^1; v^2), \ w = (w^1; w^2).$$

We consider the following boundary conditions:

$$u = 0, \quad (b, n) = 0, \quad \partial_{x_1} b^2 - \partial_{x_2} b^1 = 0 \quad \text{on } \partial D. \tag{4.6.4}$$

In the equations above ν_1 is the kinematic viscosity, ν_2 is the magnetic diffusivity (which is determined from magnetic permeability and conductivity of the fluid), the positive parameter s is defined by the relation $s = Ha^2 \nu_1 \nu_2$, where Ha is the Hartmann number, and n is the outer normal vector. The given functions $f = f(x,t)$ and $g = g(x,t)$ represent external volume forces and the curl of external current applied to the fluid. We refer to LADYZHENSKAYA/SOLONNIKOV [143], DUVAUT/LIONS [90], and SERMANGE/TEMAM[205] for the mathematical theory for the MHD equations.

The above equations are a particular case of equation (4.4.1) for some choice of spaces and operators which satisfy Assumption 4.4.1. To see this, we first note that without loss of generality we can assume that $s = 1$ in (4.6.1) (indeed, if $s \neq 1$ we can introduce a new magnetic field $b := \sqrt{s} b$ and rescale the curl of the current $g := \sqrt{s} g$). For the velocity part of the MHD equations, we use the same spaces as in Example 4.4.4. We denote them as $H_{(1)}$ and V_1:

$$H_{(1)} = \{ f \in [L^2(D)]^2 : \text{div} f = 0 \text{ in } D \text{ and} (f, n) = 0 \text{ on} \partial D \}$$

endowed with the usual L^2 scalar product and

$$V_1 \equiv [H_0^1(D)]^2 \cap H_{(1)}.$$

We denote by A the Stokes operator on $H_{(1)}$ generated by the bilinear form defined by (4.4.10) with $\nu = \nu_1$.

For the magnetic part we set $H_{(2)} = H_{(1)}$ and $V_2 = [H^1(D)]^2 \cap H_{(2)}$ and define another Stokes operator A_2 as an unbounded operator on $H_{(2)}$ generated by the form (4.4.10) with $\nu = \nu_2$ considered on the space V_2.

We can write (4.6.1)–(4.6.4) in the form (4.4.1) in the space $H = H_{(1)} \times H_{(2)}$ with $A = A_1 \times A_2$, $K \equiv 0$. We also set $V = V_1 \times V_2$ and define $B : V \times V \to V'$ by the relation

$$(B(z_1, z_2), z_3) = (\tilde{B}(u_1, u_2), u_3) - (\tilde{B}(b_1, b_2), u_3)$$
$$+ (\tilde{B}(u_1, b_2), b_3) - (\tilde{B}(b_1, u_2), b_3)$$

for $z_i = (u_i; b_i) \in V = V_1 \times V_2$, where $\tilde{B}$ is given by (4.4.11). The conditions in Assumption 4.4.1 are satisfied with $\mathscr{H} = \left(\left[L^4(D) \right]^2 \times \left[L^4(D) \right]^2 \right) \cap H$. In the same way as in Remark 4.5.5 we can conclude that (4.5.13) holds for this case.

4.6.2 2D Boussinesq model for the Bénard convection

The next example is the following coupled system of Navier-Stokes and heat equations from the Bénard convection problem (see, e.g., FOIAS/MANLEY/TEMAM [103] and the references therein). Let $D = (0, l) \times (0, 1)$ be a rectangular domain in the vertical plane, $\{e_1, e_2\}$ the standard basis in $\mathbb{R}^2$, and $x = (x^1; x^2)$ an element of $\mathbb{R}^2$. Denote by $p(x, t)$ the pressure field, f, g external forces, $u = (u^1(x, t); u^2(x, t))$ the velocity field, and $\theta = \theta(x, t)$ the temperature field satisfying the following system:

$$\partial_t u + u \nabla u - \nu \Delta u + \nabla p = \theta e_2 + f, \quad \operatorname{div} u = 0, \qquad (4.6.5)$$
$$\partial_t \theta + u \nabla \theta - u^2 - \kappa \Delta \theta = g, \qquad (4.6.6)$$

with boundary conditions

$$u = 0, \quad \theta = 0 \text{ on } x^2 = 0 \text{ and } x^2 = 1,$$
$$u, p, \theta, u_{x^1}, \theta_{x^1} \text{ are periodic in } x^1 \text{ with period } l.[12]$$

Here ν is the kinematic viscosity, and κ is the thermal diffusion coefficient. Let

$$H_{(3)} = \left\{ u \in \left[L^2(D) \right]^2 : \operatorname{div} u = 0, \ u^2|_{x^2=0} = u^2|_{x^2=1} = 0, \ u^1|_{x^1=0} = u^1|_{x^1=l} \right\}$$

and $H_{(4)} = L^2(D)$. We also denote

$$V_3 = \left\{ u \in H_{(3)} \cap \left[H^1(D) \right]^2 : u|_{x^2=0} = u|_{x^2=1} = 0, \ u \text{ is } l\text{-periodic in } x^1 \right\},$$

---

[12]Here and below this means that $\phi|_{x^1=0} = \phi|_{x^1=l}$ for the corresponding function.

$$V_4 = \left\{\theta \in H^1(D) : \quad \theta|_{x^2=0} = \theta|_{x^2=1} = 0, \ \theta \text{ is } l\text{-periodic in } x^1\right\}.$$

Let A_3 be the Stokes operator in $H_{(3)}$ generated by the bilinear form (4.4.10) considered on V_3 and A_4 be the operator in $H_{(4)}$ generated by the Dirichlet form

$$a(\theta_1, \theta_2) = \kappa \int_D \nabla\theta_1 \cdot \nabla\theta_2 \, dx, \quad \theta_1, \theta_2 \in V_4.$$

Again, the above equations are a particular case of equation (4.4.1) for the following spaces and operators which satisfy Assumption 4.4.1. We assume $H = H_{(3)} \times H_{(4)}$ and $V = V_3 \times V_4$. We set $A(u, \theta) = (A_3 u \, ; A_4\theta)$, $K(u, \theta) = -(\theta e_2 \, ; u^2)$, and define the mapping $B : V \times V \to V'$ by the relation

$$(B(z_1, z_2), z_3) = (\tilde{B}_1(u_1, u_2), u_3) + \sum_{i=1,2} \int_D u_1^i \, \partial_i \theta_2 \, \theta_3 \, dx$$

for $z_i = (u_i; \theta_i) \in V = V_3 \times V_4$, where $\tilde{B}$ is given by (4.4.11). With this notation, the Boussinesq equations for (u, θ) are a particular case of (4.4.1) with Assumption 4.4.1 for $\mathscr{H} = \left(\left[L^4(D)\right]^2 \times L^4(D)\right) \cap H$. As above, (4.5.13) holds for this case.

4.6.3   2D magnetic Bénard problem

This is the Boussinesq model coupled with a magnetic field (see GALDI/PADULA [109]). As above, let $D = (0, l) \times (0, 1)$ be a rectangular domain in the vertical plane, $\{e_1, e_2\}$ the standard basis in $\mathbb{R}^2$. We consider the equations

$$\partial_t u + u\nabla u - \nu_1 \Delta u + \nabla\left(p + \frac{s}{2}|b|^2\right) - sb\nabla b = \theta e_2 + f, \quad \operatorname{div} u = 0,$$

$$\partial_t \theta + u\nabla\theta - u^2 - \kappa\Delta\theta = f,$$

$$\partial_t b - \nu_2 \Delta b + u\nabla b - b\nabla u = h, \quad \operatorname{div} b = 0,$$

with boundary conditions

$$u = 0, \quad \theta = 0, \quad b^2 = 0, \ \partial_2 b^1 = 0 \text{ on } x^2 = 0 \text{ and } x^2 = 1,$$

$$u, p, \theta, b, u_{x^1}, \theta_{x^1}, b_{x^1} \text{ are periodic in } x^1 \text{ with period } l.$$

As for the MHD case we can assume that $s = 1$. In this case we have (4.4.1) for the variable $z = (u; \theta; b)$ with $H = H_{(3)} \times H_{(4)} \times H_{(5)}$, where $H_{(3)}$ and $H_{(4)}$ are the same as in the previous example and $H_{(5)} = H_{(3)}$. We also set $V = V_3 \times V_4 \times V_5$, where V_3 and V_4 are the same as above and $V_5 = H_{(3)} \cap \left[H^1(D)\right]^2$. The operator A

is generated by the bilinear form

$$a(z_1, z_2) = \nu_1 \sum_{j=1}^{2} \int_D \nabla u_1^j \cdot \nabla u_2^j \, dx + \kappa \int_D \nabla \theta_1 \cdot \nabla \theta_2 \, dx + \nu_2 \sum_{j=1}^{2} \int_D \nabla b_1^j \cdot \nabla b_2^j \, dx$$

for $z_i = (u_i, \theta_i, b_i) \in V$. The bilinear operator B is defined by

$$(B(z_1, z_2), z_3) = (\tilde{B}(u_1, u_2), u_3) - (\tilde{B}(b_1, b_2), u_3)$$

$$+ (\tilde{B}(u_1, b_2), b_3) - (\tilde{B}(b_1, u_2), b_3) + \sum_{i=1,2} \int_D u_1^i \, \partial_i \theta_2 \, \theta_3 \, dx$$

for $z_i = (u_i; \theta_i; b_i) \in V$, where $\tilde{B}$ is given by (4.4.11). We also set $K(u, \theta, b) = -(\theta e_2; u^2; 0)$. It is easy to check that this model is an example of equation (4.4.1) with Assumption 4.4.1, where $\mathscr{H} = \left(\left[L^4(D) \right]^2 \times L^4(D) \times \left[L^4(D) \right]^2 \right) \cap H$. The condition in (4.5.13) is also valid for this model.

4.6.4   3D Leray α-model for Navier-Stokes equations

The theory can also be applied to some 3D models. As an example we consider the 3D Leray α-model (see LERAY [149]; for recent developments of this model we refer to CHEPYZHOV/TITI/VISHIK [30], CHESKIDOV ET AL. [34] and to the references therein). In a bounded 3D domain D we consider the following equations:

$$\partial_t u - \nu \Delta u + \nu \nabla u + \nabla p = f, \tag{4.6.7}$$

$$(1 - \alpha \Delta)v = u, \quad \text{div } u = 0, \quad \text{div } v = 0 \quad \text{in } D, \tag{4.6.8}$$

$$v = u = 0 \quad \text{on } \partial D. \tag{4.6.9}$$

where $u = (u^1; u^2; u^3)$ and $v = (v^1; v^2; v^3)$ are unknown fields, and $p(x, t)$ is the pressure. In the space

$$H = \{u \in \left[L^2(D) \right]^3 : \text{div } u = 0 \text{ in } D \text{ and } (u, n) = 0 \text{ on } \partial D\}$$

problem (4.6.7)–(4.6.9) can be written in the form

$$u_t + Au + B(G_\alpha u, u) = \tilde{f},$$

where A is the corresponding 3D Stokes operator (defined similarly as in the 2D case) by the form

$$a(u_1, u_2) = \nu \sum_{j=1}^{3} \int_D \nabla u_1^j \, \nabla u_2^j \, dx$$

on $V \equiv H \cap \left[H_0^1(D)\right]^3$, $G_\alpha = \left(Id + \alpha \nu^{-1} A\right)^{-1}$ is the Green's operator, and

$$(B(u_1, u_2), u_3) = \sum_{i,j=1}^{3} \int_D u_1^j \, \partial_j u_2^i \, u_3^i \, dx, \quad u_i \in V = H \cap \left[H_0^1(D)\right]^3.$$

Note that the embedding $H^{1/2}(D) \subset L^3(D)$ with $\dim D = 3$ (see TRIEBEL [220]) implies that inequality (4.4.3) holds true for $\mathscr{H} = \left[L^3(D)\right]^3 \cap H$. Furthermore, Hölder's inequality and the embedding $H^1(D) \subset L^6(D)$ imply that for $u_1, u_2, u_3 \in V$,

$$|(B(G_\alpha u_1, u_2), u_3)| \leq C\|u_2\|_V \|G_\alpha u_1\|_{L^6(D)} \|u_3\|_{L^3(D)}$$
$$\leq C\|u_2\|_V \|G_\alpha u_1\|_V \|u_3\|_{L^3(D)}$$
$$\leq C\|u_2\|_V \|u_1\|_{L^3(D)} \|u_3\|_{L^3(D)},$$

where the last inequality comes from the fact that $A^{\frac{1}{2}} G_\alpha$ is a bounded operator on H, so that

$$\|G_\alpha u_1\|_V = \|A^{\frac{1}{2}} G_\alpha u_1\| \leq C\|u_1\| \leq C\|u_1\|_{L^3(D)}.$$

This implies Assumption 4.4.1 for $B_\alpha(u_1, u_2) := B(G_\alpha u_1, u_2)$. Condition (4.5.13) follows from the embedding $H^{1/2}(D) \subset L^3(D)$.

4.6.5   Shell model of turbulence

Let H be a set of all sequences $u = \{u_1, u_2, \ldots\}$ of complex numbers such that $\sum_n |u_n|^2 < \infty$. We consider H as a *real* Hilbert space endowed with the inner product $(\cdot, \cdot)$ and the norm $|\cdot|$ of the form

$$(u, v) = \operatorname{Re} \sum_{n=1}^{\infty} u_n v_n^*, \quad |u|^2 = \sum_{n=1}^{\infty} |u_n|^2,$$

where v_n^* denotes the complex conjugate of v_n. In this space H we consider the evolution equation (4.4.1) with $K = 0$ and with linear operator A and bilinear mapping B defined by the formulas

$$(Au)_n = \nu k_n^2 u_n, \quad n = 1, 2, \ldots, \qquad \mathscr{D}(A) = \left\{ u \in H : \sum_{n=1}^{\infty} k_n^4 |u_n|^2 < \infty \right\},$$

where $\nu > 0$, $k_n = k_0 \mu^n$ with $k_0 > 0$ and $\mu > 1$, and

$$[B(u, v)]_n = -i \left(a k_{n+1} u_{n+1}^* v_{n+2}^* + b k_n u_{n-1}^* v_{n+1}^* - a k_{n-1} u_{n-1}^* v_{n-2}^* - b k_{n-1} u_{n-2}^* v_{n-1}^* \right)$$

for $n = 1, 2, \ldots$, where a and b are real numbers (here we also assume that $u_{-1} = u_0 = v_{-1} = v_0 = 0$). This choice of A and B corresponds to what is called the GOY model (see, e.g., OHKITANI/YAMADA [173]). If we take

$$[B(u, v)]_n = -i \left(a k_{n+1} u_{n+1}^* v_{n+2} + b k_n u_{n-1}^* v_{n+1} + a k_{n-1} u_{n-1} v_{n-2} + b k_{n-1} u_{n-2} v_{n-1} \right),$$

then we obtain the Sabra shell model introduced in LVOV ET AL. [155]. In both cases the equation (4.4.1) is an infinite sequence of ODEs.

One can easily show (see BARBATO ET AL. [10] for the GOY model and CONSTANTIN/LEVANT/TITI [80] for the Sabra model) that the trilinear form

$$\langle B(u, v), w \rangle \equiv \text{Re} \sum_{n=1}^{\infty} [B(u, v)]_n w_n^*$$

possesses the property (4.4.2) and also satisfies the inequality

$$|(B(u, v), w)| \leq C \|u\| \|A^{1/2} v\| \|w\|, \quad \forall u, w \in H, \quad \forall v \in \mathscr{D}(A^{1/2}).$$

The conditions in Assumption 4.4.1 hold with $\mathscr{H} = \mathscr{D}(A^s)$ for any choice of $s \in [0, 1/4]$. For any case $\mathscr{D}(A^{1/4}) \subset \mathscr{H}$ is also in force.

We can also consider the dyadic model (see, e.g., KATZ/PAVLOVIĆ [133] and the references therein), which can be written as an infinite system of real ODEs of the form

$$\partial_t u_n + \nu \lambda^{2\alpha n} u_n - \lambda^n u_{n-1}^2 + \lambda^{n+1} u_n u_{n+1} = f_n, \quad n = 1, 2, \ldots, \tag{4.6.10}$$

where $\nu, \alpha > 0$, $\lambda > 1$, $u_0 = 0$. Simple calculations show that under the condition $\alpha \geq 1/2$ the system (4.6.10) can be written as (4.4.1) and that Assumption 4.4.1 holds for $[B(u, v)]_n = -\lambda^n u_{n-1} v_{n-1} + \lambda^{n+1} u_n v_{n+1}$ and $(Au)_n = \nu \lambda^{2\alpha n} u_n$.

4.6.6 Hopf model of turbulence

This model was suggested in 1948 by E. Hopf (see the references in HENRY [123]) as an illustration of one of the possible scenarios (which is known now as the Landau-Hopf scenario) of the turbulence appearance in fluids. The model is described by the following equations:

$$\begin{cases} u_t = \mu u_{xx} - v * v - w * w - u * 1, \\ v_t = \mu v_{xx} + v * u + v * a + w * b, \\ w_t = \mu w_{xx} + w * u - v * b + w * a, \end{cases} \tag{4.6.11}$$

where the unknown functions u, v, w are even and 2π-periodic with respect to x and the convolution $w * u$ defined as follows:

$$(w * u)(x) = \frac{1}{2\pi} \int_0^{2\pi} w(x - y)u(y) \, dy.$$

A phase space for this model is the space

$$H = \left\{ U = (u; v; w) \in \left[L_2^{loc}(\mathbb{R}) \right]^3 \; : \; U(x) = U(-x) = U(x + 2\pi) \right\}$$

endowed with the L_2-norm:

$$\|U\|^2 = \int_0^{2\pi} \left[|u(x)|^2 + |v(x)|^2 + |w(x)|^2 \right] dx, \quad U = (u; v; w).$$

The operator A is defined on $\mathcal{D}(A) = \left[H_{loc}^2(\mathbb{R}) \right]^3 \cap H$, and has the form

$$A(u; v; w) = (u - \mu u_{xx}; v - \mu v_{xx}; w - \mu u_{xx}), \quad (u; v; w) \in \mathcal{D}(A).$$

The corresponding bilinear mapping B has the form

$$B(\tilde{U}, U) = (\tilde{v} * v + \tilde{w} * w; -\tilde{v} * u; -\tilde{w} * u),$$

where $\tilde{U} = (\tilde{u}; \tilde{v}; \tilde{w})$ and where $U = (u; v; w)$. The operator K is given by

$$KU = -(u - u * 1; v + v * a + w * b; w - v * b + w * a)$$

where $U = (u; v; w)$. One can see that Assumption 4.4.1 is satisfied here. For detailed calculations relating to this model, we refer to CHUESHOV [39, Chapter 2].

Chapter 5
Second Order Evolution Equations

In this chapter we show how the general ideas developed in Chapters 2 and 3 can be applied to second order in time evolution equations with damping and source terms of various structure, whose abstract form is the following Cauchy problem in a separable Hilbert space H:

$$\mu u_{tt} + K(u)u_t + Au + B(u) = 0, \quad t > 0; \quad u|_{t=0} = u_0, \quad u_t|_{t=0} = u_1. \qquad (5.0.1)$$

We also consider by means of an example the case when the main elliptic part A is nonlinear. The model in (5.0.1) represents nonlinear wave dynamics with the damping (operator) coefficient $K(u)$, which depends on the displacement u (but not on the velocity u_t). This type of model was studied by many authors for different classes of equations; see, e.g., CHUESHOV/KOLBASIN [49], GATTI/PATA [111], PATA/ZELIK [178, 180] and the references therein. We note that there is a wide class of models with velocity-dependent damping ($D(u_t)$ in (5.0.1) instead of $K(u)u_t$), but we do not discuss them here, and refer the reader to the surveys in CHUESHOV/LASIECKA [56, 58]. We also mention that models with different types of strong (linear) damping in wave equations have been considered by many authors; see, e.g., CARVALHO/CHOLEWA [23], CHOLEWA/DLOTKO [35], KALANTAROV/ZELIK [128], PATA/ZELIK [177], and also the literature quoted in these references. This class of models corresponds to the case when $K(u) \equiv K$ is a non-negative operator independent of u and hence can be included in our framework. Under some additional hypotheses (see, e.g., LASIECKA/TRIGGIANI [145, Chapter 3] and the references therein), the corresponding linearization generates an analytical semigroup. This situation was studied by many authors (see, e.g., the discussion in LASIECKA/TRIGGIANI [145, Chapter 3]). In principle, this allows us to use the "parabolic" methods presented in Chapter 4. However, our main examples are related to the case when $K(u)$ depends nonlinearly on u, and thus it is problematic to use analyticity of the linearization.

© Springer International Publishing Switzerland 2015   219
I. Chueshov, *Dynamics of Quasi-Stable Dissipative Systems*, Universitext,
DOI 10.1007/978-3-319-22903-4_5

The main topics in this chapter deal with the existence of a compact finite-dimensional attractor for different situations. In the case when the set of equilibria is finite and hyperbolic, we show that every trajectory is attracted by some equilibrium at an exponential rate. We also consider dynamics in the inertial zero limit $\mu \to 0$, and show how the results can be extended to the case when we have the nonlocal nonlinear Kirchhoff operator instead of A in (5.0.1).

Our arguments involve the method based on quasi-stability estimates (see Section 3.4.3). We first prove that the corresponding system is asymptotically quasi-stable in the sense of Definition 3.4.15, and then we apply general theorems on properties of quasi-stable systems.

To obtain the results concerning equilibria, we rely on some type of observability inequality and use the same idea as in CHUESHOV/LASIECKA [56, Section 4.3] (see also CHUESHOV/LASIECKA [51, 58]) and the argument given in the proof of Theorem 4.3.18.

The main applications, which we keep in mind, are concerned with a class of elastic plate models with different types of boundary conditions. The results presented can also be applied to nonlinear wave equations.

5.1   Generation of a dynamical system

In this section we impose our main hypotheses and show that problem (5.0.1) generates a dynamical system. Our presentation mainly follows the ideas presented in CHUESHOV/KOLBASIN [49, 50].

5.1.1   Main hypotheses and motivation

Assumption 5.1.1. We impose the following set of hypotheses.

(A) The operator A is a linear self-adjoint positive operator densely defined on a separable Hilbert space H possessing a discrete spectrum (see Definition 4.1.1). Below we denote by $\{e_k\}$ the orthonormal basis in H consisting of the eigenvectors of the operator A:

$$Ae_k = \lambda_k e_k, \quad 0 < \lambda_1 \leq \lambda_2 \leq \cdots, \quad \lim_{k \to \infty} \lambda_k = \infty.$$

We denote by $\|\cdot\|$ and $(\cdot, \cdot)$ the norm and the scalar product in H. As in the previous chapter, we denote by H_s (with $s > 0$) the domain $\mathscr{D}(A^s)$ equipped with the graph norm $\|\cdot\|_s = \|A^s \cdot \|$. As above, H_{-s} denotes the completion of H with respect to the norm $\|\cdot\|_{-s} = \|A^{-s} \cdot \|$.

(D) For some value $\theta \in [0, 1/2]$ the damping operator $K(u)$ maps H_θ into $H_{-\theta}$ and possesses the properties:

(i) For each $u \in \mathscr{D}(A^{1/2})$ the operator $K(u)$ generates a symmetric bilinear form $b_u(v, w) = (K(u)v, v)$ on $H_\theta \times H_\theta$ and for every $\varrho > 0$ there exist $\alpha_\varrho > 0$ and $\beta_\varrho > 0$ such that[1]

$$\alpha_\varrho \|v\|_\theta^2 \le (K(u)v, v) \le \beta_\varrho \|v\|_\theta^2, \quad \forall \, \|A^{1/2}u\| \le \varrho, \ v \in H_\theta \ . \qquad (5.1.1)$$

(ii) For every $\varrho > 0$ there exists $C_\varrho > 0$ such that

$$\|K(u_1) - K(u_2)\|_{H_\theta \mapsto H_{-\theta}} \le C_\varrho \|A^{1/2}(u_1 - u_2)\|, \quad \forall \|A^{1/2}u_i\| \le \varrho, \qquad (5.1.2)$$

where $\| \cdot \|_{X \mapsto Y}$ stands for the operator norm of linear mappings from X into Y.
(iii) We also have that

$$\|[K(u_1) - K(u_2)]\psi\|_{-\theta} \le C_\varrho \|A^{1/2-\delta}(u_1 - u_2)\| \|A^l \psi\| \qquad (5.1.3)$$

for some $\delta > 0$ and for all $\|A^{1/2}u_i\| \le \varrho$ and $\psi \in \mathscr{D}(A^l)$ with some $l \ge 0$.

(B) There exists $\delta \ge 0$ such that the nonlinear operator B maps $H_{1/2-\delta}$ into $H_{-\theta}$ and is locally Lipschitz, i.e.,

$$\|B(u_1) - B(u_2)\|_{-\theta} \le L(\varrho)\|u_1 - u_2\|_{1/2-\delta}, \quad \forall \|A^{1/2}u_i\| \le \varrho. \qquad (5.1.4)$$

In addition we assume:

(i) The mapping B is weakly continuous[2] from $H_{1/2}$ into H_{-l} for some $l \ge \theta$, i.e.,

$$|(B(u_n) - B(u), \psi)| \to 0 \ \text{ for every } \psi \in H_l \qquad (5.1.5)$$

provided $u_n \to u$ weakly in $H_{1/2}$.
(ii) $B(u) = \Pi'(u)$, where $\Pi(u)$ is a C^1 functional[3] on $H_1 = \mathscr{D}(A^{1/2})$, and $'$ stands for the Fréchet derivative (see the definition in (A.5.3)). We assume that there exist $\eta < 1/2$ and $C \ge 0$ such that

$$\eta\|A^{1/2}u\|^2 + \Pi(u) + C \ge 0, \quad u \in H_{1/2} = \mathscr{D}(A^{1/2}) \ . \qquad (5.1.6)$$

Remark 5.1.2. As already mentioned, our main goal is to demonstrate the method but not to obtain the best possible result for the considered class of abstract systems. However, we note that our conditions can be relaxed in different directions; see,

---

[1] It is allowed that $\alpha_\varrho \to 0$ and $\beta_\varrho \to \infty$ as $\varrho \to \infty$.
[2] We need this property in order to make limit transitions on the Galerkin approximations in the nonlinear term. We also note that this property is valid in the case when (5.1.4) holds with $\delta > 0$.
[3] We note that $\Pi(u)$ is locally bounded on $H_{1/2}$, see Remark 4.2.21.

e.g., CHUESHOV/KOLBASIN [50] for some details and discussions. We note that our damping term $D(u, u_t) = K(u)u_t$ is positive (in the sense that $(D(u, u_t), u_t) \geq 0$, see the left inequality in (5.1.1)) but not monotone in general. Thus, we cannot apply the theory developed in HARAUX [119] and CHUESHOV/LASIECKA [56]. We also mention that problem (5.0.1) was studied in CARVALHO/CHOLEWA/DLOTKO [24] in the Banach space setting for a constant operator $K(u) = A^{2\theta}$ with $1/4 \leq \theta \leq 1/2$ (in contrast with our case of $0 \leq \theta \leq 1/2$). The situation in CARVALHO/CHOLEWA/DLOTKO [24] corresponds to the case when the linear part of the problem generates an analytic semigroup.   ∎

Our main motivation is related to plate models (with hinged boundary conditions, for definiteness). In this case the middle surface of a plate is a domain Ω in $\mathbb{R}^2$, and $u(t) = u(x, t), x \in \Omega, t > 0$, is the transverse displacement of this middle surface at point x and time t. In these models we have the following.

- $A = (-\Delta_D)^2$, where Δ_D is the Laplace operator in a bounded smooth domain Ω in $\mathbb{R}^2$ with Dirichlet boundary conditions. We then have that $H = L_2(\Omega)$ and

$$\mathscr{D}(A) = \left\{ u \in H^4(\Omega) \, : \, u = \Delta u = 0 \text{ on } \partial\Omega \right\}.$$

Here $H^\sigma(\Omega)$ is the L_2-based Sobolev space of the order σ.
- The damping operator $K(u)$ may have the form

$$K(u)u_t = \Delta \left[\sigma_0(u)\Delta u_t \right] - \text{div} \left[\sigma_1(u, \nabla u)\nabla u_t \right] + \sigma_2(u)u_t, \qquad (5.1.7)$$

where $\sigma_0(s_1)$, $\sigma_1(s_1, s_2, s_3)$ and $\sigma_2(s_1)$ are non-negative locally Lipschitz functions of $s_i \in \mathbb{R}$, $i = 1, 2, 3$, satisfying some growth conditions (for a more detailed discussion of properties of the damping functions we refer to CHUESHOV/KOLBASIN [50]). We note that every term in (5.1.7) represents a different type of damping mechanism. The first one is viscoelastic Kelvin-Voight damping, the second one represents structural damping, and the term $\sigma_2(u)u_t$ is the dynamical friction (or viscous damping). We refer to LASIECKA/TRIGGIANI [145, Chapter 3] and to the references therein for a discussion of stability properties caused by each type of damping term in the case of linear systems.
- The nonlinear feedback (elastic) force $F(u)$ may have one of the following forms (which represent different plate models):

(a) *Kirchhoff model*: $B(u)$ is the Nemytskii operator

$$u \mapsto -\kappa \cdot \text{div} \left\{ |\nabla u|^q \nabla u - \mu |\nabla u|^r \nabla u \right\} - p(x), \qquad (5.1.8)$$

where $\kappa \geq 0, q > r \geq 0, \mu \in \mathbb{R}$ are parameters, $p \in L_2(\Omega)$.
(b) *Von Karman model* (see LIONS [151] and also CHUESHOV/LASIECKA [58]):

$$B(u) = -[u, v(u) + F_0] - p(x),$$

where $F_0 \in H^4(\Omega)$ and $p \in L_2(\Omega)$ are given functions, the von Karman bracket $[u, v]$ is given by

$$[u, v] = \partial^2_{x_1} u \cdot \partial^2_{x_2} v + \partial^2_{x_2} u \cdot \partial^2_{x_1} v - 2 \cdot \partial^2_{x_1 x_2} u \cdot \partial^2_{x_1 x_2} v,$$

and the Airy stress function $v(u)$ solves the following elliptic problem:

$$\Delta^2 v(u) + [u, u] = 0 \text{ in } \Omega, \quad \frac{\partial v(u)}{\partial n} = v(u) = 0 \text{ on } \partial\Omega. \tag{5.1.9}$$

(c) *Berger model:* In this case the feedback force has the form

$$B(u) = -\left[\kappa \int_\Omega |\nabla u|^2 dx - \Gamma \right] \Delta u - p(x),$$

where $\kappa > 0$ and $\Gamma \in \mathbb{R}$ are parameters, $p \in L_2(\Omega)$; for more details and references see, e.g., CHUESHOV [39, Chapter 4] and CHUESHOV/LASIECKA [56, Chapter 7].

We do not provide full details concerning these models and refer to CHUESHOV/KOLBASIN [49, 50] and to the references therein. We also refer to CIARLET [72] for a general presentation of the plate theory.

As an example, we can also consider the following wave equation on a bounded domain Ω in $\mathbb{R}^3$ with a nonlocal damping coefficient:

$$u_{tt} + \sigma(\|u\|_{H^s(\Omega)})u_t - \Delta u + \varphi(u) = f(x), \quad u\big|_{\partial\Omega} = 0. \tag{5.1.10}$$

One can see that Assumption 5.1.1 is valid if we assume that (a) $s < 1$ and σ is a positive and locally Lipschitz function, and (b) the source term $\varphi \in C^2(\mathbb{R})$ possesses the properties

$$\liminf_{|s|\to\infty} \{\varphi(s)s^{-1}\} > -\lambda_1, \quad |\varphi''(s)| \leq C(1 + |s|), \ s \in \mathbb{R},$$

where λ_1 is the first eigenvalue of the minus Laplace operator with Dirichlet boundary conditions.

In a similar way, we can consider the wave model (5.1.10) in an arbitrary spatial dimension and with another form of the damping operator and cover the case when $s = 1$ in (5.1.10); see Section 5.4 below.

We refer to PATA/ZELIK [178, 180] for the case when the damping coefficient has the form $\sigma(u(t, x))$ with some scalar function σ and mention the recent papers SAVOSTIANOV/ZELIK [202, 203] which deal with wave models damped by linear terms of the form $(-\Delta)^{2\theta} u_t$ with $\theta \in [0, 1/2]$. The results in these papers involve specific properties of wave dynamics, and it seems that they cannot be derived at the abstract level.

5.1.2   Well-posedness: nondegenerate case

We first prove the existence and uniqueness of weak solutions to problem (5.0.1) in the case of the positive inertial parameter μ.

Definition 5.1.3. A function $u(t)$ is said to be a *weak solution* to (5.0.1) on an interval $[0, T]$ if

$$u \in L_\infty(0, T; \mathcal{D}(A^{1/2})), \quad u_t \in L_\infty(0, T; H) \cap L_2(0, T; \mathcal{D}(A^\theta))$$

and (5.0.1) is satisfied in the sense of distributions, i.e., $u(0) = u_0$ and

$$-\mu \int_0^T (u_t, v_t)dt + \int_0^T (K(u)u_t, v)dt + \int_0^T (Au, v)dt + \int_0^T (B(u), v)dt = \mu(u_1, v(0)),$$
(5.1.11)

for every $v \in \mathscr{W}_T$, where

$$\mathscr{W}_T = \left\{ v \in L_\infty(0, T; \mathcal{D}(A^{1/2})), \; v_t \in L_\infty(0, T; H), \; v(T) = 0 \right\}.$$

For a description of the functional L_p spaces involved above, see Section A.3 in the Appendix. ∎

The main statement in this section is the following assertion, which also contains some auxiliary solution properties needed for the results on asymptotic dynamics.

Theorem 5.1.4. *Let Assumption 5.1.1 be in force and* $(u_0; u_1) \in \mathscr{H} \equiv \mathcal{D}(A^{1/2}) \times H$. *Then the following assertions hold.*

1. *Problem (5.0.1) has a unique weak solution $u(t)$ on $\mathbb{R}_+$. This solution belongs to the class*

$$\mathscr{W} \equiv C(\mathbb{R}_+; \mathcal{D}(A^{1/2})) \cap C^1(\mathbb{R}_+; H),$$

and the following energy relation:

$$\mathscr{E}(u(t), u_t(t)) + \int_0^t (K(u(\tau))u_t(\tau), u_t(\tau))d\tau = \mathscr{E}(u_0, u_1)$$
(5.1.12)

holds for every $t > 0$, where the energy $\mathscr{E}$ is defined by the formula

$$\mathscr{E}(u_0, u_1) = E(u_0, u_1) + \Pi(u_0) \equiv \frac{1}{2}\left(\mu\|u_1\|^2 + \|A^{1/2}u_0\|^2\right) + \Pi(u_0).$$

Moreover, this solution $u(t)$ satisfies the estimate

$$\sup_{t \geq 0} E(u(t), u_t(t)) + \int_0^{+\infty} \|A^\theta u_t(t)\|^2 dt \leq C_R \text{ when } E(u_0, u_1) \leq R^2, \quad (5.1.13)$$

where $C_R > 0$ does not depend on μ.

2. *If $u^1(t)$ and $u^2(t)$ are two weak solutions such that $E(u^i(0), u^i_t(0)) \leq R^2$, $i = 1, 2$, then their difference $z(t) = u^1(t) - u^2(t)$ satisfies the relation*

$$E(z(t), z_t(t)) + \int_0^t \|A^\theta z_t(\tau)\|^2 d\tau \leq a_R E(z(0), z_t(0)) e^{b_R t} \quad (5.1.14)$$

for some constants $a_R, b_R > 0$ independent of μ.

Proof. To prove the existence of solutions, we use the standard Galerkin method, seeking approximations of the form

$$u^N(t) = \sum_{k=1}^N y_k(t) e_k, \quad N = 1, 2, \ldots, \quad (5.1.15)$$

that solve the finite-dimensional projections of (5.0.1):

$$\mu u^N_{tt} + P_N K(u^N) u^N_t + A u^N + P_N B(u^N) = 0, \ t > 0; \ u^N|_{t=0} = P_N u_0, \ u^N_t|_{t=0} = P_N u_1, \quad (5.1.16)$$

where P_N is the orthoprojector onto $\text{Span}\{e_k : k = 1, 2, \ldots, N\}$.

Exercise 5.1.5. Show that problem (5.1.16) has a unique local solution. Hint: Rewrite (5.1.16) as a first order equation and apply Theorem A.1.2. ∎

Multiplying (5.1.16) by u^N_t, we get that $u^N(t)$ satisfies the energy relation (5.1.12). By (5.1.6) we obtain that

$$c_0 E(u_0, u_1) - c_1 \leq \mathscr{E}(u_0, u_1) \leq C(R)$$

whenever $E(u_0, u_1) \leq R^2$. Therefore, by (5.1.1) the energy relation for $u_N(t)$ yields the a priori estimate, which implies the global existence of approximate solutions. Moreover, estimate (5.1.13) holds with constant $C(R)$ independent of N for these solutions. Using the equation for $u^N(t)$ and also the conditions (5.1.1) and (5.1.4), it can be shown in the standard way that

$$\mu^2 \int_0^T \|A^{-1/2} u^N_{tt}(t)\|^2 dt \leq C_T(R), \quad N = 1, 2, \ldots,$$

for every $T > 0$. These a priori estimates show that $(u^N; u_t^N; u_{tt}^N)$ is *-weakly compact in

$$Y_T := L_\infty(0, T; H_{1/2}) \times [L_\infty(0, T; H) \cap L_2(0, T; H_\theta)] \times L_2(0, T; H_{-1/2}), \quad \forall\, T > 0.$$

The Aubin-Dubinskii-Lions theorem (see SIMON [213, Corollary 4] and also Theorem A.3.7 in the Appendix) yields that $(u^N; u_t^N)$ is compact in $C([0, T]; H_{1/2-\varepsilon} \times H_{-\varepsilon})$ for every $\varepsilon > 0$. Thus there exists $u(t) \in L_\infty(0, T; H)$ such that

$$(u; u_t; u_{tt}) \in Y_T, \quad (u; u_t) \in C([0, T]; H_{1/2-\varepsilon} \times H_{-\varepsilon}), \quad \forall\, \varepsilon > 0,$$

and, along a subsequence, we have

- $u^N \to u$ *-weakly in $L_\infty(0, T; H_{1/2})$ and strongly in $C([0, T]; H_{1/2-\varepsilon})$;
- $u_t^N \to u_t$ *-weakly in $L_\infty(0, T; H) \cap L_2(0, T; H_\theta)$ and strongly in $C([0, T]; H_{-\varepsilon})$;
- $u_{tt}^N \to u_{tt}$ weakly in $L_2(0, T; H_{-1/2})$.

It is also clear from Lions' lemma (see, e.g., LIONS/MAGENES [152] and Lemma A.3.1 in the Appendix) that $t \mapsto (u(t); u_t(t))$ is a weakly continuous function in $\mathscr{H} = H_{1/2} \times H$. Therefore, the uniform convergence

$$\|u^N(t) - u(t)\|_{1/2-\varepsilon} + \|u_t^N(t) - u_t(t)\|_{-\varepsilon} \to 0, \quad t \in [0, T]$$

yields $(u^N(t); u_t^N(t)) \to (u(t); u_t(t))$ weakly in $H_{1/2} \times H$ for every $t \in [0, T]$. In particular, this and also (5.1.5) imply that

$$(B(u^N(t)) - B(u(t)), \psi) \to 0 \text{ for every } \psi \in H_l,\ t \in [0, T].$$

For the damping term we have that

$$(K(u^N)u_t^N - K(u)u_t, v) = (u_t^N - u_t, K(u)v) + (u_t^N, [K(u^N) - K(u)]v).$$

The weak convergence $u_t^N \to u_t$ in $L_2(0, T; H_\theta)$ yields

$$\int_0^T (u_t^N - u_t, K(u)v)\, d\tau \to 0 \text{ as } N \to \infty$$

for every $v \in \mathscr{W}_T$. As for the second term, by (5.1.3)

$$|(u_t^N, [K(u^N) - K(u)]v)| \le C\|u_t^N\|_\theta \|u^N - u\|_{1/2-\delta} \|v\|_l.$$

This implies that

$$\int_0^T (u_t^N, [K(u^N) - K(u)]v)\, d\tau \to 0 \text{ as } N \to \infty, \quad \forall\, v \in \mathscr{W}_T.$$

All of these convergence properties make it possible to show that $u(t)$ is a weak solution satisfying (5.1.13).

To obtain the energy relation in (5.1.12), we note that the function $u^n(t) = P_n u(t)$ for every $n = 1, 2, \ldots$ solves an equation of the form

$$\mu \partial_{tt} u^n + A u^n = h(t) \quad \text{with} \quad h(t) = -(P_n K(u) u_t + P_n B(u)) \in L_2(0, T; H).$$

Therefore, using the multiplier u_t we can obtain that

$$E(u(t), u_t(t)) + \int_0^t (K(u) u_t, u_t) d\tau + \int_0^t (B(u), u_t) d\tau = E(u_0, u_1) \qquad (5.1.17)$$

where the energy $E(u, u_t)$ is given by

$$E(u, u_t) = \frac{1}{2} \left[\mu \|u_t\|^2 + \|A^{1/2} u\|^2 \right].$$

In particular, this implies that $t \mapsto E(u(t), u_t(t))$ is continuous and thus the weak continuity of $t \mapsto (u(t); u_t(t))$ implies its strong continuity in $\mathcal{D}(A^{1/2}) \times H$, i.e., $u(t) \in \mathcal{W}$. Now using approximations and relation (4.2.31) in Remark 4.2.21, we can derive the energy relation in (5.1.12) from (5.1.17).

To prove (5.1.14), we note that $z(t) = u^1(t) - u^2(t)$ solves the equation

$$\mu z_{tt} + K(u^1) u_t^1 - K(u^2) u_t^2 + Az + B(u^1) - B(u^2) = 0. \qquad (5.1.18)$$

Thus, multiplying this equation by z_t and integrating from s to t, we have

$$E_z(t) + \int_s^t (K(u^1) u_t^1 - K(u^2) u_t^2, z_t) d\tau = E_z(s) - \int_s^t (B(u^1) - B(u^2), z_t) d\tau \qquad (5.1.19)$$

for any $0 \leq s < t$, where $E_z(t) = E(z(t), z_t(t))$. Using Assumption 5.1.1D(i,ii) and (5.1.13), we obtain that

$$E_z(t) + \gamma_R \int_s^t \|A^\theta z_t\|^2 d\tau \leq E_z(s) + c_R \int_s^t \left(1 + \|u_t^2\|_\theta^2\right) \|A^{1/2} z\|^2 d\tau,$$

for all $s < t$ and for some $\gamma_R, c_R > 0$. Now we can apply Gronwall's lemma to obtain (5.1.14), which, in particular, implies the uniqueness of weak solutions. □

Remark 5.1.6. Under additional hypotheses concerning K and B, we can also establish the existence of more regular solutions. We do not pursue these generalizations at the abstract level, and refer to CHUESHOV/KOLBASIN [49] and the references therein for some results in this direction. ∎

Applying Theorem 5.1.4, we obtain the following assertion.

Proposition 5.1.7. *Let Assumption 5.1.1 be in force. Then problem (5.0.1) generates a dynamical system in the space $\mathscr{H} = \mathscr{D}(A^{1/2}) \times H$ with the evolution operator S_t^μ given by*

$$S_t^\mu y = (u(t); u_t(t)), \quad \text{where } y = (u_0; u_1) \text{ and } u(t) \text{ solves (5.0.1).}$$

This system is gradient (see Definition 2.4.1) with the full energy $\mathscr{E}(u_0; u_1)$ as a strict Lyapunov function (this follows from the energy relation in (5.1.12)).

5.1.3 Well-posedness: degenerate case

One of our goals is to study asymptotic properties of the system $(\mathscr{H}, S_t^\mu)$ in the zero mass density limit $\mu \to 0$. To describe these properties we need to consider the model (5.0.1) in the degenerate case $\mu = 0$. Thus, we arrive at the problem

$$K(u)u_t + Au + B(u) = 0, \quad u|_{t=0} = u_0. \tag{5.1.20}$$

In the case when $K(u) = Id$ this equation belongs to the class of models studied in Chapter 4. Our conditions in Assumption 5.1.1 concerning A and B are very close to the hypotheses of Theorem 4.2.22. However, we cannot apply the theory developed in Chapter 4 for generic operators K. Equation (5.1.20) does not contain a naturally separated linear part, and thus it is problematic to use the idea of a mild solution for (5.1.20). On the other hand, the result established for $\mu = 0$ in this chapter can be applied to the models studied in Sections 4.2 and 4.3. Thus, in addition to the method based on mild formulation, we can use the approach presented here.

It is remarkable that many steps in the well-posedness argument for (5.1.20) repeat the corresponding argument for the second order in time model in (5.0.1). To realize this analogy, it is convenient to accept the following definition.

Definition 5.1.8. A function $u(t)$ is called a *strong solution* to (5.1.20) on an interval $[0, T]$, if

$$u \in L^\infty(0, T; H_{1/2}), \quad u_t \in L^2(0, T; H_\theta), \quad u(0) = u_0,$$

and

$$\int_0^T \left(K(u)u_t, v \right) dt + \int_0^T \left(A^{1/2}u, A^{1/2}v \right) dt + \int_0^T \left(B(u), v \right) dt = 0 \tag{5.1.21}$$

holds for all $v \in L^2(0, T; H_{1/2})$.   ∎

Although we define solutions in the variational sense, we apply the term "strong" because, from Theorem 5.1.9 below (in the case $\theta = 0$), these solutions possess values in $\mathscr{D}(A)$ for almost all $t \in [0, T]$ and satisfy (5.1.20) in H. We do not work with the standard weak solutions, because their uniqueness seems to be out of reach under the conditions imposed.

Theorem 5.1.9. *Let Assumption 5.1.1 be in force and $u_0 \in H_{1/2} = \mathscr{D}(A^{1/2})$. Then problem (5.1.20) has a unique strong solution $u(t)$ on any interval $[0, T]$ such that*

$$u(t) \in C\big([0, T]; H_{1/2}\big) \cap L^2\big(0, T; H_{1-\theta}\big) \tag{5.1.22}$$

and the following balance relation holds:

$$\Pi_*(u(t)) + \int_0^t \big(K(u(\tau))u_t(\tau), u_t(\tau)\big)d\tau = \Pi_*(u_0), \tag{5.1.23}$$

where $\Pi_*(u) = \frac{1}{2}\big\|A^{1/2}u\big\|^2 + \Pi(u)$. Moreover,

$$\sup_{t \in \mathbb{R}_+} \big\|A^{1/2}u(t)\big\|^2 + \sup_{t \in \mathbb{R}_+} \int_t^{t+1} \big\|A^{1-\theta}u(\tau)\big\|^2 d\tau + \int_0^\infty \big\|A^\theta u_t(\tau)\big\|^2 d\tau \leq c_R, \tag{5.1.24}$$

provided $u_0 \in B_{1/2}(R) \equiv \{u \in \mathscr{D}(A^{1/2}) : \|A^{1/2}u\| \leq R\}$. If $u^1(t)$ and $u^2(t)$ are strong solutions with initial data u_0^1 and u_0^2 from the ball $B_{1/2}(R)$, then

$$\|u^1(t) - u^2(t)\|_{1/2}^2 + \int_0^t \|u_t^1(\tau) - u_t^2(\tau)\|_\theta^2 d\tau \leq a_R \|u_0^1 - u_0^2\|_{1/2}^2 e^{b_R t}, \tag{5.1.25}$$

for all $t \geq 0$. Thus, problem (5.1.20) generates a dynamical system $(H_{1/2}, S_t)$ with the evolution operator defined as $S_t u_0 = u(t)$, where $u(t)$ is the solution to problem (5.1.20) with the initial condition $u(0) = u_0 \in H_{1/2}$. This system is gradient with $\Pi_*(u)$ as a strict Lyapunov function.

Proof. We adopt the approach presented in the proof of Theorem 5.1.4 and rely on some ideas presented in BABIN/VISHIK [9] and CHUESHOV [39].

Step 1: Existence. Let $u^N(t)$ be a Galerkin approximate solution of (5.1.20), i.e., a function $u^N(t)$ with values in $P_N H = \text{Span}\{e_1, \ldots, e_N\}$ solving the problem

$$P_N K(u^N)u_t^N + Au^N + P_N B(u^N) = 0, \quad u_N(0) = P_N u_0. \tag{5.1.26}$$

Since the matrix $\big((K(v)e_k, e_j)\big)_{N \times N}$ is strictly positive for every $v \in \mathscr{D}(A^{1/2})$ and, by (5.1.2), is Lipschitz with respect to v, we can guarantee the existence of $u_N(t)$ at least locally on some interval $[0, T_*)$ (see, e.g., Theorem A.1.2 in the Appendix).

Multiplying (5.1.26) by u_t^N, we obtain (5.1.23) with $u^N(t)$ instead of $u(t)$. Relation (5.1.23) leads to the following a priori estimate:

$$\|A^{1/2}u^N(t)\|^2 \le C_R, \quad t \in [0, T_*),$$

under the condition $\|A^{1/2}u^N(0)\| \le R$. By (5.1.1) we also have that

$$\left(K(u^N)u_t^N, u_t^N\right) \ge a_R\|u_t^N\|_\theta^2.$$

Thus, we arrive at an a priori estimate of the form

$$\|A^{1/2}u^N(t)\|^2 + \int_0^t \|A^\theta u_t^N(\tau)\|^2 d\tau \le C_R, \quad t \in [0, T_*),$$

which implies the existence of approximate solutions on the semi-axis $\mathbb{R}_+$ with the following uniform estimate:

$$\sup_{t\in\mathbb{R}_+} \|A^{1/2}u^N(t)\|^2 + \int_0^\infty \|A^\theta u_t^N(t)\|^2 dt \le C_R, \quad N = 1, 2, \ldots. \qquad (5.1.27)$$

Since $Au^N = -P_N\left(K(u^N)u_t^N + B(u^N)\right)$, from (5.1.27) we obtain

$$\sup_{t\in\mathbb{R}_+} \int_t^{t+1} \|u^N(\tau)\|_{1-\theta}^2 d\tau \le C_R. \qquad (5.1.28)$$

It follows from (5.1.27) and (5.1.28) that there exists

$$u(t) \in L_\infty(\mathbb{R}_+; H_{1/2}) \cap L_2^{loc}(\mathbb{R}_+; H_{1-\theta})$$

with $u_t \in L_2(\mathbb{R}_+; H_\theta)$ such that along a subsequence

$$u^N \to u(t) \quad \text{*-weakly in} \quad L_\infty(0, T; H_{1/2}) \cap L^2(0, T; H_{1-\theta}),$$

$$u_t^N \to u_t(t) \quad \text{weakly in} \quad L_2(0, T; H_\theta),$$

as $N \to \infty$ for every $T > 0$. By the Aubin-Dubinskii-Lions theorem (see Theorem A.3.7 in the Appendix),

$$u^N \to u \quad \text{in} \quad C([0, T]; H_{1/2-\epsilon}) \cap L_2(0, T; H_{1-\theta-\epsilon})), \quad \forall\epsilon > 0. \qquad (5.1.29)$$

Moreover, by Proposition A.3.3 with $V = H_{1-\theta}$, $H = H_{1/2}$, and $V = H_\theta$, one can see that $u(t)$ belongs to $C([0, T]; H_{1/2})$. It is also clear that $u^N(t)$ satisfies the variational form in (5.1.21) with

$$v \equiv v_M \in L_2(0, T; P_M\mathscr{D}(A^{1/2})),$$

where $M \leq N$. Therefore, passing to the limit (sequentially as $N \to \infty$ and $M \to \infty$) one can see that $u(t)$ is a strong solution to problem (5.1.20) satisfying (5.1.22) and (5.1.24). The subcritical estimate in (5.1.3) for K and the weak convergence in (5.1.5) for B along with (5.1.29) make it possible to pass to the limit in the nonlinear terms.

Step 2: Energy balance relation. In the following lemma we prove that any strong solution satisfies (5.1.23) and obeys some estimates which we need to prove (5.1.25).

Lemma 5.1.10. *Any strong solution $u(t)$ to (5.1.20) on an interval $[0, T]$ possesses property (5.1.22) and satisfies the energy balance relation in (5.1.23) for $t \in [0, T]$. Moreover,*

$$\|u(t)\|_{1/2}^2 + \int_0^t \|u_t(\tau)\|_\theta^2 d\tau \leq C_R, \quad t \in [0, T], \tag{5.1.30}$$

and also

$$\sup_{t \in \mathbb{R}_+} \int_t^{t+T} \|u(\tau)\|_{1-\theta}^2 d\tau \leq C_R(1 + T), \tag{5.1.31}$$

provided $\|A^{1/2}u_0\| \leq R$, where C_R does not depend on T.

Proof. One can see from (5.1.21) that $u(t) \in L_2(0, T; H_{1-\theta})$ and the relation (5.1.20) holds in $H_{-\theta}$ for almost all $t \in [0, T]$. Thus, by Proposition A.3.3 any strong solution satisfies (5.1.22). Multiplying (5.1.20) by $P_N u_t(t) \in L_2(0, T; P_N H)$, after integration we obtain that

$$\frac{1}{2}\left(\|P_N A^{1/2}u(t)\|^2 - \|P_N A^{1/2}u(0)\|^2\right) + \int_0^t \left(K(u)u_t + B(u), P_N u_t\right)d\tau = 0.$$

Since $P_N \to I$ strongly, using the smoothness property (5.1.22) and the relation $B(u) = \Pi'(u)$, we can pass to the limit $N \to \infty$ and obtain (5.1.23). Relations (5.1.30) and (5.1.31) follow from (5.1.23) in the same way as in the proof of (5.1.27) and (5.1.28). $\qquad\square$

Step 3: Uniqueness and Lipschitz property. It is obvious that the uniqueness of strong solutions follows from the Lipschitz property in (5.1.25).

Let $u^1(t)$ and $u^2(t)$ be two strong solutions to problem (5.1.20) with different initial data such that $\|A^{1/2}u^i(0)\| \leq R$. Then $z(t) = u^1(t) - u^2(t)$ solves the equation

$$\left[K(u^1)u_t^1 - K(u^2)u_t^2\right] + Az + \left[B(u_1) - B(u_2)\right] = 0.$$

In the same way as in the proof of Lemma 5.1.10, using the test function $P_N z_t$ and then passing to the limit as $N \to \infty$, we get that

$$\frac{1}{2}\left(\|A^{1/2}z(t)\|^2 - \|A^{1/2}z(0)\|^2\right)$$

$$+ \int_0^t \left(K(u^1)u_t^1 - K(u^2)u_t^2, z_t\right)d\tau + \int_0^t \left(B(u^1) - B(u^2), z_t\right)d\tau = 0.$$

By the properties (5.1.2) and (5.1.4) of B and K, this implies that

$$\|A^{1/2}z(t)\|^2 + \alpha_R \int_0^t \|z_t\|_\theta^2 d\tau \leq \|A^{1/2}z(0)\|^2 + \int_0^t \left[\|K(u^1) - K(u^2)\|_{H_\theta \mapsto H_{-\theta}}\|u_t^2\|_\theta\right.$$

$$\left. + \|B(u^1) - B(u^2)\|_{-\theta}\right]\|z_t\|_\theta d\tau$$

$$\leq \|A^{1/2}z(0)\|^2 + C_R \int_0^t \|A^{1/2}z\|\left(1 + \|u_t^2\|_\theta\right)\|z_t\|_\theta d\tau.$$

Thus,

$$\|z(t)\|_{1/2}^2 + \frac{\alpha_R}{2}\int_0^t \|z_t\|_\theta^2 d\tau \leq \|z(0)\|_{1/2}^2 + C_R \int_0^t \|z\|_{1/2}^2\left(1 + \|u^2\|_\theta^2\right)d\tau,$$

$$(5.1.32)$$

which implies (5.1.25). This completes the proof of Theorem 5.1.9.   □

5.2   Global attractors

In this section we prove the existence of global attractors for the dynamical systems $(\mathscr{H}, S_t^\mu)$ and $(H_{1/2}, S_t)$ and study their properties. As was shown in the previous section, these systems are gradient, and thus we can use the criterion of the global attractor existence for gradient systems stated in Theorem 2.4.16.

We also apply a weak quasi-stability method based on Theorem 2.2.17 and Proposition 2.2.18. To implement the method for the models with state-dependent damping coefficient, we need additional hypotheses.

Assumption 5.2.1. We assume the following.

(1) The family of operators $K(u)$ is subcritical in the sense that

$$\|K(u_1) - K(u_2)\|_{H_\theta \mapsto H_{-\theta}} \leq C_\varrho \|A^{1/2-\delta}(u_1 - u_2)\|, \quad \forall \|A^{1/2}u_i\| \leq \varrho, \quad (5.2.1)$$

for some $\delta > 0$.

(2) Also,

- either $0 \leq \theta \leq 1/2$ and B is subcritical, i.e., (5.1.4) holds with some $\delta > 0$;
- or else $0 \leq \theta < 1/2$ and the mapping $u \mapsto \Pi(u)$ is continuous with respect to convergence in $H_{1/2-\delta}$ for some $\delta > 0$.

(3) There exist $\nu < 1$ and $C > 0$ such that

$$\nu\|A^{1/2}u\|^2 + (B(u), u) + C \geq 0, \quad u \in H_{1/2}.$$

We need the last requirement to obtain the following fact.

Exercise 5.2.2. The set $\mathcal{N}_* = \{u \in H_{1/2} : Au + B(u) = 0\}$ of stationary solutions is bounded. Hint: Use the same idea as in the proof of Proposition 4.3.6.　■

5.2.1 Existence of regular attractors

Our goal is to prove the following result on attractors.

Theorem 5.2.3. *Let Assumptions* 5.1.1 *and* 5.2.1 *be in force. Then the dynamical systems* $(\mathcal{H}, S_t^\mu)$ *and* $(H_{1/2}, S_t)$ *possess compact global attractors* $\mathfrak{A}^\mu$ *and* $\mathfrak{A}$*, which have the form* $\mathfrak{A}^\mu = \mathcal{M}^u(\mathcal{N})$ *and* $\mathfrak{A} = \mathcal{M}^u(\mathcal{N}_*)$*, where*

$$\mathcal{N} = \{(u; 0) \in \mathcal{H} : u \in \mathcal{N}_*\} \quad and \quad \mathcal{N}_* = \{u \in \mathcal{D}(A^{1/2}) : Au + B(u) = 0\}$$

are the corresponding sets of equilibria for $(\mathcal{H}, S_t^\mu)$ *and* $(H_{1/2}, S_t)$*. Moreover,*

$$\mathrm{dist}_{\mathcal{H}}(S_t^\mu y, \mathcal{N}) \equiv \inf\{\|S_t^\mu y - e\|_{\mathcal{H}} : e \in \mathcal{N}\} \to 0 \; as \; t \to +\infty \qquad (5.2.2)$$

for every $y \in \mathcal{H}$ *and*

$$\mathrm{dist}_{H_{1/2}}(S_t u, \mathcal{N}_*) \equiv \inf\{\|S_t y - e\|_{1/2} : e \in \mathcal{N}_*\} \to 0 \; as \; t \to +\infty \qquad (5.2.3)$$

for every $u \in H_{1/2}$*.*

We recall (see Definition 2.3.10) that $\mathcal{M}^u(\mathcal{N})$ denotes the unstable set emanating from $\mathcal{N}$ which is a subset of $\mathcal{H}$ such that for each $z \in \mathcal{M}_+(\mathcal{N})$ there exists a full trajectory $\{y(t) : t \in \mathbb{R}\}$ satisfying $u(0) = z$ and $\mathrm{dist}_X(y(t), \mathcal{N}) \to 0$ as $t \to -\infty$.

The proof of Theorem 5.2.3 consists of several steps. We start with some preliminary remarks.

As we know from Proposition 5.1.7 and Theorem 5.1.9, both systems $(\mathcal{H}, S_t^\mu)$ and $(H_{1/2}, S_t)$ are gradient. Moreover, by Assumption 5.1.1 one can see that the corresponding Lyapunov functions satisfy the following requirements: (i) $\Phi(x)$ is bounded from above on any bounded set; (ii) the set $\Phi_R = \{x \in X : \Phi(x) \leq R\}$ is bounded for every R, where $X = \mathcal{H}$ or $X = H_{1/2}$. The sets of stationary points are bounded under Assumption 5.2.1(3); see Exercise 5.2.2.

Therefore, in order to apply Theorem 2.4.16 we need only to check that the corresponding system is asymptotically smooth. For this we use Proposition 2.2.18 based on weak quasi-stability.

We start with two preliminary lemmas.

Lemma 5.2.4. *Let $\mu \in [0, \mu_0]$ for some $\mu_0 > 0$. Under Assumption 5.1.1 there exist $T_0 > 0$ and a constant $c > 0$ independent of T and μ such that for any pair u^1 and u^2 of weak solutions to (5.0.1) (to (5.1.20) in the case when $\mu = 0$) we have the following relation:*

$$TE_z^\mu(T) + \int_0^T E_z^\mu(t)dt + \int_0^T dt \int_t^T (D(\tau), z_t)d\tau \tag{5.2.4}$$

$$\leq c \left\{ \int_0^T \|z_t(t)\|^2 dt + \int_0^T |(D(t), z_t)|dt + \int_0^T |(D(t), z)|\, dt + \Psi_T(u^1, u^2) \right\}$$

for every $T \geq T_0$, where $z(t) = u^1(t) - u^2(t)$, and the functionals[4] E_z^μ, D, and Ψ_T are defined as

$$E_z^\mu(t) = E_0^\mu(z(t), z_t(t)) = \frac{1}{2}\left(\mu\|z_t(t)\|^2 + \|A^{1/2}z(t)\|^2 \right),$$

$$D(t) = K(u^1(t))u_t^1(t) - K(u^2(t))u_t^2(t),$$

$$\Psi_T(u^1, u^2) = \left| \int_0^T (G(\tau), z_t(\tau))d\tau \right| + \left| \int_0^T (G(t), z(t))dt \right|$$

$$+ \left| \int_0^T dt \int_t^T (G(\tau), z_t(\tau))d\tau \right|$$

with $G(t) = B(u^1(t)) - B(u^2(t))$.

Proof. We use the standard arguments involving the multipliers z_t and z for (5.1.18). We refer to the proof of Lemma 3.23 in CHUESHOV/LASIECKA [56] and also to CHUESHOV/LASIECKA [58, Lemma 8.3.1], where this lemma is proved under another set of hypotheses concerning the damping operator. However, the corresponding argument does not depend on the structure of the damping operator. For self-containment, we sketch the argument.

The variable z satisfies the equation

$$\mu z_{tt} + Az + D(t) + G(t) = 0, \tag{5.2.5}$$

and hence we have the following energy relation:

$$E_z^\mu(T) + \int_t^T (D(\tau), z_t(\tau))d\tau = E_z^\mu(t) + \int_t^T (G(\tau), z_t(\tau))d\tau. \tag{5.2.6}$$

---

[4]To emphasize the uniform dependence of bounds on μ, we use the notation E_z^μ instead of E_z.

Multiplying (5.2.5) by z after integration we obtain

$$\int_0^T E_z^\mu(t)dt \le c_0\sqrt{\mu}\left(E_z^\mu(T) + E_z^\mu(0)\right) + \mu\int_0^T \|z_t(s)\|^2 ds$$

$$+ \frac{1}{2}\int_0^T |(D(s), z(s))|ds + \frac{1}{2}\int_0^T (G(s), z(s))ds. \quad (5.2.7)$$

By (5.2.6),

$$E_z^\mu(0) = E_z^\mu(T) + \int_0^T (D(\tau), z_t(\tau))d\tau - \int_0^T (G(\tau), z_t(\tau))d\tau.$$

This relation allows us to exclude $E_z^\mu(0)$ from (5.2.7). Integrating (5.2.6) from 0 to T, we obtain

$$TE_z^\mu(T) + \int_0^T dt \int_t^T (D(\tau), z_t(\tau))d\tau \le \int_0^T E_z^\mu(t)dt + \int_0^T dt \int_t^T (G(\tau), z_t(\tau))d\tau.$$

Therefore, (5.2.4) with T_0 and c depending on μ_0 follows from (5.2.7). $\qquad\square$

Remark 5.2.5. The inequality stated in Lemma 5.2.4 constitutes a common first step in the proofs of several assertions on the existence and finite dimensionality of global attractors for second order in time evolution equations (see, e.g., CHUESHOV/LASIECKA [56, 58]). In the terminology of CHUESHOV/LASIECKA [58], the inequality in (5.2.4) represents equipartition of the energy. The potential energy is reconstructed from the kinetic energy and the nonlinear quantities entering the equation (see CHUESHOV/LASIECKA [58] for some details for other models). Eventually, these quantities will need to be absorbed ("modulo" lower order terms) by the damping. The realization of this step depends heavily on the assumptions imposed on the model. Although this approach was originally designed for second order in time evolution equations (see, e.g., CHUESHOV/LASIECKA [56] and the references therein), we can also apply the same idea for the parabolic-like problem (5.1.20). $\qquad\blacksquare$

Lemma 5.2.6. *Let u^1 and u^2 be two solutions to (5.0.1) with initial data $(u_0^i; u_1^i)$. We assume that $\mu\|u_1^i\|^2 + \|u_0^i\|_{1/2}^2 \le R^2$. Then*

$$\max_{[0,T]} E_z^\mu(t) \le E_z^\mu(T) + \int_0^T |(D(t), z_t)|dt + C_R\left[\int_0^T \|z_t\|_\theta^2 dt + \int_0^T \|A^{1/2}z\|^2 dt\right],$$

with $z = u^1 - u^2$, where E_z^μ and $D(t)$ are the same as in Lemma 5.2.4.

Proof. It follows from (5.2.6) that

$$\max_{[0,T]} E_z^\mu(t) \le E_z^\mu(T) + \int_0^T |(D(\tau), z_t(\tau))| d\tau + \int_0^T |(G(\tau), z_t(\tau))| d\tau.$$

Using (5.1.4) and (5.1.13) we have

$$|(G, z_t)| \le C_R \|A^{1/2} z\| \|z_t\|_\theta \le C_R \left(\|A^{1/2} z\|^2 + \|z_t\|_\theta^2 \right).$$

Substitution in the previous formula yields the conclusion. $\qquad\square$

Now we simplify the representation in Lemma 5.2.4. Namely, we prove the following assertion.

Lemma 5.2.7. *Let $\mu \in [0, \mu_0]$ for some $\mu_0 > 0$. Under Assumptions 5.1.1 and 5.2.1, there exist $T_0 > 0$ and a constant $c > 0$ independent of T and μ such that for any pair u^1 and u^2 of weak solutions to (5.0.1) (to (5.1.20) in the case when $\mu = 0$) possessing the properties $E^\mu(u^j(0), u_t^j(0)) \le R$ we have*

$$TE_z^\mu(T) + \int_0^T E_z^\mu(t) dt \le C_{R,T} \max_{[0,T]} \|z\|^2 + C_R \int_0^T \|z_t\|_\theta^2 dt + c\Psi_T(u^1, u^2), \tag{5.2.8}$$

where $z = u^1 - u^2$, and E_z and $\Psi_T(u^1, u^2)$ are the same as in Lemma 5.2.4. The constants in (5.2.8) do not depend on $\mu \in [0, \mu_0]$.

Proof. Using (5.2.1) we obtain that

$$|(D(t), z_t)| \le C_{R,\varepsilon} \|z_t\|_\theta^2 + \varepsilon \|z\|_{1/2-\delta}^2 \|u_t^2\|_\theta^2 \tag{5.2.9}$$

for any $\varepsilon > 0$ and

$$(D(t), z_t) \ge \gamma_R \|z_t\|_\theta^2 - C_R \|z\|_{1/2-\delta}^2 \|u_t^2\|_\theta^2.$$

We also have that

$$|(D(t), z)| \le \varepsilon \|z\|_\theta^2 + C_{R,\varepsilon} \left[\|z\|_{1/2-\delta}^2 \|u_t^2\|_\theta^2 + \|z_t\|_\theta^2 \right]$$

for any $\varepsilon > 0$. Therefore, after choosing ε in an appropriate way, Lemma 5.2.4 implies

$$TE_z^\mu(T) + \int_0^T E_z^\mu(t) dt \le C_R(1 + T) \int_0^T \|z\|_{1/2-\delta}^2 \|u_t^2\|_\theta^2 dt$$

$$+ C_R \int_0^T \|z_t\|_\theta^2 dt + c\Psi_T(u^1, u^2)$$

for every $T \geq T_0$. Thus, using the interpolation (see Exercise 4.1.2(E,F))

$$\|A^{1/2-\delta}z\|^2 \leq \|A^{1/2}z\|^{2-4\delta}\|z\|^{4\delta} \leq \varepsilon\|z\|^2 + C_\varepsilon\|z\|^2, \quad 0 < \delta \leq 1/2, \ \forall \varepsilon > 0,$$

we obtain

$$TE_z^\mu(T) + \int_0^T E_z^\mu(t)dt \leq \varepsilon \int_0^T \|z\|_{1/2}^2\|u_t^2\|_\theta^2 dt$$

$$+ C_{R,\varepsilon,T}\int_0^T \|z\|^2\|u_t^2\|_\theta^2 dt + C_R\int_0^T \|z_t\|_\theta^2 dt + c\Psi_T(u^1, u^2)$$

Next, it follows from Lemma 5.2.6 and (5.2.9) that

$$\max_{[0,T]}E_z^\mu(t) \leq E_z^\mu(T) + C_{R,\epsilon}\int_0^T \|z_t\|_\theta^2 dt$$

$$+ \epsilon\int_0^T \|A^{1/2}z\|^2\|u_t^2\|^2 dt + C_R\int_0^T \|A^{1/2}z\|^2 dt.$$

We note that due to (5.1.13) (or (5.1.24) in the case $\mu = 0$),

$$\int_0^T \|A^{1/2}z\|^2\|u_t^2\|_\theta^2 dt \leq \max_{[0,T]}E_z^\mu(t) \cdot \int_0^T \|u_t^2\|_\theta^2 dt \leq C_R\max_{[0,T]}E_z^\mu(t).$$

Choosing $\epsilon = \epsilon(R)$ in an appropriate way, we have the bound

$$\max_{[0,T]}E_z^\mu(t) \leq 2E_z^\mu(T) + C_R\int_0^T \|z_t\|_\theta^2 dt + C_R\int_0^T \|A^{1/2}z\|^2 dt. \qquad (5.2.10)$$

This gives us

$$TE_z^\mu(T) + \int_0^T E_z^\mu(t)dt \leq C_{R,T}\int_0^T \|z\|^2\|u_t^2\|_\theta^2 dt$$

$$+ C_R\int_0^T \|z_t\|_\theta^2 dt + c\Psi_T(u^1, u^2) \qquad (5.2.11)$$

for $T > T_0$. Therefore, (5.2.8) follows. $\qquad \square$

Proof of Theorem 5.2.3. As we mentioned above, we need only to prove that the corresponding system is asymptotically smooth.

Let u^1 and u^2 be weak solutions to (5.0.1) (to (5.1.20) in the case when $\mu = 0$) possessing the properties $E^\mu(u^j(0), u_t^j(0)) \leq R$. We also suppose $z = u^1 - u^2$.

The case $\theta < 1/2$ with a critical force B. We recall that the criticality of the force B means that (5.1.4) holds with $\delta = 0$. In this case,

$$\left| \int_0^T (G(t), z(t)) dt \right| \leq C_R \int_0^T \|z\|_{1/2} \|z\|_\theta dt \leq C_R \int_0^T \|z\|_{1/2}^{1+2\theta} \|z\|^{1-2\theta} dt$$

$$\leq \varepsilon \int_0^T \|z\|_{1/2}^2 dt + T C_{R,\varepsilon} \max_{[0,T]} \|z\|^2.$$

Therefore by Lemma 5.2.7,

$$T E_z^\mu(T) \leq C_{R,T} \max_{[0,T]} \|z\|^2 + C_R \int_0^T \|z_t\|_\theta^2 dt + c \Psi_T^0(u^1, u^2), \tag{5.2.12}$$

where

$$\Psi_T^0(u^1, u^2) = \left| \int_0^T (G(\tau), z_t(\tau)) d\tau \right| + \left| \int_0^T dt \int_t^T (G(\tau), z_t(\tau)) d\tau \right|. \tag{5.2.13}$$

Under the conditions of Lemma 5.2.4, by Theorems 5.1.4 and 5.1.9 we have that

$$\int_0^T \|z_t\|_\theta^2 dt \leq 2 \int_0^T (\|u_t^1\|_\theta^2 + \|u_t^2\|_\theta^2) dt \leq C_R.$$

Therefore, relation (5.2.12) gives us that for any $\epsilon > 0$ there exists $T = T(R, \epsilon)$ such that

$$E_z(T) \leq \epsilon + C_{R,\epsilon} \max_{[0,T]} \|z(t)\|^2 + C_{R,\epsilon} \Psi_T^0(u_1, u_2).$$

The case $\mu > 0$. Let $\{y_n = (u_0^n; u_1^n)\}$ be a bounded sequence in the space $\mathscr{H}$, i.e., $E^\mu(u_0^n, u_1^n) \leq R^2$ for some R. Let $(u^n(t); u_t^n(t))$ be the corresponding solution to (5.0.1). Then by estimate (5.1.13) in Theorem 5.1.4 we can assume that

$$(u^n(t); u_t^n(t)) \to (u(t); u_t(t)) \quad \text{*-weakly in} \quad L_\infty(0, T; \mathscr{D}(A^{1/2}) \times H). \tag{5.2.14}$$

In particular, by the Aubin-Dubinskii-Lions theorem (see Theorem A.3.7), this implies that

$$\max_{[0,T]} \|u^n(t) - u^m(t)\|_\eta^2 \to 0 \quad \text{as } n, m \to \infty, \quad \forall 0 \leq \eta < 1/2. \tag{5.2.15}$$

Thus, to apply Proposition 2.2.18 we need only to prove

$$\liminf_{m\to\infty} \liminf_{n\to\infty} \Psi_T^0(u^n, u^m) = 0 \tag{5.2.16}$$

for every $T > 0$, where $\Psi_T^0(w, v)$ is given by (5.2.13).
    We claim that

$$\lim_{m\to\infty} \lim_{n\to\infty} \int_t^T (G^{n,m}(\tau), u_t^n(\tau) - u_t^m(\tau))d\tau = 0 \tag{5.2.17}$$

for any $t < T$, where $G^{n,m}(t) = B(u^n(t)) - B(u^m(t))$. Indeed, since

$$\int_t^T (G^{n,m}(\tau), u_t^n(\tau) - u_t^m(\tau))d\tau$$

$$= \Pi(u^n(T)) - \Pi(u^n(t)) + \Pi(u^m(T)) - \Pi(u^m(t)) - \mathscr{I}_t^T(n, m),$$

where

$$\mathscr{I}_t^T(n, m) = \int_t^T \left[(\Pi'(u^n(\tau)), u_t^m(\tau)) + (\Pi'(u^m(\tau)), u_t^n(\tau))\right] d\tau,$$

by Assumption 5.2.1(2) and (5.2.15) we have that

$$\lim_{m\to\infty} \lim_{n\to\infty} \int_t^T (G^{n,m}(\tau), u_t^n(\tau) - u_t^m(\tau))d\tau$$

$$= 2\Pi(u(T)) - 2\Pi(u(t)) - \lim_{m\to\infty} \lim_{n\to\infty} \mathscr{I}_t^T(n, m). \tag{5.2.18}$$

It follows from (5.2.15) and Assumption 5.1.1(B) that

$$\Pi'(u^n(t)) \to \Pi'(u(t)) \quad \text{*-weakly in} \quad L_\infty(0, T; H_{-\theta}).$$

Therefore, using (5.2.14) we obtain that

$$\lim_{m\to\infty} \lim_{n\to\infty} \int_t^T (\Pi'(u^n(\tau)), u_t^m(\tau))d\tau = \lim_{m\to\infty} \int_t^T (\Pi'(u(\tau)), u_t^m(\tau))d\tau$$

$$= \int_t^T (\Pi'(u(\tau)), u_t(\tau))d\tau$$

In a similar way, we also have that

$$\lim_{m\to\infty} \lim_{n\to\infty} \int_t^T (\Pi'(u^m(\tau)), u_t^n(\tau))d\tau = \lim_{m\to\infty} \int_t^T (\Pi'(u^m(\tau)), u_t(\tau))d\tau$$

$$= \int_t^T (\Pi'(u(\tau)), u_t(\tau))d\tau.$$

Therefore,

$$\lim_{m\to\infty}\lim_{n\to\infty}\mathscr{I}_t^T(n,m) = 2\int_t^T (\Pi'(u(\tau)), u_t(\tau))d\tau. \qquad (5.2.19)$$

Hence after substituting (5.2.19) in (5.2.18), we get (5.2.17).

To conclude the proof, we note that (5.2.16) follows from (5.2.13) and (5.2.17) and also from the Lebesgue dominated convergence theorem. Thus, we can apply Proposition 2.2.18 and obtain asymptotic compactness for the case $\mu > 0$.

The case $\mu = 0$. Now we have the relation

$$\|A^{1/2}z(T)\|^2 \le \epsilon + C_{R,\epsilon}\max_{[0,T]}\|z(t)\|^2 + C_{R,\epsilon}\Psi_T^0(u_1, u_2).$$

Let $\{u_0^n\}$ be a sequence in $\mathscr{D}(A^{1/2})$ such that $\|A^{1/2}u_0^n\| \le R$ for some R. Let $u^n(t)$ be the corresponding solution to (5.1.20). In this case by estimate (5.1.24) in Theorem 5.1.9 we can assume that $(u^n(t); u_t^n(t))$ is a *-weakly convergent sequence in

$$\left[L_\infty(0, T; H_{1/2}) \cap L_2(0, T; H_{1-\theta}))\right] \times L_2(0, T; H_\theta).$$

In particular, this implies the strong convergence of $u^n(t)$ in

$$C([0, T]; \mathscr{D}(A^{1/2-\delta})) \cap L_2(0, T; \mathscr{D}(A^{1-\theta-\delta})), \quad \forall \delta > 0.$$

These convergence properties allow us to apply the same argument as in the case $\mu > 0$ to prove that $\lim_{n\to\infty}\lim_{m\to\infty}\Psi_T^0(T, u^n, u^m) = 0$. This implies the asymptotic smoothness in the case $\mu = 0$.

The case $\theta \le 1/2$ with a subcritical force B. Now the case $\theta = 1/2$ is included at the expense that B is subcritical in the sense that (5.1.4) holds with $\delta > 0$.

The subcritical estimate in (5.1.4) with $\delta > 0$ yields

$$|\Psi_T| \le C_R\int_0^T \|z_t\|_\theta^2 dt + \varepsilon\int_0^T \|z\|_{1/2}^2 dt + C_{R,\varepsilon,T}\int_0^T \|z\|^2 dt \quad \text{for every } \varepsilon > 0.$$

Therefore, by (5.2.8)

$$TE_z(T) + \int_0^T E_z(t)dt \le C_R\int_0^T \|z_t\|_\theta^2 dt + C_{R,T}\max_{[0,T]}|z(t)\|^2. \qquad (5.2.20)$$

Thus, as above, we obtain

$$E_z(T) \le \epsilon + C_{R,\epsilon}\max_{[0,T]}\|z(t)\|^2 \quad \text{for } T \ge T_\varepsilon.$$

This implies the desired conclusion on asymptotic compactness.   ∎

□

5.2.2 Rate of stabilization to equilibria

Using the same idea as in CHUESHOV/LASIECKA [51, 56, 58], we can establish the following result on convergence of individual solutions to equilibria at an exponential rate.

Theorem 5.2.8 (Rate of stabilization). *In addition to Assumption 5.1.1 and 5.2.1, we assume that $B(u)$ is Fréchet differentiable[5] and its derivative $B'(u)$ possesses the properties*

$$\|B'(u)w\|_{-1/2} \le C_R\|w\|_{1/2}, \quad w \in H_{-1/2}, \tag{5.2.21}$$

and

$$\|[B'(u) - B'(v)]w\|_{-1/2} \le C_R\|u - v\|_{1/2-\delta} \cdot \|w\|_{1/2}, \ w \in H_{1/2}, \tag{5.2.22}$$

for any $u, v \in H_{1/2}$ such that $\|u\|_{1/2} \le R$ and $\|v\|_{1/2} \le R$ with $\delta > 0$. Let the set

$$\mathcal{N}_* = \{u \in \mathcal{D}(A^{1/2}) : Au + B(u) = 0\}$$

of stationary solutions be finite and all equilibria be hyperbolic in the sense that the equation $Au + B'(\phi)u = 0$ has only a trivial solution for each $\phi \in \mathcal{N}_*$. Then

- **Case $\mu > 0$:** *For any $y = (u_0; u_1) \in \mathcal{H}$ there exist a stationary solution $\phi \in \mathcal{N}_*$ and constants $\gamma > 0$, $C > 0$ such that*

$$\mu\|u_t(t)\|^2 + \|u(t) - \phi\|_{1/2}^2 \le Ce^{-\gamma t}, \quad t > 0, \tag{5.2.23}$$

  where $(u(t); u_t(t)) = S_t^\mu y$ is the flow generated by (5.0.1) on $\mathcal{H}$.
- **Case $\mu = 0$:** *For any $u_0 \in H_{1/2}$ there exist a stationary solution $\phi \in \mathcal{N}_*$ and constants $\gamma > 0$, $C > 0$ such that*

$$\|S_t u_0 - \phi\|_{1/2}^2 \le Ce^{-\gamma t}, \quad t > 0, \tag{5.2.24}$$

  where S_t is the evolution operator for (5.1.20) on $H_{1/2}$.

As we mentioned in Chapter 4, rate stabilization theorems are well known for different classes of gradient systems, and several approaches to this question are available (see, e.g., BABIN/VISHIK [9] and also CHUESHOV/LASIECKA [51, 56, 58]). Here we use the method developed in CHUESHOV/LASIECKA [51, 56] (see also the discussion in CHUESHOV/LASIECKA [58]), and we use (in contrast with BABIN/VISHIK [9]) the static form of the hyperbolicity condition. A parabolic realization of this method was already applied in the proof of Theorem 4.3.18.

---

[5] See Section A.5 in the Appendix for the definitions.

Proof. Let $u(t)$ be a solution (5.0.1) for some fixed $\mu \in [0, \mu_0]$. Since $\mathcal{N}_*$ is finite by (5.2.2) and (5.2.3) in Theorem 5.2.3, we have that there exists a stationary solution $\phi \in \mathcal{N}_*$ such that

$$\mu \|u_t(t)\|^2 + \|u(t) - \phi\|_{1/2}^2 \to 0, \quad t \to \infty. \tag{5.2.25}$$

Thus, we need only to prove that $u(t)$ tends to ϕ at the stated rate.

We can assume that

$$\sup_{t \geq 0} \left\{ \mu \|u_t(t)\|^2 + \|u(t)\|_{1/2}^2 \right\} \leq R \quad \text{for some } R > 0.$$

The function $z(t) = u(t) - \phi$ satisfies the following equation:

$$\mu z_{tt}(t) + K(\phi + z(t))z_t(t) + Az(t) + B(\phi + z(t)) - B(\phi) = 0, \ t > 0. \tag{5.2.26}$$

Let $\tilde{\mathscr{E}}(t) = E_z^\mu(t) + \Phi(t)$, where $E_z(t)$ is the same as in Lemma 5.2.4, and

$$\Phi(t) = \Pi(\phi + z(t)) - \Pi(\phi) - (B(\phi), z(t)) \equiv \int_0^1 (B(\phi + \lambda z) - B(\phi), z)d\lambda.$$

One can see that

$$\tilde{\mathscr{E}}(t) + \int_0^t (K(\phi + z(\tau))z_t(\tau), z_t(\tau))d\tau = \tilde{\mathscr{E}}(0). \tag{5.2.27}$$

In particular, we have that $\tilde{\mathscr{E}}(t)$ is non-increasing. Moreover, since $(z; z_t) \to 0$ in $H_{1/2} \times H$ as $t \to +\infty$, we have that $\tilde{\mathscr{E}}(t) \to 0$ when $t \to +\infty$. Thus $\tilde{\mathscr{E}}(t) \geq 0$ for all $t \geq 0$. It is also clear from (5.1.4) that

$$|\tilde{\mathscr{E}}(t) - E_z^\mu(t)| \leq C_R \|z(t)\|_{1/2-\delta}\|z(t)\|_\theta \leq \varepsilon \|z(t)\|_{1/2}^2 + C_{R,\varepsilon}\|z(t)\|^2, \quad \forall \varepsilon > 0, \tag{5.2.28}$$

under the conditions imposed on B in Assumptions 5.1.1 and 5.2.1.

Using the multiplier z in (5.2.26), we obtain

$$\mu \frac{d}{dt}(z_t, z) - \mu \|z_t\|^2 + (K(\phi + z)z_t, z) + \|A^{1/2}z\|^2 + (B(\phi + z) - B(\phi), z) = 0.$$

Therefore,

$$\int_0^T \|A^{1/2}z\|^2 dt \leq c_0(E_z^\mu(T) + E_z^\mu(0)) + \mu \int_0^T \|z_t\|^2 dt$$

$$+ C_R \int_0^T \|z_t\|_\theta \|z\|_\theta dt + C_R \int_0^T \|z_t\|_{1/2-\delta}\|z\|_\theta dt,$$

where $\delta = 0$ is allowed when $\theta < 1/2$. The constants may depend on μ_0. This implies that

$$\int_0^T E_z^\mu(t)dt \leq 2c_0(E_z^\mu(T) + E_z^\mu(0)) + C_R \int_0^T \|z_t\|_\theta^2 dt + C_R \int_0^T \|z\|^2 dt.$$

By (5.2.28),

$$\tilde{\mathscr{E}}(t) \leq 2E_z^\mu(t) + C_R\|z(t)\|^2 \quad \text{and} \quad E_z^\mu(t) \leq 2\tilde{\mathscr{E}}(t) + C_R\|z(t)\|^2$$

for all $t \in \mathbb{R}_+$. Since

$$\tilde{\mathscr{E}}(0) = \tilde{\mathscr{E}}(T) + \int_0^T (K(\phi + z(\tau))z_t(\tau), z_t(\tau))d\tau \leq \tilde{\mathscr{E}}(T) + C_R \int_0^T \|z_t(\tau)\|_\theta^2 d\tau$$

and

$$T\tilde{\mathscr{E}}(T) \leq \int_0^T \tilde{\mathscr{E}}(t)dt \leq 2\int_0^T E_z^\mu(t)dt + C_R \int_0^T \|z(t)\|^2 dt,$$

we have that

$$T\tilde{\mathscr{E}}(T) + 2\int_0^T \tilde{\mathscr{E}}(t)dt \leq c_0\tilde{\mathscr{E}}(T) + C_R \int_0^T \|z_t(t)\|_\theta^2 dt + C_{R,T} \max_{[0,T]} \|z(t)\|^2.$$

This implies that

$$TE_z^\mu(T) + \int_0^T E_z^\mu(t)dt \leq C_R \int_0^T \|z_t(t)\|_\theta^2 dt + C_{R,T} \max_{[0,T]} \|z(t)\|^2 \tag{5.2.29}$$

for $T \geq T_0$ with some $T_0 > 0$.

A parabolic analog of the following lemma with $K(u) = \mathrm{Id}$ and $\theta = 0$ was already proved in Chapter 4 (see Lemma 4.3.19). Here we apply the same idea.

Lemma 5.2.9. *Let $z(t)$ be a weak solution to (5.2.26) such that*

$$\int_{T-1}^T E_z^\mu(t)dt \leq \delta \quad \text{and} \quad \sup_{t \in \mathbb{R}_+} E_z^\mu(t) \leq \varrho \tag{5.2.30}$$

with some $\delta, \varrho > 0$ and $T > 1$. Then there exists $\delta_0 > 0$ such that

$$\max_{[0,T]} \|z(t)\|^2 \leq C \int_0^T \|z_t(t)\|_\theta^2 dt \tag{5.2.31}$$

for every $0 < \delta \leq \delta_0$, where the constant C may depend on δ, ϱ, and T.

Proof. Assume that (5.2.31) is not true. Then for some $\delta > 0$ small enough there exists a sequence of solutions $\{z^n(t)\}$ satisfying (5.2.30) and such that

$$\lim_{n \to \infty} \left\{ \max_{[0,T]} \|z^n(t)\|^2 \left[\int_0^T \|z_t^n(t)\|_\theta^2 dt \right]^{-1} \right\} = \infty. \tag{5.2.32}$$

By (5.2.30), $\max_{[0,T]} \|z^n(t)\|^2 \le C_\varrho$ for all n. Thus, (5.2.32) implies that

$$\lim_{n \to \infty} \int_0^T \|z_t^n(t)\|_\theta^2 dt = 0. \tag{5.2.33}$$

Therefore, we can assume that there exists $z^* \in H_{1/2}$ such that

$$z^n \to z^* \quad \text{*-weakly in } L_\infty(0, T; H_{1/2}). \tag{5.2.34}$$

It follows from (5.1.1) and (5.2.33) that for $u^n(t) = \phi + z^n(t)$ we have the relation

$$\lim_{n \to \infty} \int_0^T |(K(\phi + z^n(t))z_t^n(t), \psi(t))| dt = 0 \quad \text{for any } \psi \in L_2(0, T; H_\theta).$$

This allows us to conclude that $u^* = \phi + z^* \in H_{1/2}$ solves the problem $Au + B(u) = 0$. From (5.2.30) we have that $\|A^{1/2}(u^* - \phi)\|^2 \le 2\delta$. If we choose $\delta_0 > 0$ such that $\|A^{1/2}(\phi_1 - \phi_2)\|^2 > 2\delta_0$ for every pair ϕ_1 and ϕ_2 of stationary solutions (we can do this because the set $\mathcal{N}_*$ is finite), then we can conclude that $u^* = \phi$, provided $\delta \le \delta_0$. Thus, we have $z^* = 0$ in (5.2.34).

Now we normalize the sequence z^n by defining

$$\hat{z}^n \equiv c_n^{-1} z^n \quad \text{with } c_n = \max_{[0,T]} \|z^n(t)\|,$$

where we account only for a suitable subsequence of nonzero terms in c_n. By the compactness embedding properties (see Theorem A.3.7), it is clear from (5.2.33) and (5.2.34) with $z^* = 0$ that $c_n \to 0$ as $n \to \infty$. By (5.2.32),

$$\int_0^T \|\hat{z}_t^n(t)\|_\theta^2 dt \to 0 \quad \text{as } n \to \infty. \tag{5.2.35}$$

Relations (5.2.29) and (5.2.35) imply the following uniform estimate:

$$\sup_{t \in [0,T]} \{ \mu \|\hat{z}_t^n(t)\|^2 + \|A^{1/2} \hat{z}^n(t)\|^2 \} \le C, \quad n = 1, 2, \dots.$$

Thus, we can suppose that there exists $\hat{z}^* \in H_1$ such that

$$(\hat{z}^n; \sqrt{\mu} \hat{z}_t^n) \to (\hat{z}^*; 0) \quad \text{*-weakly in } L_\infty(0, T; H_{1/2} \times H), \quad \mu \ge 0. \tag{5.2.36}$$

The function $\hat{z}^n$ satisfies the equation

$$\mu \hat{z}^n_{tt} + \frac{1}{c_n} K(\phi + z^n) z^n_t + A \hat{z}^n + \frac{1}{c_n} \left[B(\phi + z^n) - B(\phi) \right] = 0. \qquad (5.2.37)$$

As above, from (5.1.1) and (5.2.35) we conclude that

$$\frac{1}{c_n} \int_0^T |(K(\phi + z^n) z^n_t, \psi)| dt \to 0 \quad \text{as} \quad n \to \infty \quad \text{for any} \ \psi \in L_2(0, T; H_\theta).$$

It also follows from (5.2.21) and (5.2.22) that

$$\frac{1}{c_n} \left[B(\phi + z^n) - B(\phi) \right] \to B'(\phi) \hat{z}^* \quad \text{weakly in} \ L_2(0, T; H_{-1/2}).$$

Therefore, after the limit transition in (5.2.37) we conclude that $\hat{z}^*$ satisfies the equation $A \hat{z}^* + B'(\phi) \hat{z}^* = 0$ and, by hyperbolicity of ϕ, we have that $\hat{z}^* = 0$. Thus, (5.2.36) (we use also (5.2.35) in the case $\mu = 0$) and the Aubin-Dubinskii-Lions theorem (see Theorem A.3.7) imply that $\max_{[0,T]} \|\hat{z}^n\| \to 0$ as $n \to \infty$, which is impossible. □

Remark 5.2.10. In the parabolic case ($\mu = 0$) with $\theta < 1/2$, we can avoid the hypotheses in (5.2.21) and (5.2.22) concerning the derivative B'. To do this, we can use the same idea as in the proof of Lemma 4.3.19. ∎

Completion of the proof of Theorem 5.2.8. By (5.2.25) we can choose $T_0 > 0$ such that (5.2.30) holds with $\delta \le \delta_0$ and $T > T_0$. From (5.2.28) we have that $\tilde{\mathscr{E}}(T) \le C_R E_z(T)$. The energy relation in (5.2.27) and the lower bound in (5.1.1) yield that

$$\int_0^T \|z_t(t)\|_\theta^2 dt \le C_R \left[\tilde{\mathscr{E}}(0) - \tilde{\mathscr{E}}(T) \right]. \qquad (5.2.38)$$

Therefore, Lemma 5.2.9, relation (5.2.29), and the energy relation in (5.2.27) imply that $\tilde{\mathscr{E}}(T) \le \gamma_R \tilde{\mathscr{E}}(0)$ for some $0 < \gamma_R < 1$. This implies that $\tilde{\mathscr{E}}(mT) \le \gamma_R^m \tilde{\mathscr{E}}(0)$ for $m = 1, 2, \ldots$. By (5.2.28), (5.2.31), and (5.2.38) we have

$$E_z(mT) \le 2\tilde{\mathscr{E}}(mT) + C_R \max_{[mT,(m+1)T]} \|z(t)\|^2 \le C_R \tilde{\mathscr{E}}(mT), \quad m = 1, 2, \ldots.$$

Thus, $E_z(mT) \le C_R \gamma_R^m$ for $m = 1, 2, \ldots$. Now using (5.1.14) and (5.1.25), we obtain (5.2.23) and (5.2.24). The proof is complete. □

5.3   Properties of the attractor

Our main goal in this section is to demonstrate how the quasi-stability method developed in Chapter 3 can be applied to the systems generated by second order evolution equations with a state-dependent damping coefficient. To cover the critical case of the force term B, we need some structural hypotheses concerning B. We impose them in Assumption 5.3.1 below.

5.3.1   Finite dimension and smoothness

The proof of finiteness of the fractal dimension of the global attractors for $(\mathcal{H}, S_t^\mu)$ and $(H_{1/2}, S_t)$ requires the following additional hypotheses concerning the nonlinear force $B(u)$ in the critical case.

Assumption 5.3.1. Assume that

- $B(u) = \Pi'(u)$ with the functional $\Pi : \mathscr{D}(A^{1/2}) \mapsto \mathbb{R}$ which is a (Fréchet) C^3 mapping. See Section A.5 in the Appendix for the definition.
- The second $\Pi^{(2)}(u)$ and the third $\Pi^{(3)}(u)$ derivatives of $\Pi(u)$ satisfy the conditions[6]

$$\left|\langle \Pi^{(2)}(u); v, v\rangle\right| \le C_\rho \|A^\sigma v\|^2, \quad v \in \mathscr{D}(A^{1/2}), \tag{5.3.1}$$

for some $\sigma < 1/2$, and

$$\left|\langle \Pi^{(3)}(u); v_1, v_2, v_3\rangle\right| \le C_\rho \|A^{1/2}v_1\|\,\|A^{1/2}v_2\|\,\|v_3\|_\theta, \quad v_i \in \mathscr{D}(A^{1/2}), \tag{5.3.2}$$

for all $u \in \mathscr{D}(A^{1/2})$ such that $\|A^{1/2}u\| \le \rho$, where $\rho > 0$ is arbitrary and C_ρ is a positive constant. Here $\langle \Pi^{(k)}(u); v_1, \ldots, v_k\rangle$ denotes the value of the derivative $\Pi^{(k)}(u)$ on elements $v_1, \ldots, v_k$.

This assumption concerning nonlinear feedback force $B(u)$ appeared earlier for systems with monotone velocity-dependent damping (see CHUESHOV/LASIECKA [56, p. 98]) to cover the case of critical nonlinearities.

Recall (see Proposition A.5.3 in the Appendix) that the derivatives $\Pi^{(k)}(u)$ of the functional Π are *symmetric* k-linear continuous forms on $\mathscr{D}(A^{1/2})$. Moreover, if $\Pi \in C^3$, then $(B(u), v) \equiv \langle \Pi'(u); v\rangle$ is a C^2 functional for every fixed $v \in \mathscr{D}(A^{1/2})$, and the following Taylor's expansion holds:

---

[6] Another version of the condition in (5.3.1) is possible; see (5.4.45) below.

$$(B(u + w) - B(u), v) = \langle \Pi^{(2)}(u); w, v \rangle + \int_0^1 (1 - \lambda) \langle \Pi^{(3)}(u + \lambda w); w, w, v \rangle d\lambda \tag{5.3.3}$$

for any $u, v \in \mathscr{D}(A^{1/2})$. See Theorem A.5.4.

Let us assume that $u(t)$ and $z(t)$ belong to the class $C^1([a, b]; \mathscr{D}(A^{1/2}))$ for some interval $[a, b] \subseteq \mathbb{R}$. Then, by the differentiation rule for composition of mappings (Proposition A.5.2) and using the symmetry of the form $\Pi^{(2)}(u)$, we have that

$$\frac{d}{dt} \langle \Pi^{(2)}(u); z, z \rangle = \langle \Pi^{(3)}(u); u_t, z, z \rangle + 2 \langle \Pi^{(2)}(u); z, z_t \rangle.$$

Therefore, from (5.3.3) we obtain the representation

$$(B(u(t) + z(t)) - B(u(t)), z_t(t)) = \frac{d}{dt} Q(t) + R(t), \quad t \in [a, b] \subseteq \mathbb{R}, \tag{5.3.4}$$

with

$$Q(t) = \frac{1}{2} \langle \Pi^{(2)}(u(t)); z(t), z(t) \rangle \tag{5.3.5}$$

and

$$R(t) = -\frac{1}{2} \langle \Pi^{(3)}(u); u_t, z, z \rangle + \int_0^1 (1 - \lambda) \langle \Pi^{(3)}(u + \lambda z); z, z, z_t \rangle d\lambda. \tag{5.3.6}$$

The representation in (5.3.4) leads to the following assertion which, in fact, is proved in [56] (see (4.38), p. 99).

Proposition 5.3.2. *Let Assumptions 5.1.1 and 5.3.1 be in force. Assume that functions $(u^1; u_t^1)$ and $(u^2; u_t^2)$ from $C(\mathbb{R}_+; \mathscr{D}(A^{1/2})) \times L_2^{loc}(\mathbb{R}_+; H)$ possess the property*

$$\max_{s \in \mathbb{R}_+} \left\{ \|A^{1/2} u^1(s)\|^2 + \|A^{1/2} u^2(s)\|^2 \right\} \leq R^2 \tag{5.3.7}$$

for some $R > 0$. Let

$$G_{u^1 u^2}(t) = B(u^1(t)) - B(u^2(t)). \tag{5.3.8}$$

Then for any $\varepsilon > 0$ and $T > 0$ there exist $a(R)$ and $b(R)$ independent of $\mu \in (0, \mu_0]$ such that for $z(t) = u^1(t) - u^2(t)$ we have the following relation:

$$\sup_{t \in [0,T]} \left| \int_{s+t}^{s+T} (G_{u^1 u^2}(\tau), z_t(\tau)) d\tau \right| \leq \varepsilon \int_s^{s+T} \|A^{1/2} z(\tau)\|^2 d\tau \tag{5.3.9}$$

$$+ \frac{a(R)}{\epsilon} \int_s^{s+T} d(\tau)\|A^{1/2}z(\tau)\|^2 d\tau + b(R) \sup_{\tau \in [0,T]} \|A^\sigma z(s+\tau)\|^2$$

for all $s \geq 0$ and $\epsilon > 0$, where $\sigma < 1/2$ and

$$d(t) \equiv d(t; u_1, u_2) = \|u_t^1(t)\|_\theta^2 + \|u_t^2(t)\|_\theta^2. \tag{5.3.10}$$

Proof. Using (5.3.4) and also (5.3.1) and (5.3.2), we have

$$\left| \int_{s+t}^{s+T} (G_{u^1 u^2}, z_t) d\tau \right| \leq |Q(s+T)| + |Q(s+t)| + \int_{s+t}^{s+T} |R(\tau)| d\tau$$

$$\leq C_R \sup_{[s,s+T]} \|A^\sigma z(\tau)\|^2 + C_R \int_{s+t}^{s+T} (\|u_t^1\|_\theta + \|u_t^2|_\theta) \|A^{1/2}z\|^2 d\tau,$$

which implies (5.3.9).   □

Remark 5.3.3. If the nonlinear force $B(u)$ is subcritical; i.e., instead of (5.1.4) with $\delta = 0$ we have

$$\|B(u_1) - B(u_2)\|_{-\theta} \leq L(\varrho)\|A^\sigma(u_1 - u_2)\|, \quad \forall \|A^{1/2}u_i\| \leq \varrho, \tag{5.3.11}$$

for some $\sigma < 1/2$, then we can avoid the hypotheses listed in Assumption 5.3.1. Indeed, under the condition in (5.3.7), relation (5.3.11) implies that

$$\sup_{t \in [0,T]} \left| \int_{s+t}^{s+T} (G_{u^1 u^2}(\tau), z_t(\tau)) d\tau \right|$$

$$\leq \varepsilon \int_s^{s+T} \|z_t(\tau)\|_\theta^2 d\tau + \frac{b(R)T}{\epsilon} \sup_{\tau \in [0,T]} \|A^\sigma z(s+\tau)\|^2 \tag{5.3.12}$$

for all $s \geq 0$ and $\epsilon > 0$. As we see below, this inequality can be used in the next proposition instead of (5.3.9).   ∎

The following proposition establishes the quasi-stability estimate for the dynamical systems $(\mathscr{H}, S_t^\mu)$ and $(H_{1/2}, S_t)$.

Proposition 5.3.4 (Quasi-stability estimate). *Let $\mu \in [0, \mu_0]$. In addition to Assumptions 5.1.1 and 5.2.1, we suppose that **either** Assumption 5.3.1 **or else** relation (5.3.11) holds. Let u^i, $i = 1, 2$ be two solutions either to (5.0.1) ($\mu > 0$) with the initial data such that $\|A^{1/2}u_0^i\|^2 + \mu\|u_1^i\|^2 \leq R^2$ or to (5.1.20) with $\|A^{1/2}u_0^i\| \leq R$, $i = 1, 2$. Let $z = u_1 - u_2$. Then there exist $C(R), \gamma(R) > 0$ independent of $\mu \in [0, \mu_0]$ such that*

$$E_z^\mu(t) \leq C(R)E_z^\mu(0)e^{-\gamma(R)t} + C(R)\max_{[0,t]}\|z(\tau)\|^2, \quad \forall t > 0, \tag{5.3.13}$$

where $E_z^\mu(t) = \frac{1}{2}\left(\mu\|z(t)\|^2 + \|A^{1/2}z(t)\|^2\right)$.

Proof. Using Lemma 5.2.7 and either estimate (5.3.9) from Proposition 5.3.2 or relation (5.3.12) from Remark 5.3.3, we obtain that

$$
TE_z^\mu(T) + \int_0^T E_z^\mu(\tau)d\tau
$$

$$
\leq C_R \int_0^T \|z_t\|_\theta^2 dt + C_{R,T} \left[\sup_{[0,T]} \|A^\sigma z(\tau)\|^2 + \int_0^T d(\tau)\|A^{1/2}z(\tau)\|^2 d\tau \right], \qquad (5.3.14)
$$

where $d(t)$ is defined in (5.3.10).

From (5.2.6) we have that

$$
\int_0^T \|z_t\|_\theta^2 dt \leq C_R \left[E_z(0) - E_z(T) + \int_0^T \|A^{1/2}z\|^2 \|u_t^2\|_\theta^2 dt + \left| \int_0^T (G_{u_1 u_2}, z_t)dt \right| \right]
$$

Therefore, using either (5.3.9) or (5.3.12) we obtain that

$$
\int_0^T \|z_t\|_\theta^2 dt \leq C_R E_z(0) + C_{R,\epsilon} \int_0^T d(t)\|A^{1/2}z\|^2 dt
$$

$$
+ C_{R,T,\epsilon} \max_{[0,T]} \|A^\sigma z(t)\|^2 + \epsilon \int_0^T \|A^{1/2}z\|^2 dt,
$$

where $d(t)$ is given by (5.3.10). Inserting this estimate for z_t in (5.3.14), we get

$$
TE_z^\mu(T) + \int_0^T \|z_t\|_\theta^2 d\tau + \int_0^T E_z^\mu(\tau)d\tau \leq
$$

$$
C_R E_z^\mu(0) + C_{R,T} \left[\sup_{[0,T]} \|A^\sigma z(\tau)\|^2 + \int_0^T d(\tau)\|A^{1/2}z(\tau)\|^2 d\tau \right]. \qquad (5.3.15)
$$

This inequality holds for $\sigma = 0$ as well. Indeed, by interpolation,

$$
\sup_{[0,T]} \|A^\sigma z\|^2 \leq \epsilon \sup_{[0,T]} \|A^{1/2}z\|^2 + C_\epsilon \sup_{[0,T]} \|z\|^2, \qquad \forall \epsilon > 0.
$$

The term $\epsilon \sup \|A^{1/2}z\|^2$ is controlled by (5.2.10); therefore, after an appropriate choice of ϵ, we obtain (5.3.15) with $\sigma = 0$. As a consequence,

$$TE_z^\mu(T) \le C_R E_z^\mu(0) + C_{R,T}\sup_{[0,T]}\|z(t)\|^2 + C_{R,T}\int_0^T d(t)E_z^\mu(t)dt$$

for T large enough. This implies

$$E_z^\mu(T) \le \varkappa_R E_z^\mu(0) + C_{R,T}\int_0^T \left[\|u_t^1\|_\theta^2 + \|u_t^2\|_\theta^2\right]E_z^\mu(t)dt + C_{R,T}\sup_{[0,T]}\|z(t)\|^2,$$

(5.3.16)

where $\varkappa_R < 1$ and $T \ge T_0(R)$.

To conclude the proof of Proposition 5.3.4, we apply the same argument as in CHUESHOV/LASIECKA [56, p. 62] (see also CHUESHOV/LASIECKA [58, p. 414]). First we note that (5.3.16) yields

$$E_z^\mu((m+1)T) \le \varkappa_R E_z^\mu(mT) + c_{R,T}b_m, \quad m = 0, 1, 2, \ldots,$$

with $\varkappa_R < 1$, where

$$b_m \equiv \sup_{[mT,(m+1)T]}\|z(t)\|^2 + \int_{mT}^{(m+1)T}\left(\|u_t^1\|_\theta^2 + \|u_t^2\|_\theta^2\right)E_z^\mu(\tau)d\tau.$$

This yields

$$E_z(mT) \le \varkappa_R^m E_z(0) + c\sum_{l=1}^m \kappa_R^{m-l}b_{l-1}.$$

Because $\varkappa_R < 1$, we can see that there exists $\gamma > 0$ such that

$$E_z^\mu(t) \le C_1 e^{-\gamma t}E_z(0)$$

$$+C_2\left\{\sup_{[0,t]}\|z(\tau)\|^2 + \int_0^t e^{-\gamma(t-\tau)}d(\tau; u^1, u^2)E_z^\mu(\tau)d\tau\right\}$$

for all $t \ge 0$, where

$$d(\tau; u^1, u^2) = \|u_t^1(\tau)\|_\theta^2 + \|u_t^2(\tau)\|_\theta^2,$$

Therefore, applying Gronwall's lemma to the function $E_z^\mu(t)e^{\gamma t}$, we find that

$$E_z(t) \le \left[C_1 E_z(0)e^{-\gamma t} + C_2\sup_{[0,t]}\|z(\tau)\|^2\right]\exp\left\{C_2\int_0^t d(\tau; u^1, u^2)\,d\tau\right\}.$$

Now using the finiteness of the dissipation integrals in (5.1.13) and (5.1.24), we obtain estimate (5.3.13). This concludes the proof of Proposition 5.3.4.   □

Now we are in position to prove our main result on the finiteness of the fractal dimension of the attractors.

Theorem 5.3.5. *Let Assumptions 5.1.1 and 5.2.1 be in force. Concerning the forcing term $B(u)$, we assume that either $B(u)$ is subcritical (i.e., (5.3.11) holds) or $B(u)$ satisfies Assumption 5.3.1. Then*

Case $\mu > 0$:

1. *The global attractor $\mathfrak{A}^\mu$ of the system $(\mathscr{H}, S_t^\mu)$ has a finite fractal dimension $\dim_f \mathfrak{A}^\mu$ with upper bound independent of $\mu \in (0, \mu_0]$.*
2. *The global attractor $\mathfrak{A}^\mu$ lies in $\mathscr{D}(A^{1-\theta}) \times \mathscr{D}(A^{1/2})$ and for any full trajectory $(u(t); u_t(t))$ from the attractor we have that*

$$\|A^{1-\theta}u(t)\|^2 + \int_{-\infty}^{\infty} \|u_t\|_\theta^2 d\tau + \mu\|A^{1/2}u_t(t)\|^2 + \mu^2\|u_{tt}(t)\|^2 \le R_0^2 \qquad (5.3.17)$$

for all $t \in \mathbb{R}$. Moreover, the (uniform) Hölder condition

$$\frac{1}{h}\left[\|A^{1/2}(u(t+h) - u(t))\|^2 + \mu\|u_t(t+h) - u_t(t)\|^2\right] \le R_0^2 \qquad (5.3.18)$$

holds for every $t \in \mathbb{R}$ and $h > 0$. Here R_0 does not depend on μ.
3. *Assume in addition that $K : H_{1/2} \mapsto \mathscr{L}(H_\theta, H_{-\theta})$ and $B : H_{1/2} \mapsto H_{-\theta}$ are C^1 Fréchet smooth mappings such that their derivatives satisfy the relations*

$$\|\langle K'(u); v\rangle w\|_{-\theta} \le C_\varrho \|A^{1/2}v\|\|w\|_\theta, \quad v \in \mathscr{D}(A^{1/2}), \ w \in H_\theta, \qquad (5.3.19)$$

and there exist $\delta > 0$ such that

$$\|\langle B'(u); v\rangle\|_{-\theta} \le C_\varrho \|A^{1/2-\delta}v\|, \quad v \in \mathscr{D}(A^{1/2}), \qquad (5.3.20)$$

for any $u \in \mathscr{D}(A^{1/2})$ such that $\|A^{1/2}u\| \le \varrho$, where ϱ is arbitrary. Here we denote by $\langle \Phi'(u); v\rangle$ the value of the derivative $\Phi'(u)$ on the element v. Then the uniform (in μ) bounds in (5.3.17) and (5.3.18) can be improved in the following way:

$$\|A^{1/2}u_t(t)\|^2 + \mu\|u_{tt}(t)\|^2 \le R_0^2, \quad \forall t \in \mathbb{R}. \qquad (5.3.21)$$

Case $\mu = 0$:

4. *The global attractor $\mathfrak{A}$ of the system $(H_{1/2}, S_t)$ has a finite fractal dimension $\dim_f \mathfrak{A}$.*
5. *For any full trajectory $u(t)$ from the attractor $\mathfrak{A}$ we have that*

$$\|A^{1/2}u(t)\|^2 + \int_{-\infty}^{\infty} \|u_t\|_{\theta}^2 d\tau \le c_0, \quad \|A^{1/2}(u(t+h) - u(t))\| \le c_0 h^{1/2}$$

$$(5.3.22)$$

for all $t \in \mathbb{R}$ and $h > 0$.

Proof. We apply the quasi-stability method developed in Section 3.4.3. Its realization is based on the estimate in (5.3.13).

The case $\mu > 0$: To obtain the independence of a bound for the dimension, it is convenient to introduce the scaled dynamical system $(\mathcal{H}, \tilde{S}_t^{\mu})$ with the evolution operator $\tilde{S}_t^{\mu}$ defined by the formula

$$\tilde{S}_t^{\mu} = M_{\mu} \circ S_t^{\mu} \circ M_{\mu}^{-1} \quad \text{with} \quad M_{\mu} = \begin{pmatrix} 1 & 0 \\ 0 & \sqrt{\mu} \end{pmatrix}. \qquad (5.3.23)$$

It is clear that $(\mathcal{H}, \tilde{S}_t^{\mu})$ possesses a global attractor $\tilde{\mathfrak{A}}^{\mu}$ and $\tilde{\mathfrak{A}}^{\mu} = M_{\mu}\mathfrak{A}^{\mu}$, where $\mathfrak{A}^{\mu}$ is the global attractor for $(\mathcal{H}, S_t^{\mu})$.

It follows from (5.1.14) and (5.3.13) that

$$\|\tilde{S}_t^{\mu} y_1 - \tilde{S}_t^{\mu} y_2\|_{\mathcal{H}}^2 \le a_R \|y_1 - y_2\|_{\mathcal{H}}^2 e^{b_R t} \qquad (5.3.24)$$

and

$$\|\tilde{S}_t^{\mu} y_1 - \tilde{S}_t^{\mu} y_2\|_{\mathcal{H}}^2 \le c_R \|y_1 - y_2\|_{\mathcal{H}}^2 e^{-\gamma(R)t} + c_R \max_{[0,t]} \|u_1^{\mu}(\tau) - u_2^{\mu}(\tau)\|^2 \qquad (5.3.25)$$

for all $t > 0$, where $y_i = (u_i^0; u_i^1)$ is such that $\|y_i\|_{\mathcal{H}}^2 \equiv \|u_i^1\|^2 + \|A^{1/2}u_i^0\|^2 < R^2$ and the positive constants in (5.3.24) and (5.3.25) do not depend on μ. We also use the notation $\tilde{S}_t^{\mu} y_1 = (u_i^{\mu}(t); \sqrt{\mu}\partial_t u_i^{\mu}(t))$, where $u_i^{\mu}(t)$ solves (5.0.1) with initial data $y_i = (u_i^0; \mu^{-1/2} u_i^1)$.

Relations (5.3.24) and (5.3.25) mean that the system $(\mathcal{H}, \tilde{S}_t^{\mu})$ is asymptotically quasi-stable (see Definition 3.4.15), and thus by Proposition 3.4.17 this system is quasi-stable at some time moment. Therefore, to prove the finiteness (with a uniform bound) of the fractal dimension of the global attractor $\tilde{\mathfrak{A}}^{\mu}$, we apply Theorem 3.4.18 and the bound for its dimension given in Theorem 3.4.5, see (3.4.3). Since the constants in (5.3.24) and (5.3.25) do not depend on μ, we can conclude that there is a bound for $\dim_f^{\mathcal{H}} \tilde{\mathfrak{A}}^{\mu}$ which does not depend on μ.

To prove the second statement of Theorem 5.3.5, we note that by Theorem 5.2.3 $\mathfrak{A}^{\mu} = \mathcal{M}^u(\mathcal{N})$. Since the system is gradient and the energy functional $\mathscr{E}(u, v)$ is a strict Lyapunov function, we conclude that

$$\sup_{(u;v) \in \mathfrak{A}^{\mu}} \mathscr{E}(u, v) \le \sup_{(u;v) \in \mathcal{N}} \mathscr{E}(u, v) = \sup_{u \in \mathcal{N}_*} \left\{ \frac{1}{2}\|A^{1/2}u\|^2 + \Pi(u) \right\} \le c_0,$$

where c_0 does not depend on μ. Therefore by (5.1.13),

$$\|A^{1/2}u(t)\|^2 + \int_{-\infty}^{\infty} \|u_t\|_\theta^2 d\tau + \mu\|u_t(t)\|^2 \le R^2, \quad \forall t \in \mathbb{R}, \qquad (5.3.26)$$

for any full trajectory $(u(t); u_t(t))$ from $\mathfrak{A}^\mu$. Applying the quasi-stability estimate in (5.3.13) to this and the shifted trajectory $(u(t+h); u_t(t+h))$, in the same way as we did in the proof of Theorem 3.4.19, we can conclude that

$$\frac{1}{h}\left[\|A^{1/2}(u(t+h) - u(t))\|^2 + \mu\|u_t(t+h) - u_t(t)\|^2\right]$$

$$\le C_R \sup_{-\infty < \tau < t} \int_\tau^{\tau+h} \|u_t(\xi)\|^2 d\xi. \qquad (5.3.27)$$

Using the estimate $\|u_t(t)\|^2 \le R^2/\mu$ (which follows from (5.3.26)) and also the relation

$$\|A^{1-\theta}u(t)\|^2 \le C\left[\mu^2\|u_{tt}\|^2 + \|K(u)u_t\|_{-\theta}^2 + \|B(u)\|_{-\theta}^2\right]$$

which follows from (5.0.1), we can easily obtain (5.3.17). Relation (5.3.27) with the integral bound for $\|u_t\|^2$ in (5.3.26) yields (5.3.18).

To prove the third statement in Theorem 5.3.5, we restrict ourselves to the case of small intervals $(0, \mu_0]$. To prove the desired smoothness we consider the equation for $w = u_t$ on the attractor $\mathfrak{A}^\mu$:

$$\mu w_{tt} + K(u(t))w_t + Aw + \langle K'(u(t)); w \rangle u_t + \langle B'(u); w \rangle = 0. \qquad (5.3.28)$$

We multiply (5.3.28) by w_t in H. Using (5.3.19) and (5.3.20) we have that

$$\frac{d}{dt}E(w, w_t) + (K(u(t))w_t, w_t) \le \eta\left(\|w_t\|_\theta^2 + \|A^{1/2}w\|^2\right)$$

$$+ C_{R,\eta}\|u_t(t)\|_\theta^2\|A^{1/2}w\|^2 + C_{R,\eta}\|w\|_\theta^2 \qquad (5.3.29)$$

for any $\eta > 0$, where

$$E(w, w_t) = \frac{1}{2}\left[\mu\|w_t\|^2 + \|A^{1/2}w\|^2\right].$$

This implies that

$$\frac{d}{dt}E(w, w_t) + \alpha_0\|w_t\|_\theta^2 \le \frac{1}{4}\|A^{1/2}w\|^2 + C_0\|u_t(t)\|_\theta^2\left(1 + \|A^{1/2}w\|^2\right) \qquad (5.3.30)$$

for the trajectory $(u(t); u_t(t))$ from $\mathfrak{A}^\mu$. Multiplying (5.3.28) by w and using (5.3.19) and (5.3.20), one can also see that

$$\mu \frac{d}{dt}(w, w_t) \leq -\frac{1}{2}\|A^{1/2}w\|^2 + \mu\|w_t\|^2 + \epsilon\|w_t\|_\theta^2$$
$$+ C_{1,\epsilon}\|u_t(t)\|_\theta^2 \cdot \|A^{1/2}w\|^2 + C_2\|w\|_\theta^2$$

for $\mu > 0$ with arbitrary $\epsilon > 0$. Therefore, with an appropriate choice of $\mu_0 > 0$ and $\epsilon > 0$ (both small), the function

$$V(t) = \frac{1}{2}\left[\mu\|w_t\|^2 + \|A^{1/2}w\|^2\right] + \mu(w, w_t)$$

satisfies the relations

$$\frac{1}{2}\left(\mu\|w_t\|^2 + \|A^{1/2}w\|^2\right) \leq V(t) \leq 2(\mu\|w_t\|^2 + \|A^{1/2}w\|^2)$$

and

$$\frac{d}{dt}V(t) + c_0 V(t) \leq c_1\|u_t(t)\|_\theta^2 \cdot V(t) + c_2\|u_t(t)\|_\theta^2,$$

where the constants c_i do not depend on μ. Using the boundedness of the integral term in (5.3.26), by a Gronwall-type argument we conclude that

$$V(t) \leq C_1 e^{-\gamma(t-s)}V(s) + C_2, \quad t > s,$$

with the constants independent of μ. In the limit $s \to -\infty$ this yields

$$\mu\|u_{tt}(t)\|^2 + \|A^{1/2}u_t(t)\|^2 \leq C, \quad t \in \mathbb{R},$$

which implies (5.3.21).

     The case $\mu = 0$: We first note that by (5.1.24) on the attractor $\mathfrak{A}$ we have

$$\sup_{t \in \mathbb{R}}\|A^{1/2}u(t)\|^2 + \sup_{t \in \mathbb{R}}\int_t^{t+1}\|A^{1-\theta}u(\tau)\|^2 d\tau + \int_{-\infty}^\infty\|A^\theta u_t(\tau)\|^2 d\tau \leq C. \quad (5.3.31)$$

For dimension we use the same quasi-stability idea as above.

     Let u_0^i, $i = 1, 2$, be two elements from the attractor $\mathfrak{A}$. We denote $u^i(t) = S_t u_0^i$ and $z(t) = u^1(t) - u^2(t)$. By (5.3.13) with $\mu = 0$ we have

$$\|A^{1/2}z(t)\|^2 \leq C_1\|A^{1/2}z(0)\|^2 e^{-\gamma_0 t} + C_2 \max_{[0,t]}\|z(\tau)\|^2, \quad \forall t > 0. \quad (5.3.32)$$

This implies that $(H_{1/2}, S_t)$ is quasi-stable on the attractor $\mathfrak{A}$ with $Z = W_1(0, T)$, where

$$W_1(0, T) = \left\{ z(t) : |z|^2_{W_1(0,T)} \equiv \int_0^T \left(\|A^{1/2}z(t)\|^2 + |z_t(t)\|^2 \right) dt < \infty \right\}. \qquad (5.3.33)$$

In Definition 3.4.1 we also choose the operator $K : H_{1/2} \mapsto Z$ according to the formula

$$Ku_0 = \{u(t) : t \in [0, T]\}, \quad \text{where } u(t) = S_t u_0,$$

and suppose $n_Z(u) = C_2 \max_{[0,T]} \|u(\tau)\|^2$. By (5.1.25) the operator K is a Lipschitz mapping from $H_{1/2}$ into Z. The seminorm n_Z is compact on Z by the compactness Theorem A.3.7. Thus, we can apply Theorem 3.4.5 to prove the finiteness of the fractal dimension of $\mathfrak{A}$.

To obtain (5.3.22) we use (5.3.31) and the quasi-stability estimate in (5.3.32) for the trajectories $u(t)$ and $u(t + h)$, i.e., with $z(t) = u(t + h) - u(t)$. □

Remark 5.3.6. Using Theorem 3.4.7 one can show the existence of a (generalized) fractal exponential attractor for the systems considered. We can also establish criteria for the existence of finite families of determining functionals. For this we can use the idea presented in Theorem 3.4.20. However, we do not pursue these directions for the model considered in (5.0.1). The corresponding arguments are standard. We refer to CHUESHOV/LASIECKA [56, 58] for similar approaches and results for second order in time models with other damping mechanisms. ■

5.3.2 Upper semicontinuity of attractors

Now we show that the attractors $\mathfrak{A}^\mu$ for problem (5.0.1) in some sense converge to the attractor of the system generated by (5.1.20) as $\mu \to 0$. If $\theta < 1/2$ and the conditions in (5.3.19) and (5.3.20) are valid, we can apply abstract Theorem 2.3.31. However, we prefer to establish a slightly more general result to demonstrate a direct approach to upper semicontinuity.

Theorem 5.3.7. *Let Assumptions 5.1.1 and 5.2.1 be in force and $\theta < 1/2$. Assume in addition that the forcing term $B(u)$ either is subcritical (i.e., (5.3.11) holds) or satisfies Assumption 5.3.1. Let $\mathfrak{A}$ be the attractor of the system $(H_{1/2}, S_t)$ generated by (5.1.20). Then*

$$\lim_{\mu \to 0} \left[\sup \left\{ \text{dist}_{\mathcal{H}}(y, \tilde{\mathfrak{A}}) : y \in \tilde{\mathfrak{A}}^\mu \right\} \right] = 0, \qquad (5.3.34)$$

where $\tilde{\mathfrak{A}} = \{(u; 0) : u \in \mathfrak{A}\}$ and $\tilde{\mathfrak{A}}^\mu$ is the global attractor for the dynamical system $(\mathcal{H}, \tilde{S}_t^\mu)$ with the evolution operator $\tilde{S}_t^\mu$ given by (5.3.23).

If we assume in addition that (5.3.19) and (5.3.20) hold, then

$$\lim_{\mu \to 0} \left[\{ \mathrm{dist}_{\mathscr{H}}(y, \mathfrak{A}^*) : y \in \mathfrak{A}^\mu \} \right] = 0, \qquad (5.3.35)$$

where $\mathfrak{A}^\mu$ is the attractor of system $(\mathscr{H}, S_t^\mu)$ generated by (5.0.1) and

$$\mathfrak{A}^* = \{ (u; v) : u \in \mathfrak{A}, \ v = K^{-1}(u)[Au + F(u)] \}.$$

Proof. We first note that by the compactness Theorem A.3.7 in the Appendix it follows from (5.3.17) and (5.3.18) that for any trajectory

$$\{ y^\mu(t) = (u^\mu(t); \sqrt{\mu} u_t^\mu(t)) : t \in \mathbb{R} \} \subset \tilde{\mathfrak{A}}^\mu$$

the family $\{ y^\mu(t) \}_{\mu > 0}$ is relatively compact in the space

$$C([a, b]; \mathscr{D}(A^{1-\theta-\eta}) \times \mathscr{D}(A^{1/2-\eta})), \quad \forall a < b, \ \eta > 0.$$

Moreover, we have that $\sqrt{\mu} u_t^\mu \to 0$ in $L^2(\mathbb{R}; H_\theta)$ as $\mu \to 0$.

These observations allow us to obtain (5.3.34) by the standard contradiction argument (see BABIN/VISHIK [9], CHUESHOV [39], RAUGEL [188], for instance). Indeed, assume that (5.3.34) is not valid. Then there exist sequences $\mu_n \to 0$ and $y^n \in \tilde{\mathfrak{A}}^{\mu_n}$ such that

$$\mathrm{dist}_{\mathscr{H}}(y^n, \tilde{\mathfrak{A}}) \geq \delta, \ \ n = 1, 2, \ldots \qquad (5.3.36)$$

Let $\gamma_n = \{ y^n(t) \}$ be a full trajectory for $(\mathscr{H}, \tilde{S}_t^\mu)$ in the attractor $\tilde{\mathfrak{A}}^{\mu_n}$ such that $y^n(0) = y^n$. Such a trajectory exists due to Exercise 2.3.4(B). By the observations made we can assume that $y^n(t) = (u^n(t); \sqrt{\mu_n} u_t^n(t))$ is convergent in the space

$$C([a, b]; H_{1-\theta-\eta} \times H_{1/2-\eta}), \quad \forall a < b, \ \eta > 0.$$

Thus, there exists $u \in C_b(\mathbb{R}; H_{1-\theta-\eta})$ such that

$$\max_{[a,b]} \left[\| u^n(t) - u(t) \|_{1/2+\eta}^2 + \mu_n \| u_t^n \|_{1/2-\eta}^2 \right] \to 0, \ \ n \to \infty, \qquad (5.3.37)$$

for every interval $[a, b]$ and for all $\eta > 0$ small enough. Moreover, we have that $u_t \in L_2(\mathbb{R}; H_\theta)$ and $u_t^n \to u_t$ weakly in $L_2(\mathbb{R}; H_\theta)$. Passing to the limit in the variational form of equation (5.0.1), we can see that $u(t)$ is a strong solution to problem (5.1.20) on $\mathbb{R}$ which is bounded in $H_{1/2}$. Therefore, applying Exercise 2.3.4(B) again, we obtain that $\{ u(t) \}$ is a full trajectory which lies in the attractor $\mathfrak{A}$. Thus, (5.3.37) implies that $y^n \to (u(0); 0) \in \tilde{\mathfrak{A}}$, which contradicts (5.3.36).

To prove (5.3.35) we note that in this case from (5.3.21) we additionally have that $\mu\|u_{tt}^{\mu}(t)\| \to 0$ uniformly in t. This observation makes it possible to obtain (5.3.35) in the same way as (5.3.34). □

5.3.3 A model with noncompact global attractor

We conclude this section with an example which shows that a second order in time model with a degenerate damping coefficient may possess a global attractor which is not compact in the phase space. This means that the requirement of boundedness (instead of compactness) in Definition 2.3.1 of a global attractor is reasonable.

We consider the following infinite-dimensional version of the Krasovskii system (see Exercises 1.8.22 and 2.1.10):

$$u_{tt} + Au + k(\|u_t(t)\|^2 + \|A^{1/2}u(t)\|^2)u_t = 0, \quad u\big|_{t=0} = u_0, \quad u_t\big|_{t=0} = u_1, \quad (5.3.38)$$

in a separable Hilbert space H, where, as above, A is a positive operator with a discrete spectrum. The intensity k of the damping is determined by the energy of the system. We suppose that $k(\cdot)$ is a scalar bounded Lipschitz function on $\mathbb{R}_+$ such that $k(1) = 0$ and $k(s)$ is strictly increasing for $s > 1$. On the interval $[0, 1]$ the function $k(s)$ can be arbitrary. For instance, we can take

$$k(s) = k_0\left(1 - \frac{2}{1+s}\right) \quad \text{or} \quad k(s) = \begin{cases} 0, & \text{if } 0 \le s \le 1; \\ k_0\left(1 - s^{-1}\right), & \text{if } s > 1. \end{cases}$$

Similar to Definition 5.1.3, we introduce the notion of a weak solution to (5.3.38) on an interval $[0, T]$ which we define as a function

$$u \in L_\infty(0, T; \mathscr{D}(A^{1/2})), \quad u_t \in L_\infty(0, T; H)$$

such that $u(0) = u_0$ and

$$-\int_0^T (u_t, v_t)dt + \int_0^T (k(\|u_t\|^2 + \|A^{1/2}u\|^2))u_t, v)dt + \int_0^T (Au, v)dt = (u_1, v(0)),$$

for every $v \in \mathscr{W}_T$, where

$$\mathscr{W}_T = \left\{v \in L_\infty(0, T; \mathscr{D}(A^{1/2})), \ v_t \in L_\infty(0, T; H), \ v(T) = 0\right\}.$$

For a description of the functional spaces above, see Section A.3 in the Appendix.

Proposition 5.3.8. *For every* $(u_0; u_1) \in \mathcal{H} \equiv H_{1/2} \times H$, *problem* (5.3.38) *has a unique weak solution* $u(t)$ *on* $\mathbb{R}_+$. *This solution belongs to the class*

$$C(\mathbb{R}_+; H_{1/2}) \cap C^1(\mathbb{R}_+; H)$$

and the energy relation

$$E(u(t), u_t(t)) + \int_0^t (k(2E(u(\tau), u_t(\tau)))\|u_t(\tau)\|^2 d\tau = E(u_0, u_1) \qquad (5.3.39)$$

holds for every $t > 0$, *where the energy* E *is defined by the formula*

$$E(u_0, u_1) = \frac{1}{2} \left(\|u_1\|^2 + \|A^{1/2}u_0\|^2 \right).$$

Moreover, for every pair $u^1(t)$ *and* $u^2(t)$ *of weak solutions such that*

$$E(u^i(0), u_t^i(0)) \leq R^2, \quad i = 1, 2,$$

we have that $z(t) = u^1(t) - u^2(t)$ *satisfies the relation*

$$E(z(t), z_t(t)) \leq b_{T,R} E(z(0), z_t(0)), \quad t \in [0, T], \quad \forall T > 0, \qquad (5.3.40)$$

for some constant $b_{T,R} > 0$.

Proof. We cannot directly refer here to the standard compactness method as in Theorem 5.1.4, because the nonlinear term in (5.3.38) is critical and cannot be defined as a weakly continuous mapping from $H_{1/2} \times H$ in some negative order space. However, we can use the structure of the problem, which provides an important invariance property.

Let P_N be the orthoprojector onto Span $\{e_1, \ldots, e_N\}$, where $\{e_k\}$ is the eigenbasis for A. It is easy to see that for every initial data $(u_0; u_1) \in \mathcal{H}_N \equiv P_N H_{1/2} \times P_N H$ there exists a solution u to (5.3.38) which takes its values $P_N H_{1/2}$. In fact, this means that the Galerkin approximate solutions satisfy the equation in (5.3.38). Using this observation, we take arbitrary $(u_0; u_1) \in \mathcal{H}$ and denote by u^N the solution to (5.3.38) with the initial data $(P_N u_0; P_N u_1)$. Every solution u^N satisfies the energy relation in (5.3.39) with $(P_N u_0; P_N u_1)$ instead of $(u_0; u_1)$. This implies that

$$\|u_t^N(t)\|^2 + \|A^{1/2}u^N(t)\|^2 \leq \|P_N u_1\|^2 + \|A^{1/2}P_N u_0\|^2 + 2k_0 \int_0^t \|u_t^N(\tau)\|^2 d\tau,$$

where $k_0 = -\inf\{k(s) : s \geq 0]\}$. Thus, Gronwall's lemma yields

$$\|u_t^N(t)\|^2 + \|A^{1/2}u^N(t)\|^2 \leq \left[\|P_N u_1\|^2 + \|A^{1/2}P_N u_0\|^2 \right] e^{2k_0 t} \leq 2E(u_0, u_1)e^{2k_0 t}.$$

Now we consider the difference $z^{N,M} = u^N - u^M$ for different N and M. This function solves the equation

$$z_{tt}^{N,M} + Au + k(\|u_t^N(t)\|^2 + \|A^{1/2}u^N(t)\|^2)z_t^{N,M}$$
$$= -\left[k(\|u_t^N(t)\|^2 + \|A^{1/2}u^N(t)\|^2) - k(\|u_t^M(t)\|^2 + \|A^{1/2}M^N(t)\|^2)\right]u_t^M.$$

Multiplying this equation by $z_t^{N,M}$ in H and using the notation

$$Z^{N,M} = (z^{N,M}; z_t^{N,M}) \quad \text{and} \quad U^N = (u^N; u_t^N)$$

and the relations

$$\|Z\|_{\mathscr{H}}^2 = \|z_t\|^2 + \|A^{1/2}z\|^2 = 2E(z, z_t),$$

we obtain t

$$\frac{1}{2}\frac{d}{dt}\|Z^{N,M}(t)\|_{\mathscr{H}}^2 \le k_0\|Z^{N,M}\|_{\mathscr{H}}^2 + C\|u_t^M\| \cdot \|z_t^{N,M}\| \left(\|U^N\|_{\mathscr{H}} + \|U^M\|_{\mathscr{H}}\right)\|Z^{N,M}\|_{\mathscr{H}}$$
$$\le C_T(E(u_0; u_1))\|Z^{N,M}\|_{\mathscr{H}}^2.$$

Thus, Gronwall's lemma yields

$$\max_{[0,T]}\left[\|z_t^{N,M}(t)\|^2 + \|A^{1/2}z^{N,M}(t)\|^2\right]$$
$$\le C_T(u_0, u_1)\left[\|(P_N - P_M)u_1\|^2 + \|A^{1/2}(P_N - P_M)u_0\|^2\right].$$

This implies that $\{u^N(t)\}$ is a Cauchy sequence in

$$C([0, T]; H_{1/2}) \cap C^1([0, T]; H), \quad \forall T > 0.$$

This allows us to pass to the limit $N \to \infty$ and prove the existence of weak solutions possessing the required smoothness. The uniqueness of weak solutions and the Lipschitz property in (5.3.40) can be achieved in the standard way.   □

By Proposition 5.3.8, problem (5.3.38) generates an evolution operator in the space $\mathscr{H} = H_{1/2} \times H$ by the formula $S_t(u_0; u_1) = (u(t); u_t(t))$, where $u(t)$ solves (5.3.38).

We note that the change of the time variable $t \mapsto -t$ in equation (5.3.38) gives us the same problem but with damping function $-k(s)$ instead of $k(s)$ in (5.3.38). Since $k(s)$ is bounded, we can use the same argument as in Proposition 5.3.8 to prove well-posedness for the inverse time problem. This allows us to state that the evolution operator S_t is invertible and thus $\{S_t\}$ is an evolution group. This observation is important below, where we discuss the invariance properties of sets in $\mathscr{H}$.

Our main result concerning the system $(\mathcal{H}, S_t)$ is the following assertion.

Proposition 5.3.9. *The global attractor of the system* $(\mathcal{H}, S_t)$ *generated by the equation in (5.3.38) is the ball*

$$\mathcal{B} = \left\{ U_0 \equiv (u_0; u_1) \in \mathcal{H} : \|U_0\|_{\mathcal{H}}^2 \equiv \|u_1\|^2 + \left\| A^{1/2} u_0 \right\|^2 \le 1 \right\}.$$

Proof. The linear problem

$$u_{tt} + Au = 0, \quad u\big|_{t=0} = u_0, \quad u_t\big|_{t=0} = u_1,$$

has a unique solution $u(t)$ such that $E(u(t); u_t(t)) = E(u_0, u_1)$ for all $t \in \mathbb{R}$. Thus, by the uniqueness statement in Proposition 5.3.8, due to the property $k(1) = 0$, this implies that the unit sphere $\partial\mathcal{B}$ is a strictly invariant set with respect to S_t. Therefore, one can show that both sets $\mathcal{B}$ and Closure$(\mathcal{H} \setminus \mathcal{B})$ are also *strictly* invariant.

It is also easy to see from the energy relation that any ball

$$\mathcal{B}_R = \left\{ U_0 \equiv (u_0; u_1) \in \mathcal{H} : \|U_0\|_{\mathcal{H}}^2 \equiv \|u_1\|^2 + \left\| A^{1/2} u_0 \right\|^2 \le R^2 \right\}$$

is a forward invariant set for every $R > 1$. Thus, we need only to prove that $S_t \mathcal{B}_R$ goes to $\mathcal{B}$ uniformly in $u \in \mathcal{B}_R$ for every $R > 1$.

First we show that for every $U_0 \notin \mathcal{B}$ we have $\|S_t U_0\|_{\mathcal{H}} \to 1$ as $t \to \infty$. Assume that this is not true. By the energy relation and the invariance of Closure$(\mathcal{H} \setminus \mathcal{B})$ the function $t \mapsto \|S_t U_0\|_{\mathcal{H}}$ is non-increasing. This means that there exists $R_0 > 1$ such that

$$\|S_t U_0\|_{\mathcal{H}} \ge R_0, \quad \forall\, t > 0, \quad \text{and} \quad \lim_{t \to \infty} \|S_t U_0\|_{\mathcal{H}} = R_0.$$

In this case the damping coefficient satisfies the inequality

$$k(\|S_t U_0\|_{\mathcal{H}}^2) \ge k(R_0^2) \equiv k_* > 0 \quad \text{for all } t \ge 0.$$

This implies that

$$\frac{d}{dt}\|S_t U_0\|_{\mathcal{H}}^2 + 2k_*\|u_t\|^2 \le 0,$$

where $S_t U_0 = (u(t); u_t(t))$. We also have that

$$\frac{d}{dt}(u, u_t) = \|u_t\|^2 - \left\| A^{1/2} u \right\|^2 - k(\|S_t U_0\|_{\mathcal{H}}^2)(u_t, u)$$

$$\le \left(1 + \frac{k_1^2}{2\lambda_1} \right) \|u_t\|^2 - \frac{1}{2}\left\| A^{1/2} u \right\|^2,$$

where $k_1 = \sup\{k(s) : s \geq 1\}$. The relations above allow us to state that there are parameters $\gamma > 0$ and $\eta > 0$ such that the function

$$V(t) = \|S_t U_0\|_{\mathcal{H}}^2 + \eta(u, u_t)$$

possesses the properties

$$\frac{d}{dt} V(t) + \gamma V(t) \leq 0 \quad \text{and} \quad \frac{1}{2} \|S_t U_0\|_{\mathcal{H}}^2 \leq V(t) \leq 2\|S_t U_0\|_{\mathcal{H}}^2.$$

This implies

$$\|S_t U_0\|_{\mathcal{H}}^2 \leq 4\|U_0\|_{\mathcal{H}}^2 e^{-\gamma t} \quad \text{as long as} \quad \|S_t U_0\|_{\mathcal{H}}^2 \geq R_0 > 1, \tag{5.3.41}$$

which contradicts the initial guess that $\|S_t U_0\|_{\mathcal{H}} \geq R_0 > 1$ for all $t > 0$. Thus,

$$\lim_{t \to \infty} \|S_t U_0\|_{\mathcal{H}} = 1.$$

One can show that this convergence is uniform with respect to $U_0 \in \mathscr{B}_R \setminus \text{interior}(\mathscr{B})$. The point is that the relation in (5.3.41) yields that for any $R_1 \geq R_2 > 1$ there exists $t_* = t_*(R_1, R_2) > 0$ such that $S_t \mathscr{B}_{R_1} \subset \mathscr{B}_{R_2}$ for all $t \geq t_*$.

To conclude the proof of convergence $S_t \mathscr{B}_R \to \mathscr{B}$, we note that

$$\text{dist}_{\mathcal{H}}(S_t U_0, \mathscr{B}) \leq \left\| S_t U_0 - \frac{S_t U_0}{\|S_t U_0\|_{\mathcal{H}}} \right\|_{\mathcal{H}} = \|S_t U_0\|_{\mathcal{H}} - 1$$

for every $U_0 \in \mathscr{B}_R \setminus \text{interior}(\mathscr{B})$. Therefore,

$$\text{dist}_{\mathcal{H}}(S_t U_0, \mathscr{B}) \to 0 \quad \text{as} \quad t \to \infty$$

uniformly for every $U_0 \in \mathscr{B}_R \setminus \text{interior}(\mathscr{B})$. This means that the ball $\mathscr{B}$ is a global attractor for the system generated by (5.3.38). $\qquad\qquad \square$

5.4 Kirchhoff wave models with a structural nonlinear damping

In this section our main goal is to illustrate the previously developed methods by means of an abstract second order evolution equation with nonlinearity in the main part. As a model we choose the following Cauchy problem:

$$\begin{cases} u_{tt} + \phi(\|A^{1/2}u\|^2)Au + \sigma(\|A^{1/2}u\|^2)A^{2\theta}u_t + B(u) = 0, & t > 0, \\ u(0) = u_0, \quad u_t(0) = u_1, \end{cases} \tag{5.4.1}$$

with $\theta \in [1/4, 1/2)$ under the following set of hypotheses.

Assumption 5.4.1. (i) The operator A is a linear positive self-adjoint operator with domain $\mathscr{D}(A)$ in a separable Hilbert space H (we denote by $\|\cdot\|$ and $(\cdot, \cdot)$ the norm and the inner product in this space). We assume that A has a discrete spectrum (see Definition 4.1.1):

$$Ae_k = \lambda_k e_k, \quad 0 < \lambda_1 \le \lambda_2 \le \cdots, \quad \lim_{k \to \infty} \lambda_k = \infty.$$

As above, we denote by P_N the orthoprojector onto $\text{Span}\{e_k : k = 1, 2, \ldots, N\}$ and by H_s (with $s > 0$) the domain $\mathscr{D}(A^s)$ equipped with the graph norm $\|\cdot\|_s = \|A^s \cdot \|$. In this case H_{-s} denotes the completion of H with respect to the norm $\|\cdot\|_{-s} = \|A^{-s} \cdot \|$.

(ii) The damping (σ) and the stiffness (ϕ) factors are positive C^1 functions on the semi-axis $\mathbb{R}_+ = [0, +\infty)$. Moreover, we assume that $\theta \in [1/4, 1/2)$ and

$$\Phi(s) \equiv \int_0^s \phi(\xi)d\xi \to +\infty \quad \text{as} \quad s \to +\infty.^7 \tag{5.4.2}$$

(iii) The nonlinear operator B maps $H_{1/2}$ into $H_{-\theta}$ and is locally Lipschitz, i.e.,

$$\|B(u_1) - B(u_2)\|_{-\theta} \le L(\varrho)\|u_1 - u_2\|_{1/2}, \quad \forall \|A^{1/2}u_i\| \le \varrho. \tag{5.4.3}$$

We also assume that **(a)** $B(u) = \Pi'(u)$, where $\Pi(u)$ is a C^1 functional on $H_{1/2} = \mathscr{D}(A^{1/2})$, and $'$ stands for the Fréchet derivative, and **(b)** there exist $\eta < 1/2$ and $C \ge 0$ such that

$$\eta\Phi(\|A^{1/2}u\|^2) + \Pi(u) + C \ge 0, \quad u \in H_{1/2} = \mathscr{D}(A^{1/2}). \tag{5.4.4}$$

Our main motivating example is the case when $H = L_2(\Omega)$, where Ω is a bounded smooth domain in $\mathbb{R}^d$, A is the minus Dirichlet Laplace operator, and $B(u)$ is a Nemytskii operator generated by an appropriate C^1 function f (see Section 5.4.4 for more details). This kind of second order evolution model goes back to G. Kirchhoff ($d = 1, \phi(s) = \varphi_0 + \varphi_1 s, \sigma(s) \equiv 0, f(u) \equiv 0$) and has been studied by many authors under different types of hypotheses starting with the pioneering papers BERNSTEIN [12], LIONS [153], POHOZHAEV [184] (see also the literature cited in the survey MEDEIROS/FERREL/MENEZES [163] and the discussion below).

Remark 5.4.2. We concentrate on the case when $1/4 \le \theta < 1/2$. The case of strong damping ($\theta = 1/2$) requires a separate consideration and has been studied in CHUESHOV [42]. This paper covers the case of possibly non-positive stiffness coefficients ϕ and supercritical source terms B (in the sense that $B(u) \notin H_{-1/2}$ is

---

[7] This coercive behavior holds if $\liminf_{s \to +\infty}\{s\phi(s)\} > 0$, for instance. The standard example is $\phi(s) = \phi_0 + \phi_1 s^\alpha$ with $\phi_0 > 0, \phi_1 \ge 0$, and $\alpha \ge 1$. However, we can also take a decreasing $\phi(s)$ like $(1 + s)^{-1}$.

allowed for some $u \in H_{1/2}$). This became possible due to the fact that the damping operator has the same regularity order as the main elastic part. For the case $0 \leq \theta < 1/4$, to the best of our knowledge, the uniqueness theorem for this model is not established for initial data from the natural energy space $\mathscr{H} = H_{1/2} \times H$ (see the discussion in MEDEIROS/MILLA MIRANDA [164]). Thus, we are not able to apply here the approaches from the theory of single-valued dynamical systems. ∎

In this section we study well-posedness and long-time dynamics for general nonlinear stiffness and damping coefficients. Then we show the existence of a compact global attractor possessing some regularity and structural properties. Under some additional conditions concerning the source term B, we prove that the attractor has a finite fractal dimension. Moreover, in this case we prove the existence of a fractal exponential attractor and give conditions for the existence of finite sets of determining functionals. As in the case of the constant stiffness coefficient, we rely on the quasi-stability method developed in Chapter 3. As an application of the results obtained in Section 5.4.4 we consider the Kirchhoff wave equation in a bounded domain in $\mathbb{R}^d$ with $d \geq 1$. Our presentation is partially based on the paper of CHUESHOV [40].

5.4.1 Well-posedness

Well-posedness issues for Kirchhoff-type models like (5.4.1) have been studied intensively in past years. The primary focus was the case when $\theta = 0$ or $\theta = 1/2$, the damping coefficient $\sigma \geq 0$ is a constant, and the source term $B(u)$ is either absent or subcritical. For the case $\theta = 0$ we refer to GHISI [112], MATSUYAMA/IKEHATA [162], ONO [174], TANIGUCHI [214] and also to the survey MEDEIROS/FERREL/MENEZES [163]. In these papers the authors have studied sets of initial data for which solutions exist and are unique. The papers of MATSUYAMA/IKEHATA [162] and TANIGUCHI [214] also consider the case of nonlinear viscous damping. For the case $\theta = 1/2$ of strong (Kelvin-Voigt) damping the global well-posedness results are available in the literature (see, e.g., AUTUORI/PUCCI/SALVATORI [6], CAVALCANTI ET AL. [27], KALAN-TAROV/ZELIK [128], MEDEIROS/MILLA MIRANDA [164], NAKAO/ZHIJIAN [170], ONO [175] and the references therein). All these publications assume that the damping coefficient $\sigma(s) \equiv \sigma_0 > 0$ is a constant.

The case of a structural damping with $0 < \theta < 1/2$ was considered in MEDEIROS/MILLA MIRANDA [164] with the constant damping coefficient σ and $B \equiv 0$. The main result in that paper states the existence of weak solutions for $\theta \in (0, 1/2]$ with initial data $(u_0; u_1)$ from $H_{\max\{1/4,\theta\}} \times H$ and their uniqueness for $\theta \in [1/4, 1/2]$. We also mention the recent paper LAZO [147], which deals with nonlinear damping of the form $\sigma(\|A^\alpha u\|^2)A^\alpha u_t$ with $0 < \alpha \leq 1$. The main result of LAZO [147] states only the existence of weak solutions for uniformly positive ϕ and σ in the case when $B(u) \equiv 0$.

Our main result in this section is Theorem 5.4.4 on the well-posedness of problem (5.4.1). This theorem also contains some auxiliary properties of weak solutions which we need for the results on the asymptotic dynamics. We start with the following notion which repeats Definition 5.1.3 with obvious changes.

Definition 5.4.3. A function $u(t)$ is said to be a weak solution to (5.4.1) on an interval $[0, T]$ if

$$u \in L_\infty(0, T; H_{1/2}), \quad u_t \in L_\infty(0, T; H) \cap L_2(0, T; H_\theta) \tag{5.4.5}$$

and (5.4.1) is satisfied in the sense of distributions, i.e., $u(0) = u_0$ and

$$-\int_0^T (u_t, v_t)dt + \int_0^T \sigma(\|A^{1/2}u(t)\|^2)(A^\theta u_t, A^\theta v)dt$$

$$+ \int_0^T \phi(\|A^{1/2}u(t)\|^2)(Au, v)dt + \int_0^T (B(u), v)dt = (u_1, v(0)), \tag{5.4.6}$$

for every $v \in \mathcal{W}_T$, where

$$\mathcal{W}_T = \left\{ v \in L_\infty(0, T; \mathcal{D}(A^{1/2})), v_t \in L_\infty(0, T; H), \ v(T) = 0 \right\}.$$

∎

Theorem 5.4.4 (Well-posedness). *Let Assumption 5.4.1 be in force and $(u_0; u_1) \in \mathcal{H} = H_{1/2} \times H$. Then for every $T > 0$ problem (5.4.1) has a unique weak solution $u(t)$ on $[0, T]$. This solution possesses the following properties:*

1. *The function $t \mapsto (u(t); u_t(t))$ is strongly continuous in $\mathcal{H}$. Moreover,*

$$u \in L_2(0, T; H_{1-\theta}), \quad \forall T > 0, \tag{5.4.7}$$

$$u_{tt} \in L_\infty(0, T; H_{-2\theta}) \cap L_2(0, T; H_{-\theta}), \quad \forall T > 0, \tag{5.4.8}$$

and there exists a constant $C_R > 0$ such that

$$\|u_{tt}(t)\|^2_{-2\theta} + \|u_t(t)\|^2 + \|u(t)\|^2_{1/2} + \int_0^t \|u_t(\tau)\|^2_\theta d\tau$$

$$+ \frac{1}{1+t}\int_0^t \left[\|u(\tau)\|^2_{1-\theta} + \|u_{tt}(\tau)\|^2_{-\theta}\right] d\tau \le C_R \tag{5.4.9}$$

for every $t \in \mathbb{R}_+$ and initial data such that $\|(u_0; u_1)\|_{\mathcal{H}} \le R$.
2. *The energy identity*

$$\mathcal{E}(u(t), u_t(t)) + \int_s^t \sigma(\|u(\tau)\|^2_{1/2})\|u_t(\tau)\|^2_\theta d\tau = \mathcal{E}(u(s), u_t(s)) \tag{5.4.10}$$

holds for every $t > s \geq 0$, where the energy $\mathscr{E}$ is defined by the relation

$$\mathscr{E}(u_0, u_1) = \frac{1}{2}\left[\|u_1\|^2 + \Phi\left(\|A^{1/2}u_0\|^2\right)\right] + \Pi(u_0), \quad (u_0; u_1) \in \mathscr{H},$$

with Φ and Π defined in Assumption 5.4.1.

3. *If $u^1(t)$ and $u^2(t)$ are two weak solutions such that $\|(u^i(0); u^i_t(0))\|_{\mathscr{H}} \leq R$, $i = 1, 2$, then there exists $b_{R,T} > 0$ such that the difference $z(t) = u^1(t) - u^2(t)$ satisfies the relation*

$$\|z_t(t)\|^2 + \|z(t)\|^2_{1/2} + \int_0^t \|z_t(\tau)\|^2_\theta d\tau \leq b_{R,T}\left(\|z_t(0)\|^2 + \|z(0)\|^2_1\right) \qquad (5.4.11)$$

for all $t \in [0, T]$.

Proof. As in Section 5.1.2, to prove the existence of solutions, we use the standard Galerkin method.

We assume that $\|(u_0; u_1)\|_{\mathscr{H}} \leq R$ for some $R > 0$ and seek approximate solutions $u^N(t)$, $N = 1, 2, \ldots$, with values in $P_N H$ which solve the corresponding finite-dimensional projections of (5.4.1). Such solutions exist (at least locally), and after multiplication of the corresponding projection of (5.4.1) by $u^N_t(t)$ we get that $u^N(t)$ satisfies the energy relation in (5.4.10). By the coercivity requirements in (5.4.2) and (5.4.4), we conclude that

$$\|(u^N(t); u^N_t(t))\|_{\mathscr{H}} \leq C_R \quad \text{for all } t \in [0, T], \quad N = 1, 2, 3 \ldots. \qquad (5.4.12)$$

Since $\sigma(s) > 0$, this implies that $\sigma(\|u^N(t)\|^2_{1/2}) \geq \sigma_R > 0$ for all $t \in [0, T]$. Therefore, the energy relation (5.4.10) for u^N yields that

$$\int_0^T \|u^N_t(t)\|^2_\theta dt \leq C(R), \quad N = 1, 2, \ldots, \quad \text{for any } T > 0. \qquad (5.4.13)$$

The next a priori estimate involves the fact that $\theta \geq 1/4$. We use the multiplier $A^{1-2\theta}u^N$ to obtain

$$\frac{d}{dt}\left[(A^{1-2\theta}u^N, u^N_t) + \frac{1}{2}\Sigma(\|u^N\|^2_{1/2})\right] + \phi(\|u^N\|^2_{1/2})\|u^N\|^2_{1-\theta}$$
$$= -(B(u^N), A^{1-2\theta}u^N) + \|u^N_t\|^2_{1/2-\theta},$$

where $\Sigma(s) = \int_0^s \sigma(\xi)d\xi$. It follows from (5.4.12) (and a posteriori from the requirements in (5.4.2) and (5.4.4)) that $\phi(\|u^N\|^2_{1/2}) \geq \phi_R > 0$ and

$$|(B(w), A^{1-2\theta}u)| \leq \|B(w)\|_{-\theta}\|u\|_{1-\theta} \leq \varepsilon\|u\|^2_{1-\theta} + C_\varepsilon\|B(w)\|^2_{-\theta} \qquad (5.4.14)$$

for all $w \in H_{1/2}$ and $u \in H_{1-\theta}$. Therefore, using (5.4.12), (5.4.3), and the relation

$$\|u_t^N\|_{1/2-\theta}^2 \leq \lambda_1^{1-4\theta}\|u_t^N\|_\theta^2, \quad \theta \geq 1/4,$$

we have

$$\frac{d}{dt}\left[(A^{1-2\theta}u^N, u_t^N) + \frac{1}{2}\Sigma(\|u^N\|_{1/2}^2)\right] + \frac{\phi_R}{2}\|u^N\|_{1-\theta}^2 \leq \lambda_1^{1-4\theta}\|u_t^N\|_\theta^2 + C(R),$$

for $t \in [0, T]$. Since $|(A^{1-2\theta}u^N, u_t^N)| \leq \lambda_1^{1/2-2\theta}\|A^{1/2}u^N\|\|u_t^N\| \leq C_R$ in our case (when $\theta \geq 1/4$), this and (5.4.13) imply the following a priori estimate:

$$\int_0^T \|u^N(t)\|_{1-\theta}^2 dt \leq C(R)(1+T), \quad N = 1, 2, \ldots, \quad \text{for any } T > 0. \tag{5.4.15}$$

The above a priori estimates show that $(u^N; u_t^N)$ is *-weakly compact in

$$W_T \equiv \left[L_\infty(0, T; H_{1/2}) \cap L_2(0, T; H_{1-\theta}))\right] \times \left[L_\infty(0, T; H) \cap L_2(0, T; H_\theta)\right] \tag{5.4.16}$$

for every $T > 0$. Moreover, using the equation for $u^N(t)$ we see that

$$\sup_{t\in[0,T]} \|u_{tt}^N(t)\|_{-2\theta}^2 \leq C_R, \quad N = 1, 2, \ldots, \tag{5.4.17}$$

and also

$$\int_0^T \|u_{tt}^N(t)\|_{-\theta}^2 dt \leq C_R \int_0^T \left[1 + \|u^N(t)\|_{1-\theta}^2 + \|u_t^N(t)\|_\theta^2\right] dt \leq C_R(1+T)$$

for $N = 1, 2, \ldots$. Thus, the Aubin-Dubinskii-Lions compactness theorem (see Section A.3.3 in the Appendix) yields that $(u^N; u_t^N)$ is also compact in

$$[C(0, T; H_{1/2-\epsilon}) \cap L_2(0, T; H_{1-\theta-\epsilon})] \times [C(0, T; H_{-\epsilon}) \cap L_2(0, T; H_{\theta-\epsilon})]$$

for every $\epsilon > 0$. Hence there exists an element $(u; u_t)$ in W_T (see (5.4.16)) such that (along a subsequence) the following convergence holds:

$$\max_{[0,T]} \|u^N(t) - u(t)\|_{1/2-\epsilon}^2$$

$$+ \int_0^T \|u^N(t) - u(t)\|_{1-\theta-\epsilon}^2 dt + \int_0^T \|u_t^N(t) - u_t(t)\|_{\theta-\epsilon}^2 dt \to 0$$

as $N \to \infty$. In particular, due to the fact that $\theta < 1/2$, this implies that

$$\int_0^T \|B(u^N(t)) - B(u(t))\|_{-\theta}^2 \le C_R \int_0^T \|u^N(t) - u(t)\|_{1/2}^2 dt \to 0 \quad \text{as } N \to \infty,$$

and also

$$\int_0^T |\psi(\|u^N(t)\|_{1/2}^2) - \psi(\|u(t)\|_{1/2}^2)|^2 \le C_R \int_0^T \|u^N(t) - u(t)\|_{1/2}^2 dt \to 0 \quad \text{as } N \to \infty,$$

where ψ is either ϕ or σ. These observations allow us to make a limit transition in nonlinear terms and prove the existence of weak solutions. One can see that these solutions possess the properties stated in (5.4.7), (5.4.8), and (5.4.9).

To prove the uniqueness statement, we first show that *any* weak solution $u(t)$ satisfies (5.4.7). Indeed, since

$$P_N u \in \{v \in C^1(0, T, P_N H) : v_{tt} \in L_\infty(0, T, P_N H)\} \quad \text{and} \quad P_N H \subset \mathscr{D}(A),$$

we can use the element $A^{1-2\theta} P_N u(t)$ as a multiplier for (5.4.1). This yields

$$\frac{d}{dt}(A^{1-2\theta} P_N u, P_N u_t) + \sigma(\|u\|_{1/2}^2)(AP_N u, P_N u_t)$$

$$+ \phi(\|u\|_{1/2}^2)\|P_N u\|_{1-\theta}^2 + (B(u), A^{1-2\theta} P_N u) = \|P_N u_t\|_{1/2-\theta}^2.$$

Therefore, using (5.4.5), (5.4.14), and also the relation

$$|(Aw, v)| \le \varepsilon \|w\|_{1-\theta}^2 + C_\varepsilon \|v\|_\theta^2, \quad \forall \varepsilon > 0, \ 0 \le \theta < 1,$$

we obtain the uniform estimate

$$\int_0^T \|P_N u(t)\|_{1-\theta}^2 dt \le C_T, \quad N = 1, 2, \ldots$$

This implies (5.4.7) for the solution $u(t)$.

We also note that

$$\|u(t)\|_{1/2}^2 - \|u(s)\|_{1/2}^2 = 2 \int_s^t (A^\theta u_t(\tau), A^{1-\theta} u(\tau)) d\tau$$

for almost all $t, s \in [0, T]$, where $u(t)$ is an arbitrary weak solution to (5.4.1). Thus, by (5.4.5) and (5.4.7) we can assume that $t \mapsto \|u(t)\|_{1/2}^2$ is an absolutely continuous function.

Now we prove that (5.4.11) holds for every pair $u^1(t)$ and $u^2(t)$ of weak solutions.

One can see that $z(t) = u^1(t) - u^2(t)$ solves the equation

$$z_{tt} + \frac{1}{2}\sigma_{12}(t)A^{2\theta}z_t + \frac{1}{2}\phi_{12}(t)Az + G(u^1, u^2; t) = 0, \tag{5.4.18}$$

where $\sigma_{12}(t) = \sigma_1(t) + \sigma_2(t)$ and $\phi_{12}(t) = \phi_1(t) + \phi_2(t)$ with

$$\sigma_i(t) = \sigma(\|u^i(t)\|_{1/2}^2), \quad \phi_i(t) = \phi(\|u^i(t)\|_{1/2}^2), \quad i = 1, 2, \tag{5.4.19}$$

and

$$G(u^1, u^2; t) = \frac{1}{2}\left\{[\sigma_1(t) - \sigma_2(t)]A^{2\theta}(u_t^1 + u_t^2) + [\phi_1(t) - \phi_2(t)]A(u^1 + u^2)\right\}$$
$$+ B(u^1) - B(u^2). \tag{5.4.20}$$

It is clear that $G \in L_2^{loc}(\mathbb{R}_+; H_{-\theta})$. Since $t \mapsto \|u(t)\|_{1/2}^2$ is an absolutely continuous function, the same property holds for $\phi_{12}(t)$. Therefore, we can multiply equation (5.4.18) by $P_N z_t$ and obtain that

$$\frac{1}{2}\frac{d}{dt}\left[\|P_N z_t\|^2 + \frac{1}{2}\phi_{12}(t)\|P_N z\|_{1/2}^2\right] + \frac{1}{2}\sigma_{12}(t)\|P_N z_t\|_\theta^2 + G^N(t) = \frac{1}{4}\phi_{12}'(t)\|P_N z\|_{1/2}^2. \tag{5.4.21}$$

Here

$$\phi_{12}'(t) = \phi_1'(t) + \phi_2'(t) \quad \text{with} \quad \phi_i'(t) = 2\phi'(\|u^i(t)\|_{1/2}^2)(Au^i, u_t^i), \quad i = 1, 2,$$

and

$$G^N(t) \equiv (G(u^1, u^2; t), P_N z_t) = H_1^N(t) + H_2^N(t) + H_3^N(t), \tag{5.4.22}$$

where

$$H_1^N(t) = \frac{1}{2}[\sigma_1(t) - \sigma_2(t)](A^{2\theta}(u_t^1 + u_t^2), P_N z_t),$$

$$H_2^N(t) = \frac{1}{2}[\phi_1(t) - \phi_2(t)](A(u^1 + u^2), P_N z_t)$$

$$H_3^N(t) = (B(u^1) - B(u^2), P_N z_t).$$

One can see that

$$|\phi_{12}'(t)| \leq C_R\left(\|u_t^1\|_\theta\|u^1\|_{1-\theta} + \|u_t^2\|_\theta\|u^2\|_{1-\theta}\right). \tag{5.4.23}$$

Since

$$\phi_1(t) - \phi_2(t) = 2(A(u^1 + u^2), z) \cdot \tilde{\phi}_{12}(t), \tag{5.4.24}$$

where $\tilde{\phi}_{12}$ is given by

$$\tilde{\phi}_{12}(t) = \frac{1}{2} \int_0^1 \phi'(\lambda \|u^1(t)\|_{1/2}^2 + (1-\lambda)\|u^2(t)\|_{1/2}^2) d\lambda \qquad (5.4.25)$$

and a similar relation holds true for $\sigma_1(t) - \sigma_2(t)$, we have that

$$|H_1^N(t)| \le \epsilon \|P_N z_t\|_\theta^2 + C_{\epsilon,R} \left(\|u_t^1\|_\theta^2 + \|u_t^2\|_\theta^2 \right) \|z\|_{1/2}^2,$$

$$|H_2^N(t)| \le \epsilon \|P_N z_t\|_\theta^2 + C_{\epsilon,R} \left(\|u^1\|_{1-\theta}^2 + \|u^2\|_{1-\theta}^2 \right) \|z\|_{1/2}^2,$$

$$|H_3^N(t)| \le \epsilon \|P_N z_t\|_\theta^2 + C_{\epsilon,R} \|z\|_{1/2}^2,$$

with the constant $C_{\epsilon,R}$ independent of N. This implies

$$\frac{d}{dt}\left[\|P_N z_t\|^2 + \frac{1}{2}\phi_{12}(t)\|P_N z\|_{1/2}^2 \right] + a_R\|P_N z_t\|_\theta^2 \le b_R \left(1 + d_{12}(t)\right) \|z\|_{1/2}^2, \qquad (5.4.26)$$

where $d_{12}(t) = \|u_t^1\|_\theta^2 + \|u_t^2\|_\theta^2 + \|u^1\|_{1-\theta}^2 + \|u^2\|_{1-\theta}^2$. By (5.4.5) and (5.4.7),

$$\int_0^t d_{12}(\tau)d\tau \le C_{R,T}, \quad t \in [0,T].$$

Therefore, writing (5.4.26) in integral form after the limit transition $N \to \infty$ via a Gronwall's-type argument we obtain (5.4.11).

By (5.4.11) we have the uniqueness statement for weak solutions.

The continuity in $\mathscr{H}$ and the energy relation for weak solutions follow from the corresponding properties of solutions to some nonautonomous linear problem which we state in the following assertion.

Lemma 5.4.5. *Assume that $\phi(s)$ and $\sigma(s)$ are strictly positive absolutely continuous scalar functions and $\phi' \in L_1^{loc}(\mathbb{R}_+)$. Let $f \in L_2^{loc}(\mathbb{R}_+, H_{-\theta})$ and $(u_0; u_1) \in \mathscr{H} = H_{1/2} \times H$. Then the linear Cauchy problem*

$$u_{tt} + \phi(t)Au + \sigma(t)A^{2\theta}u_t = f(t), \quad t > 0, \quad u(0) = u_0, \quad u_t(0) = u_1, \qquad (5.4.27)$$

has a unique weak solution $u(t)$. This solution possesses the properties

$$(u(t); u_t(t)) \in \mathscr{V} \equiv C(\mathbb{R}_+; H_{1/2} \times H) \cap L_2^{loc}(\mathbb{R}_+; H_{1-\theta} \times H_\theta), \qquad (5.4.28)$$

and

$$\|u_t(t)\|^2 + \|u(t)\|_{1/2}^2 + \int_0^t \left[\|u(\tau)\|_{1-\theta}^2 + \|u_t(\tau)\|_\theta^2 \right] d\tau$$

$$\le C_T \left[\|u_1\|^2 + \|u_0\|_{1/2}^2 + \int_0^t \|f(\tau)\|_{-\theta}^2 d\tau \right] \qquad (5.4.29)$$

for every $t \in [0, T]$, and satisfies the energy relation

$$E_\phi(t; u(t), u_t(t)) + \int_0^t \sigma(\tau) \|u_t(\tau)\|_\theta^2 d\tau$$

$$= E_\phi(0; u_0, u_1) + \frac{1}{2} \int_0^t \phi'(\tau) \|u(\tau)\|_{1/2}^2 d\tau + \int_0^t (f(\tau), u_t(\tau)) d\tau, \qquad (5.4.30)$$

where the energy $E_\phi(t; u(t), u_t(t))$ is defined by the relation

$$E_\phi(t; u(t), u_t(t)) = \frac{1}{2} \left[\|u_t(t)\|^2 + \phi(t) \|u(t)\|_{1/2}^2 \right].$$

Proof. We use the Galerkin method. Approximate solutions $u^N(t)$ with values in $P_N H$ exist and, since problem (5.4.27) is linear, the differences $u^{N,M}(t) = u^N(t) - u^M(t)$ satisfy the energy relation of the form

$$\frac{d}{dt} E_\phi(t; u^{N,M}(t), u_t^{N,M}(t)) + \sigma(t) \|u_t^{N,M}(t)\|_\theta^2$$

$$= \frac{1}{2} \phi'(t) \|u^{N,M}(t)\|_{1/2}^2 + (f(t), u_t^{N,M}(t)).$$

This implies that

$$\sup_{t \in [0,T]} \left\{ \|u_t^{N,M}(t)\|^2 + \|u^{N,M}(t)\|_{1/2}^2 \right\} + \int_0^T \|u_t^{N,M}(\tau)\|_\theta^2 d\tau$$

$$\leq C_T \left[\|(P_N - P_M)u_1\|^2 + \|P_N - P_M)u_0\|_{1/2}^2 + \int_0^t \|(P_N - P_M)f(\tau)\|_{-\theta}^2 d\tau \right]$$

$$(5.4.31)$$

for every $T > 0$. Moreover, using the multiplier $A^{1-2\theta} u^{N,M}$ we can also find that

$$\int_0^T \|u^{N,M}\|_{1-\theta}^2 d\tau \leq C_T \left[\|(P_N - P_M)u_1\|^2 + \|P_N - P_M)u_0\|_{1/2}^2 \right.$$

$$\left. + \int_0^t \|P_N - P_M)f(\tau)\|_{-\theta}^2 d\tau \right]. \qquad (5.4.32)$$

Thus, by (5.4.31) and (5.4.32) the approximate solutions $(u^N; u_t^N)$ converge to a weak solution in the space $\mathcal{V}$ defined in (5.4.28). It is also clear that relation (5.4.29) holds and the energy relation in (5.4.30) is satisfied. $\square$

Remark 5.4.6. We note that in the case of "frozen" coefficients ϕ and σ (i.e., when $\phi(t) \equiv \phi_0 > 0$ and $\sigma(t) \equiv \sigma_0 > 0$), problem (5.4.27) generates an analytic semigroup (see, e.g., LASIECKA/TRIGGIANI [145, Chapter 3] and the references therein). In particular, this allows us to apply general results (see LUNARDI [154], for instance) on evolution families generated by time-dependent sectorial operators to obtain further regularity properties of solutions to (5.4.27). ∎

Lemma 5.4.5 implies the corresponding continuity properties for the original nonlinear problem and also the energy relation (5.4.10). For this we consider $u(t)$ as a solution to (5.4.27) with $\phi(t) = \phi(\|u(t)\|_{1/2}^2)$, $\sigma(t) = \sigma(\|u(t)\|_{1/2}^2)$, and $f(t) = -B(u(t))$. To obtain the energy relation we also use the obvious equality

$$\phi'(t)\|u(t)\|_{1/2}^2 = \frac{d}{dt}\left\{ \Phi(t)\|u(t)\|_{1/2}^2 - \Phi(\|u(t)\|_{1/2}^2)\right\}.$$

The proof of Theorem 5.4.4 is complete. □

Remark 5.4.7. It follows from (5.4.9) that

$$\sup_{t\in\mathbb{R}_+}\left\{\|u_{tt}(t)\|_{-2\theta}^2 + \|u_t(t)\|_\theta^2 + \|u(t)\|_{1/2}^2\right\} + \int_0^\infty \|u_t(\tau)\|_\theta^2 d\tau \leq C_R \qquad (5.4.33)$$

for any initial data such that $\|(u_0, u_1)\|_{\mathscr{H}} \leq R$. The finiteness of the dissipation integral in (5.4.33) is important for the study of long-time dynamics. ∎

Remark 5.4.8. In the case when $0 < \theta < 1/4$, from the energy relations for approximate solutions $u^N(t)$ we can obtain the same a priori estimate as in (5.4.12) and (5.4.13). Then using the multiplier $A^{2\theta}u^N$ we find that

$$\frac{d}{dt}(A^{2\theta}u^N, u_t^N) + \sigma(\|u^N\|_{1/2}^2)(A^{2\theta}u^N, A^{2\theta}u_t^N)$$
$$+ \phi(\|u^N\|_{1/2}^2)\|u^N\|_{1/2+\theta}^2 + (B(u^N), A^{2\theta}u^N) = \|u_t^N\|_\theta^2.$$

This and also estimates (5.4.12) and (5.4.13) yield the relation

$$\int_0^T \|u^N(t)\|_{1/2+\theta}^2 dt \leq C(R)(1 + T), \quad N = 1, 2, \ldots, \quad \text{for any } T > 0.$$

This bound allows us to make the limit transition in nonlinear terms for $0 < \theta < 1/4$ and to prove the existence of weak solutions $u(t)$ such that

$$(u; u_t) \in L_\infty(0, T; H_{1/2} \times H) \cap L_2(0, T; H_{1/2+\theta} \times H_\theta).$$

We do not know how to prove the uniqueness of weak solutions in the case $0 < \theta < 1/4$. The point is that the estimate in (5.4.15) is possible in the case $\theta \geq 1/4$ only. ∎

5.4.2 Smooth solutions

Now we prove additional smoothness of weak solutions. To do this, we need additional requirements concerning the nonlinear force $B(u)$.

Assumption 5.4.9. We assume that $B : H_{1/2} \mapsto H_{-\theta}$ has a Fréchet derivative $B'(u)$ which is a linear operator from $H_{1/2}$ into $H_{-\theta}$ for each $u \in H_{1/2}$ and there exists $c_\rho \geq 0$ such that

$$(B'(u)v, v) \geq -c_\rho \|v\|_\theta^2 \quad \text{and} \quad \|A^{-2\theta} B'(u)v\| \leq c_\rho \|v\|_\theta$$

for every $u, v \in H_{1/2}$, $\|u\|_{1/2} \leq \rho$.

Proposition 5.4.10 (Regularity). *Let Assumptions 5.4.1 and 5.4.9 be in force. Then for every weak solution $u(t)$ with initial data $(u_0; u_1) \in \mathscr{H}$ we have the following additional regularity:*

$$(u; u_t) \in L_\infty(a, T; H_{1-\theta} \times H_\theta), \quad u_{tt} \in L_\infty(a, T; \times H_{-\theta}) \cap L_2(a, T; H)$$

for every $0 < a < T$. Moreover, there exists $c_{R,T} > 0$ such that

$$\|u_{tt}(t)\|_{-\theta}^2 + \|u_t(t)\|_\theta^2 + \|u(t)\|_{1-\theta}^2 + \int_t^{t+1} \|u_{tt}(\tau)\|^2 d\tau \leq \frac{c_{R,T}}{t^2} \tag{5.4.34}$$

for every $t \in (0, T]$ under the condition $\|(u_0; u_1)\|_{\mathscr{H}} \leq R$.
If we assume in addition that $(u_0; u_1) \in H_{1-\theta} \times H_\theta$, then

$$\|u_{tt}(t)\|_{-\theta}^2 + \|u_t(t)\|_\theta^2 + \|u(t)\|_{1-\theta}^2 + \int_0^t \|u_{tt}(\tau)\|^2 d\tau \leq c_{R,T} \left(1 + \|u_1\|_\theta^2 + \|u_0\|_{1-\theta}^2\right) \tag{5.4.35}$$

for every $t \in (0, T]$.

Proof. The argument below can be justified by considering Galerkin approximations.

Let $u(t)$ be a solution such that $\|(u(t); u_t(t))\|_{\mathscr{H}} \leq R$ for $t \in [0, T]$. Formal differentiation gives that $v = u_t(t)$ solves the equation

$$v_{tt} + \sigma(\|u\|_{1/2}^2)A^{2\theta} v_t + \phi(\|u\|_{1/2}^2)Av + B'(u)v + G_*(u, u_t; t) = 0, \tag{5.4.36}$$

where

$$G_*(u, u_t; t) = 2\left[\sigma'(\|u\|_{1/2}^2)A^{2\theta} u_t + \phi'(\|u\|_{1/2}^2)Au\right](Au, u_t).$$

Thus, multiplying equation (5.4.36) by v and using (5.4.9), we have that

$$\frac{d}{dt}\left[(v, v_t) + \frac{1}{2}\sigma(\|u\|_{1/2}^2)\|v\|_\theta^2\right] + \phi(\|u\|_{1/2}^2)\|v\|_{1/2}^2 + (B'(u)v, v)$$
$$\leq \|v_t\|^2 + C_R\left[|(Au, v)|^2 + |(Au, v)|\|v\|_\theta^2\right].$$

This implies that

$$\frac{d}{dt}\left[(v, v_t) + \frac{1}{2}\sigma(\|u\|_{1/2}^2)\|v\|_\theta^2\right] + \phi_R\|v\|_{1/2}^2 +$$
$$+ (B'(u)v, v) \leq \|v_t\|^2 + C_R\left[\|u\|_{1-\theta}^2 + \|u_t\|_\theta\|u\|_{1-\theta}\right]\|v\|_\theta^2.$$

Using the multiplier $A^{-2\theta}v_t$ in (5.4.36) we obtain

$$\frac{1}{2}\frac{d}{dt}\left[\|v_t\|_{-\theta}^2 + \phi(\|u\|_{1/2}^2)\|v\|_{1/2-\theta}^2\right] + \sigma(\|u\|_{1/2}^2)\|v_t\|^2$$
$$\leq C_R\left[\|v\|_\theta\|v_t\|\|u\|_{1-\theta} + \|v\|_{1/2-\theta}^2\|u_t\|_\theta\|u\|_{1-\theta}\right] + |(A^{-2\theta}B'(u)v, v_t)|.$$

Here we also use the fact that $\theta \geq 1/4$. Let

$$\Psi_*(t) = \frac{1}{2}\left[\|v_t\|_{-\theta}^2 + \phi(\|u\|_{1/2}^2)\|v\|_{1/2-\theta}^2\right] + \eta\left[(v, v_t) + \frac{1}{2}\sigma(\|u\|_{1/2}^2)\|v\|_\theta^2\right]$$

for $\eta > 0$ small enough. It is obvious that there exists $\eta_0 > 0$ such that

$$\eta a_R\left[\|v_t\|_{-\theta}^2 + \|v\|_\theta^2\right] \leq \Psi_*(t) \leq b_R\left[\|v_t\|_{-\theta}^2 + \|v\|_\theta^2\right],$$

for $0 < \eta \leq \eta_0$. Moreover,

$$\frac{d}{dt}\Psi_*(t) + \eta(B'(u)v, v) + c_R\|v_t\|^2$$
$$\leq C_R(\|u_t\|_\theta^2 + \|u\|_{1-\theta}^2)\|v\|_\theta^2 + |(A^{-2\theta}B'(u)v, v_t)|$$

Thus, by Assumption 5.4.9,

$$\frac{d}{dt}\Psi_*(t) + c_R\|v_t\|^2 \leq C_R(1 + \|u_t\|_\theta^2 + \|u\|_{1-\theta}^2)\|v\|_\theta^2. \tag{5.4.37}$$

Using the finiteness of the integrals in (5.4.9) and a Gronwall's-type argument, we obtain that

$$\|u_{tt}(t)\|_{-\theta}^2 + \|u_t(t)\|_\theta^2 + \int_0^t \|u_{tt}(\tau)\|^2 d\tau \leq C_{R,T}\left(\|u_{tt}(0)\|_{-\theta}^2 + \|u_t(0)\|_\theta^2\right) \tag{5.4.38}$$

for $t \in (0, T]$. It follows from (5.4.1) and (5.4.9) that

$$\|u_{tt}(t)\|_{-\theta}^2 \le C_R \left(1 + \|u(t)\|_{1-\theta}^2 + \|u_t(t)\|_\theta^2\right), \tag{5.4.39}$$

$$\|u(t)\|_{1-\theta}^2 \le C_R \left(1 + \|u_{tt}(t)\|_{-\theta}^2 + \|u_t(t)\|_\theta^2\right) \tag{5.4.40}$$

for all $t \ge 0$ under the condition $\|(u_0; u_1)\|_{\mathscr{H}} \le R$. Therefore, (5.4.38) implies (5.4.35).

To obtain (5.4.34) we multiply (5.4.37) by t^2. This yields

$$\frac{d}{dt}[t^2 \Psi_*] + \alpha_R t^2 \|v_t\|^2 \le C_R \left(1 + \|u_t\|_\theta^2 + \|u\|_{1-\theta}^2\right) [t^2 \Psi_*] + 2tb_R \left[\|v_t\|_{-\theta}^2 + \|v\|_\theta^2\right]. \tag{5.4.41}$$

One can see that

$$2t\|v\|_\theta^2 \le 1 + t^2 \|u_t\|_\theta^2 \|v\|_\theta^2 \le C_R[1 + \|u_t\|_\theta^2 (t^2 \Psi_*)], \quad t \in [0, T],$$

and

$$\|v_t\|_{-\theta}^2 \le C\|v_t\|\|u_{tt}\|_{-2\theta}, \quad t \in [0, T].$$

The relation

$$A^{-2\theta} u_{tt} = -\sigma(\|u\|_{1/2}^2)u_t - \phi(\|u\|_{1/2}^2)A^{1-2\theta}u - A^{-2\theta}B(u)$$

yields $\|A^{-2\theta} u_{tt}\| \le C_R$. Therefore,

$$t\|v_t\|_{-\theta}^2 \le C_R t\|v_t\| \le \varepsilon t^2 \|v_t\|^2 + C_{R,\varepsilon}, \quad t \in [0, T].$$

Thus, from (5.4.41) we have

$$\frac{d}{dt}(t^2 \Psi_*) \le C_R + C_R \left(1 + \|u_t\|_\theta^2 + \|u\|_{1-\theta}^2\right) [t^2 \Psi_*].$$

Using the finiteness of the integrals in (5.4.9) and also relation (5.4.40), we obtain (5.4.34). □

In further considerations we need the following properties of stationary solutions.

Proposition 5.4.11. *In addition to Assumption 5.4.1 we assume that*

$$\liminf_{\|u\|_{1/2} \to \infty} \left\{\phi(\|u\|_{1/2}^2)\|u\|_{1/2}^2 + (B(u), u)\right\} > 0. \tag{5.4.42}$$

Then the set

$$\mathcal{N}_* = \left\{ u \in H_{1/2} \;:\; \phi(\|u\|_{1/2}^2)Au + B(u) = 0 \right\} \qquad (5.4.43)$$

of stationary solutions is a nonempty compact set in $H_{1/2}$. Moreover, $\mathcal{N}_*$ is bounded in $H_{1-\theta}$.

Proof. One can see from (5.4.42) and from Lemma 4.3.7 on the "acute angle" that there exists a sequence $\{u^N\}$ of approximate stationary solutions (with values in $P_N H$) which is bounded in $H_{1/2}$: $\|u^N\|_{1/2} \le R$ for some $R > 0$. From the equation for u^N we have

$$c_R \|u^N\|_{1-\theta}^2 \le \phi(\|u^N\|_{1/2}^2)\|u^N\|_{1-\theta}^2 \le \|B(u^N)\|_{-\theta}\|u^N\|_{1-\theta} \le C_R \|u^N\|_{1-\theta}$$

for positive c_R and C_R. This implies that $\|u^N\|_{1-\theta} \le C_R$. Since $H_{1-\theta}$ is compactly embedded in $H_{1/2}$, we can make the limit transition along a subsequence of $\{u^N\}$ and conclude the proof. $\qquad\square$

Remark 5.4.12. The condition in (5.4.42) holds true, if $s\phi(s) \to +\infty$ as $s \to +\infty$ and $(B(u), u) \ge -c$ for all $u \in H_{1/2}$, for instance. Another possibility for (5.4.42) in the case of the wave model in $\mathbb{R}^d$ is shown in Section 5.4.4 below. $\qquad\blacksquare$

5.4.3 Long-time dynamics

In this section we deal with a global attractor for the dynamical system generated by (5.4.1). There are many papers on stabilization to zero equilibrium for Kirchhoff-type models (see, e.g., AUTUORI/PUCCI/SALVATORI [6], CAVALCANTI ET AL. [27], MATSUYAMA/IKEHATA [162], MEDEIROS/FERREL/MENEZES [163], MEDEIROS/MILLA MIRANDA [164], TANIGUCHI [214] and the references therein) and only a few recent results devoted to (nontrivial) attractors for systems like (5.4.1). We refer to NAKAO [169] for studies of local attractors in the case of viscous ($\theta = 0$) linear damping. The *global* attractors were studied only in the case of a strong damping ($\theta = 1/2$); see CHUESHOV [42] and also FAN/ZHOU [96], NAKAO/ZHIJIAN [170], YANG ET AL. [225–227] in the case $\sigma = const > 0$ possibly perturbed by nonlinear viscous damping terms.

We concentrate on the case $1/4 \le \theta < 1/2$ with nontrivial stiffness ϕ and damping σ factors.

By Theorem 5.4.4 problem (5.4.1) generates an evolution semigroup S_t in the space $\mathcal{H} = H_{1/2} \times H$ by the formula

$$S_t y = (u(t); u_t(t)), \text{ where } y = (u_0; u_1) \in \mathcal{H} \text{ and } u(t) \text{ solves (5.4.1).} \qquad (5.4.44)$$

One can see from the energy relation in (5.4.10) that the dynamical system $(\mathcal{H}, S_t)$ is gradient on $\mathcal{H}$ (see Definition 2.4.1). The full energy $\mathscr{E}(u_0; u_1)$ is a strict

Lyapunov function. Thus, we can use the criterion of the existence of global attractors for gradient systems (see Theorem 2.4.16).

To prove finiteness of the fractal dimension of global attractors, we use the quasi-stability method based on an appropriate stabilizability estimate. For this we need additional hypotheses concerning nonlinear force $B(u)$. In fact, they require some smoothness of the potential mapping $B(\cdot)$ (see Section A.5 for the basic definitions) and are almost the same[8] as those in Assumption 5.3.1.

Assumption 5.4.13. Assume that

- $B(u) = \Pi'(u)$ with the functional $\Pi : H_{1/2} \mapsto \mathbb{R}$, which is a Fréchet C^3 mapping.
- The second $\Pi^{(2)}(u)$ and the third $\Pi^{(3)}(u)$ Fréchet derivatives of $\Pi(u)$ satisfy the conditions

$$\left|\langle \Pi^{(2)}(u); v, v\rangle\right| \le \varepsilon\|v\|_{1/2}^2 + C_{\varrho,\varepsilon}\|v\|^2, \quad v \in H_{1/2}, \ \forall\, \varepsilon > 0, \qquad (5.4.45)$$

and

$$\left|\langle \Pi^{(3)}(u); v_1, v_2, v_3\rangle\right| \le C_\rho\|v_1\|_{1/2}\|v_2\|_{1/2}\|v_3\|_\theta, \quad v_i \in H_1, \qquad (5.4.46)$$

for all $u \in H_{1-\theta}$ such that $\|u\|_{1-\theta} \le \rho$, where $\rho > 0$ is arbitrary and C_ρ is a positive constant. Here $\langle \Pi^{(k)}(u); v_1, \ldots, v_k\rangle$ denotes the value of the derivative $\Pi^{(k)}(u)$ on elements $v_1, \ldots, v_k$. For details concerning the Fréchet calculus we refer to Section A.5 in the Appendix.

Assume now that $u^1(t)$ and $u^2(t)$ belong to the class $C^1([a, b]; H_1)$ for some interval $[a, b] \subset \mathbb{R}$. Let $z(t) = u^1(t) - u^2(t)$. Then, by the same argument as in Section 5.3.1, we can obtain the representation

$$(B(u^1(t)) - B(u^2(t)), z_t(t)) = \frac{d}{dt}Q(t) + R(t), \quad t \in [a, b] \subset \mathbb{R}, \qquad (5.4.47)$$

with

$$Q(t) = \frac{1}{2}\langle \Pi^{(2)}(u^2(t)); z(t), z(t)\rangle \qquad (5.4.48)$$

and

$$R(t) = -\frac{1}{2}\langle \Pi^{(3)}(u^2); u_t^2, z, z\rangle + \int_0^1 (1 - \lambda)\langle \Pi^{(3)}(u^2 + \lambda z); z, z, z_t\rangle d\lambda.$$

---

[8] The main difference is that relation (5.4.45) and (5.4.46) hold for $u \in H_{1-\theta}$ in the present case.

If we assume in addition that $\|u^i(t)\|_{1-\theta} \leq R$ for all $t \in [a,b]$, $i = 1,2$, then by Assumption 5.4.13 we have the estimates

$$|Q(t)| \leq \varepsilon\|z(t)\|_{1/2}^2 + C_{R,\varepsilon}\|z(t)\|^2, \quad \forall \varepsilon > 0, \tag{5.4.49}$$

and

$$|R(t)| \leq C_R(\|u_t^1\|_\theta + \|u_t^2\|_\theta)\|z(t)\|_{1/2}^2. \tag{5.4.50}$$

Our main result in this section is the following theorem.

Theorem 5.4.14. *Let Assumptions 5.4.1 and 5.4.9 be in force and relation (5.4.42) be valid. Then the following assertions hold.*

(1) *The evolution semigroup S_t possesses a compact global attractor $\mathfrak{A}$ in $\mathscr{H}$ which is a bounded set in $H_{1-\theta} \times H_\theta$. Moreover,*

$$\sup_{t\in\mathbb{R}} \left(\|u(t)\|_{1-\theta}^2 + \|u_t(t)\|_\theta^2 + \|u_{tt}(t)\|_{-\theta}^2 + \int_t^{t+1}\|u_{tt}(\tau)\|^2 d\tau\right) \leq C_{\mathfrak{A}} \tag{5.4.51}$$

for any full trajectory $\gamma = \{(u(t); u_t(t)) : t \in \mathbb{R}\}$ from the attractor $\mathfrak{A}$. Moreover,

$$\mathfrak{A} = \mathcal{M}^u(\mathscr{N}), \quad \text{where} \quad \mathscr{N} = \{(u;0) \in \mathscr{H} : \phi(\|A^{1/2}u\|^2)Au + B(u) = 0\} \tag{5.4.52}$$

and $\text{dist}_{\mathscr{H}}(y, \mathscr{N}) \to 0$ as $t \to \infty$ for any $y \in \mathscr{H}$.

(2) *Assume in addition that either $B(u)$ is subcritical [9] on $H_{1-\theta}$ in the sense that there exists $\delta > 0$ such that*

$$\|B(u_1) - B(u_2)\|_{-\theta} \leq L(\varrho)\|u_1 - u_2\|_{1/2-\delta}, \quad \forall \|u_i\|_{1-\theta} \leq \varrho, \tag{5.4.53}$$

or else Assumption 5.4.13 holds. Then the attractor $\mathfrak{A}$ has a finite fractal dimension and the following hold.

(a) *Any trajectory $\gamma = \{(u(t); u_t(t)) : t \in \mathbb{R}\}$ from the attractor $\mathfrak{A}$ possesses the properties*

$$(u; u_t; u_{tt}) \in L_\infty(\mathbb{R}; H_{1-\theta} \times H_{1/2} \times H) \tag{5.4.54}$$

and there is $R > 0$ such that

$$\sup_{\gamma\subset\mathfrak{A}} \sup_{t\in\mathbb{R}} \left(\|u(t)\|_{1-\theta}^2 + \|u_t(t)\|_{1/2}^2 + \|u_{tt}(t)\|^2\right) \leq R^2. \tag{5.4.55}$$

---

[9] As we will see in Section 5.4.4, the mapping B can be subcritical on $H_{1-\theta}$ and critical (or even supercritical) on the potential energy space $H_{1/2}$.

(b) *There exists a fractal exponential attractor* $\mathfrak{A}_{exp}$ *in* $\mathcal{H}$.

(c) *Let* $\mathcal{L} = \{l_j : j = 1,\ldots,N\}$ *be a finite set of functionals on* $H_{1/2}$ *with the completeness defect* $\epsilon_{\mathcal{L}} = \epsilon_{\mathcal{L}}(H_{1/2}, H)$. *Then there exists* $\varepsilon_0 > 0$ *such that under the condition* $\epsilon_{\mathcal{L}} \leq \varepsilon_0$ *the set* $\mathcal{L}$ *is (asymptotically) determining in the sense that the property*

$$\lim_{t\to\infty} \max_{j} |l_j(u^1(t) - u^2(t))| = 0$$

for a pair of weak solutions $u^1(t)$ *and* $u^1(t)$ *implies that*

$$\lim_{t\to\infty} \left\{ \|u_t^1(t) - u_t^2(t)\|^2 + \|u^1(t) - u^2(t)\|_{1/2}^2 \right\} = 0.$$

Proof of Theorem 5.4.14. Let

$$\mathscr{E}_R = \{(u_0; u_1) \in \mathcal{H} : \mathscr{E}(u_0, u_1) \leq R\}, \tag{5.4.56}$$

where $\mathscr{E}(u_0, u_1)$ is the energy defined in the statement of Theorem 5.4.4. It follows from Assumption 5.4.1 that $\Pi(u)$ is bounded on every bounded set (see Remark 4.2.21), and for every bounded set $B \subset \mathcal{H}$ there exists R_B such that $B \subset \mathscr{E}_{R_B}$. By (5.4.2) and (5.4.4) the set $\mathscr{E}_R$ is bounded for every R, and by the energy relation in (5.4.10) $\mathscr{E}_R$ is a positively invariant set in $\mathcal{H}$. This implies that $\gamma_0(B) = \bigcup_{t\geq 0} S_t B$ is bounded for every bounded set B. The smoothness result stated in Proposition 5.4.10 implies that $\mathscr{E}_R^{(1)} := S_1 \mathscr{E}_R$ is a compact set for every $R > 0$. Therefore, S_t is asymptotically smooth (even compact) on every set $\mathscr{E}_R$. Thus, the first part of the theorem follows from Theorem 2.4.16 and Proposition 5.4.10.

To prove the second part of Theorem 5.4.14, we first note that the set $\mathscr{E}_R$ given by (5.4.56) is absorbing for R large enough; i.e., there exists R_* such that for any bounded set $B \subset \mathcal{H}$ one can find $t_B \geq 0$ such that $S_t B \subset \mathscr{E}_{R_*}$ for all $t \geq t_B$. This follows from the existence of a global attractor. By Proposition 5.4.10 the set $\mathscr{E}_{R_*}^{(1)} = S_1 \mathscr{E}_{R_*}$ is bounded in $H_{1-\theta} \times H_\theta$. Since $\mathscr{E}_{R_*}$ is a positively invariant absorbing set, the same property is true for $\mathscr{E}_{R_*}^{(1)}$. We use this observation in the proof of a quasi-stability property of S_t in the energy space $\mathcal{H}$, which is stated in the following assertion.

Proposition 5.4.15 (Quasi-stability). *Let the hypotheses of the second part of Theorem 5.4.14 be in force. Assume that* $u^1(t)$ *and* $u^2(t)$ *are two weak solutions such that*

$$\|u^i(t)\|_{1-\theta}^2 + \|u_t^i(t)\|_\theta^2 \leq R^2 \text{ for all } t > 0 \text{ and for some } R > 0. \tag{5.4.57}$$

Then the difference $z(t) = u^1(t) - u^2(t)$ satisfies the relation

$$\|z_t(t)\|^2 + \|z(t)\|_{1/2}^2 \leq a_R \left(\|z_t(0)\|^2 + \|z(0)\|_{1/2}^2 \right) e^{-\gamma_R t} + b_R \int_0^t e^{-\gamma_R(t-\tau)} \|z(\tau)\|^2 d\tau,$$

$$(5.4.58)$$

where a_R, b_R, γ_R are positive constants.

Proof. We start with the following lemma.

Lemma 5.4.16. *Let $u^1(t)$ and $u^2(t)$ be two weak solutions to (5.4.1) satisfying (5.4.57). Then for $z(t) = u^1(t) - u^2(t)$ we have the relation[10]*

$$\frac{d}{dt}(z, z_t) + c_R \|z\|_{1/2}^2 \leq c_0 \|z_t\|_\theta^2 + C_R \|z\|^2 \tag{5.4.59}$$

for all $t \in [0, T]$, where $c_0 = 1 + \lambda_1^{-\theta}$.

Proof. The difference $z(t) = u^1(t) - u^2(t)$ solves (5.4.18). Therefore, multiplying equation (5.4.18) by z in H yields

$$\frac{d}{dt}(z, z_t) + \frac{1}{2}\sigma_{12}(t) \cdot (A^{2\theta} z, z_t) + \frac{1}{2}\phi_{12}(t) \cdot \|z\|_{1/2}^2 + (G(u^1, u^2, t), z) = \|z_t\|^2,$$

where

$$\sigma_{12}(t) = \sigma_1(t) + \sigma_2(t), \quad \phi_{12}(t) = \phi_1(t) + \phi_2(t)$$

with $\sigma_i(t)$ and $\phi_i(t)$ given by (5.4.19), and $G(u^1, u^2, t)$ is defined by (5.4.20).
   Using (5.4.24) and (5.4.49) under the condition (5.4.57), we have that

$$|\phi_1(t) - \phi_2(t)| \leq C_R \|z\|_\theta,$$

and hence

$$|[\phi_1(t) - \phi_2(t)](A(u^1 + u^2), z)| \leq C_R \|z\|_\theta^2 \leq \epsilon \|z\|_{1/2}^2 + C_{\epsilon, R} \|z\|^2, \quad \forall \epsilon > 0.$$

Similarly,

$$|[\sigma_1(t) - \sigma_2(t)](A^{2\theta}(u_t^1 + u_t^2), z)| \leq C_R \|z\|_\theta^2 \leq \epsilon \|z\|_{1/2}^2 + C_{\epsilon, R} \|z\|^2, \quad \forall \epsilon > 0.$$

We also have that

$$|(B(u^1) - B(u^2), z)| \leq \|B(u^1) - B(u^2)\|_{-\theta} \|z\|_\theta \leq \epsilon \|z\|_{1/2}^2 + C_{\epsilon, R} \|z\|^2$$

---

[10] It is easy to show that the product (z, z_t) is absolutely continuous with respect to t for every pair of weak solutions, and thus the relation in (5.4.59) has meaning.

and

$$|(A^{2\theta} z_t, z)| \leq \|z_t\|_\theta \|z\|_\theta \leq \varepsilon \left(\|z_t\|_\theta^2 + \|z\|_{1/2}^2 \right) + C_\varepsilon \|z\|^2.$$

Thus, using the structure (5.4.20) of the term $G(u^1, u^2; t)$ in (5.4.18) we obtain (5.4.59).   □

Now we multiply (5.4.18) by z_t and obtain that

$$\frac{1}{2} \frac{d}{dt} \left[\|z_t\|^2 + \frac{1}{2} \phi_{12}(t) \|z\|_{1/2}^2 \right] + \frac{1}{2} \sigma_{12}(t) \|z_t\|_\theta^2 + G(t) = \frac{1}{4} \phi'_{12}(t) \|z\|_{1/2}^2 \qquad (5.4.60)$$

with

$$G(t) \equiv (G(u^1, u^2; t), z_t) = H_1(t) + H_2(t) + H_3(t),$$

where

$$H_1(t) = \frac{1}{2} [\sigma_1(t) - \sigma_2(t)] (A^{2\theta}(u_t^1 + u_t^2), z_t),$$

$$H_2(t) = \frac{1}{2} [\phi_1(t) - \phi_2(t)] (A(u^1 + u^2), z_t)$$

$$H_3(t) = (B(u^1) - B(u^2), z_t).$$

One can see that under the condition in (5.4.57),

$$|H_i(t)| \leq \epsilon (\|z_t\|_\theta^2 + \|z\|_{1/2}^2) + C_{\epsilon,R} \|z\|^2, \quad \forall \epsilon > 0, \quad i = 1, 2.$$

If B is subcritical on $H_{1-\theta}$ in the sense of (5.4.53), then the estimate for $H_3(t)$ is direct and the same as that for H_1 and H_2:

$$|H_3(t)| \leq C_R \|z_t\|_\theta \|z\|_{1/2-\delta} \leq \varepsilon \left(\|z_t\|_\theta^2 + \|z\|_{1/2}^2 \right) + C_{R,\varepsilon} \|z\|^2, \quad \forall \epsilon > 0.$$

Therefore, in the argument below we concentrate on the case described in Assumption 5.4.13. In this case we use relation (5.4.47) and introduce the energy-type functional

$$E_*(t) = \frac{1}{2} \left[\|z_t\|^2 + \frac{1}{2} \phi_{12}(t) \|z\|_{1/2}^2 + Q(t) \right],$$

where $Q(t)$ is given by (5.4.48). From (5.4.60) and (5.4.50) using the calculations above, we obviously have

$$\frac{d}{dt} E_*(t) + c_R \|z_t\|_\theta^2 \leq [\varepsilon + C_{R,\varepsilon}(\|u_t^1\|_\theta^2 + \|u_t^2\|_\theta^2)] \|z\|_{1/2}^2 + C_{R,\varepsilon} \|z\|^2.$$

Therefore, using Lemma 5.4.16 and relation (5.4.49), we obtain that the function

$$W_*(t) = E_*(t) + \eta(z, z_t) + K\|z(t)\|^2, \quad \eta > 0,$$

with appropriate $K > 0$ and $\eta > 0$ small enough, satisfies the relations

$$a_R \left(\|z_t(t)\|^2 + \|z(t)\|_{1/2}^2 \right) \le W_*(t) \le b_R \left(\|z_t(t)\|^2 + \|z(t)\|_{1/2}^2 \right)$$

and

$$\frac{d}{dt} W_*(t) + c_R W_*(t) \le C_R(\|u_t^1\|_\theta^2 + \|u_t^2\|_\theta^2) W_*(t) + C_R\|z(t)\|^2$$

with positive constants. Thus, the finiteness of the dissipation integral in (5.4.33) and the standard Gronwall's argument imply the result in (5.4.58) in the case when Assumption 5.4.13 holds. In the subcritical case (5.4.53) we use the same argument but for the functional E_* without the term containing Q. □

Completion of the proof of Theorem 5.4.14. We first note that Proposition 5.4.15 means that the semigroup S_t is (asymptotically) quasi-stable in the sense of Definition 3.4.15 on the positively invariant absorbing set $\mathscr{E}_{R_*}^{(1)} = S_1 \mathscr{E}_{R_*}$, where $\mathscr{E}_R$ is defined in (5.4.56) and R_* is sufficiently large.

To obtain the result on regularity stated in (5.4.54) and (5.4.55), we first apply Theorem 3.4.19, which gives us

$$\sup_{t\in\mathbb{R}} \left(\|u_t(t)\|_{1/2}^2 + \|u_{tt}(t)\|^2 \right) \le C_\mathfrak{A}$$

for any trajectory $\gamma = \{(u(t); u_t(t)) : t \in \mathbb{R}\} \subset \mathfrak{A}$. Therefore, applying (5.4.51) we obtain (5.4.54) and (5.4.55).

By (5.4.35) any weak solution $u(t)$ possesses the property

$$\|u(t)\|_{1-\theta}^2 + \|u_t(t)\|_\theta^2 + \int_t^{t+1} \|u_{tt}(\tau)\|^2 d\tau \le C_{R_*,T} \text{ for } t \in [0, T], \ \forall T > 0,$$

provided $(u_0; u_1) \in \mathscr{E}_{R_*}^{(1)}$. This implies that $t \mapsto (u(t); u_t(t)) \equiv S_t y$ is a $1/4$-Hölder continuous function with values in $\mathscr{H}$ for every $y \in \mathscr{E}_{R_*}^{(1)}$. Indeed, using the interpolation inequality (4.1.2) with $\mu = 1/2, \alpha = 1 - \theta, \beta = \theta$, we have

$$\|u(t+h) - u(t)\|_{1/2} \le [\|u(t+h)\|_{1-\theta} + \|u(t)\|_{1-\theta}]^{1/2} \|u(t+h) - u(t)\|_\theta^{1/2}$$

$$\le C_{R_*,T} \left| \int_t^{t+h} \|u_t(\tau)\|_\theta d\tau \right|^{1/2} \le C|h|^{1/4},$$

and similarly for $\|u_t(t+h) - u_t(t)\|$. Since $\mathscr{E}_{R_*}^{(1)}$ is a positively invariant absorbing set, the existence of a fractal exponential attractor follows from Theorem 3.4.18.

To prove the statement concerning determining functionals, we use the same idea as in the proof of Theorem 3.4.20.

□

5.4.4 An application

In a bounded smooth domain $\Omega \subset \mathbb{R}^d$, $d = 1,2,3$, we consider the following Kirchhoff wave model with a structural nonlinear damping:

$$\begin{cases} u_{tt} + \sigma(\|\nabla u\|^2)[-\Delta_D]^{2\theta} u_t - \phi(\|\nabla u\|^2)\Delta u + f(u) = 0, \quad x \in \Omega, \ t > 0, \\ u|_{\partial\Omega} = 0, \quad u(0) = u_0, \quad \partial_t u(0) = u_1. \end{cases}$$

(5.4.61)

Here $[-\Delta_D]^{2\theta}$ is the power of a (positive self-adjoint) operator $-\Delta_D$ generated by the Laplacian Δ with Dirichlet boundary conditions and $\|\cdot\|$ is the L_2-norm. We can represent this problem in the abstract form (5.4.1) by setting $H = L_2(\Omega)$, $A = -\Delta_D$ and with $B(u)$ the Nemytskii operator generated by the function $f(u)$. We note that the presence of the parameter θ in the model allows us to control the "strength" of the dissipative mechanism between the viscoelastic Kelvin-Voigt damping ($\theta = 1/2$) and the dynamical friction ($\theta = 0$). We do not know whether *all* values of $0 < \theta < 1/2$ can be realized in real physical situations.

One can show (see CHUESHOV [40] for details) that Assumption 5.4.1 is satisfied if we impose the following hypotheses.

- The damping (σ) and the stiffness (ϕ) factors are positive C^1 functions on the semi-axis $\mathbb{R}_+ = [0, +\infty)$. Moreover,

$$\int_0^s \phi(\xi)d\xi \to +\infty \quad \text{as } s \to +\infty,$$

and $1/4 < \theta < 1/2$.
- $f(u)$ is a C^1 function such that $f(0) = 0$ (without loss of generality) and (a) if $d = 1$, then f is arbitrary; (b) if $d = 2$, then

$$|f'(u)| \leq C(1 + |u|^{p-1}) \quad \text{for some } p \geq 1;$$

(c) if $d = 3$, then

$$|f'(u)| \leq C(1 + |u|^{p-1}) \quad \text{with some } 1 \leq p \leq p_* < 2 + \frac{3}{4\theta - 1}.$$

- $\hat{\mu}_\phi \lambda_1 + \mu_f > 0$, where λ_1 is the first eigenvalue of the operator $-\Delta_D$ and

$$\hat{\mu}_\phi := \liminf_{s \to +\infty} \phi(s) > 0 \quad \text{and} \quad \mu_f := \liminf_{|s| \to \infty} \{s^{-1} f(s)\} > 0$$

(if $\hat{\mu}_\phi = +\infty$, then $\mu_f > -\infty$ can be arbitrary).

These hypotheses imply the statement of Theorem 5.1.4 for the model in (5.4.61). Moreover, this is the subcritical case, i.e., (5.4.53) holds. To employ Assumption 5.4.9, we need to assume only that

$$\inf \{f'(u) \ : \ u \in \mathbb{R}\} > -\infty.$$

Therefore, we have that both Proposition 5.4.10 and Theorem 5.4.14 hold true for the Kirchhoff wave model in (5.4.61). For more details concerning this example, we refer to [40].

Chapter 6
Delay Equations in Infinite-Dimensional Spaces

Our main goal in this chapter is to demonstrate how the method developed in Chapters 2 and 3 can be applied to study qualitative dynamics of abstract evolution equations containing delay terms. These equations naturally arise in various applications, such as viscoelasticity, nuclear reactors, heat flow, neural networks, combustion, interaction of species, microbiology, and many others. The theory of delay, or more generally, functional differential equations has been developed by many authors (see, e.g., the discussion and the references in the monographs DIEKMANN ET AL. [84], HALE [115] and WU [224]).

The general theory of delay equations in infinite-dimensional spaces started with FITZGIBBON [101] and TRAVIS/WEBB [218] at the abstract level and was developed in the last decades mainly for parabolic-type models with constant and time-dependent delays (see, e.g., the monograph WU [224] and the survey RUESS [200]). Abstract approaches for C-type (FITZGIBBON [101], TRAVIS/WEBB [218]) and L_p-type (KUNISCH/SCHAPPACHER [138]) phase spaces are available.

In this chapter we deal with two classes of models.

One of them is represented by a first order equation whose linear stationary part is a positive self-adjoint operator. In fact we consider some delay perturbations of the model studied in Sections 4.2 and 4.3. We deal with two types of perturbations. The first type is represented by Lipschitz mappings defined on the natural "history" space. The perturbations of the second type are more singular and include a case of a discrete state-dependent delay.

The second class of models is represented by second order in time evolution equations, which are similar to the ones considered in Chapter 5, but perturbed by delay terms. The hypotheses imposed on the delay forces in the latter case allow us to cover an important class of models with state-dependent delay.

In all cases, to study the long-time dynamics we rely on the quasi-stability method. In the delay case this method was applied earlier in

© Springer International Publishing Switzerland 2015
I. Chueshov, *Dynamics of Quasi-Stable Dissipative Systems*, Universitext,
DOI 10.1007/978-3-319-22903-4_6

CHUESHOV/LASIECKA/WEBSTER [63, 64] and CHUESHOV/REZOUNENKO [66] for second order models and in CHUESHOV/REZOUNENKO [67] for parabolic equations.

6.1   Models with parabolic main part and Lipschitz delay term

We consider abstract evolution equations of the form

$$u_t + Au = B(u^t), \quad u(s)\big|_{s \in [-h,0]} = u^0(s) = \varphi(s), \tag{6.1.1}$$

in a Hilbert space H, where A is a positive self-adjoint operator and $B(u^t)$ is a nonlinear mapping which is defined on pieces $u^t := \{u(t + s) : s \in [-h, 0]\}$ of an unknown function u and has its values in H, $\varphi : [-h, 0] \mapsto H$ is a given (initial) function.

Here and below h represents the (maximal) delay effect. We deal with arbitrary (but fixed) $0 < h < +\infty$. The case $h = +\infty$ is beyond the scope of the theory developed here.

Our main motivating example is a reaction-diffusion equation with (discrete) delay time h in the reaction term of the form

$$u_t(x, t) = \Delta u(x, t) + f(u(x, t - h)), \quad x \in \Omega \subset \mathbb{R}^d, \ t > 0.$$

We note that in order to state the problem rigorously for all time moments $t > 0$ we need to know prehistory, i.e., the values of the concentration $u(x, t)$ for $t \in [-h, 0]$. This leads to a Cauchy problem of the form (6.1.1) with initial data imposed on the interval $[-h, 0]$ and with nonlinear term B which is determined at given time t by the values of solutions in the time interval $t - h, t]$.

To obtain our well-posedness result below, we mainly follow the method suggested in TRAVIS/WEBB [219]; see also FITZGIBBON [101], TRAVIS/WEBB [218] and WU [224].

6.1.1   Well-posedness and generation of a dynamical system

Similar to Chapter 4 we impose the following hypotheses.

Assumption 6.1.1. We assume that H is a separable Hilbert space with the norm $\| \cdot \|$ and the inner product $(\cdot, \cdot)$ and

(A) A is a linear positive self-adjoint operator with discrete spectrum on H (see Definition 4.1.1). As in Section 4.1 we consider the scale of spaces H_s generated by powers A^s of the operator A.

(B) B is a (nonlinear) continuous[1] mapping from $\mathscr{C}_\alpha$ into H with $0 \le \alpha < 1$, where $\mathscr{C}_\alpha = C([-h, 0], H_\alpha)$ is a Banach space endowed with the norm

$$|v|_{\mathscr{C}_\alpha} \equiv \sup\{\| A^\alpha v(\theta) \|: \theta \in [-h, 0]\}.$$

We also assume that B is bounded on bounded sets in $\mathscr{C}_\alpha$, i.e.,

$$\forall R > 0, \ \exists C_R : \ \|B(v)\| \le C_R \ \text{for all} \ |v|_{\mathscr{C}_\alpha} \le R.$$

As in Chapter 4 we introduce the following definition.

Definition 6.1.2 (Mild solution). A function $u \in C([-h, T); H_\alpha)$ is said to be a *mild solution* to (6.1.1) on an interval $[0, T)$ if $u(t) = \varphi(t)$ for all $t \in [-h, 0]$ and

$$u(t) = e^{-tA}u(0) + \int_0^t e^{-(t-\tau)A} B(u^\tau)\, d\tau, \quad t \in [0, T). \tag{6.1.2}$$

Similarly we can define this notion for the closed interval $[0, T]$. Here and below we denote by u^t an element in $\mathscr{C}_\alpha = C([-h, 0]; H_\alpha)$ of the form $u^t(\theta) \equiv u(t + \theta)$, $\theta \in [-h, 0]$. ∎

In the standard way (see TRAVIS/WEBB [219]) we can prove the following result on the existence of local solutions, which is the main ingredient of the well-posedness result stated later.

Proposition 6.1.3 (Local existence). *Let Assumption 6.1.1 be in force. Then for every $\varphi \in \mathscr{C}_\alpha$ we can find $0 < T_{\max} \le \infty$ such that*

(a) *there is a mild solution $u(t)$ to problem (6.1.1) defined on $[-h, T_{\max})$;*
(b) *we have that either $T_{\max} = \infty$ or $\lim_{t \to T_{\max}-0} |u(t)|_{\mathscr{C}_\alpha} = \infty$.*

Proof. We use the same idea as in TRAVIS/WEBB [219] (see also WU [224, Section 2.2]) and, in contrast with the non-delay result presented in Theorem 4.2.3, we involve Schauder's fixed point theorem (see, e.g., ZEIDLER [231, Volume I, Chapter 2]) instead of the contraction principle. The point is that we now do not assume any Lipschitz conditions concerning B. In some sense this proposition is an infinite-dimensional analog of Theorem A.1.2.

Let $W_T = C([-h, T], H_\alpha)$ endowed with the corresponding sup-norm. For a given $\varphi \in \mathscr{C}_\alpha$ we define an element in W_T by the formula

$$\phi(t) = \begin{cases} \varphi(t), & \text{if } t \in [-h, 0]; \\ e^{-At}\varphi(0), & \text{if } t \in (0, T]. \end{cases}$$

---

[1] In contrast with Assumption 4.2.1, at this point we do not assume any Lipschitz properties for $B(u)$. See the comments in Remark 6.1.4.

It is obvious that $|\phi|_{W_T} \leq |\varphi|_{\mathscr{C}_\alpha}$. Now we assume that $|\varphi|_{\mathscr{C}_\alpha} \leq \rho$ for some $\rho > 0$, and on the ball

$$D_R = \{v \in W_T \; : \; |u|_{W_T} \leq R\} \subset W_T$$

we consider the mapping

$$\mathscr{B}[v](t) = \begin{cases} 0, & \text{if } t \in [-h, 0]; \\ \displaystyle\int_0^t e^{-(t-\tau)A} B(\phi^\tau + v^\tau)\, d\tau, & \text{if } t \in (0, T]. \end{cases}$$

If v is a fixed point for $\mathscr{B}$, then the function $u = \phi + v$ is a mild solution.
  For every $0 \leq t_1 \leq t_2 \leq T$ we have that

$$\mathscr{B}[v](t_2) - \mathscr{B}[v](t_1) = \int_{t_1}^{t_2} e^{-(t_2-\tau)A} B(\phi^\tau + v^\tau)\, d\tau$$

$$+ \int_0^{t_1} e^{-(t_1-\tau)A} \left[e^{-(t_2-t_1)A} - 1 \right] B(\phi^\tau + v^\tau)\, d\tau.$$

Due to (4.1.8), (4.1.9), and (4.1.10), this implies that

$$\|\mathscr{B}[v](t_2) - \mathscr{B}[v](t_1)\|_\alpha \leq \left(\frac{\alpha}{e}\right)^\alpha |t_2 - t_1|^{1-\alpha}(1-\alpha)^{-1} C_{R+\rho}$$

$$+ |t_2 - t_1|^\beta [1 + \varkappa_{\alpha+\beta}] \lambda_1^{-(1-\alpha-\beta)} C_{R+\rho}$$

for every $0 \leq \beta < 1 - \alpha$, where $C_r = \max\{\|B(v)\| \; : \; |v|_{\mathscr{C}_\alpha} \leq r\}$ and

$$\varkappa_\alpha = \alpha^\alpha \int_0^\infty \xi^{-\alpha} e^{-\xi}\, d\xi.$$

Thus $\mathscr{B}[v]$ is Hölder in the sense that

$$|\mathscr{B}[v]|_{H_\alpha^\beta(0,T)} \leq C_{\alpha,\beta}(R, \rho) < \infty, \quad v \in D_R, \tag{6.1.3}$$

where

$$|v|_{H_\alpha^\beta(0,T)} \equiv \sup \left\{ \frac{\|v(t_1) - v(t_2)\|_\alpha}{|t_1 - t_2|^\beta} \; \middle| \; \begin{array}{c} t_1, t_2 \in [0, T] \\ |t_1 - t_2| \leq 1 \end{array} \right\}.$$

In particular, this means that $\mathscr{B}[u] \in W_T$ and thus $\mathscr{B} : D_R \mapsto W_T$. It is also easy to see that this mapping is continuous in the W_T-topology.

For all $t \in [-h, T]$ using (4.1.9) we have that

$$\|\mathscr{B}[u](t)\|_\beta \leq \left(\frac{\beta}{e}\right)^\beta T^{1-\beta}(1-\beta)^{-1} C_{R+\rho}, \quad u \in D_R, \tag{6.1.4}$$

for every $0 \leq \beta < 1$. Thus taking T such that

$$\left(\frac{\alpha}{e}\right)^\alpha T^{1-\alpha}(1-\alpha)^{-1} C_{R+\rho} \leq R,$$

we obtain that the ball D_R is invariant with respect to $\mathscr{B}$.

Now we show that the closure $\overline{\mathscr{B}(D_R)}$ of $\mathscr{B}(D_R)$ is compact in the W_T-topology. Indeed, applying (6.1.4) with $\beta > \alpha$ we obtain that the set

$$\mathscr{B}_t(D_R) = \{v(t) \, : \, v \in \mathscr{B}(D_R)\}$$

belongs to a bounded set in H_β for each $t \in [-h, T]$. By Proposition 4.1.6 H_β is compactly embedded into H_α. Therefore $\mathscr{B}_t(D_R)$ lies in a compact set of H_α for every t. By (6.1.3) $\mathscr{B}(D_R)$ is an equicontinuous collection on $[-h, T]$. Thus we can apply the Arzelà-Ascoli theorem (see Lemma A.3.5 in the Appendix) to show that $\mathscr{B}(D_R)$ is compact in W_T. Therefore by Schauder's fixed point theorem[2] $\mathscr{B}$ has a fixed point in D_R which defines a solution.

To complete the proof we use the same argument as in Theorem 4.2.3.

If a solution exists on a closed interval $[0, T]$, then by the previous argument we can extend it on the interval $[0, T + \delta]$, where δ depends on an upper bound for $|u^T|_{\mathscr{C}_\alpha}$, where, as above, $u^T = \{u(T + \theta) \, : \, \theta \in [-h, 0]\}$. This means that there is a maximal existence interval $[0, T_{max})$. If $T_{max} < +\infty$ and

$$\lim_{t \to T_{max}-0} |u^t|_{\mathscr{C}_\alpha} = +\infty \text{ is not true,}$$

then there exists a sequence $T_n \to T_{max} - 0$ and a number R_* such that $|u^{T_n}|_{\mathscr{C}_\alpha} \leq R_*$ for all $n = 1, 2, \ldots$. Thus using u^{T_n} as an initial data we can extend the solution to an interval $[0, T_n + \delta]$ for some $\delta > 0$, which does not depend on n. Since $T_n \to T_{max}$, this means that we are able to extend the solution beyond T_{max}.

This completes the proof of Proposition 6.1.3. $\qquad\qquad\square$

Remark 6.1.4. The argument above can be applied in the non-delay case considered in Section 4.2. In this case Proposition 6.1.3 presents another way to obtain the local existence result stated in Theorem 4.2.3 even without assuming the Lipschitz

---

[2] The theorem asserts (see, e.g., ZEIDLER [231, Volume I, Chapter 2]) that if K is a convex subset of a Banach space and T is a continuous mapping of K into itself such that $T(K)$ is contained in a compact subset of K, then T has a fixed point.

property (4.2.2). The principal role is played by the fact that the linear part of the problem generates an analytic semigroup with smoothening properties (see Section 4.1).   ∎

Using the non-explosion condition in Proposition 6.1.3 we can provide the following result on the existence of global solutions.

Exercise 6.1.5. Let the hypotheses of Proposition 6.1.3 be in force. Assume that the nonlinearity $B(v)$ is linearly bounded, i.e., there exist constants C_1 and C_2 such that

$$\|B(v)\| \leq C_1 + C_2 |v|_{\mathscr{C}_\alpha}, \quad \forall\, v \in \mathscr{C}_\alpha.$$

Then every local solution can be extended on the whole $\mathbb{R}_+$. Hint: Use (6.1.2), (4.1.9), and the step-by-step procedure as in Section 4.2 or in the proof of Theorem 6.1.6 below.   ∎

Now we are in position to establish the main result of the section.

Theorem 6.1.6 (Well-posedness). *Let Assumption 6.1.1 be in force. Assume in addition that B is a (nonlinear) locally Lipschitz mapping from $\mathscr{C}_\alpha$ into H with $0 \leq \alpha < 1$, i.e., we assume that for every $\rho > 0$ there exists L_ρ such that*

$$\|B(v_1) - B(v_2)\| \leq L_\rho |v_1 - v_2|_{\mathscr{C}_\alpha}, \quad v_i \in \mathscr{C}_\alpha, \quad |v_i|_{\mathscr{C}_\alpha} \leq \rho, \quad i = 1, 2. \qquad (6.1.5)$$

Then the local solution given by Proposition 6.1.3 is unique. Moreover, any two mild solutions $u_1(t)$ and $u_2(t)$ with initial data u_1^0 and u_2^0 on the joint interval $[0, T]$ of the existence admit the estimate

$$\| u_1(t) - u_2(t) \|_\alpha \leq C_T(R) |u_1^0 - u_2^0|_{\mathscr{C}_\alpha}, \quad t \in [0, T], \qquad (6.1.6)$$

under the condition $\sup_{[-h,T]} \|u_j(t)\|_\alpha \leq R$.

If $B(u)$ is globally Lipschitz, i.e., (6.1.5) is satisfied with $L_\rho \equiv L$ independent of ρ, then for every $u^0 \in \mathscr{C}_\alpha$ there exists a unique mild solution to (6.1.1) for every interval $[0, T]$. In this case (6.1.6) can be written in the form

$$\| u_1(t) - u_2(t) \|_\alpha \leq A e^{\omega t} |u_1^0 - u_2^0|_{\mathscr{C}_\alpha}, \quad t > 0, \qquad (6.1.7)$$

for every pair of mild solutions $u_1(t)$ and $u_2(t)$, where $A > 0$ and $\omega \geq 0$ are constants.

Proof. We need to show the Lipschitz properties in (6.1.6) and (6.1.7) only. We first note that (6.1.2) and also (6.1.5) and (4.1.9) imply that

$$\|u_1(t) - u_2(t)\|_\alpha \leq \|u_1^0(0) - u_2^0(0)\|_\alpha + C_{R,\alpha} \int_0^t \frac{1}{(t - \tau)^\alpha} |u_1^\tau - u_2^\tau|_{\mathscr{C}_\alpha}\, d\tau.$$

This yields

$$\max_{[0,\tilde{T}]} \|u_1^t - u_2^t\|_{\mathscr{C}_\alpha} \le |u_1^0 - u_2^0|_{\mathscr{C}_\alpha} + C_{R,\alpha}\tilde{T}^{1-\alpha}(1-\alpha)^{-1} \max_{[0,\tilde{T}]} \|u_1^t - u_2^t\|_{\mathscr{C}_\alpha}$$

for every $\tilde{T} \le T$. Thus we can choose $\tilde{T} = \tilde{T}(R,\alpha)$ such that

$$\max_{[0,\tilde{T}]} \|u_1^t - u_2^t\|_{\mathscr{C}_\alpha} \le 2|u_1^0 - u_2^0|_{\mathscr{C}_\alpha}. \tag{6.1.8}$$

Therefore, applying the step-by-step procedure we obtain (6.1.6). In the globally Lipschitz case the constant $C_{R,\alpha}$ does not depend on R. This implies the same property for $\tilde{T}$. Thus (6.1.8) yields (6.1.7). $\qquad\square$

Now we impose a hypothesis which guarantees the global solvability of problem (6.1.1) in the locally Lipschitz case. This set of hypotheses is a delay version of Assumption 4.2.20.

Assumption 6.1.7. Assumption 6.1.1 and relation (6.1.5) hold with $\alpha = 1/2$. Moreover we assume that

$$B(v) = -B_0(v(0)) + B_1(v),$$

where $B_1 : \mathscr{C}_{1/2} \mapsto H$ is *linearly* bounded, i.e.,

$$\|B_1(v)\| \le c_1 + c_2|v|_{\mathscr{C}_{1/2}}, \quad v \in \mathscr{C}_{1/2},$$

and $B_0 : H_{1/2} \mapsto H$ is a potential operator on the space $H_{1/2}$; i.e., there exists a Frechét differentiable functional $\Pi(u)$ on $H_{1/2}$ such that $B_0(u) = \Pi'(u)$ in the sense of relation (4.2.30). See also Section A.5 in the Appendix.

Roughly speaking, we assume that the nonlinearity is split into a non-delay potential part and a globally Lipschitz delay term.

Theorem 6.1.8. *Let Assumption 6.1.7 be in force. Assume that the functional $\Pi(u)$ possesses the property: there exist $\beta < 1/2$ and $\gamma \ge 0$ such that*

$$\beta\|A^{1/2}u\|^2 + \Pi(u) + \gamma \ge 0, \quad \forall u \in H_{1/2}. \tag{6.1.9}$$

Then for every $\varphi \in \mathscr{C}_{1/2}$ problem (6.1.1) has a unique mild solution on $\mathbb{R}_+$ lying in the space $C([-h,+\infty); H_{1/2})$.

Proof. The following argument is formal. To justify it we can use calculations with the projection $P_N u$ as in the end of the proof of Theorem 4.2.22.

If we multiply (6.1.1) by u_t, then we obtain that

$$\|u_t(t)\|^2 + \frac{d}{dt}\left[\frac{1}{2}\|A^{1/2}u(t)\|^2 + \Pi(u(t))\right]$$

$$= (B_1(u^t), u_t(t)) \le \frac{1}{2}\|u_t(t)\|^2 + c_1^2 + c_2^2|u^t|_{\mathscr{C}_{1/2}}^2.$$

It is easy to see that

$$W(u) := \frac{1}{2}\|A^{1/2}u\|^2 + \Pi(u) + \gamma \geq \left(\frac{1}{2} - \beta\right)\|A^{1/2}u\|^2$$

and satisfies the inequality

$$W(u(t)) \leq W(u^0(0)) + c_1^2 T + a_2^2 \int_0^t |u^\tau|^2_{\mathscr{C}_{1/2}} d\tau$$

on any interval $[0, T]$ with some $a_i > 0$. Therefore,

$$\max_{s \in [0,t]} \|A^{1/2}u(s)\|^2 \leq C_\rho(T) + C_1 \int_0^t \max_{s \in [0,\tau]} \|A^{1/2}u(t)\|^2 d\tau$$

on any interval $[0, T]$ under the condition that $|u^0|_{\mathscr{C}_{1/2}} \leq \rho$. This relation allows us to apply Gronwall's lemma and use the non-explosion criterion stated in Proposition 6.1.3. This yields the desired conclusion.   □

Remark 6.1.9. As in Section 4.2.3 we can apply the Galerkin method to study the problem in (6.1.1). Under the conditions of Theorem 6.1.6 we can prove the corresponding analog of Theorem 4.2.19 on convergence of approximations and give an alternative argument in the proof of Theorem 6.1.8. We do not pursue these issues for the model discussed now, and postpone them to Section 6.2, which is devoted to a state-dependent delay.   ∎

The results above allow us to construct a dynamical system. Indeed, let the conditions of Theorem 6.1.6 be in force. Assume in addition that for every initial data $\varphi \in \mathscr{C}_\alpha$ the corresponding solution $u(t)$ to (6.1.1) exists globally (e.g., the hypotheses of Theorem 6.1.8 are valid). Following the standard procedure (see, e.g., [115] or [224]) we can define a family of mappings $S_t : \mathscr{C}_\alpha \mapsto \mathscr{C}_\alpha$ by the formula

$$[S_t\varphi](\theta) = u(t + \theta), \quad \varphi \in \mathscr{C}_\alpha, \tag{6.1.10}$$

where $u(t)$ solves (6.1.1) with the initial data φ. One can see from Theorem 6.1.6 that

- for each $t \in \mathbb{R}_+$ the mapping S_t is continuous on $\mathscr{C}_\alpha$;
- the family $\{S_t\}_{t \in \mathbb{R}_+}$ satisfies the semigroup property;
- the function $t \mapsto S_t\varphi$ is continuous in $\mathscr{C}_\alpha$ for every $\varphi \in \mathscr{C}_\alpha$.

Thus problem (6.1.1) generates (see Definition 1.1.1) a dynamical system $(\mathscr{C}_\alpha, S_t)$ with phase space $\mathscr{C}_\alpha$ and evolution operator S_t.

6.1.2 Global and exponential attractors

Now we consider the long-time dynamics of the dynamical system $(\mathscr{C}_\alpha, S_t)$ generated by (6.1.1). Our first goal is to show that the system $(\mathscr{C}_\alpha, S_t)$ possesses some compactness and quasi-stability properties. To do this, in the space $\mathscr{C}_\alpha$ we introduce the following linear set:

$$\mathscr{Y}_\beta = \left\{ v \in \mathscr{C}_\alpha \; : \; |v|_{\mathscr{Y}_\beta} < \infty \right\},$$

where $\beta \in (\alpha, 1)$ and

$$|v|_{\mathscr{Y}_\beta} = \max_{\theta \in [-h,0]} \|A^\beta v(\theta)\| + \max_{\theta_1, \theta_2 \in [-h,0]} \frac{\|A^\alpha [v(\theta_1) - v(\theta_2)]\|}{|\theta_1 - \theta_2|^{\beta - \alpha}}.$$

This space $\mathscr{Y}_\beta$ can be written as

$$\mathscr{Y}_\beta = C([-h,0]; H_\beta) \cap C^{\beta - \alpha}([-h,0]; H_\alpha),$$

where $C^\nu([-h,0]; H_\alpha)$ denotes the corresponding Hölder space (see Section A.3.1 in the Appendix). By the Arzelà-Ascoli theorem (see Lemma A.3.5), $\mathscr{Y}_\beta$ is a Banach space which is compactly embedded in $\mathscr{C}_\alpha$ for $\alpha < \beta$.

Proposition 6.1.10 (Conditional compactness and quasi-stability). *Let the hypotheses of Theorem 6.1.6 be in force. This means that (a) Assumption 6.1.1 is valid, and (b) $B : \mathscr{C}_\alpha \mapsto H$ is a locally Lipschitz mapping from with $0 \leq \alpha < 1$ and for every $\rho > 0$ there exists L_ρ such that (6.1.5) holds. Assume that the problem (6.1.1) generates a dynamical system $(\mathscr{C}_\alpha, S_t)$. Let D be a forward invariant bounded set in $\mathscr{C}_\alpha$. Then*

(1) *For every $t > h$ the set $S_t D$ is bounded in $\mathscr{Y}_\beta$ for arbitrary $\beta \in (\alpha, 1)$. Moreover, for every $\delta > 0$ there exists R_δ such that*

$$S_t D \subset B_\beta = \{ u \in \mathscr{Y}_\beta \; : \; |u|_{\mathscr{Y}_\beta} \leq R_\delta \} \quad \text{for all } t \geq \delta + h. \tag{6.1.11}$$

In particular, this means that the system $(\mathscr{C}_\alpha, S_t)$ is conditionally compact and thus asymptotically smooth (see Definition 2.2.1).

(2) *The mapping S_t is Lipschitz from D into $\mathscr{Y}_\beta$. Moreover, for every $h < a < b < +\infty$ there exists a constant $M_D(a, b)$ such that*

$$|S_t u^0 - S_t \tilde{u}^0|_{\mathscr{Y}_\beta} \leq M_D(a, b)|u^0 - \tilde{u}^0|_{\mathscr{C}_\alpha}, \quad t \in [a, b], \quad u^0, \tilde{u}^0 \in D. \tag{6.1.12}$$

In particular, this means that the system $(\mathscr{C}_\alpha, S_t)$ is quasi-stable at any time from the interval $[a, b]$ (see Exercise 3.4.2).

Proof. Let $\beta \in (\alpha, 1)$ and $u(t) = S_t u^0$ be a solution to (6.1.1). It follows from (6.1.2) and (4.1.9) that

$$\|u(t)\|_\beta \le \left(\frac{\beta - \alpha}{e(t-s)}\right)^{\beta-\alpha} \|u(s)\|_\alpha + \int_s^t \left(\frac{\beta}{e(t-\tau)}\right)^\beta \|B(u^\tau)\| \, d\tau$$

for all $t > s > 0$. Since $S_t u^0 \in D$ for all $t \ge 0$ we have that $\|u(t)\|_\alpha \le C_D$ for all $t \ge 0$. Thus,

$$\|u(t)\|_\beta \le \left(\frac{\beta - \alpha}{e(t-s)}\right)^{\beta-\alpha} C_D + K_D(B) \left(\frac{\beta}{e}\right)^\beta \frac{|t-s|^{1-\beta}}{1-\beta} \tag{6.1.13}$$

for all $t > s \ge 0$, where $K_D(B) = \sup\{\|B(v)\| : v \in D\}$. Taking $s = t - \delta$ we obtain that

$$S_t D \subset \{u \in \mathscr{C}_\beta : \|u\|_\beta \le R_\delta^*\} \quad \text{for all } t \ge \delta + h,$$

where

$$R_\delta^* = \left(\frac{\beta - \alpha}{e\delta}\right)^{\beta-\alpha} C_D + K_D(B) \left(\frac{\beta}{e}\right)^\beta \frac{\delta^{1-\beta}}{1-\beta}.$$

As in the proof of equicontinuity in Proposition 6.1.3, using the representation

$$u(t_2) - u(t_1) = \left[e^{-(t_2-t_1)A} - 1\right] u(t_1) + \int_{t_1}^{t_2} e^{-(t_2-\tau)A} B(u^\tau) \, d\tau, \quad t_2 \ge t_1 > 0,$$

and the Hölder continuity of the operator exponent in relation (4.1.8) we can show that

$$\|u(t_2) - u(t_1)\|_\alpha \le |t_2 - t_1|^{\beta-\alpha} \|u(t_1)\|_\beta + K_D(B) \left(\frac{\alpha}{e}\right)^\alpha |t_2 - t_1|^{1-\alpha}(1-\alpha)^{-1}$$

for every $\alpha \le \beta < 1$ and $t_2, t_1 > 0$. This implies that

$$\sup\left\{ \frac{\|u(t_2) - u(t_1)\|_\alpha}{|t_2 - t_1|^{\beta-\alpha}} \;\middle|\; \begin{array}{l} t_1, t_2 \in [\delta, T] \\ |t_1 - t_2| \le 1 \end{array} \right\} \le C_{D,\delta} \quad \text{for every } \delta > 0.$$

Therefore (6.1.11) follows.

Now we prove the second part of the statement. Let $u(t)$ and $u_*(t)$ be two solutions with initial data from D and $w(t) = u(t) - u_*(t)$. Using (4.1.9) one can see that there exists a constant $C_D > 0$ such that

$$\|w(t)\|_\beta \le \left(\frac{\beta - \alpha}{e(t-s)}\right)^{\beta-\alpha} \|w(s)\|_\alpha + C_D \int_s^t \left(\frac{\beta}{t-\tau}\right)^\beta |w^\tau|_{\mathscr{C}_\alpha} \, d\tau \tag{6.1.14}$$

for all $t > s \ge 0$. Using (6.1.6) we obtain that

$$\|w(t)\|_\beta \leq C_D(a,b)|w^s|_{\mathscr{C}_\alpha}, \quad t \in [s+a, s+b], \text{ for all } s \geq 0 \text{ and } a > 0. \quad (6.1.15)$$

Let $t_1, t_2 \in [a, b]$ and $t_2 > t_1$. Using the representation

$$w(t_2) = e^{-(t_2-t_1)A}w(t_1) + \int_{t_1}^{t_2} e^{-(t_2-\tau)A}[B(u^\tau) - [B(u^\tau_*)]\,d\tau,$$

one can see that

$$\|w(t_2) - w(t_1)\|_\alpha \leq \| \left[e^{-(t_2-t_1)A} - 1\right]w(t_1)\|_\alpha + C_D \int_{t_1}^{t_2} \left(\frac{\alpha}{t-\tau}\right)^\alpha |w^\tau|_{\mathscr{C}_\alpha}\,d\tau.$$

Therefore from (4.1.8) and (6.1.6) we obtain

$$\|w(t_2) - w(t_1)\|_\alpha \leq |t_2 - t_1|^{\beta-\alpha}\|w(t_1)\|_\beta + C_D(b)\left\{\int_{t_1}^{t_2} \left(\frac{\alpha}{t_2-\tau}\right)^\alpha d\tau\right\}|w^0|_{\mathscr{C}_\alpha}$$

Thus (6.1.15) with $s = 0$ yields

$$\|w(t_2) - w(t_1)\|_\alpha \leq C_D(a,b)\left[|t_2 - t_1|^{\beta-\alpha} + |t_2 - t_1|^{1-\alpha}\right]|w^0|_{\mathscr{C}_\alpha}$$

for all $t_1, t_2 \in [a, b]$. This and (6.1.15) with $s = 0$ give (6.1.12) and conclude the proof. □

Exercise 6.1.11. Let the hypotheses of Proposition 6.1.10 be in force. Show that any bounded semitrajectory for $(\mathscr{C}_\alpha, S_t)$ is a relatively compact set. ■

Proposition 6.1.10 allows us to state the existence of global and exponential attractors for $(\mathscr{C}_\alpha, S_t)$ under the condition that this system is dissipative.[3]

Theorem 6.1.12 (Global attractor). *Let (a) Assumption 6.1.1 be valid, and (b)* $B : \mathscr{C}_\alpha \mapsto H$ *be a locally Lipschitz mapping with* $0 \leq \alpha < 1$ *and for every* $\rho > 0$ *there exists* L_ρ *such that (6.1.5) holds. Assume that problem (6.1.1) generates a dissipative dynamical system* $(\mathscr{C}_\alpha, S_t)$. *Then this system possesses a compact global attractor. This attractor is a bounded set in the space* $\mathscr{Y}_\beta$ *for every* $\beta \in (\alpha, 1)$ *and has a finite fractal dimension.*

Proof. Since $(\mathscr{C}_\alpha, S_t)$ is dissipative, by Proposition 6.1.10 the system $(\mathscr{C}_\alpha, S_t)$ is compact (see Exercise 2.2.3). Thus the existence of a compact global attractor follows from Theorem 2.3.5. It is also clear from Proposition 6.1.10 that the attractor is bounded in $\mathscr{Y}_\beta$. To prove finiteness of the fractal dimension we use Theorem 3.4.5. □

The next outcome of Proposition 6.1.10 is the existence of a fractal exponential attractor.

---

[3] We refer to Theorem 6.1.15 below for sufficient conditions of dissipativity.

Theorem 6.1.13 (Fractal exponential attractor). *Under the hypotheses of Theorem 6.1.12 the system $(\mathscr{C}_\alpha, S_t)$ generated by (6.1.1) possesses a fractal exponential attractor $\mathfrak{A}_{exp}$ whose dimension is finite in $\mathscr{C}_\alpha$.*

Proof. By Proposition 6.1.10 the system is quasi-stable at every time $t_* > h$ on any (forward invariant) absorbing set. Thus we can apply Theorem 3.4.7. Due to (6.1.11) there exists a forward invariant absorbing set which belongs to $\mathscr{Y}_\beta$ for $\beta > \alpha$. Thus every trajectory from this set is Hölder continuous in $\mathscr{C}_\alpha$. Thus $\dim_f \mathfrak{A}_{exp}$ is finite in $\mathscr{C}_\alpha$. ☐

Using Proposition 6.1.10 we can also obtain a result on determining functionals for every pair of bounded solutions, even without assuming dissipativity of the system.

Theorem 6.1.14 (Determining functionals). *Let the hypotheses of Theorem 6.1.6 be in force, that is (a) Assumption 6.1.1 is valid, and (b) $B : \mathscr{C}_\alpha \mapsto H$ is a locally Lipschitz mapping with $0 \leq \alpha < 1$ and for every $\rho > 0$ there exists L_ρ such that (6.1.5) holds. Let $u(t)$ and $u_*(t)$ be two solutions to problem (6.1.1) (with different initial data) on $\mathbb{R}_+$ such that*

$$\limsup_{t \to +\infty} (\|u(t)\|_\alpha + \|u_*(t)\|_\alpha) < R \text{ for some } R > 0. \tag{6.1.16}$$

Let $\mathscr{L} = \{l_j : j = 1, \ldots, N\}$ be a set of functionals on H_β for some $\beta \in (\alpha, 1)$ with the completeness defect $\varepsilon_{\mathscr{L}}(\beta, \alpha) \equiv \varepsilon_{\mathscr{L}}(H_\beta, H_\alpha)$. Then there exists $\varepsilon_0 = \varepsilon_0(R)$ such that under the condition $\varepsilon_{\mathscr{L}}(\beta, \alpha) < \varepsilon_0$ the set $\mathscr{L}$ is asymptotically determining, i.e., the property

$$l_j(u(t)) - l_j(u_*(t)) \to 0 \text{ as } t \to +\infty \text{ for all } j = 1, \ldots, N$$

implies that $\|u(t) - u_*(t)\|_\alpha \to 0$ as $t \to +\infty$.

Proof. By (6.1.16) there exists $s_0 \in \mathbb{R}_+$ such that

$$\|u(t)\|_\alpha + \|u_*(t)\|_\alpha \leq R \text{ for all } t \geq s_0.$$

Let $w(t) = u(t) - u_*(t)$. Using the property of the completeness defect stated in Proposition 3.3.4 and also (6.1.15), we can conclude that

$$\|w(t)\|_\alpha \leq \varepsilon_{\mathscr{L}}(\beta, \alpha)\|w(t)\|_\beta + C_{\mathscr{L}} \max_j |l_j(w(t))|$$

$$\varepsilon_{\mathscr{L}}(\beta, \alpha)C_R(a, b)\|w^s\|_{\mathscr{C}_\alpha} + C_{\mathscr{L}} \max_j |l_j(w(t))|$$

for all $0 < a \leq t - s \leq b$ with $s \geq s_0$. This yields

$$\|w^t\|_{\mathscr{C}_\alpha} \leq \varepsilon_{\mathscr{L}}(\beta, \alpha)C_R(a, b)\|w^s\|_{\mathscr{C}_\alpha} + C_{\mathscr{L}} \max_{\theta \in [-h, 0]} \max_j |l_j(w(t + \theta))|$$

for all $0 < a + h + s \le t \le s + b$ with $s \ge s_0$. If we take $a = h$, $b = 2h$, and $t = s + 2h$ with $s = s_0 + 2(m-1)h$, then we obtain

$$\|w^{s_0+2mh}\|_{\mathscr{C}_\alpha} \le q\|w^{s_0+2(m-1)h}\|_{\mathscr{C}_\alpha} + C_{\mathscr{L}} \max_{\theta \in [-h,0]} \max_j |l_j(w(s_0 + 2mh + \theta))|$$

for every $m = 1, 2, \ldots$, where $q := \varepsilon_{\mathscr{L}}(\beta, \alpha)C_R(a,b)$. If we choose $\varepsilon_{\mathscr{L}}(\beta, \alpha)$ sufficiently small such that $q < 1$, then after iterations in the same way as was done in the proof of Theorem 3.4.12 we can conclude that $\|w(t)\|_\alpha \to 0$ as $t \to +\infty$. $\quad\square$

Now we give conditions under which equation (6.1.1) generates a *dissipative* dynamical system.

Theorem 6.1.15 (Dissipativity). *Assume that* $B(v) = -B_0(v(0)) + B_1(v)$ *and Assumption 6.1.7 is in force. Assume in addition that*

- *the potential* $\Pi(u)$ *is bounded from below,*[4] *i.e., there exists* $\gamma \ge 0$ *such that*

$$\Pi(u) \ge -\gamma, \quad \forall\, u \in H_{1/2}; \tag{6.1.17}$$

- *there exist* $\delta > 0$ *and* $c \ge 0$ *such that*

$$- (B_0(u), u) \le \frac{1}{2}\|A^{1/2}u\|^2 - \delta\Pi(u) + c, \quad \forall\, u \in H_{1/2}; \tag{6.1.18}$$

- *there exist* $\varkappa > 0$ *and* $c \ge 0$ *such that*

$$\|B_1(v)\|^2 \le c + \varkappa \int_{-h}^0 \|A^{1/2}v(\theta)\|^2\sigma(d\theta) \quad \text{for all } v \in \mathscr{C}_{1/2}, \tag{6.1.19}$$

where $\sigma(d\theta)$ *is a Borel measure on* $[-h, 0]$ *such that* $\sigma([-h, 0]) = 1$.

Then problem (6.1.1) generates a dissipative dynamical system $(H_{1/2}, S_t)$ *provided* $\varkappa < \lambda_1(2\lambda_1 + 8)^{-1}$.

Remark 6.1.16 (Admissible structure of the delay term). Condition (6.1.19) concerning the delay term B_1 admits both point and distributed delays. For instance, we can consider

$$B_1(v) = g\left(\sum_{i=1}^m c_i A^{\beta_i}v(-h_i) + \int_{-h_0}^0 A^{\beta_0}v(-\theta)f(\theta)d\theta\right), \quad v \in \mathscr{C}_{1/2},$$

---

[4] We can relax this condition by changing (6.1.17) into (6.1.9). However, this requires some additional calculations and leads, which is more important, to a smaller interval of admissible values of the intensity parameter $\varkappa$ in the bound for the delay term in (6.1.19). We do not pursue this case and leave the details for the readers.

where g is a globally Lipschitz mapping on H, $c_j \in \mathbb{R}$, $\beta_i \in [0, 1/2]$, $h_i \geq 0$ are fixed constants, and $f \in L_1(-h_0, 0)$ is a real function. The corresponding constant $\varkappa$ in (6.1.19) can be controlled by the Lipschitz constant of g and also by the parameters $\mu_1 = \sum_{i=1}^{m} |c_j|$ and $\mu_2 = \|f\|_{L_1(-h_0, 0)}$. Moreover, there are no restrictions on the parameters μ_1 and μ_2 in the case when the mapping g is sublinear, i.e., it satisfies the inequality

$$\|g(u)\| \leq a_1 + a_2 \|u\|^{\gamma}, \quad \forall u \in H, \quad \text{with } 0 \leq \gamma < 1.$$

In this case for every $\varkappa > 0$ we can find $c = c_{\varkappa}$ such that (6.1.19) is valid. We also note that the restriction concerning the intensity parameter $\varkappa$ in the statement of Theorem 6.1.15 is not surprising. We refer to Exercise 6.1.17 below, which demonstrates different types of behaviors depending on the intensity of a delay term. See also Remark 6.3.12 below, where a similar effect is discussed for second order in time models.   ∎

Exercise 6.1.17. Consider the following delay ODE:

$$\dot{x} + x - \varkappa \cdot x(t - 1) = 0. \tag{6.1.20}$$

Show that

(A) If $\varkappa > 1$, then there exists $\lambda_* > 0$ such that $x(t) = e^{\lambda_* t}$ solves (6.1.20) on $\mathbb{R}_+$. Thus (6.1.20) has unbounded solutions.
(B) If $0 < \varkappa < 1$, then any solution to (6.1.20) is bounded on $\mathbb{R}_+$. Hint: Show that the function

$$V(t) = \frac{1}{2} \left([x(t)]^2 + \int_{t-1}^{t} [x(\tau)]^2 d\tau \right)$$

does not increase along solutions.   ∎

Proof of Theorem 6.1.15. The following calculations can be justified on Galerkin approximations.

Multiplying equation (6.1.1) by u we obtain that

$$\frac{1}{2}\frac{d}{dt}\|u(t)\|^2 + \|A^{1/2}u(t)\|^2 + (B_0(u(t)), u(t)) = (B_1(u^t), u(t))$$

$$\leq \eta\|A^{1/2}u(t)\|^2 + \frac{c}{4\eta\lambda_1} + \frac{\varkappa}{4\eta\lambda_1}\int_{-h}^{0}\|A^{1/2}u(t+\theta)\|^2\sigma(d\theta) \tag{6.1.21}$$

for every $\eta > 0$. Using the multiplier u_t as in the proof of Theorem 6.1.8, we also have

$$\|u_t(t)\|^2 + \frac{d}{dt}\left[\frac{1}{2}\|A^{1/2}u(t)\|^2 + \Pi(u(t))\right] = (B_1(u^t), u_t(t))$$

$$\leq \|u_t(t)\|^2 + \frac{c}{4} + \frac{\varkappa}{4}\int_{-h}^0 \|A^{1/2}u(t+\theta)\|^2\sigma(d\theta). \qquad (6.1.22)$$

Now we consider the function

$$\mathscr{W}(t) = \frac{1}{2}\|u(t)\|^2 + \frac{1}{2}\|A^{1/2}u(t)\|^2 + \Pi(u(t)) + \gamma + \mu\mathscr{W}_0(t),$$

where μ is a positive parameter and

$$\mathscr{W}_0(t) = \frac{1}{h}\int_{-h}^0 ds \int_{t+s}^t \|A^{1/2}u(\theta)\|^2 d\theta + \int_{-h}^0 \sigma(ds)\int_{t+s}^t \|A^{1/2}u(\theta)\|^2 d\theta$$

We note that the main idea behind inclusion of the additional delay term $\mathscr{W}_0$ is to compensate the contribution from $B_1(u^t)$. This idea goes back to the considerations in [115] and was already used in infinite dimensions (see, e.g., CHUESHOV/LASIE-CKA [58, p. 480] and also CHUESHOV/LASIECKA/WEBSTER [63], CHUESHOV/RE-ZOUNENKO [66, 67]). The corresponding compensator is model-dependent. Below in Sections 6.2 and 6.3 we demonstrate this effect for other models.

It is clear that

$$0 \leq \mathscr{W}_0(t) \leq 2\int_{t-h}^t \|A^{1/2}u(s)\|^2 ds \leq 2h|u^t|^2_{\mathscr{C}_{1/2}}$$

and

$$\frac{\mathscr{W}_0(t)}{dt} = 2\|A^{1/2}u(t)\|^2 - \frac{1}{h}\int_{-h}^0 \|A^{1/2}u(t+s)\|^2 ds - \int_{-h}^0 \|A^{1/2}u(t+s)\|^2\sigma(ds).$$

Since by Remark 4.2.21 $\Pi(u)$ is bounded on every bounded set, we conclude from (6.1.17) that

$$\frac{1}{2}\|A^{1/2}u(t)\|^2 \leq \mathscr{W}(t) \leq \phi(\|A^{1/2}u\|) + 2\mu h|u^t|^2_{\mathscr{C}_{1/2}}, \qquad (6.1.23)$$

where $\phi(r) = \gamma + ar^2 + \sup\{|\Pi(u)| : \|u\|_{1/2} \leq r\}$ for some $a > 0$. It also follows from (6.1.21) and (6.1.22) that

$$\frac{d}{dt}\mathscr{W}(t) + \nu\mathscr{W}(t) \leq -(1 - \eta - 2\mu)\|A^{1/2}u(t)\|^2 - (B_0(u(t)), u(t))$$

$$+ \frac{\nu}{2}\left[\|u(t)\|^2 + \|A^{1/2}u(t)\|^2 + 2\Pi(u(t))\right]$$

$$+ \left[-\frac{\mu}{h} + 2\nu\mu \right] \int_{-h}^{0} \|A^{1/2}u(t+\theta)\|^2 d\theta$$

$$+ \left[-\mu + \frac{\varkappa}{4\eta\lambda_1} + \frac{\varkappa}{4} \right] \int_{-h}^{0} \|A^{1/2}u(t+\theta)\|^2 \sigma(d\theta) + \gamma\nu + C_\eta$$

By (6.1.18) this implies that

$$\frac{d}{dt}\mathscr{W}(t) + \nu\mathscr{W}(t) \leq - \left(\frac{1}{2} - \eta - 2\mu - \frac{\nu}{2}[\lambda_1^{-1} + 1] \right) \|A^{1/2}u(t)\|^2$$

$$- (\delta - \nu)\Pi(u(t)) + \left[-\frac{\mu}{h} + 2\nu\mu \right] \int_{-h}^{0} \|A^{1/2}u(t+\theta)\|^2 d\theta$$

$$+ \left[-\mu + \frac{\varkappa}{4\eta\lambda_1} + \frac{\varkappa}{4} \right] \int_{-h}^{0} \|A^{1/2}u(t+\theta)\|^2 \sigma(d\theta) + \gamma\nu + C_\eta.$$

This yields

$$\frac{d}{dt}\mathscr{W}(t) + \nu\mathscr{W}(t) \leq b, \quad t > 0, \tag{6.1.24}$$

for some $\nu, b > 0$ provided that

$$\frac{1}{2} - \eta - 2\mu - \frac{\nu}{2}[\lambda_1^{-1} + 1] \geq 0, \quad \delta - \nu \geq 0,$$

and

$$-\frac{\mu}{h} + 2\nu\mu \leq 0, \quad -\mu + \frac{\varkappa}{4\eta\lambda_1} + \frac{\varkappa}{4} \leq 0.$$

These relations hold with $\eta = 1/4$ and with $\nu > 0$ small enough if we demand that

$$\frac{1}{4} - 2\mu > 0, \quad -\mu + \frac{\varkappa}{\lambda_1} + \frac{\varkappa}{4} < 0.$$

Thus under the condition $\varkappa < \lambda_1(2\lambda_1 + 8)^{-1}$ we can find appropriate μ and prove dissipativity using (6.1.23), (6.1.24) and also the observation made in Exercise 2.1.3.

□

Using Theorems 6.1.12 and 6.1.13 we can derive from Theorem 6.1.15 the following assertion.

Corollary 6.1.18 (Global and exponential attractors). *Let the hypotheses of Theorem 6.1.15 be in force and $(\mathscr{C}_{1/2}, S_t)$ be the system generated by (6.1.1). Then*

- *The system $(\mathscr{C}_{1/2}, S_t)$ possesses a compact global attractor. This attractor is a bounded set in the space $\mathscr{Y}_\beta$ for every $\beta \in (1/2, 1)$ and has a finite fractal dimension.*
- *The system $(\mathscr{C}_{1/2}, S_t)$ possesses a fractal exponential attractor whose dimension is finite in $\mathscr{C}_{1/2}$.*

Proof. Theorem 6.1.15 guarantees the dissipativity of $(\mathscr{C}_{1/2}, S_t)$. Thus the result follows from Theorems 6.1.12 and 6.1.13. $\qquad\square$

6.1.3 Application: reaction-diffusion (heat) equation with delay

In a bounded domain $\Omega \subset \mathbb{R}^d$ we consider the following problem:

$$u_t(x, t) - \Delta u(x, t) + f_0(u(x, t)) = f_1(u(x, t - h_1), \nabla u(x, t - h_2)) \qquad (6.1.25)$$

endowed with boundary and initial conditions of the form

$$u\big|_{\partial\Omega} = 0, \quad u\big|_{t \in [-\max\{h_1, h_2\}, 0]} = \varphi(t). \qquad (6.1.26)$$

We assume that $f_1 : \mathbb{R}^{1+d} \mapsto \mathbb{R}$ is globally Lipschitz and $f_0 : \mathbb{R}^1 \mapsto \mathbb{R}$ satisfies the inequality

$$|f_0(u) - f_0(v)| \le C(1 + |u|^r + |v|^r)|u - v|, \qquad (6.1.27)$$

where $r \in [0, +\infty)$ when $d \le 2$ and $r \le 2(d-2)^{-1}$ for $d \ge 3$.

We consider (6.1.25) in the space $H = L_2(\Omega)$ and assume $A = -\Delta$ on the domain

$$\mathscr{D}(A) = H^2(\Omega) \cap H_0^1(\Omega) \equiv \{u \in L_2(\Omega) : \partial_{x_i x_j} u \in L_2(\Omega), \ u\big|_{\partial\Omega} = 0\},$$

where we use the notation $H^s(\Omega)$ for the Sobolev space of order s (ADAMS [1]) and $H_0^s(\Omega)$ denotes the closure of $C_0^\infty(\Omega)$ in $H^s(\Omega)$. It is well known that $\mathscr{D}(A^{1/2}) = H_0^1(\Omega)$. The nonlinear mapping B is defined by the relation

$$B(u^t) = -B_0(u(t)) + B_1(u^t),$$

where

$$[B_0(u)](x) = f_0(u(x)), \quad u \in H_0^1(\Omega),$$

and the definition of B_1 is obvious. As was seen in Section 4.2.5 the mapping B_0 is locally Lipschitz from $H_0^1(\Omega)$ into $L^2(\Omega)$. The condition in (6.1.9) is satisfied when

$$\liminf_{|s|\to\infty} \frac{f_0(s)}{s} > -\lambda_1, \tag{6.1.28}$$

where λ_1 is a first eigenvalue of the operator $-\Delta$ with Dirichlet boundary conditions. Thus equation (6.1.25) defines a dynamical system in the space

$$\mathscr{C} = C([-\max\{h_1, h_2\}, 0]; H_0^1(\Omega)).$$

To guarantee the hypotheses of Theorem 6.1.15 it is sufficient to assume in addition that

$$uf_0(u) \geq \alpha |u|^{r+2} - \beta \quad \text{for some } \alpha > 0 \text{ and } \beta \geq 0.$$

The condition in (6.1.19) concerning the delay term is valid with discrete measure σ concentrated at $\{-h_1, -h_2\}$ and with the parameter $\varkappa$ defined by the Lipschitz constant of f_1.

6.2 Parabolic problems with state-dependent delay: a case study

As we mentioned at the beginning of this chapter, the general theory of delay systems was mainly developed in the case of constant delays. On the other hand, it is clear that the constancy of the delay is just an extra assumption made to simplify the study, but it is not really well-motivated by real-world models. To describe a process more naturally, a new class of *state-dependent delay* models was introduced and studied during the last decades. As mentioned in the survey of HARTUNG ET AL. [121], the discussion of differential equations with such delays goes back to 1806 when Poisson studied a geometrical problem.[5] However, the theory of (ordinary) differential equations with state-dependent delay was developed only recently (see, e.g., KRISZTIN/ARINO [137], MALLET-PARET ET AL. [160], WALTHER [222] and also the survey HARTUNG ET AL. [121] and the references therein). Partial differential equations with state-dependent delay have been essentially less investigated; see the discussions in the papers REZOUNENKO [190, 191] devoted to the parabolic case.

The simplest case of a state-dependent delay is a delay explicitly given by a real-valued function $\eta : \mathbb{R} \to \mathbb{R}_+$ which depends on the value $x(t)$ at the reference time t but not on previous values of the solution $\{x(\tau), \tau \leq t\}$. This leads to terms of the form $f(x(t - \eta(x(t))))$ in the model considered. Even in this case non-uniqueness can appear (see the scalar ODE example constructed in 1963 by

---

[5]We refer to WALTHER [223] for a modern and detailed discussion of Poisson's example with state-dependent damping.

DRIVER [86]). The standard way for general models to avoid non-uniqueness in the case of infinite-dimensional dynamics is to consider smoother (narrower) classes of solutions. However in this case the existence problem may become critical. The main task is to find a good balance between these two issues.

In this section we deal with a certain abstract parabolic problem with a state-dependent delay term of a rather general structure. Our considerations are based on the paper CHUESHOV/REZOUNENKO [67] and motivated by several biological models; see the discussion and the references in BRITTON [18], GOURLEY/SO/WU [113] and REZOUNENKO/ZAGALAK [194]. We note that in the context of population dynamics, delays arise frequently as the maturation time, and this time is a function of the total population. Similarly, in the modeling of infectious disease transmission or in the modeling of immune response, the delay is due to the time required to accumulate an appropriate dosage of infection or antigen concentration.

As for previous topics of this book, we first discuss well-posedness of the problem with a concentration on variational-type solutions. Then we deal with the existence of a global *finite-dimensional* attractor and consider exponential attractors. The main difficulty we face is related to the fact that the corresponding delay term is not Lipschitz on the natural energy balance space. This circumstance makes it impossible to prove that the evolution operators S_t we construct are continuous mappings on the phase space for t small. We have continuity of the evolution operators for relatively large times only.

6.2.1 Model description

We deal with the model in (6.1.1) with a special choice of the nonlinear (delay) term B. More precisely, we take $B(u^t) = F(u^t) - G(u(t))$. Formally this form is the same as the one postulated in Assumption 6.1.7. However we prefer to use a different notation because our hypotheses concerning the delay term and the potential part are different. Thus we consider the dynamics of abstract evolution delay equations of the form

$$u_t(t) + Au(t) + G(u(t)) = F(u^t), \quad t > 0, \qquad (6.2.1)$$

in some Hilbert space H. Here A is a linear and G is a nonlinear operator, and the term $F(u^t)$ represents a delay effect in the dynamics. As in the previous section, the history segment (the state) is denoted by $u^t \equiv u^t(\theta) \equiv u(t + \theta)$ for $\theta \in [-h, 0]$.

Assumption 6.2.1 (Basic hypotheses). We assume that:

(A) A is a positive operator with a discrete spectrum in a separable Hilbert space H with a dense domain $\mathscr{D}(A) \subset H$ (see Definition 4.1.1). As above we suppose $H_s = \mathscr{D}(A^s)$ for $s \geq 0$ and H_s is the completion of H with respect to the norm

$\|A^s \cdot \|$ when $s < 0$ (see Section 4.1). Here and below, $\| \cdot \|$ is the norm of H, and $(\cdot, \cdot)$ is the corresponding scalar product. We denote by $\|v\|_s = \|A^s v\|$ the norm in H_s.

(F) The delay term $F(u^t)$ has the form $F(u_t) \equiv F_0(u(t - \eta(u^t)))$, where **(a)** F_0 : $H_\alpha \mapsto H_\alpha$ is globally Lipschitz for $\alpha = 0$ and $\alpha = -1/2$, i.e., there exists $L_F > 0$ such that

$$\|F_0(v) - F_0(u)\|_\alpha \leq L_F \|v - u\|_\alpha, \quad v, u \in H_\alpha, \quad \alpha = 0, -1/2; \quad (6.2.2)$$

and **(b)** $\eta : \mathscr{C} \equiv C([-h, 0]; H) \mapsto [0, h] \subset \mathbb{R}$ is globally Lipschitz:

$$|\eta(\phi) - \eta(\psi)| \leq L_\eta |\phi - \psi|_\mathscr{C}, \quad \forall \phi, \psi \in \mathscr{C}, \quad (6.2.3)$$

where $|v|_\mathscr{C} \equiv \sup\{\|v(\theta)\| : \theta \in [-h, 0]\}$ is the norm in the space $\mathscr{C}$.

(G) $G : H_{1/2} \mapsto H$ is locally Lipschitz, i.e.,

$$\|G(v) - G(u)\| \leq L_G(R)\|v - u\|_{1/2}, \quad v, u \in H_{1/2}, \quad \|v\|_{1/2}, \|u\|_{1/2} \leq R, \quad (6.2.4)$$

where $L_G : \mathbb{R}_+ \to \mathbb{R}_+$ is a non-decreasing function. In addition we assume that G is a potential mapping, which means that there exists a (Frechét differentiable) functional $\Pi(u) : H_{1/2} \to \mathbb{R}$ such that $G(u) = \Pi'(u)$ in the sense

$$\lim_{\|v\|_{1/2} \to 0} \|v\|_{1/2}^{-1}\big[\Pi(u + v) - \Pi(u) + (G(u), v)\big] = 0.$$

Moreover, we assume that **(a)** there exist positive constants c_1 and c_2 such that

$$(G(u), Au) \geq -c_1\|A^{\frac{1}{2}}u\|^2 - c_2, \quad u \in \mathscr{D}(A); \quad (6.2.5)$$

and **(b)** there exist $\delta > 0$ and $m \geq 0$ such that $G : H_{1/2-\delta} \mapsto H_{-m}$ is continuous.

Our main motivating example of a system with discrete state-dependent delay is the following one:

$$u_t(t, x) - \Delta u(t, x) + g(u(t, x)) = d(x) - f\left(K[u(t - \eta(u^t), \cdot)](x)\right), \quad x \in \Omega, \ t > 0, \quad (6.2.6)$$

in a bounded domain $\Omega \subset \mathbb{R}^n$, where $K : L^2(\Omega) \to L^2(\Omega)$ is a bounded operator and $f : \mathbb{R} \to \mathbb{R}$ stands for a Lipschitz function. The function

$$\eta : C([-h, 0]; L^2(\Omega)) \to [0, h] \subset \mathbb{R}_+$$

denotes *a state-dependent discrete delay*. The Nemytskii operator $u \mapsto g(u)$ with a C^1 function g represents a nonlinear (non-delayed) reaction term and $d(x)$

describes sources. The form of the delay term is motivated by models in population dynamics where function f is a birth function (it could be $f(s) = c_1 s \cdot e^{-c_2 s}$, with $c_1, c_2 > 0$) and the delay η represents the maturity age. For more detailed discussions and further examples (e.g., the diffusive Nicholson blowflies equation, the Mackey-Glass equation — a diffusive model of blood cell production, and the Lasota-Wazewska-Czyzewska model in hematology) with state-dependent delay, we refer to GOURLEY/SO/WU [113] and REZOUNENKO/ZAGALAK [194] and to the references therein. Several special cases of the model in (6.2.6) were studied in REZOUNENKO [191–193] and REZOUNENKO/ZAGALAK [194]). We note that if we equip (6.2.6) with the Dirichlet boundary condition, then the dissipativity property in (6.2.5) holds provided $g \in C^1(\mathbb{R})$, $g(0) = 0$, and the derivative $g'(s)$ is bounded from below. This follows by standard integration by parts. Thus population dynamics models with nonlinear sink/source feedback terms can be included in the framework of this section. For this kind of a biological model, but with a state-*independent* delay, we refer to WU [224].

We equip the equation (6.2.1) with the initial condition

$$u(\theta) = \varphi(\theta), \quad \theta \in [-h, 0], \tag{6.2.7}$$

and for initial data φ consider the space

$$\mathscr{L} \equiv \left\{ \varphi \in C([-h, 0]; H) \mid \mathrm{Lip}_{[-h,0]}(A^{-\frac{1}{2}}\varphi) < +\infty; \; \varphi(0) \in \mathscr{D}(A^{\frac{1}{2}}) \right\}, \tag{6.2.8}$$

where

$$\mathrm{Lip}_{[a,b]}(\varphi) \equiv \sup_{s \neq t} \left\{ \frac{||\varphi(s) - \varphi(t)||}{|s - t|} : s, t \in [a, b], \; s \neq t \right\}$$

denotes the corresponding Lipschitz constant. One can show that all elements from $\mathscr{L}$ are absolutely continuous functions φ on $[-h, 0]$ with values in $H_{-1/2}$. The latter means that there exists a derivative $\varphi_t \in L_\infty(-h, 0; H_{-1/2})$ such that

$$\varphi(s) = \varphi(0) - \int_s^0 \varphi_t(\xi) d\xi, \quad s \in [-h, 0].$$

Moreover, one can see that

$$\mathrm{Lip}_{[-h,0]}(A^{-\frac{1}{2}}\varphi) = \mathrm{ess\,sup} \left\{ ||A^{-\frac{1}{2}}\varphi_s(s)|| : s \in [-h, 0] \right\} \equiv |\varphi_t|_{L_\infty(-h,0;H_{-1/2})}.$$

We equip the space $\mathscr{L}$ with the natural norm

$$|\varphi|_{\mathscr{L}} \equiv \max_{s \in [-h,0]} ||\varphi(s)|| + \mathrm{Lip}_{[-h,0]}(A^{-\frac{1}{2}}\varphi) + ||A^{\frac{1}{2}}\varphi(0)||. \tag{6.2.9}$$

We note that the delay term $F(\varphi) \equiv F_0(\varphi(-\eta(\varphi)))$ in (6.2.1) is well defined for every $\varphi \in \mathscr{C}$ and possesses the property (see (6.2.2) for $\alpha = 0$):

$$||F(\varphi)|| \le c_1 + c_2|\varphi|_{\mathscr{C}}, \quad \varphi \in \mathscr{C}, \tag{6.2.10}$$

with $c_1 = ||F(0)||$ and $c_2 = L_F$. One can see that F is continuous on $\mathscr{C}$, but it is not Lipschitz on this space. We can only show that the delay term F satisfies the inequality

$$||F(\varphi) - F(\psi)||_{-1/2} \le L_F \left(1 + L_\eta \text{Lip}_{[-h,0]}(A^{-\frac{1}{2}}\varphi)\right) |\varphi - \psi|_{\mathscr{C}} \tag{6.2.11}$$

for every $\varphi \in \mathscr{L}$ and $\psi \in \mathscr{C}$. Using the terminology of [160] we can call this mapping F "almost Lipschitz" from $\mathscr{C}$ into $H_{-1/2}$. See also the discussion in [121].

Remark 6.2.2. We can also include in (6.2.1) a delay term $M(u^t)$ which is defined by a globally Lipschitz function from $C([-h, 0]; H_{1/2})$ into H. We will not pursue this generalization because our main goal in this section is state-dependent delay models. Parabolic-type delay equations with those globally Lipschitz $M(u^t)$ were discussed in Section 6.1.   ∎

6.2.2 Well-posedness

In contrast with Section 6.1, based on the mild formulation of the problem we now consider variational-type solutions which possess additional smoothness. The main reason for this is a singularity of the delay term on the "standard" phase space.

We introduce the following definition.

Definition 6.2.3 (Strong solution). A vector function

$$u(t) \in C([-h, T]; H) \cap C([0, T]; H_{1/2}) \cap L_2(0, T; H_1) \tag{6.2.12}$$

is said to be a (strong) solution to the problem defined by (6.2.1) and (6.2.7) on $[0, T]$ if

(a) $u(\theta) = \varphi(\theta)$ for $\theta \in [-h, 0]$;
(b) $\forall v \in L_2(0, T; H)$ such that $v_t \in L_2(0, T; H_{-1})$ and $v(T) = 0$ we have that

$$-\int_0^T (u(t), v_t(t))\, dt + \int_0^T (Au(t), v(t))\, dt$$

$$+ \int_0^T (-F(u^t) + G(u(t)), v(t))\, dt = (\varphi(0), v(0)). \tag{6.2.13}$$

                                                                     ∎

Remark 6.2.4. Let $u(t)$ be a strong solution on an interval $[0, T]$ with some $\varphi \in \mathscr{C}$. Then it follows from (6.2.12) and also from (6.2.4) and (6.2.10) that

$$F(u^t) - G(u(t)) \in L_\infty(0, T; H).$$

This allows us to conclude from (6.2.12) and (6.2.13) that

$$u_t(t) \in L_\infty(0, T; H_{-1/2}) \cap L_2(0, T; H). \tag{6.2.14}$$

Moreover, the relation in (6.2.13) implies that $u(t)$ satisfies (6.2.1) for almost all $t \in [0, T]$ as an equality in H. In particular, this implies that $u(t)$ solves the integral equation in (6.1.2) with $B(u^t) = F(u^t) - G(u(t))$, i.e., $u(t)$ is a mild solution to (6.2.1) and (6.2.7) as well. We also note that relations (6.2.12) and (6.2.14) yield

$$u^t \in \mathscr{L} \text{ for every } t \in [0, T] \text{ and } \max_{[0,T]} |u^t|_{\mathscr{L}} < +\infty \tag{6.2.15}$$

for every strong solution $u(t)$ with initial data φ from the space $\mathscr{L}$ which is defined in (6.2.8). ∎

We have the following theorem on the existence and uniqueness of solutions.

Theorem 6.2.5. *Let Assumption 6.2.1 be in force. Assume that $\varphi \in \mathscr{L}$, see (6.2.8). Then the initial value problem defined by (6.2.1) and (6.2.7) has a unique strong solution on any time interval $[0, T]$. This solution possesses the property*

$$u_t(t) \in C([0, T]; H_{-1/2}) \cap L_2(0, T; H) \tag{6.2.16}$$

and satisfies the estimate

$$||A^{-1/2}u_t(t)||^2 + ||A^{1/2}u(t)||^2 + \int_0^t \left[||u_t(\tau)||^2 + ||Au(\tau)||^2 \right] d\tau \leq C_T(R) \tag{6.2.17}$$

for all $t \in [0, T]$ and $||A^{1/2}\varphi(0)||^2 + |\varphi|_{\mathscr{C}}^2 \leq R^2$. Moreover, for every two strong solutions u^1 and u^2 with initial data φ^1 and φ^2 from $\mathscr{L}$ we have that

$$\sup_{\tau \in [0,t]} ||u^1(\tau) - u^2(\tau)||^2 + \int_0^t ||A^{1/2}(u^1(\tau) - u^2(\tau))||^2 d\tau \leq C_R(T)|\varphi^1 - \varphi^2|_{\mathscr{C}}^2 \tag{6.2.18}$$

for every $t \in [0, T]$ and for all φ^i such that $|\varphi^i|_{\mathscr{L}} \leq R$.

Proof. To prove the existence we use the standard compactness method (LIONS [151]) based on Galerkin approximations with respect to the eigenbasis $\{e_k\}$ of the operator A.

We define a Galerkin approximate solution of order N by the formula

$$u^N = u^N(t) = \sum_{k=1}^{N} g_{k,N}(t) e_k,$$

where the functions $g_{k,N}$ are defined on $[-h, T]$, absolutely continuous on $[0, T]$, and are such that the following equations are satisfied:

$$\begin{cases} (u_t^N + A u^N - F([u^N]^t) + G(u^N), e_k) = 0, & t > 0, \\ (u^N(\theta), e_k) = (\varphi(\theta), e_k), & \forall \theta \in [-h, 0], \quad \forall k = 1, \ldots, N. \end{cases} \tag{6.2.19}$$

The equation in (6.2.19) is a system of delay differential equations in $\mathbb{R}^N \simeq P_N H$, where P_N is the orthogonal projection onto the subspace Span $\{e_1, \ldots, e_N\}$. Hence, we can apply a finite-dimensional analog of Proposition 6.1.3 (see also HARTUNG ET AL. [121] for the purely ODE argument) to get the *local* existence of solutions to (6.2.19).

Next, we derive an a priori estimate which allows us to extend solutions u^N to (6.2.19) on an arbitrary time interval $[0, T]$. We also use it for the compactness of the set of approximate solutions.

We multiply the first equation in (6.2.19) by $\lambda_k g_{k,N}$ and sum for $k = 1, \ldots, N$ to get

$$\frac{1}{2} \frac{d}{dt} \|A^{1/2} u^N(t)\|^2 + \|A u^N(t)\|^2 + (-F([u^N]^t) + G(u^N(t)), A u^N(t)) = 0.$$

Due to (6.2.10) and (6.2.5) this implies that

$$\frac{d}{dt} \left[\|A^{1/2} u^N(t)\|^2 + \int_0^t \|A u^N(\tau)\|^2 d\tau \right] \leq c \left[1 + |[u^N]^t|_{\mathscr{C}}^2 + \|A^{1/2} u^N(t)\|^2 \right]$$

$$\leq c_0 \left[1 + |\varphi|_{\mathscr{C}}^2 \right] + c_1 \max_{\tau \in [0,t]} \|A^{1/2} u^N(\tau)\|^2.$$

Integrating the last inequality we can easily see that the function

$$\Psi(t) = \max_{\tau \in [0,t]} \|A^{1/2} u^N(\tau)\|^2 + \int_0^t \|A u^N(\tau)\|^2 d\tau$$

satisfies the inequality

$$\Psi(t) \leq 2 \|A^{1/2} \varphi(0)\|^2 + 2t c_0 \left[1 + |\varphi|_{\mathscr{C}}^2 \right] + 2c_1 \int_0^t \Psi(\tau) d\tau.$$

Therefore Gronwall's lemma gives us the a priori estimate

$$\|A^{1/2} u^N(t)\|^2 + \int_0^t \|A u^N(\tau)\|^2 d\tau \leq 2e^{at} \left[\|A^{1/2} \varphi(0)\|^2 + b \left[1 + |\varphi|_{\mathscr{C}}^2 \right] \right], \tag{6.2.20}$$

for all t from an existence interval, where a and b are positive constants. This a priori estimate allows us to extend approximate solutions on every time interval $[0, T]$ such that (6.2.20) remains true for every $t \in [0, T]$.

Now we establish additional a priori bounds. Using (6.2.20), (6.2.4), and (6.2.10), from the first equation in (6.2.19) we obtain that

$$\|u_t^N(t) + Au^N(t)\| \le \|F([u^N]^t)\| + \|G(u^N(t))\| \le C(R, T), \quad t \in [0, T],$$

provided $\|A^{1/2}\varphi(0)\|^2 + |\varphi|_{\mathscr{C}}^2 \le R^2$. Thus by (6.2.20) we obtain the estimate

$$\|A^{1/2}u^N(t)\|^2 + \int_0^t \left[\|u_t^N(\tau)\|^2 + \|Au^N(\tau)\|^2 \right] d\tau \le C_T(R) \qquad (6.2.21)$$

for all $t \in [0, T]$ and $\|A^{1/2}\varphi(0)\|^2 + |\varphi|_{\mathscr{C}}^2 \le R^2$. It also follows from (6.2.19) that

$$\sup_{t \in [0,T]} \|A^{-1/2}u_t^N(t)\|^2 \le C_T(R). \qquad (6.2.22)$$

Thus

$$\{u^N\}_{N=1}^\infty \text{ is a bounded set in } W_1 \equiv L_\infty(0, T; H_{1/2}) \cap L_2(0, T; D(A)),$$

and

$$\{u_t^N\}_{N=1}^\infty \text{ is a bounded set in } W_2 \equiv L_\infty(0, T; H_{-1/2}) \cap L_2(0, T; H).$$

Hence, there exist a subsequence $\{(u^k; u_t^k)\}$ and an element $(u; u_t) \in W_1 \times W_2$ such that

$$\{(u^k; u_t^k)\} \text{ *-weakly converges to } (u; u_t) \text{ in } W_1 \times W_2.$$

By the Aubin-Dubinskii-Lions theorem (see the Appendix, Section A.3.3) we also have

$$u^k \to u \text{ in } C([0, T]; H_{1/2-\delta})) \cap L_2(0, T; H_{1-\delta}) \text{ as } k \to \infty.$$

Now the proof that any *-weak limit $u(t)$ is a solution is standard. To make the limit transition in the nonlinear terms F and G we use relation (6.2.11) and Assumption 6.2.1(**Gb**).

The property $u(t) \in C([0, T]; H_{1/2})$ follows from the well-known continuous embedding (see Proposition A.3.3 in the Appendix)

$$\{u \in L_2(0, T; H_1) : u_t \in L_2(0, T; H)\} \subset C([0, T]; H_{1/2}).$$

The continuity of u_t in $H_{-1/2}$ follows from equation (6.2.1) and from the continuity of u in $H_{1/2}$. Thus the existence of strong solutions is proved. It is easy to see from (6.2.21) and (6.2.22) that the strong solution constructed satisfies (6.2.17).

Now we prove the uniqueness.

Let u^1 and u^2 be two solutions (at this point we do not assume that they have the same initial data). Then the difference

$$z = u^1 - u^2 \in C([0, T]; H_{1/2}) \cap L_2(0, T; H_1)$$

is a strong solution to the linear parabolic-type (non-delay) equation

$$z_t(t) + Az(t) = f(t), \quad t > 0, \tag{6.2.23}$$

with

$$f(t) \equiv F([u^1]^t) - F([u^2]^t) + G(u^2(t)) - G(u^1(t)).$$

By Remark 6.2.4, $f \in L_\infty(0, T; H)$. From (6.2.4) and (6.2.11) and using (6.2.15) we also have that

$$\|G(u^2(t)) - G(u^1(t))\| \le L_G(\varrho)\|z(t)\|_{1/2}, \quad t \in [0, T],$$

and

$$\|A^{-1/2}(F([u^1]^t) - F([u^2]^t))\| \le L_F(1 + L_\eta\varrho)|z^t|_{\mathscr{C}}, \quad t \in [0, T],$$

for every $\varrho \ge \max_{[0,T]} \{|[u^1]^t|_{\mathscr{L}} + |[u^2]^t|_{\mathscr{L}}\}$. Therefore

$$|(f(t), z(t))| \le L_F(1 + L_\eta\varrho)|z^t|_{\mathscr{C}}\|z(t)\|_{1/2} + L_G(\varrho)\|z(t)\|_{1/2}\|z(t)\|$$

$$\le \frac{1}{2}\|z(t)\|^2_{1/2} + C(\varrho)|z^t|_{\mathscr{C}}.$$

The observations made in Remark 6.2.4 allow us to use the standard multiplier z in (6.2.23). Thus we can obtain that

$$\frac{d}{dt}\|z(t)\|^2 + \|A^{1/2}z(t)\|^2 \le C(\varrho)\|z^t\|^2_{\mathscr{C}} \le C(\varrho)\left[|\varphi^1 - \varphi^2|^2_{\mathscr{C}} + \sup_{\tau\in[0,t]} \|z(\tau)\|^2 \right]$$

for every $\varrho \ge \max_{[0,T]} \{|[u^1]^t|_{\mathscr{L}} + |[u^2]^t|_{\mathscr{L}}\}$. Applying Gronwall's lemma we obtain

$$\sup_{\tau\in[0,t]} \|u^1(\tau) - u^2(\tau)\|^2 + \int_0^t \|A^{1/2}(u^1(\tau) - u^2(\tau))\|^2 d\tau \le C(\varrho)|\varphi^1 - \varphi^2|^2_{\mathscr{C}}$$

$$\tag{6.2.24}$$

for all $t \in [0, T]$. This implies the uniqueness of strong solutions.

As a by-product the uniqueness yields that *any* strong solution satisfies (6.2.17). Therefore we can apply (6.2.24) with some $\varrho = \varrho(R, T)$ to obtain (6.2.18).

Thus the proof of Theorem 6.2.5 is complete.   □

Theorem 6.2.5 allows us to define an evolution semigroup S_t on the space $\mathscr{L}$ (see (6.2.8)) by the formula

$$S_t\varphi \equiv u_t, \quad t \geq 0, \tag{6.2.25}$$

where $u(t)$ is the unique solution to problem (6.2.1) and (6.2.7). We note that (6.2.18) implies that S_t is *almost* locally Lipschitz on $\mathscr{C}$, i.e.,

$$|S_t\varphi^1 - S_t\varphi^2|_{\mathscr{C}} \leq C_R(T)|\varphi^1 - \varphi^2|_{\mathscr{C}} \text{ for every } \varphi^i \in \mathscr{L}, \ |\varphi^i|_{\mathscr{L}} \leq R, \ t \in [0, T]$$

However, it seems that a similar bound is not true in the space $\mathscr{L}$. We can only guarantee that $\varphi \mapsto S_t\varphi$ is a continuous mapping on $\mathscr{L}$ for $t > h$. Moreover, the following assertion shows that the mapping $\varphi \mapsto S_t\varphi$ is even $\frac{1}{2}$-Hölder on $\mathscr{L}$ with respect to φ when $t > h$.

Proposition 6.2.6 (Dependence on initial data in the space $\mathscr{L}$). *Assume that the hypotheses of Theorem 6.2.5 are in force. Let u^1 and u^2 be two solutions on $[0, T]$ with initial data φ^1 and φ^2 from $\mathscr{L}$. Then the difference $z = u^1 - u^2$ satisfies the estimate*

$$(t - h)\left[\|A^{-1/2}\dot{z}(t)\|^2 + \|A^{1/2}z(t)\|^2\right]$$

$$+ \int_h^t (\tau - h)\left[\|z_t(\tau)\|^2 + \|Az(\tau)\|^2\right] d\tau \leq C_T(R)|\varphi^1 - \varphi^2|_{\mathscr{C}} \tag{6.2.26}$$

for all $t \in [h, T]$ and for all initial data φ^i such that $|\varphi^i|_{\mathscr{L}} \leq R$. This implies that for every $t > h$ the evolution semigroup S_t is $\frac{1}{2}$-Hölder continuous in the norm of $\mathscr{L}$. In the case when $t \in (0, h]$ we can guarantee the closedness of the evolution operator S_t only. This means[6] (see, e.g., PATA/ZELIK [179]) that the properties $\varphi_n \to \varphi$ and $S_t\varphi_n \to \psi$ in the norm of $\mathscr{L}$ as $n \to \infty$ imply that $S_t\varphi = \psi$.

Proof. Let P_N be the orthoprojector on Span $\{e_1, \ldots, e_N\}$. Multiplying (6.2.23) by $P_N Az$ and using (6.2.17) and (6.2.4) we obtain that

$$\frac{d}{dt}\|P_N A^{1/2}z(t)\|^2 + \|P_N Az(t)\|^2 \leq \|F([u^2]^t) - F([u^1]^t)\|^2 + C_R(T)\|A^{1/2}z(t)\|^2$$

---

[6]We mention that any continuous mapping is closed, and a mapping can be closed but not continuous. See examples in [179] and also in Remark 1.1.6.

for all $t > 0$. From (6.2.17), (6.2.2) and (6.2.3) we also have that

$$\|F([u^2]^t) - F([u^1]^t)\|^2 \leq 2L_F^2 \left[\left| \int_{t-\eta([u^1]^t)}^{t-\eta([u^2]^t)} \|u_t^2(\xi)\| d\xi \right|^2 + |[u^2]^t - [u^1]^t|_{\mathscr{C}}^2 \right]$$

$$\leq 2L_F^2 \left[|\eta([u^1]^t) - \eta([u^2]^t)| \int_{t-h}^{t} \|u_t^2(\xi)\|^2 d\xi \right.$$

$$\left. + |[u^2]^t - [u^1]^t|_{\mathscr{C}}^2 \right] \leq C_T(R)|[u^2]^t - [u^1]^t|_{\mathscr{C}} \qquad (6.2.27)$$

for every $t \geq h$. Therefore

$$\frac{d}{dt} \|P_N A^{1/2} z(t)\|^2 + \|P_N A z(t)\|^2 \leq C_T(R) \left[\max_{[0,t]} \|z(s)\|^2 + \|A^{1/2} z(t)\|^2 \right]^{1/2}, \quad t \geq h.$$

Integrating over interval $[\tau, t]$ with $\tau \geq h$ and using (6.2.18), after the limit transition $N \to \infty$ we obtain that

$$\|A^{1/2} z(t)\|^2 + \int_{\tau}^{t} \|A z(\xi)\|^2 d\xi \leq \|A^{1/2} z(\tau)\|^2 + C_T(R)|\varphi^1 - \varphi^2|_{\mathscr{C}} \qquad (6.2.28)$$

for all $t \geq \tau \geq h$. Now we integrate (6.2.28) with respect to τ over $[h, t]$, change the order of integration, and use (6.2.18) to get

$$(t-h)\|A^{1/2} z(t)\|^2 + \int_{h}^{t} (\xi - h)\|A z(\xi)\|^2 d\xi \leq C_T(R)|\varphi^1 - \varphi^2|_{\mathscr{C}}, \quad t \geq h.$$

Using the expression for z_t from (6.2.23) and also the bounds in (6.2.18) and (6.2.27) we have that

$$\|z_t(t) + A z(t)\|^2 + \|A^{-1/2} z_t(t)\|^2 \leq C_T(R) \left[\|A^{1/2} z(t)\|^2 + |\varphi^1 - \varphi^2|_{\mathscr{C}} \right], \quad t \geq h.$$

This implies (6.2.26), which yields the $\frac{1}{2}$-Hölder continuity of the evolution semigroup S_t in the norm of $\mathscr{L}$ for $t > h$.

The closedness of S_t for $t \in (0, h]$ follows from (6.2.18). Indeed, the continuity in a weaker topology on bounded sequences in $\mathscr{L}$ allows us to identify the limit of $S_t \varphi_n$ with $S_t(\lim \varphi_n)$. □

Remark 6.2.7. From (6.2.27) we can obtain a $\frac{1}{2}$-Hölder continuity relation like (6.2.26) *for all* $t \geq 0$ if we assume in addition that one of initial data φ^i possesses the property $\varphi_t^i \in L_2(-h, 0; H)$. In this case the argument above leads to the relation

$$\|A^{-1/2} z_t(t)\|^2 + \|A^{1/2} z(t)\|^2 + \int_{0}^{t} \left[\|z_t(\tau)\|^2 + \|A z(\tau)\|^2 \right] d\tau$$

$$\leq C_T(R) \left[\|A^{1/2}(\varphi^1(0) - \varphi^2(0))\| + |\varphi^1 - \varphi^2|_{\mathscr{C}} \right] \qquad (6.2.29)$$

for all $t \in [0, T]$ and for all initial data φ^i such that $|\varphi^i|_{\mathscr{L}} + |\varphi^i_t|_{L_2(-h,0;H)} \leq R$. Moreover, one can also see that the set

$$\mathscr{L}_0 = \{\varphi \in \mathscr{L} : \varphi_t \in L_2(-h, 0; H)\} \tag{6.2.30}$$

is forward invariant with respect to S_t. Thus $\varphi \mapsto S_t\varphi$ is a $\frac{1}{2}$-Hölder continuous mapping for each $t \geq 0$ on the Banach space $\mathscr{L}_0$ endowed with the norm

$$|\varphi|_{\mathscr{L}_0} = |\varphi|_{\mathscr{L}} + |\varphi_t|_{L^2(-h,0;H)}.$$

Hence a dynamical (in the classical sense, see Definition 1.1.1) system $(\mathscr{L}_0, S_t)$ arises. However we prefer to avoid property $\varphi_t \in L_2(-h, 0; H)$ in the description of the phase space. Our goal is long-time dynamics, and the existence of limiting objects requires some compactness properties. Unfortunately, we cannot guarantee these properties in the space $\mathscr{L}_0$ without serious restrictions concerning the delay term.   ∎

Remark 6.2.8. We have a similar problem to that above with the *time* continuity of the evolution operator S_t. It is clear from (6.2.12) and (6.2.16) that $t \mapsto S_t\varphi$ is continuous in $\mathscr{L}$ for every $\varphi \in \mathscr{L}$ when $t > h$. To guarantee the continuity $t \mapsto S_t\varphi$ *for all* $t \geq 0$ we must make a further restriction[7] on the initial data. The main restriction is a compatibility condition at time $t = 0$. To describe this condition we introduce the following (complete) metric space:

$$Y \equiv \left\{\varphi \in \mathscr{C} \equiv C([-h, 0]; H) \,\middle|\, \begin{array}{l} \varphi \in C^1([-h, 0]; H_{-1/2}); \\ \varphi(0) \in H_{1/2}; \\ \varphi_t(0) + A\varphi(0) + G(\varphi(0)) = F(\varphi) \end{array}\right\} \tag{6.2.31}$$

Here the compatibility condition $\varphi_t(0) + A\varphi(0) + G(\varphi(0)) = F(\varphi)$ is understood as an equality in $H_{-1/2}$. The distance in Y is given by the relation

$$\mathrm{dist}_Y(\varphi, \psi) = \|A^{1/2}(\varphi(0) - \psi(0))\|$$
$$+ \max_{[-h,0]} \left\{\|A^{-1/2}(\varphi_t(\theta) - \psi_t(\theta))\| + \|\varphi(\theta) - \psi(\theta)\|\right\}. \tag{6.2.32}$$

One can see that Y is a *closed* subset in the space $\mathscr{L}$ and the topology generated by the metric dist_Y coincides with the induced topology of $\mathscr{L}$; see (6.2.9).   ∎

In the following assertion we collect several dynamical properties of the evolution semigroup S_t which are direct consequences of Theorem 6.2.5, Proposition 6.2.6, and Remark 6.2.8.

---

[7] We refer to the discussions in REZOUNENKO [192] and REZOUNENKO/ZAGALAK [194] for the related PDE models.

Proposition 6.2.9. *Under the conditions of Theorem 6.2.5 problem (6.2.1) generates an evolution semigroup S_t of closed mappings on $\mathscr{L}$ such that*

(a) *$S_t \mathscr{L} \subset Y$ for every $t \geq h$ and the set $S_t B$ is bounded in Y for each $t \geq h$ when B is bounded in the space $\mathscr{L}$;*
(b) *the set Y is forward invariant: $S_t Y \subset Y$ for all $t > 0$;*
(c) *the mapping $\varphi \mapsto S_t \varphi$ is $\frac{1}{2}$-Hölder continuous on $\mathscr{L}$ (and hence on Y) for all $t > h$;*
(d) *the trajectories $t \mapsto S_t \varphi$ are continuous for $t > h$ and $\varphi \in \mathscr{L}$. If $\varphi \in Y$, then these trajectories are continuous for all $t \geq 0$.*

6.2.3   Long-time dynamics: hypotheses and statement

We impose the following (standard) hypotheses (see, e.g., TEMAM [216]) concerning the nonlinear (non-delayed) sink/source term G.

Assumption 6.2.10. The nonlinear mapping $G : H_{1/2} \to H$ is potential and has the form

$$G(u) = \Pi'(u) \quad \text{with} \quad \Pi(u) = \Pi_0(u) + \Pi_1(u),$$

where $\Pi_0(u) \geq 0$ is bounded on bounded sets in $H_{1/2}$ and $\Pi_1(u)$ satisfies the property

$$\forall\, \eta > 0 \,\exists\, C_\eta > 0 : \ |\Pi_1(u)| \leq \eta \left(||A^{1/2}u||^2 + \Pi_0(u) \right) + C_\eta, \ u \in H_{1/2}. \tag{6.2.33}$$

Moreover, we assume that

(a) there are constants $\nu \in [0, 1), c_4, c_5 > 0$ such that

$$- (u, G(u)) \leq \nu ||A^{1/2}u||^2 - c_4 \Pi_0(u) + c_5, \ u \in H_{1/2}; \tag{6.2.34}$$

(b) for every $\tilde{\eta} > 0$ there exists $C_{\tilde{\eta}} > 0$ such that

$$||u||^2 \leq C_{\tilde{\eta}} + \tilde{\eta} \left(||A^{1/2}u||^2 + \Pi_0(u) \right), \ u \in H_{1/2}. \tag{6.2.35}$$

In the case of parabolic models like (6.2.6), examples of functions $g(u)$ such that the corresponding Nemytskii operator satisfies Assumptions 6.2.1(G) and 6.2.10 can be found in [9] and [216]. The simplest one is $g(u) = u^3 + a_1 u^2 + a_2 u$ with arbitrary $a_1, a_2 \in \mathbb{R}$ in the case when Ω is a 3D domain.

Theorem 6.2.11 (Global and exponential attractors). *Let Assumptions 6.2.1 and 6.2.10 be in force. Suppose that S_t is the evolution semigroup generated in $\mathscr{L}$ by (6.2.1) and (6.2.7). Then there exists $\ell_0 > 0$ such that this semigroup possesses a* *compact connected global attractor* $\mathfrak{A}$ *provided* $m_F h < \ell_0$, *where* h *is the delay time and* m_F *is the linear growth constant for* F_0 *in* H *defined by the relation*

$$m_F = \limsup_{\|u\| \to +\infty} \frac{\|F_0(u)\|}{\|u\|}. \qquad (6.2.36)$$

Moreover, for every $0 < \beta \leq 1$, $0 < \alpha \leq 1/2$, $\alpha < \beta$, *this attractor belongs to the set*

$$D_{\alpha,\beta}^R = \left\{ \varphi \in Y \middle| \begin{array}{l} |A^{1-\beta}\varphi|_C + |A^{-\beta}\varphi_t|_C + \mathrm{Hld}_\alpha(A^{1-\beta}\varphi) + \mathrm{Hld}_\alpha(A^{-\beta}\varphi_t) \\ + \left[\int_{-h}^0 \left(\|A\varphi(\theta)\|^2 + \|\varphi_t(\theta)\|^2 \right) d\theta \right]^{1/2} \leq R \end{array} \right\}$$

$$(6.2.37)$$

for some $R = R(\alpha, \beta)$, *where the Hölder seminorm* $\mathrm{Hld}_\alpha(\psi)$ *is given by*

$$\mathrm{Hld}_\alpha(\psi) = \sup \left\{ \frac{\|\psi(t_1) - \psi(t_2)\|}{|t_1 - t_2|^\alpha} : t_1 \neq t_2, \ t_1, t_2 \in [-h, 0] \right\}.$$

Assume in addition that there exist $\gamma, \delta \in (0, 1/2]$ *such that*

(i) *the mapping* F_0 *is globally Lipschitz from* $H_{-\gamma}$ *into* $H_{-1/2+\delta}$, *i.e.,*

$$\|F_0(u) - F_0(v)\|_{-1/2+\delta} \leq c\|u - v\|_{-\gamma}, \quad u, v \in H_{-\gamma}; \qquad (6.2.38)$$

(ii) *the mapping* G *is almost locally Lipschitz from* $H_{1/2-\gamma}$ *into* $H_{-1/2+\delta}$ *in the sense that*

$$\|G(u) - G(v)\|_{-1/2+\delta} \leq c(R)\|u - v\|_{1/2-\gamma} \qquad (6.2.39)$$

for all $u, v \in H_{1-\beta}$ *such that* $\|u\|_{1-\beta}, \|v\|_{1-\beta} \leq R$ *with some* $0 < \beta < 1/2$.

Then:

(A) *The global attractor* $\mathfrak{A}$ *has finite fractal dimension.*
(B) *There exists a fractal exponential attractor* $\mathfrak{A}_{\exp}$.

Remark 6.2.12. It follows from the statement of Theorem 6.2.11 that the global attractor $\mathfrak{A}$ is a bounded set in the Hölder-type space

$$C^\alpha([-h, 0] : H_{1-\alpha-\delta}) \cap C^{1+\alpha}([-h, 0] : H_{-\alpha-\delta}), \quad \forall \alpha \in [0, 1/2], \ \delta > 0.$$

Concerning Hölder spaces, we refer to Section A.3.1 in the Appendix.   ■

The following subsections are devoted to the proof of Theorem 6.2.11.

6.2.4 Proof of the existence of a global attractor

To prove the existence of a global attractor we first show that the evolution operator possesses a compact absorbing set in $\mathscr{L}$. Obviously the same is true in the space Y. Since $S_t Y \subset Y$ and S_t is continuous on Y, we can apply the standard existence result given by Theorem 2.3.5.

We start with the existence of a bounded absorbing set.

Proposition 6.2.13 (Bounded dissipativity). *Assume that $u(t)$ solves (6.2.1) and (6.2.7) with $\varphi \in \mathscr{L}$. Then one can find $\ell_0 > 0$ such that for every delay time h satisfying the inequality $m_F h \leq \ell_0$ the following property holds: there exists R_* such that for every bounded set B in $\mathscr{L}$ there is t_B such that*

$$||A^{-1/2}u_t(t)||^2 + ||A^{1/2}u(t)||^2 + \int_t^{t+1} \left[||u_t(\tau)||^2 + ||Au(\tau)||^2 \right] d\tau \leq R_*^2 \quad (6.2.40)$$

for all $t \geq t_B$ and for all initial data $\varphi \in B$. This yields that the evolution semigroup S_t is dissipative on $\mathscr{L}$ provided $m_F h < \ell_0$.

Proof. We use the Lyapunov method to get the result. For this we consider the following functional:

$$\tilde{V}(t) \equiv \frac{1}{2} \left[||u(t)||^2 + ||A^{1/2}u(t)||^2 \right] + \Pi(u(t)) + \frac{\mu}{h} \int_0^h \left\{ \int_{t-s}^t ||u_t(\xi)||^2 d\xi \right\} ds,$$

defined on strong solutions $u(t)$ for $t \geq h$. The positive parameter μ will be chosen later. As in the proof of Theorem 6.1.15 the main idea behind inclusion of an additional integral term in $\tilde{V}$ is to find a compensator for the delay term in (6.2.1).

One can see from (6.2.33) that there exist $0 < c_0 < 1/2$ and $c, c_1 > 0$ such that

$$c_0 \left[||A^{1/2}u(t)||^2 + \Pi_0(u(t)) \right] - c \leq \tilde{V}(t)$$

$$\leq c_1 \left[||A^{1/2}u(t)||^2 + \Pi_0(u(t)) \right] + \mu \int_0^h ||u_t(t-\xi)||^2 d\xi + c. \quad (6.2.41)$$

We consider the time derivative of $\tilde{V}$ along a solution. One can easily check that

$$\frac{d}{dt}\tilde{V}(t) = (u(t), u_t(t)) + (Au(t), u_t(t)) + (G(u(t)), u_t(t))$$

$$+ \frac{\mu}{h} \int_0^h \left\{ ||u_t(t)||^2 - ||u_t(t-s)||^2 \right\} ds$$

$$= (u_t(t) + Au(t) + G(u(t)), u_t(t)) - (u_t(t), u_t(t))$$

$$+ (u(t), u_t(t)) + \mu ||u_t(t)||^2 - \frac{\mu}{h} \int_0^h ||u_t(t-\xi)||^2 d\xi.$$

Using (6.2.1) and also the relation

$$(u(t), u_t(t)) = -(Au(t), u(t)) + (F(u^t) - G(u(t)), u(t))$$

we find that

$$\frac{d}{dt}\tilde{V}(t) = (F(u^t), u_t(t) + u(t)) - (1 - \mu)||u_t(t)||^2$$

$$- \frac{\mu}{h}\int_0^h ||u_t(t - \xi)||^2 d\xi - ||A^{1/2}u(t)||^2 - (G(u(t)), u(t)).$$

By the definition of m_F in (6.2.36), for any number M_F greater than m_F we can find $C(M_F)$ such that

$$||F(u^t)|| = ||F_0(u(t - \eta(u^t)))|| \leq M_F||u(t - \eta(u^t))|| + C(M_F).$$

Therefore

$$||F(u^t)|| \leq M_F||u(t - \eta(u^t)) - u(t)|| + M_F||u(t)|| + C(M_F)$$

$$= M_F\left\|\int_{t-\eta(u^t)}^t u_t(\theta)d\theta\right\| + M_F||u(t)|| + C(M_F),$$

and thus

$$||F(u^t)|| \leq M_F \cdot \left[||u(t)|| + \int_0^h ||u_t(t - \xi)||d\xi\right] + C(M_F), \quad t \geq h.$$

Since

$$\int_0^h ||u_t(t - \xi)||d\xi \leq h^{1/2}\left(\int_0^h ||u_t(t - \xi)||^2 d\xi\right)^{1/2},$$

we have that

$$|(F(u^t), u_t(t))| \leq \frac{1}{2}||u_t(t)||^2 + c_1 M_F^2||u(t)||^2$$

$$+ c_2 M_F^2 h\int_0^h ||u_t(t - \xi)||^2 d\xi + C(M_F), \quad t \geq h.$$

In a similar way

$$|(F(u^t), u(t)))| \leq c_1 M_F^2 h\int_0^h ||u_t(t - \xi)||^2 d\xi + C(M_F)(1 + ||u(t)||^2).$$

Thus

$$|(F(u'), u_t(t) + u(t)))| \leq$$

$$\frac{1}{2}\|u_t(t)\|^2 + c_0 M_F^2 h \int_0^h \|u_t(t - \xi)\|^2 d\xi + c_1(M_F)(1 + \|u(t)\|^2).$$

The relations in (6.2.34) with $\nu \in [0, 1)$ and (6.2.35) with $\tilde{\eta} > 0$ small enough yield

$$c_1(M_F)(1 + \|u\|^2) - \|A^{1/2}u\|^2 - (u, G(u)) \leq -a_0 \left[\|A^{1/2}u\|^2 + \Pi_0(u)\right] + a_1(M_F)$$

for some $a_i > 0$ with a_0 independent of M_F. Thus it follows from the relations above that

$$\frac{d}{dt}\tilde{V}(t) \leq -\left(\frac{1}{2} - \mu\right)\|u_t(t)\|^2 - a_0 \left[\|A^{1/2}u\|^2 + \Pi_0(u)\right]$$

$$+ a_1(M_F) + \left[-\frac{\mu}{h} + a_2 M_F^2 h\right]\int_0^h \|u_t(t - \xi)\|^2 d\xi$$

for some $a_i > 0$. Hence using the right inequality in (6.2.41) we arrive at the relation

$$\frac{d}{dt}\tilde{V}(t) + \gamma\tilde{V}(t) \leq -\left(\frac{1}{2} - \mu\right)\|u_t(t)\|^2 - (a_0 - \gamma c_1)\left[\|A^{1/2}u\|^2 + \Pi_0(u)\right]$$

$$+ \left[-\frac{\mu}{h} + \mu\gamma + a_2 M_F^2 h\right]\int_0^h \|u_t(t - \xi)\|^2 d\xi + a_1(M_F).$$

$$(6.2.42)$$

Therefore taking $\mu = 1/4$ and fixing $0 < \gamma \leq a_0 c_1^{-1}$ we obtain that

$$\frac{d}{dt}\tilde{V}(t) + \gamma\tilde{V}(t) + \frac{1}{4}\|u_t(t)\|^2 \leq C, \quad t \geq h, \qquad (6.2.43)$$

provided $\gamma h + 4a_2 M_F^2 h^2 \leq 1$. Thus under the condition $4a_2 m_F^2 h^2 < 1$ we can choose $\gamma \in (0, a_0 c_1^{-1}]$ and $M_F > m_F$ such that (6.2.43) holds. In particular,

$$\frac{d}{dt}\tilde{V}(t) + \gamma\tilde{V}(t) \leq C, \quad t \geq h.$$

This yields

$$\tilde{V}(t) \leq \tilde{V}(h)e^{-\gamma(t-h)} + \frac{C}{\gamma}(1 - e^{-\gamma(t-h)}), \quad t \geq h, \qquad (6.2.44)$$

when $m_F h < \ell_0$. Using (6.2.41) and (6.2.17) we can conclude that $|\tilde{V}(h)| \leq C_B$ for all initial data from a bounded set B in $\mathscr{L}$. Hence (see (6.2.1)) there exists R such that for every initial data from a bounded set B in $\mathscr{L}$

$$\|A^{1/2}u(t)\| + \|A^{-1/2}u_t(t)\| + \|u_t(t) + Au(t)\| \leq R \quad \text{for all } t \geq t_B.$$

Moreover, it follows from (6.2.43) that

$$\int_t^{t+1} \|u_t(\tau)\|^2 d\tau \leq C_R \quad \text{for all } t \geq t_B.$$

These relations imply (6.2.40) and will allow us to complete the proof of Proposition 6.3.11. $\square$

Remark 6.2.14. If the mapping F_0 has a sublinear growth in H, i.e., there exists $\beta < 1$ such that

$$\|F_0(u)\| \leq c_1 + c_2 \|u\|^\beta, \quad u \in H,$$

then the linear growth parameter m_F given by (6.2.36) is zero. Thus in this case we have no restrictions concerning h in the statement of Proposition 6.2.13. In particular, this is true in the case of bounded mappings F_0. Moreover, in the latter case the argument can be simplified substantially (we can use a Lyapunov-type function without delay terms as was done in REZOUNENKO [192] and REZOUNENKO/ZAGALAK [194] for the case of parabolic model (6.2.6) with bounded f). ∎

We use Proposition 6.2.13 to obtain the following assertion, which means that the evolution semigroup S_t is (ultimately) compact.

Proposition 6.2.15 (Compact dissipativity). *As in Proposition 6.2.13 we assume that $m_F h < \ell_0$. Then the evolution operator S_t possesses a compact absorbing set. More precisely, for every $0 < \beta \leq 1$, $0 < \alpha \leq 1/2$, $\alpha < \beta$, the set $D_{\alpha,\beta}^R$ given by (6.2.37) is absorbing for some R. This set $D_{\alpha,\beta}^R$ is compact in Y provided $0 < \alpha < \beta < 1/2$.*

Proof. We first note that the compactness of $D_{\alpha,\beta}^R$ in $Y \subset \mathscr{L}$ for $0 < \alpha < \beta < 1/2$ follows from the Arzelà-Ascoli theorem (see, e.g., Lemma A.3.5 in the Appendix).

Now we show that $D_{\alpha,\beta}^R$ is absorbing.

Using the mild form of the problem, the bound in (6.2.10), and then the expression for u_t from (6.2.1) one can show that

$$\|A^{1-\delta}u(t)\| + \|A^{-\delta}u_t(t)\| \leq C_{R_*}(\delta) \quad \text{for all } t \geq t_B, \tag{6.2.45}$$

for every $\delta > 0$, where $u(t)$ is a solution possessing property (6.2.40).

Now we consider the difference $u(t_1) - u(t_2)$ with $t_1 > t_2$. Namely, using the mild form we obtain

$$||A^{1-\beta}(u(t_1) - u(t_2))|| \leq ||A^{1-\beta}(e^{-A(t_1-t_2)} - 1)u(t_2)||$$
$$+ \int_{t_2}^{t_1} ||A^{1-\beta}e^{-A(t_1-\tau)}|| \cdot (||F(u^\tau)|| + ||G(u(\tau))||)\, d\tau.$$

Since (see relation (4.1.8) and Exercise 4.1.10)

$$||A^{-\alpha}(1 - e^{-At})|| \leq t^\alpha \quad \text{and} \quad ||A^\alpha e^{-At}|| \leq \left(\frac{\alpha}{t}\right)^\alpha e^{-\alpha}$$

for all $t > 0$ and $0 \leq \alpha \leq 1$, we obtain

$$||A^{1-\beta}(u(t_1) - u(t_2))|| \leq |t_1 - t_2|^\alpha ||A^{1-\beta+\alpha}u(t_2)||$$
$$+ c_\beta \int_{t_2}^{t_1} \frac{1}{|t_1 - \tau|^{1-\beta}} \left[C_{R_*} + c|u^\tau|_\mathscr{C} \right] d\tau$$

for $t_1 > t_2 \geq t_B$. Thus for every $0 < \alpha < \beta \leq 1$ we have

$$||A^{1-\beta}(u(t_1) - u(t_2))|| \leq C_{R_*}|t_1 - t_2|^\alpha \quad \text{for all} \;\; t_i \geq t_B, |t_1 - t_2| \leq 1. \qquad (6.2.46)$$

Similarly to (6.2.27), using (6.2.46) with $\beta = 1$ and $\alpha = 1/2$ we obtain that

$$||F(u^{t_1}) - F(u^{t_2})|| \leq L_F \left| \int_{t_1-\eta(u^{t_1})}^{t_2-\eta(u^{t_2})} ||u_t(\xi)|| d\xi \right|$$
$$\leq C_{R_*} \left[|t_1 - t_2| + |u^{t_1} - u^{t_2}|_\mathscr{C}^2 \right]^{1/2} \leq C_{R_*} |t_1 - t_2|^{1/2}$$

for every $t_1, t_2 \geq t_B \geq h$. Thus from (6.2.1) and (6.2.46) we obtain

$$||A^{-\beta}(u_t(t_1) - u_t(t_2))|| \leq C_{R_*}|t_1 - t_2|^\alpha \quad \text{for all} \;\; t_i \geq t_B, \;\; |t_1 - t_2| \leq 1,$$

for every $0 < \alpha < 1/2$. This implies that the set $D_{\alpha,\beta}^R$ given by (6.2.37) is absorbing for some R provided $0 < \beta \leq 1$ and $0 < \alpha \leq 1/2, \alpha < \beta$. $\qquad \square$

To conclude the proof of the existence of a compact connected global attractor, we apply[8] Proposition 6.2.15 and the existence result given by Theorem 2.3.5.

---

[8] Another way is to apply the existence result due to PATA/ZELIK [179] for closed semigroups. See CHUESHOV/REZOUNENKO [67] for details.

6.2.5 Long-time dynamics: dimension and exponential attractor

In our situation we can assume that there exists a *forward invariant* closed absorbing set D_0 which belongs to $D^R_{\alpha,\beta}$ for an appropriate choice of the parameters (see Proposition 6.2.15). We also note that the restriction of S_t on D_0 is *continuous* in both t and initial data in the topology induced by $\mathscr{L}$ (see (6.2.9)). Thus a dynamical system (S_t, D_0) in the classical sense arises. Therefore we can apply the quasi-stability method.

Proposition 6.2.16 (Quasi-stability). *Let Assumptions* 6.2.1 *and* 6.2.10 *be in force. Assume that* (6.2.38) *and* (6.2.39) *are valid. Let D_0 be a forward invariant closed absorbing set D_0 which belongs to $D^R_{\alpha,\beta}$. Then*

$$|S_t\varphi^1 - S_t\varphi^2|_{\mathscr{L}} \leq C_R e^{-\lambda_1 t}\left[||\varphi^1(0) - \varphi^2(0)||_{1/2} + |\varphi^1 - \varphi^2|_{\mathscr{C}}\right]$$
$$+ C_R \max_{s\in[0,t]} ||A^{1/2-\gamma}(u^1(s) - u^2(s))||, \quad t \geq h, \tag{6.2.47}$$

for every $\varphi^i \in D_0$, where $u^i(t) = (S_t\varphi^i)(\theta)|_{\theta=0}$ and $\gamma \in (0, 1/2]$ is the parameter in (6.2.38) *and* (6.2.39).

Proof. Using the mild form presentation for $u^i(t)$ and (6.2.39) we obtain that

$$||A^{1/2}(u^1(t) - u^2(t))|| \leq e^{-\lambda_1 t}||A^{1/2}(u^1(0) - u^2(0))||$$
$$+ \int_0^t ||A^{1-\delta}e^{-A(t-\tau)}|| \cdot \mathscr{Q}(\tau; u^1, u^2)\, d\tau,$$

where

$$\mathscr{Q}(\tau; u^1, u^2) = C||A^{-1/2+\delta}\left[F([u^1]^\tau) - F([u^2]^\tau)\right]|| + C_R||u^1(\tau) - u^2(\tau)||_{1/2-\gamma}$$

and $\gamma, \delta > 0$ are parameters from (6.2.38) and (6.2.39). As in (6.2.27), using (6.2.38) we also have that

$$||A^{-1/2+\delta}\left[F([u^2]^t) - F([u^1]^t)\right]|| \leq C\left|\int_{t-\eta([u^1]^t)}^{t-\eta([u^2]^t)} ||A^{-\gamma}u^2_t(\xi)||\,d\xi\right|$$
$$+ C|[u^2 - u^1]^t|_{\mathscr{C}} \leq C(R)\max_{\theta\in[-h,0]} ||u^2(t+\theta) - u^1(t+\theta)||$$

for every $t \geq 0$. Therefore

$$||A^{1/2}(u^1(t) - u^2(t))|| \leq c_1 e^{-\lambda_1 t}\left[||A^{1/2}(\varphi^1(0) - \varphi^2(0))|| + |\varphi^1 - \varphi^2|_{\mathscr{C}}\right]$$
$$+ c_2(R)\max_{s\in[0,t]} ||A^{1/2-\gamma}(u^1(s) - u^2(s))||. \tag{6.2.48}$$

Using (6.2.1), (6.2.4), and (6.2.11) we obtain that

$$||A^{-1/2}(u_t^1(t) - u_t^2(t))|| \leq C(R) \left[||A^{1/2}(u^1(t) - u^2(t))|| + |[u^2 - u^1]'|_{\mathscr{C}}\right]$$

Thus

$$||A^{-1/2}(u_t^1(t) - u_t^2(t))|| \leq c_1 e^{-\lambda_1 t} \left[||A^{1/2}(\varphi^1(0) - \varphi^2(0))|| + |\varphi^1 - \varphi^2|_{\mathscr{C}}\right]$$
$$+ c_2(R) \max_{s \in [0,t]} ||A^{1/2-\gamma}(u^1(s) - u^2(s))||, \quad t \geq h. \qquad (6.2.49)$$

Relations (6.2.48) and (6.2.49) imply (6.2.47). The proof of Proposition 6.2.16 is complete.   □

In order to prove finite dimensionality of the attractor $\mathfrak{A}$ we apply Theorem 3.1.21 on the attractor with an appropriate choice of operators and spaces. Indeed, let $T > 0$ be chosen such that $\eta \equiv C_R e^{-\lambda_1 T} < 1$ where C_R is the constant from (6.2.47). We define the Lipschitz mapping

$$K : D_0 \mapsto Z_{[0,T]} \equiv C^1([0,T]; H_{-1/2}) \cap C([0,T]; H_{1/2})$$

by the rule $K\varphi = u(t), t \in [0,T]$, where u is the unique solution of (6.2.1) and (6.2.7) with initial function $\varphi \in D_0$. The seminorm $n_Z(u) \equiv \max_{s \in [0,T]} ||A^{1/2-\gamma} u(s)||$ is compact on $Z_{[0,T]}$ due to the compact embedding of $Z_{[0,T]}$ into $C([0,T]; H_{1/2-\gamma}))$ by the Arzelà-Ascoli theorem (see Lemma A.3.5).

If we take

$$X \equiv \left\{\varphi \in C^1([-h,0]; H_{-1/2}) \cap C([-h,0]; H) \,\big|\, \varphi(0) \in H_{1/2}\right\}$$

equipped with the norm (6.2.32) and suppose that $V = S_T$, then the (discrete) quasi-stability inequality in (3.1.15) is valid on D_0. Hence we can apply Theorem 3.1.21 with $V = S_T$ and $M = \mathfrak{A}$. Thus $\dim_f \mathfrak{A}$ is finite (in X and thus in $\mathscr{L}$).

To prove the existence of a fractal exponential attractor we can use Theorem 3.4.7. For this we need only to note that $t \mapsto S_t \varphi$ is α-Hölder on the absorbing set D_0:

$$|S_{t_1}\varphi - S_{t_2}\varphi|_X \leq C_{D_0}|t_1 - t_2|^\alpha, \quad t_1, t_2 \in [0,T], \quad \varphi \in D_0.$$

This follows from the fact that D_0 is included in the set $D_{\alpha,\beta}^R$ given by (6.2.37) with $\beta \leq 1/2$.

Thus the proof of Theorem 6.2.11 is complete.

6.3 Second order in time evolution equations with delay

Now we consider the dynamics of second order in time equations with delay of the form

$$u_{tt}(t) + ku_t(t) + Au(t) + B(u(t)) + M(u^t) = 0, \quad t > 0, \tag{6.3.1}$$

in a Hilbert space H. Here, as above, A is a linear and $B(\cdot)$ is a nonlinear operator, and $M(u^t)$ represents a (nonlinear) delay effect in the dynamics. All these objects will be specified later. Our consideration is based mainly on ideas and some results established in CHUESHOV/REZOUNENKO [66].

The main model we keep in mind is a nonlinear plate equation of the form

$$u_{tt}(t, x) + ku_t(t, x) + \Delta^2 u(t, x) + f(u(t, x)) + au(t - \tau[u(t)], x) = 0, \tag{6.3.2}$$

in a smooth bounded domain $\Omega \subset \mathbb{R}^2$ with some boundary conditions on $\partial\Omega$. Here τ is a mapping defined on solutions with values in some interval $[0, h]$, and k and a are constants. The term $au(t - \tau[u(t)], x)$ models the effect of the Winkler-type foundation (see SELVADURAI [204] or VLASOV/LEONTIEV [221]) with state-dependent delay response. The nonlinear force F can be of Kirchhoff, Berger, or von Karman type (see Section 6.3.6). Our abstract model also covers the wave equation with state-dependent delay (see the discussion in Section 6.3.6).

Plate equations with *linear* delay terms have previously been studied mainly in Hilbert L_2-type spaces on a lag interval (see, e.g., BOUTET DE MONVEL ET AL. [17], CHUESHOV [36], CHUESHOV/LASIECKA/WEBSTER [63], CHUESHOV/REZOUNENKO [65] and the references therein). However this L_2-type situation does not satisfactorily cover state-dependent delays of the form described above. In this case the delay term in (6.3.2) is not even locally Lipschitz, and thus difficulties related to uniqueness may arise. The desire to have the Lipschitz property for this type of delay term leads naturally to C-type spaces. Moreover, in our approach we employ the special structure of second order in time systems and take into account natural "displacement-velocity" compatibility from the very beginning.

We also note that some results (mainly, the existence and uniqueness) for general second order in time PDEs with delay are available. Most of them are based on a reformulation of the problem as a first order system and application of the theory of such systems (see, e.g., FITZGIBBON [101]). We also mention the papers GARRIDO-ATIENZA/REAL [110] and KARTSATOS/MARKOV [131], which involve the theory of m-accretive (see SHOWALTER [210], for instance) operators.

The main result in this section states that (6.3.1) generates a dynamical system in some space of C^1 functions on the delay time interval and possesses a compact global attractor of finite fractal dimension. We also establish the existence of a fractal exponential attractor. Again, to achieve these results we involve the method of quasi-stability estimates presented in Chapter 3. The main results are illustrated by plate and wave models.

6.3.1 Well-posedness and generation of a dynamical system

In this subsection we introduce our basic hypotheses and prove a well-posedness result. The main outcome is the fact that problem (6.3.1) generates a dynamical system in an appropriate space of C^1 functions.

Assumption 6.3.1. Let H be a separable Hilbert space H with the norm $\|\cdot\|$ and the inner product $(\cdot,\cdot)$. With reference to (6.3.1) we assume:

(A) A is a positive operator with a discrete spectrum on H with a dense domain $\mathscr{D}(A)$ (see Definition 4.1.1).

(B) The nonlinear (non-delayed) mapping $B : \mathscr{D}(A^{1/2}) \to H$ is locally Lipschitz, i.e., for any $R > 0$ there is $L_R > 0$ such that for any u^1, u^2 with $\|A^{1/2}u^i\| \leq R$, one has

$$\|B(u^1) - B(u^2)\| \leq L_R\|A^{1/2}(u^1 - u^2)\|.$$

(M) Consider the space

$$W \equiv C([-h,0]; \mathscr{D}(A^{1/2})) \cap C^1([-h,0]; H), \tag{6.3.3}$$

endowed with the norm

$$|\varphi|_W = \max_{\theta\in[-h,0]} \|A^{1/2}\varphi(\theta)\| + \max_{\theta\in[-h,0]} \|\partial_\theta\varphi(\theta)\|$$

and assume that the nonlinear delay term M maps the space W into H and is locally Lipschitz:

$$\|M(\varphi_1) - M(\varphi_2)\| \leq C_\varrho|\varphi_1 - \varphi_2|_W$$

for every $\varphi_1, \varphi_2 \in W$, $|\varphi_j|_W \leq \varrho, j = 1, 2$.

As in Section 6.1 we also use the spaces $\mathscr{C}_\alpha = C([-h,0]; \mathscr{D}(A^\alpha))$ endowed with the norm

$$|v|_{\mathscr{C}_\alpha} \equiv \sup\{\| A^\alpha v(\theta) \|: \theta \in [-h,0]\}.$$

With this notation the norm in W can be written in the form

$$|\varphi|_W = |\varphi|_{\mathscr{C}_{1/2}} + |\partial_\theta\varphi|_{\mathscr{C}_0}.$$

Below we write $\mathscr{C} = \mathscr{C}_0$. We also recall that $z^t \equiv z^t(\theta) \equiv z(t + \theta)$, $\theta \in [-h,0]$, denotes the element of $C([-h,0]; H)$, while $h > 0$ presents the (maximal) retardation time. As in the previous sections we also use the scale of the spaces H_s generated by the operator A and equipped with the norms $\|u\|_s = \|A^s u\|$.

We supply equation (6.3.1) with the following initial data:

$$u^0 = u^0(\theta) \equiv u(\theta) = \varphi(\theta), \quad \text{for } \theta \in [-h, 0], \ \varphi \in W. \tag{6.3.4}$$

We can rewrite equation (6.3.1) as the first order differential equation

$$\frac{d}{dt}U(t) + \mathscr{A}U(t) = \mathscr{N}(U^t), \quad t > 0, \tag{6.3.5}$$

in the space $\mathscr{H} = H_{1/2} \times H$, where $U(t) = (u(t); u_t(t))$. Here the operator $\mathscr{A}$ and the map $\mathscr{N}$ are defined by

$$\mathscr{A}U = (-v; Au + kv) \text{ for } \quad U = (u; v) \in \mathscr{D}(\mathscr{A}) \equiv \mathscr{D}(A) \times \mathscr{D}(A^{1/2}),$$

$$\mathscr{N}(\Phi) = (0; B(\varphi(0)) + M(\varphi)) \text{ for } \quad \Phi = (\varphi; \partial_\theta \varphi), \ \varphi \in W. \tag{6.3.6}$$

The operator $\mathscr{A}$ generates the exponentially stable C_0-semigroup $e^{-\mathscr{A}t}$ in $\mathscr{H}$; see, e.g., CHUESHOV [39] or TEMAM [216].

The representation in (6.3.5) motivates the following definition.

Definition 6.3.2. A *mild solution* to (6.3.1) and (6.3.4) on an interval $[0, T]$ is defined as a function

$$u \in C([-h, T]; \mathscr{D}(A^{1/2})) \cap C^1([-h, T]; H),$$

such that $u(\theta) = \varphi(\theta)$ for $\theta \in [-h, 0]$ and $U(t) \equiv (u(t); u_t(t))$ satisfies the relation

$$U(t) = e^{-t\mathscr{A}}U(0) + \int_0^t e^{-(t-s)\mathscr{A}}\mathscr{N}(U^s)ds, \quad t \in [0, T]. \tag{6.3.7}$$

Similarly we can also define a mild solution on the semi-interval $[0, T)$. Below $U(t)$ is also occasionally called a mild solution. ∎

Remark 6.3.3. We can also consider the equation in (6.3.7) for U which belongs to the class $C([-h, T]; \mathscr{H})$. In this case both Definitions 6.3.2 and 6.1.2 look the same, the only difference being in the structure of the corresponding linear semigroup. However we prefer to restrict our consideration to the *subspace* in $C([-h, T]; \mathscr{H})$ consisting of pairs of the form $(u; u_t)$. Thus in contrast with FITZGIBBON [101] we implement "displacement-velocity" compatibility at the level of the notion of solutions. As we will see below, this allows us to include for consideration rather general state-dependent delay terms. ∎

One can prove the following local result.

Proposition 6.3.4. *Let Assumption 6.3.1 be valid. Then for any $\varphi \in W$ there exist $T_\varphi > 0$ and a unique mild solution $U(t) \equiv (u(t); u_t(t))$ of (6.3.1) and (6.3.4) on the semi-interval $[0, T_\varphi)$. Solutions continuously depend on initial function $\varphi \in W$.*

Proof. The argument for the local existence and uniqueness of a mild solution is standard (see, e.g., FITZGIBBON [101]) and uses the Banach fixed point theorem for a contraction mapping in the space $C([-h, T]; \mathscr{D}(A^{1/2})) \cap C^1([-h, T]; H)$ with appropriately small T. We note that in contrast with parabolic-type models (see Proposition 6.1.3) the semigroup $e^{-t\mathscr{A}}$ is not compact. Therefore we cannot guarantee the compactness of the corresponding integral-type mapping $\mathscr{B}$ and apply Schauder's fixed point theorem. This is why we need Lipschitz conditions for B and M in Assumption 6.3.1.   □

To obtain a global well-posedness result we need additional hypotheses concerning B and M.

Assumption 6.3.5. We assume the following properties.

(B)′ The nonlinear mapping $B : H_{1/2} \to H$ is potential, i.e., it has the form

$$B(u) = \Pi'(u),$$

where $\Pi'(u)$ denotes the Fréchet derivative[9] of a C^1 functional $\Pi(u) : H_{1/2} \to \mathbb{R}$. Moreover, we assume that $\Pi(u) = \Pi_0(u) + \Pi_1(u)$, where $\Pi_0(u) \geq 0$ is bounded on bounded sets in $H_{1/2}$ and $\Pi_1(u)$ satisfies the property

$$\forall \eta > 0 \, \exists C_\eta > 0 : \; |\Pi_1(u)| \leq \eta \left(||A^{1/2}u||^2 + \Pi_0(u)\right) + C_\eta, \;\; u \in H_{1/2}. \tag{6.3.8}$$

(M) The nonlinear delay term $M : W \to H$ satisfies the linear growth condition

$$||M(\varphi)|| \leq M_0 + M_1 \left\{ \max_{\theta \in [-h,0]} ||A^{1/2}\varphi(\theta)|| + \max_{\theta \in [-h,0]} ||\partial_\theta \varphi(\theta)|| \right\} \tag{6.3.9}$$

for all $\varphi \in W$ and for some $M_j \geq 0$.

As is well-documented in CHUESHOV/LASIECKA [56, 58], the second order models with nonlinearities satisfying Assumption 6.3.5(B) arise in many applications (see also the discussion in Section 6.3.6). We also emphasize that the force M may contain non-delay terms; i.e., it is allowed that

$$M(\varphi) = B^*(\varphi(0)) + \bar{M}(\varphi), \;\; \varphi \in W,$$

where $\bar{M}$ obeys the conditions above concerning M and $B^* : H_{1/2} \to H$ is Lipschitz and linearly bounded.

We have the following well-posedness result.

---

[9]See the definition in Section A.5.

Theorem 6.3.6 (Well-posedness). *Let Assumptions 6.3.1 and 6.3.5 be in force. Then for any $\varphi \in W$ there exists a unique global mild solution $U(t) \equiv (u(t); u_t(t))$ to (6.3.1) and (6.3.4) on the semi-axis $[0, +\infty)$. This solution satisfies the energy equality*

$$\mathscr{E}(u(t), u_t(t)) + k \int_0^t ||u_t(s)||^2 ds = \mathscr{E}(u(0), u_t(0)) - \int_0^t (M(u^s), u_t(s)) ds.$$

(6.3.10)

Here we denote

$$\mathscr{E}(u, v) \equiv E(u, v) + \Pi_1(u) \text{ with } E(u, v) \equiv \frac{1}{2} \left(||v||^2 + ||A^{1/2}u||^2 \right) + \Pi_0(u).$$

(6.3.11)

Moreover, for any $\varrho > 0$ and $T > 0$ there exists $C_{\varrho,T}$ such that

$$||A^{1/2}(u(t) - \tilde{u}(t))|| + ||u_t(t) - \tilde{u}_t(t)|| \leq C_{\varrho,T}|\varphi - \tilde{\varphi}|_W, \quad t \in [0, T], \quad (6.3.12)$$

for any pair $u(t)$ and $\tilde{u}(t)$ of mild solutions with initial data φ and $\tilde{\varphi}$ such that $|\varphi|_W, |\tilde{\varphi}|_W \leq \varrho$.

Proof. The local existence and uniqueness of mild solutions are given by Proposition 6.3.4. Let $U = (u; u_t)$ be a mild solution to (6.3.1) and (6.3.4) on the (maximal) semi-interval $[-h, T_\varphi)$ and

$$f^u(t) \equiv B(u(t)) + M(u^t) \in C([0, T_\varphi); H).$$

It is clear that we can consider $(u(t); u_t(t))$ as a mild solution of the linear *non-delayed* equation

$$v_{tt}(t) + Av(t) + kv_t(t) + f^u(t) = 0, \quad t \in [0, T_\varphi). \quad (6.3.13)$$

Therefore (see Chapter 5) one can show that $u(t)$ satisfies an energy relation of the form

$$E_0(u(t), u_t(t)) + k \int_0^t ||u_t(s)||^2 ds = E_0(u(0), u_t(0)) - \int_0^t (f^u(s), u_t(s)) ds$$

(6.3.14)

for all $0 \leq t < T_\varphi$, where $E_0(u, v) = \frac{1}{2} \left(||A^{1/2}u||^2 + ||v||^2 \right)$. Using the structure of f^u, after some calculations (first performed on smooth functions) we can show that

$$\int_0^t (f^u(s), u_t(s)) ds = \Pi(u(t)) - \Pi(u(0)) + \int_0^t (M(u^s), u_t(s)) ds.$$

Therefore (6.3.14) yields (6.3.10) for every $t < T_\varphi$.

Using (6.3.9) and (6.3.10) we obtain

$$\mathscr{E}(u(t), u_t(t)) + \frac{k}{2} \int_0^t ||u_t(s)||^2 ds \leq \mathscr{E}(u(0), u_t(0)) + c_1 \int_0^t \left(1 + |u^s|_W^2 \right) ds.$$

(6.3.15)

One can see that

$$|u^s|_W = \max_{\theta \in [-h,0]} ||A^{1/2}u(s+\theta)|| + \max_{\theta \in [-h,0]} ||u_t(s+\theta)||$$

$$\leq |\varphi|_W + 2\sqrt{2} \max_{\sigma \in [0,s]} \left[E(u(\sigma), u_t(\sigma))\right]^{1/2} \tag{6.3.16}$$

for every $s \in [0, T_\varphi)$. It follows from (6.3.8) that there exists a constant $c > 0$ such that

$$\frac{1}{2}E(u,v) - c \leq \mathscr{E}(u,v) \leq 2E(u,v) + c, \quad u \in \mathscr{D}(A^{1/2}), \; v \in H. \tag{6.3.17}$$

Therefore we use (6.3.17) and (6.3.16) to continue (see (6.3.15)) as follows:

$$\max_{\sigma \in [0,t]} E(u(\sigma), u_t(\sigma))$$

$$\leq c\left(1 + t + E(u(0), u_t(0)) + t \cdot |\varphi|_W^2 + \int_0^t \max_{\sigma \in [0,s]} E(u(\sigma), u_t(\sigma)) \, ds\right).$$

The application of Gronwall's lemma yields the following (a priori) estimate:

$$\max_{\sigma \in [0,t]} E(u(\sigma), u_t(\sigma)) \leq C\left(1 + E(u(0), u_t(0)) + |\varphi|_W^2\right) \cdot e^{at}, \quad a > 0, \; t < T_\varphi,$$

which allows us, in the standard way, to extend the solution on the semi-axis $\mathbb{R}_+$.

To prove (6.3.12) we use the fact that the difference $w(t) = u(t) - \tilde{u}(t)$ solves the problem in (6.3.13) with

$$f^u(t) = B(u(t)) + M(u^t) - B(\tilde{u}(t)) - M(\tilde{u}^t).$$

This completes the proof of Theorem 6.3.6.   □

Using Theorem 6.3.6 we can define an *evolution operator* $S_t : W \to W$ for all $t \geq 0$ by the formula $S_t\varphi = u^t$, where $u(t)$ is the mild solution to (6.3.1) and (6.3.4), satisfying $u^0 = \varphi$. This operator satisfies the semigroup property and generates a dynamical system (W, S_t) with the phase space W defined in (6.3.3).

We conclude this subsection with the following remarks.

Remark 6.3.7 (Smooth solutions). Assume in addition that $B(u)$ is Frechét differentiable on $H_{1/2}$ and M is "locally almost Lipschitz" on $\mathscr{C}$ in the sense that

$$\|M(\varphi_1) - M(\varphi_2)\| \leq C_\varrho |\varphi_1 - \varphi_2|_\mathscr{C} \tag{6.3.18}$$

for every $\varphi_1, \varphi_2 \in W$, $|\varphi_j|_W \leq \varrho$, $j = 1, 2$. Then the smoothness of the initial data φ and some compatibility conditions imply that the solutions are C^2 smooth on $[-h, +\infty)$. Indeed, one can show (see [66]) that the set

$$\mathscr{L} = \left\{ \varphi \in C^2([-h,0]; H) \, \middle| \, \begin{array}{l} \varphi \in C^1([-h,0]; H_{1/2}) \cap C([-h,0]; H_1), \\ \varphi_{tt}(0) + k\varphi_t(0) + A\varphi(0) + B(\varphi(0)) + M(\varphi) = 0. \end{array} \right\}$$

(6.3.19)

is forward invariant with respect to the flow S_t subset in W. Thus the dynamics is defined in a smoother space. The set $\mathscr{L}$ is an analog to the solution manifold used in WALTHER [222] for the ODE case and in REZOUNENKO/ZAGALAK [194] for the parabolic PDE case as a well-posedness class. See CHUESHOV/REZOUNENKO [66] for more details. ∎

Remark 6.3.8 (Finite-dimensional case). The well-posedness results in Theorem 6.3.6 can also be applied in the ODE case when $H = \mathbb{R}^n$, A is a symmetric $n \times n$ matrix A, and the nonlinear mappings $B : \mathbb{R}^n \to \mathbb{R}^n$, $M : C([-h,0]; \mathbb{R}^n) \to \mathbb{R}^n$ satisfy appropriate requirements. The space of initial states is $W = C^1([-h,0]; \mathbb{R}^n)$. In contrast with the solution manifold method suggested in WALTHER [222] (see also HARTUNG ET AL. [121]), this approach to well-posedness does not assume any nonlinear compatibility conditions and is based on the natural (linear) "position-velocity" compatibility. This provides us with an alternative point of view on dynamics and leads to a simpler well-posedness argument compared to the method of the solution manifold. For a more detailed discussion we refer to CHUESHOV/REZOUNENKO [66]. ∎

6.3.2 Asymptotic properties: dissipativity

Now we begin to study the long-time dynamics of the system (W, S_t) generated by mild solutions to problem (6.3.1). To do this, we need to impose additional hypotheses.

Assumption 6.3.9. We assume the following.

(B) The nonlinear term $B : H_{1/2} \mapsto H$ has the potential $\Pi(u) = \Pi_0(u) + \Pi_1(u)$ satisfying (6.3.8) and also **(a)** there are constants $\nu \in [0,1), c_1, c_2 > 0$ such that

$$- (u, B(u)) \le \nu \|A^{1/2}u\|^2 - c_1 \Pi_0(u) + c_2, \quad u \in H_{1/2};$$

(6.3.20)

(b) for every $\eta > 0$ there exists $C_\eta > 0$ such that

$$\|u\|^2 \le C_\eta + \eta \left(\|A^{1/2}u\|^2 + \Pi_0(u) \right), \quad u \in H_{1/2}.$$

(6.3.21)

(M) The nonlinear delay term $M : W \mapsto H$ possesses the property

$$\|M(u^t)\|^2 \le g_0 + g_1 \|A^{1/2-\delta}u(t)\|^2 + g_2(h) \int_{t-h}^t \|u_t(s)\|^2 \, ds$$

(6.3.22)

with the parameters $g_0, g_1 > 0$, $\delta \in (0, 1/2]$ independent of h and the factor $g_2(h)$ such that $hg_2(h) \to 0$ as $h \to 0$.

Remark 6.3.10. Concerning the nonlinear (non-delayed) term B, our assumptions are motivated by nonlinear plate models and are the same as in [56] and [58, Chapter 8]. We have already involved similar requirements in the case of parabolic models with state-dependent delay (see Assumption 6.2.10). We also point out that the requirements in Assumption 6.3.9 imply the hypotheses of Assumption 6.3.5.

For the delay term, our main example is a *discrete state-dependent delay* force $M : W \mapsto H$ of the form $M(u^t) = G(u(t - \tau(u^t)))$, where τ maps W into the interval $[0, h]$ and G is a globally Lipschitz mapping from H into itself. In this case the term $M(u^t)$ can be written in the form

$$M(u^t) = G(u(t - \tau(u^t))) \equiv G\left(u(t) - \int_{t-\tau(u^t)}^{t} u_t(s)\, ds \right). \tag{6.3.23}$$

Thus we have that

$$||M(u^t)|| \leq ||G(0)|| + L_G \left[||u(t)|| + \int_{t-h}^{t} ||u_t(s)||\, ds \right],$$

where L_G is the Lipschitz constant of the mapping G. This yields (6.3.22) with $g_0 = 4||G(0)||^2$, $g_1 = 4L_G^2$, and $g_2(h) = 2L_G^2 h$. We also note that $M(u^t)$ in the form (6.3.23) satisfies the Lipschitz condition in Assumption 6.3.1(M) if we assume that τ is locally Lipschitz on W:

$$|\tau(\varphi_1) - \tau(\varphi_2)| \leq C_\varrho |\varphi_1 - \varphi_2|_W$$

for every $\varphi_1, \varphi_2 \in W$, $|\varphi_j|_W \leq \varrho, j = 1, 2$. Indeed, from (6.3.23) we have that

$$||M(u^s) - M(\tilde{u}^s)|| \leq L_G ||u(s - \tau(u^s)) - u(s - \tau(\tilde{u}^s))||$$
$$+ L_G ||u(s - \tau(\tilde{u}^s)) - \tilde{u}(s - \tau(\tilde{u}^s))||$$
$$\leq \varrho L_G |\tau(u^s) - \tau(\tilde{u}^s)| + L_G \max_{\theta \in [-h, 0]} ||u(s + \theta) - \tilde{u}(s + \theta)||$$
$$\leq (1 + \varrho C_\varrho) L_G |u^s - \tilde{u}^s|_W$$

for all $u^s, \tilde{u}^s \in W$, $|u^s|_W, ||\tilde{u}^s|_W \leq \varrho$.

Instead of the structure presented in (6.3.23) we can take a delay term of the form

$$M(u^t) = \sum_{k=1}^{N} G_k(u(t - \tau_k(u^t))),$$

or even consider an integral version of this sum and add a non-delay subcritical force $B^*(u(t))$ with linear growth. We can also include velocity terms with a (distributed) state-dependent delay of the form

$$\int_{-h}^{0} r(\theta, u^t) u_t(t + \theta) \, d\theta,$$

where $r : [-h, 0] \times W \mapsto H$ is measurable in the first variable and globally Lipschitz with respect to the second variable and satisfies appropriate properties. However, for the sake of transparency, we do not pursue these generalizations. ∎

Our first step in the study of the qualitative behavior of the system (W, S_t) is the following (ultimate) dissipativity property.

Proposition 6.3.11 (Dissipativity). *Let Assumptions 6.3.1 and 6.3.9 be valid. Then for any k_0 there exists $h_0 = h(k_0) > 0$ such that for every*

$$(k; h) \in [k_0, +\infty) \times (0, h_0]$$

the system (W, S_t) is dissipative, i.e., there exists $R > 0$ such that for every $\varrho > 0$ we can find $t_\varrho > 0$ such that

$$|S_t \varphi|_W \leq R \ \text{for all} \quad \varphi \in W, \quad |\varphi|_W \leq \varrho, \quad t \geq t_\varrho.$$

Moreover for every fixed $k_0 > 0$ the dissipativity radius R is independent of $k \geq k_0$ and the delay time $h \in (0, h_0]$. Thus the dynamical system (W, S_t) is dissipative uniformly in $k \geq k_0$ and $h \leq h_0$.

Remark 6.3.12. **(1)** The dissipativity property can be written in the form

$$||u_t(t)||^2 + ||A^{1/2} u(t)||^2 \leq R^2 \quad \text{for all} \quad t \geq t_\varrho,$$

provided the initial function $\varphi \in W$ possesses the property $|\varphi|_W \leq \varrho$. We can also show in the standard way (see Exercise 2.1.6) that there exists a bounded *forward invariant* absorbing set $\mathscr{B}$ in W which belongs to the ball $\{\varphi \in W : |\varphi|_W \leq R\}$ with radius R independent of $k \in [k_0, +\infty)$ and $h \in (0, h_0]$.

(2) As we see in the proof below, by increasing the low bound k_0 for the damping interval, we can increase the corresponding admissible interval for h. This fact is compatible with the observation that a large time lag may destabilize the system. For instance, it is known from COOKE/GROSSMAN [81] that for the delayed 1D ODE

$$\ddot{u}(t) + k\dot{u}(t) + au(t) + u(t - \tau) = 0$$

with $a > 1$ and $2a > k^2$, there exist positive numbers $\tau_* < \tau^*$ such that the zero solution is stable for all $\tau < \tau_*$ and unstable when $\tau > \tau^*$. This example also demonstrates the role of a large damping. Indeed, if $k^2 > 2a > 2$, then (see COOKE/GROSSMAN [81]) the zero solution is stable for *all* $\tau \geq 0$. Thus a large time delay requires a sufficiently large damping coefficient to stabilize this system.

∎

Proof. We use the following functional:

$$\tilde{V}(t) \equiv \mathscr{E}(u(t), u_t(t)) + \gamma(u(t), u_t(t)) + \frac{\mu}{h} \int_0^h \left\{ \int_{t-s}^t ||u_t(\xi)||^2 d\xi \right\} ds.$$

Here $\mathscr{E}$ is defined in (6.3.11) and the positive parameters γ and μ will be chosen later.

As in the proof of Theorem 6.1.15 (see also Proposition 6.2.13), the main idea behind inclusion of an additional delay term in $\tilde{V}$ is to find an appropriate compensator for $M(u^t)$. The compensator is determined by the structure of the mapping M (see (6.3.22)). For second order in time infinite-dimensional models this idea was applied in CHUESHOV/LASIECKA [58, p. 480] and CHUESHOV/LASIECKA/WEBSTER [63] in the study of a flow-plate interaction model which contains a linear constant delay term with critical spatial regularity. The corresponding compensator has a different form in the latter case and thus it is model-dependent.

One can see from (6.3.8) that there is $0 < \gamma_0 < 1$ such that

$$\frac{1}{2} E(u(t), u_t(t)) - c \leq \tilde{V}(t) \leq 2E(u(t), u_t(t)) + \mu \int_0^h ||u_t(t-\xi)||^2 d\xi + c. \quad (6.3.24)$$

for every $0 < \gamma \leq \gamma_0$, where c does not depend on k.

Let us consider the time derivative of $\tilde{V}$ along a solution. One can see that

$$\frac{d}{dt}(u(t), u_t(t)) = ||u_t(t)||^2 - k(u(t), u_t(t))$$

$$- ||A^{1/2}u(t)||^2 - (u, B(u)) - (u, M(u^t)). \quad (6.3.25)$$

Combining (6.3.25) with the energy relation in (6.3.10) and using the estimate $k(u, u_t) \leq k^2 ||u_t||^2 + \frac{1}{4} ||u||^2$ we get

$$\frac{d}{dt} \tilde{V}(t) \leq - (k - \gamma(1 + k^2) - \mu)||u_t(t)||^2 + M(u^t), u_t(t))$$

$$- \gamma \left(-\frac{1}{4}||u(t)||^2 + ||A^{1/2}u(t)||^2 + (u, B(u)) + (u, M(u^t)) \right)$$

$$- \frac{\mu}{h} \int_0^h ||u_t(t - \xi)||^2 d\xi.$$

Using the inequality $|(M(u^t), u_t(t))| \leq \frac{1}{4}k||u_t(t)||^2 + \frac{1}{k}||M(u^t)||^2$ and also estimate (6.3.22) we obtain that

$$|(M(u^t), u_t(t))| \leq \frac{1}{4}k||u_t(t)||^2$$

$$+ \frac{c_0}{k}\left[1 + ||A^{1/2-\delta}u(t)||^2\right] + \frac{2g_2(h)}{k} \int_0^h ||u_t(t - \xi)||^2 d\xi,$$

where $c_0 > 0$ does not depend on k.

In a similar way,

$$\frac{1}{4}\|u(t)\|^2 + |(u(t), M(u^t))|$$

$$\leq g_2(h) \int_0^h \|u_t(t-\xi)\|^2 d\xi + C_{g_0,g_1}(1 + \|A^{1/2-\delta}u(t)\|^2).$$

The relations in (6.3.20) and (6.3.21) with small enough $\eta > 0$ and an interpolation inequality (see Exercise 4.1.2) of the form

$$\|A^{1/2-\delta}u\|^2 \leq \varepsilon\|A^{1/2}u\|^2 + C_\varepsilon\|u\|^2, \quad u \in \mathscr{D}(A^{1/2}), \quad \forall \varepsilon > 0, \tag{6.3.26}$$

yield

$$C_{g_0,g_1}(1 + \|A^{1/2-\delta}u\|^2) - \|A^{1/2}u\|^2 - (u, B(u)) \leq -3a_0 E(u, u_t) + \|u_t\|^2 + a_1$$

for some $a_i > 0$. Thus it follows from the relations above that

$$\frac{d}{dt}\tilde{V}(t) \leq -\left(\frac{3}{4}k - \gamma(2 + k^2) - \mu\right)\|u_t(t)\|^2$$

$$+\frac{c_0}{k}\left[1 + \|A^{1/2-\delta}u(t)\|^2\right] + \gamma\left(-3a_0 E(u(t), u_t(t)) + a_1\right)$$

$$+\left[-\frac{\mu}{h} + \left(\frac{2}{k} + \gamma\right)g_2(h)\right]\int_0^h \|u_t(t-\xi)\|^2 d\xi.$$

Using (6.3.21) and (6.3.26) we find that

$$\frac{c_0}{k}\|A^{1/2-\delta}u\|^2 \leq \frac{\varepsilon}{k}\left[\|A^{1/2}u\|^2 + \Pi_0(u)\right] + \frac{1}{k}b\left(\frac{1}{\varepsilon}\right), \quad \forall \varepsilon > 0,$$

where $b(s)$ is a non-decreasing function. Taking $\varepsilon = \gamma a_0 k$ we obtain

$$\frac{c_0}{k}\|A^{1/2-\delta}u(t)\|^2 \leq \gamma a_0 E(u(t), u_t(t)) + \frac{1}{k}b\left(\frac{1}{\gamma a_0 k}\right),$$

where $b(s)$ is a non-decreasing function. Thus using (6.3.24) and rescaling the function $b(s)$ we arrive at the relation

$$\frac{d}{dt}\tilde{V}(t) + \gamma a_0\tilde{V}(t) \leq -\left(\frac{3}{4}k - \gamma(2 + k^2) - \mu\right)\|u_t(t)\|^2 + \gamma\left[\tilde{a} + \frac{1}{\gamma k}b\left(\frac{1}{\gamma k}\right)\right],$$

$$+\left[-\frac{\mu}{h} + \mu\gamma a_0 + \left(\frac{2}{k} + \gamma\right)g_2(h)\right]\int_0^h \|u_t(t-\xi)\|^2 d\xi.$$

$$\tag{6.3.27}$$

Take $\mu = \frac{k}{4}$ and $\gamma = \frac{\sigma k}{4+2k^2}$, where $0 < \sigma < 1$ is chosen such that $\gamma \leq \gamma_0$ for all $k > 0$ (the bound γ_0 arises in (6.3.24)). Assume also that h satisfies the inequality

$$-\frac{k}{4h} + \frac{\gamma k}{4}a_0 + \left(\frac{2}{k} + \gamma\right) g_2(h) \leq 0. \tag{6.3.28}$$

Then (6.3.27) implies that

$$\frac{d}{dt}\tilde{V}(t) + \gamma a_0 \tilde{V}(t) \leq \gamma \left[\tilde{a} + \frac{1}{\gamma k}b\left(\frac{1}{\gamma k}\right)\right]. \tag{6.3.29}$$

One can see that there is $\sigma_0 = \sigma_0(k_0)$ such that $\sigma_0 \leq \gamma k \leq \sigma/2$ for all $k \geq k_0$. Therefore from (6.3.29) we obtain that

$$\tilde{V}(t) \leq \tilde{V}(0)e^{-\gamma a_0 t} + \frac{1}{a_0}(1 - e^{-\gamma a_0 t})\left[\tilde{a} + \frac{1}{\sigma_0}b\left(\frac{1}{\sigma_0}\right)\right], \tag{6.3.30}$$

provided

$$-\frac{k_0}{4h} + \frac{1}{8}a_0 + g_2(h)\left(\frac{2}{k_0} + \frac{1}{2}\right) \leq 0. \tag{6.3.31}$$

Here we used (6.3.28) and properties $\gamma k < 1/2, \gamma < 1/2$, which follow from the choice of γ. The relation in (6.3.31) can be written in the form

$$h\left[a_0 \frac{k_0}{k_0 + 4} + 4g_2(h)\right] \leq \frac{2k_0^2}{k_0 + 4}.$$

This is true if we assume, for instance, that

$$a_0 h \leq k_0 \quad \text{and} \quad h g_2(h) \leq \frac{1}{4}\frac{k_0^2}{k_0 + 4}.$$

Under this condition relation (6.3.30) implies the desired (uniform in k and h) dissipativity property and completes the proof of Proposition 6.3.11.   □

6.3.3 Asymptotic properties: quasi-stability

In this section we show that the system (W, S_t) generated by the delay equation in (6.3.1) possesses some asymptotic quasi-stability property. As we have seen before, quasi-stability leads to several important conclusions concerning the global long-time dynamics of the system.

Quasi-stability requires additional hypotheses concerning the system.

Assumption 6.3.13.

(B) We assume that the nonlinear (non-delayed) mapping $B : H_{1/2} \to H$ satisfies one of the following conditions:[10]

   (a) **either** it is *subcritical*, i.e., there is positive η such that for any $R > 0$ there exists $L_B(R) > 0$ such that

$$||B(u) - B(\tilde{u})|| \leq L_B(R)||A^{1/2-\eta}(u - \tilde{u})||, \qquad (6.3.32)$$

   for all $u, \tilde{u} \in \mathscr{D}(A^{1/2})$ with the properties $||A^{1/2}u||, ||A^{1/2}\tilde{u}|| \leq R$;
   (b) **or else** it is *critical*, i.e., (6.3.32) holds with $\eta = 0$, and the damping parameter k is large enough.

(M) There exists $\delta > 0$ such that the delay term M satisfies the subcritical local Lipschitz property, i.e., for any $\varrho > 0$ there exists $L_M(\varrho) > 0$ such that

$$||M(\varphi) - M(\tilde{\varphi})|| \leq L_M(\varrho) \max_{\theta \in [-h,0]} ||A^{1/2-\delta}(\varphi(\theta) - \tilde{\varphi}(\theta))||, \qquad (6.3.33)$$

for any φ and $\tilde{\varphi}$ such that $||\varphi||_W, ||\tilde{\varphi}||_W \leq \varrho$.

As in Remark 6.3.10 one can see that (6.3.33) holds for M given by (6.3.23) if we assume that

$$|\tau(\varphi) - \tau(\tilde{\varphi})| \leq L_\tau(\varrho) \max_{\theta \in [-h,0]} ||A^{1/2-\delta}(\varphi(\theta) - \tilde{\varphi}(\theta))||. \qquad (6.3.34)$$

The following theorem is the main step in the proof of quasi-stability of the system (W, S_t).

Theorem 6.3.14 (Quasi-stability inequality). *Let Assumptions 6.3.1, 6.3.5, and 6.3.13 be in force. Then there exist positive constants $C_1(R)$, $\tilde{\lambda}$, and $C_2(R)$ such that for any two solutions $u(t)$ and $\tilde{u}(t)$ with initial data φ and $\tilde{\varphi}$ possessing the properties*

$$||u_t(t)||^2 + ||A^{1/2}u(t)||^2 \leq R^2, \quad ||\tilde{u}_t(t)||^2 + ||A^{1/2}\tilde{u}(t)||^2 \leq R^2 \quad \text{for all } t \geq -h,$$
$$(6.3.35)$$

the following quasi-stability estimate holds:

$$||u_t(t) - \tilde{u}_t(t)||^2 + ||A^{1/2}(u(t) - \tilde{u}(t))||^2$$
$$\leq C_1(R)e^{-\tilde{\lambda}t}|\varphi - \tilde{\varphi}|_W^2 + C_2(R) \max_{\xi \in [0,t]} ||A^{1/2-\delta}(u(\xi) - \tilde{u}(\xi))||^2 \qquad (6.3.36)$$

with some $\delta > 0$. In the critical case $k \geq k_0(R)$ for some $k_0(R) > 0$.

---

[10] We distinguish the cases of critical and subcritical nonlinearities.

We emphasize that Theorem 6.3.14 does not involve Assumption 6.3.9 and deals only with pairs of uniformly bounded solutions. However, if the conditions in Assumption 6.3.9 are valid, then by Proposition 6.3.11 and Remark 6.3.12(1) there exists a bounded forward invariant absorbing set. Thus under the conditions of Proposition 6.3.11 we can apply Theorem 6.3.14 on this set.

Before proving Theorem 6.3.14, we note that its main consequence is the following assertion, which states quasi-stability of the system (W, S_t) in the sense of Definition 3.4.1.

Theorem 6.3.15 (Quasi-stability). *Let Assumptions 6.3.1, 6.3.5, and 6.3.13 be in force. Then for every bounded forward invariant set $\mathscr{B}$ there exists $T = T_{\mathscr{B}}$ such that the system (W, S_t) is quasi-stable at the time T on $\mathscr{B}$.*

Proof. We assume that $\mathscr{B}$ lies in the ball $\{\varphi : |\varphi|_W \le R\}$ and choose $T > h$ in (6.3.36) such that $q = C_1(R)e^{-\tilde{\lambda}T} < 1$. As the space Z we take

$$Z = Z_T \equiv C([0, T]; H_{1/2}) \cap C^1([0, T]; H)$$

and define the seminorm

$$n_Z(u) = C_2(R) \max_{\xi \in [0,T]} \|A^{1/2-\delta}u(\xi)\|^2.$$

By the Arzelà-Ascoli theorem (see Lemma A.3.5) this seminorm is compact on Z. Thus we obtain (3.4.1) with $X = W$ and $K : W \mapsto Z$ given by the relation

$$K[u^0](t) = u(t), \quad t \in [0, T],$$

where u is a solution to (6.3.1) and (6.3.4) with initial data $u^0 \in W$.   □

6.3.3.1 Proof of Theorem 6.3.14

We split the proof into two cases and start with the simplest one.

Subcritical case: We rely on the mild solution form (6.3.7) of the problem and follow the line of argument given in CHUESHOV/LASIECKA [58, p. 479-480] with modifications which are necessary for the case of state-dependent delay force M. As in Chapter 5, here we can also use the multipliers method. However for completeness we will demonstrate the constant variation method. The multipliers method is presented below in the case of the critical force B.

Let us consider two solutions $U = (u; u_t)$ and $\tilde{U} = (\tilde{u}; \tilde{u}_t)$ to (6.3.1) possessing property (6.3.35). Using (6.3.7) and exponential stability of the semigroup $e^{-\mathscr{A}t}$ in the space $\mathscr{H} = \mathscr{D}(A^{1/2}) \times H$ we have that

$$||U(t) - \tilde{U}(t)||_{\mathscr{H}} \leq Ce^{-\tilde{\lambda}t}||U(0) - \tilde{U}(0)||_{\mathscr{H}}$$

$$+ C \int_0^t e^{-\tilde{\lambda}(t-s)}||\mathscr{N}(U^s) - \mathscr{N}(\tilde{U}^s)||_{\mathscr{H}} \, ds, \quad t > 0,$$

$$(6.3.37)$$

with $\tilde{\lambda}, C > 0$, where $\mathscr{N}$ is given by (6.3.6). Since

$$||\mathscr{N}(U^s) - \mathscr{N}(\tilde{U}^s)||_{\mathscr{H}} \leq ||B(u(t)) - B(\tilde{u}(t))|| + ||M(u^t) - M(\tilde{u}^t)||,$$

using properties (6.3.32) and (6.3.33) we obtain

$$||\mathscr{N}(U^s) - \mathscr{N}(\tilde{U}^s)||_{\mathscr{H}} \leq C(R) \max_{\theta \in [-h,0]} ||A^{1/2-\delta}(u(s+\theta) - \tilde{u}(s+\theta))||$$

for some $\delta > 0$. Thus (6.3.37) yields

$$||U(t) - \tilde{U}(t)||_{\mathscr{H}} \leq Ce^{-\tilde{\lambda}t}||U(0) - \tilde{U}(0)||_{\mathscr{H}} + C(R)\, I(t, u - \tilde{u}) \qquad (6.3.38)$$

for $t > 0$, where

$$I(t, z) = \int_0^t e^{-\tilde{\lambda}(t-s)} \max_{\ell \in [-h,0]} ||A^{1/2-\delta}z(s+\ell)|| \, ds \text{ with } z(s) = u(s) - \tilde{u}(s).$$

Now we split $I(t, z)$ as $I(t, z) = I^1(t, z) + I^2(t, z)$, where

$$I^1(t, z) \equiv \int_0^h e^{-\tilde{\lambda}(t-s)} \max_{\ell \in [-h,0]} ||A^{1/2-\delta}z(s+\ell)|| \, ds \leq C_{R,h}|z^0|_W \int_0^h e^{-\tilde{\lambda}(t-s)} \, ds$$

$$= C_{R,h}|z^0|_W \cdot e^{-\tilde{\lambda}t}(e^{\tilde{\lambda}h} - 1)\tilde{\lambda}^{-1}$$

and

$$I^2(t, z) \equiv \int_h^t e^{-\tilde{\lambda}(t-s)} \max_{\ell \in [-h,0]} ||A^{1/2-\delta}z(s+\ell)|| \, ds$$

$$\leq \int_0^t e^{-\tilde{\lambda}(t-s)} ds \max_{\xi \in [0,t]} ||A^{1/2-\delta}z(\xi)|| = (1 - e^{-\tilde{\lambda}t})\tilde{\lambda}^{-1} \cdot \max_{\xi \in [0,t]} ||A^{1/2-\delta}z(\xi)||.$$

Thus (6.3.38) yields the desired estimate in (6.3.36) for the subcritical case.

Critical case with large damping: We use the same idea as in Section 5.3.1 and partially follow the line of the arguments of CHUESHOV/LASIECKA [56, p. 85, Theorem 3.58].

Let u and $\tilde{u}$ be solutions satisfying (6.3.35). Then $z = u - \tilde{u}$ solves the equation

$$z_{tt}(t) + Az(t) + kz_t(t) = -B_{1,2}(t) - M_{1,2}(t) \tag{6.3.39}$$

with

$$B_{1,2}(t) \equiv B(u(t)) - B(\tilde{u}(t)); \quad M_{1,2}(t) \equiv M(u^t) - M(\tilde{u}^t).$$

We multiply the last equation by $z_t(t)$ and integrate over $[t, T]$:

$$E_z(T) - E_z(t) + k \int_t^T \|z_t(s)\|^2 \, ds$$

$$= - \int_t^T (B_{1,2}(s), z_t(s)) \, ds - \int_t^T (M_{1,2}(s), z_t(s)) \, ds. \tag{6.3.40}$$

Here we denote $E_z(t) \equiv \frac{1}{2}(\|z_t(t)\|^2 + \|A^{1/2}z(t)\|^2)$.

One can check that there is a constant $C_R > 0$ such that

$$|(B_{1,2}(t), z_t(t))| \leq \varepsilon \|A^{1/2}z(t)\|^2 + \frac{C_R}{\varepsilon}\|z_t(t)\|^2, \quad \forall \varepsilon > 0.$$

Similarly, using Assumption 6.3.13(M), we have

$$|(M_{1,2}(t), z_t(t))| \leq \max_{\theta \in [-h,0]} \|A^{1/2-\delta}z(t + \theta)\|^2 + C_R\|z_t(t)\|^2.$$

Hence, from (6.3.40) we get

$$\left| E_z(T) - E_z(t) + k \int_t^T \|z_t(s)\|^2 \, ds \right| \leq \varepsilon \int_t^T \|A^{1/2}z(s)\|^2 \, ds$$

$$+ \int_t^T \max_{\theta \in [-h,0]} \|A^{1/2-\delta}z(s + \theta)\|^2 \, ds + C_R\left(1 + \frac{1}{\varepsilon}\right) \int_t^T \|z_t(s)\|^2 \, ds \tag{6.3.41}$$

for every $\varepsilon > 0$. Below we choose k (assume that it is) big enough to satisfy (see the the last term in (6.3.41))

$$C_R\left(1 + \frac{1}{\varepsilon}\right) < \frac{k}{2} \quad \text{for all} \quad k \geq k_0. \tag{6.3.42}$$

This choice is made to simplify the estimates only (the final choice of k_0 will be made after the choice of ε).

Now we multiply (6.3.39) by $z(t)$ and integrate over $[0, T]$, using integration by parts. This yields

$$(z_t(T), z(T)) - (z_t(0), z(0)) + \int_0^T ||A^{1/2}z(s)||^2\, ds + k \int_0^T (z_t(s), z(s))\, ds$$

$$\leq \int_0^T ||z_t(s)||^2\, ds + \frac{1}{2} \int_0^T ||A^{1/2}z(s)||^2\, ds + \widetilde{C}_R \int_0^T ||z(s)||^2\, ds$$

$$+ \int_0^T \max_{\theta \in [-h,0]} ||A^{1/2-\delta}z(s+\theta)||^2\, ds.$$

Hence, using the definition of E_z and the relation

$$k \int_0^T (z_t(s), z(s))\, ds \leq \frac{1}{2} \int_0^T ||z_t(s)||^2\, ds + \frac{k^2}{2} \int_0^T ||z(s)||^2\, ds,$$

we obtain

$$\frac{1}{2} \int_0^T ||A^{1/2}z(s)||^2\, ds \leq \frac{3}{2} \int_0^T ||z_t(s)||^2\, ds + C(E_z(0) + E_z(T))$$

$$+ \widetilde{C}_R(k) \int_0^T \max_{\theta \in [-h,0]} ||A^{1/2-\delta}z(s+\theta)||^2\, ds. \tag{6.3.43}$$

From (6.3.41) with $t = 0$ and using (6.3.42) we get

$$E_z(0) \leq E_z(T) + \frac{3k}{2} \int_0^T ||z_t(s)||^2\, ds + \varepsilon \int_0^T ||A^{1/2}z(s)||^2\, ds$$

$$+ \int_0^T \max_{\theta \in [-h,0]} ||A^{1/2-\delta}z(s+\theta)||^2\, ds. \tag{6.3.44}$$

It also follows from (6.3.41) with the help of integration over $[0, T]$ (we use (6.3.42) again) that

$$TE_z(T) \leq \int_0^T E_z(s)\, ds + \varepsilon T \int_0^T ||A^{1/2}z(s)||^2\, ds$$

$$+ T \int_0^T \max_{\theta \in [-h,0]} ||A^{1/2-\delta}z(s+\theta)||^2\, ds. \tag{6.3.45}$$

Another consequence of (6.3.41) and (6.3.42) is

$$\frac{k}{2} \int_0^T ||z_t(s)||^2\, ds \leq E_z(0) + \varepsilon \int_0^T ||A^{1/2}z(s)||^2\, ds$$

$$+ \int_0^T \max_{\theta \in [-h,0]} ||A^{1/2-\delta}z(s+\theta)||^2\, ds. \tag{6.3.46}$$

Considering the sum of (6.3.46) and (6.3.43) and assuming that $k \geq 4$, we can get

$$
\int_0^T E_z(s) \, ds \leq C(E_z(0) + E_z(T))
$$

$$
+ c_0 \varepsilon \int_0^T ||A^{1/2} z(s)||^2 \, ds + C_{R,k}^* \int_0^T \max_{\theta \in [-h,0]} ||A^{1/2-\delta} z(s + \theta)||^2 \, ds. \qquad (6.3.47)
$$

Now to both sides of (6.3.47) we add the value $\frac{1}{2} T E_z(T)$ and use (6.3.45):

$$
\frac{1}{2} \int_0^T E_z(s) \, ds + \frac{1}{2} T E_z(T)
$$

$$
\leq c_0 \varepsilon (1 + T) \int_0^T ||A^{1/2} z(s)||^2 \, ds + C(E_z(0) + E_z(T))
$$

$$
+ C_{R,k}(1 + T) \int_0^T \max_{\theta \in [-h,0]} ||A^{1/2-\delta} z(s + \theta)||^2 \, ds. \qquad (6.3.48)
$$

Next we evaluate $E_z(0) + E_z(T)$. Using (6.3.44) we have

$$
E_z(0) + E_z(T) \leq 2 E_z(T) + \frac{3k}{2} \int_0^T ||z_t(s)||^2 \, ds + \varepsilon \int_0^T ||A^{1/2} z(s)||^2 \, ds
$$

$$
+ \int_0^T \max_{\theta \in [-h,0]} ||A^{1/2-\delta} z(s + \theta)||^2 \, ds.
$$

Substituting this into (6.3.48) we obtain

$$
\frac{1}{2} \int_0^T E_z(s) \, ds + \left(\frac{1}{2} T - 2C \right) E_z(T) \leq c_0 k \int_0^T ||z_t(s)||^2 \, ds
$$

$$
+ (1 + T) \left[c_1 \varepsilon \int_0^T ||A^{1/2} z(s)||^2 ds + \widetilde{C}_R(k) \int_0^T \max_{\theta \in [-h,0]} ||A^{1/2-\delta} z(s + \theta)||^2 ds \right].
$$

Choosing T such that

$$
\frac{1}{2} T - 2C > 1, \qquad (6.3.49)
$$

we get

$$
E_z(T) + \frac{1}{2} \int_0^T E_z(s) \, ds \leq c_1 \varepsilon (1 + T) \int_0^T ||A^{1/2} z(s)||^2 \, ds
$$

$$
+ (1 + T) \widetilde{C}_R(k) \int_0^T \max_{\theta \in [-h,0]} ||A^{1/2-\delta} z(s + \theta)||^2 \, ds + c_0 k \int_0^T ||z_t(s)||^2 \, ds.
$$

$$
(6.3.50)
$$

To estimate the last term in (6.3.50) we use (6.3.41) with $t = 0$ (remembering (6.3.42)) to get

$$\frac{k}{2}\int_0^T ||z_t(s)||^2\, ds \leq E_z(0) - E_z(T) + \varepsilon \int_0^T ||A^{1/2}z(s)||^2\, ds$$

$$+ \int_0^T \max_{\theta\in[-h,0]} ||A^{1/2-\delta}z(s+\theta)||^2\, ds.$$

So, we can rewrite (6.3.50) as

$$E_z(T) + \frac{1}{2}\int_0^T E_z(s)\, ds$$

$$\leq 2c_0(E_z(0) - E_z(T)) + c_1\varepsilon\,(1+T)\int_0^T ||A^{1/2}z(s)||^2\, ds$$

$$+ (1+T)\,\widetilde{C_R}(k)\int_0^T \max_{\theta\in[-h,0]} ||A^{1/2-\delta}z(s+\theta)||^2\, ds. \tag{6.3.51}$$

Since $||A^{1/2}z(s)||^2 \leq 2E_z(s)$, the choice of small $\varepsilon > 0$ to satisfy

$$c_1\varepsilon\,(1+T) < \frac{1}{2} \tag{6.3.52}$$

simplifies (6.3.51) as follows:

$$E_z(T) \leq \widetilde{c_0}(E_z(0) - E_z(T)) + (1+T)\,\widetilde{C_R}(k)\int_0^T \max_{\theta\in[-h,0]} ||A^{1/2-\delta}z(s+\theta)||^2\, ds.$$

The last step is

$$E_z(T) \leq \frac{\widetilde{c_0}}{1+\widetilde{c_0}}E_z(0) + \widetilde{C_R}(T,k)\int_0^T \max_{\theta\in[-h,0]} ||A^{1/2-\delta}z(s+\theta)||^2\, ds.$$

Since $\gamma \equiv \frac{\widetilde{c_0}}{1+\widetilde{c_0}} < 1$, this means that there is $\omega > 0$ such that

$$E_z(T) \leq e^{-\omega T}E_z(0) + C_{R,T,k}\int_0^T \max_{\theta\in[-h,0]} ||A^{1/2-\delta}z(s+\theta)||^2\, ds. \tag{6.3.53}$$

Note that the parameters are chosen in the following order. First we choose $T > h$ to satisfy (6.3.49), next we choose small $\varepsilon > 0$ to satisfy (6.3.52), and finally we choose $k \geq 4$ big enough to satisfy (6.3.42).

Now using the same step-by-step procedure $(mT \mapsto (m+1)T)$ as in Proposition 5.3.4, we can derive the conclusion in (6.3.36) from the relation in (6.3.53) written on the interval $[mT, (m+1)T]$. Thus the proof of Theorem 6.3.14 is complete.

6.3.4 Global and exponential attractors

In this section, relying on Proposition 6.3.11 and Theorem 6.3.15, we establish the existence of a global attractor and study its properties.

The main consequence of dissipativity and quasi-stability given by Proposition 6.3.11 and Theorem 6.3.14 is the following theorem.

Theorem 6.3.16 (Global attractor). *Let Assumptions* 6.3.1, 6.3.9, *and* 6.3.13 *be in force. Then the dynamical system* (W, S_t) *generated by* (6.3.1) *possesses the compact global attractor* $\mathfrak{A}$ *of finite fractal dimension. Moreover, for any full trajectory* $\{u(t) : t \in \mathbb{R}\}$ *such that* $u^t \in \mathfrak{A}$ *for all* $t \in \mathbb{R}$ *we have that*

$$u_{tt} \in L_\infty(\mathbb{R}, H), \quad u_t \in L_\infty(\mathbb{R}, H_{1/2}) \quad u \in L_\infty(\mathbb{R}, H_1) \tag{6.3.54}$$

and

$$\|u_{tt}(t)\| + \|A^{1/2}u_t(t)\| + \|Au(t)\| \le R_*, \quad \forall t \in \mathbb{R}. \tag{6.3.55}$$

Proof. By Theorem 6.3.15 the system (W, S_t) is quasi-stable. Due to Proposition 6.3.11 this system is dissipative. Therefore to prove the existence of a compact global attractor $\mathfrak{A}$ we can use Proposition 3.4.3 and Corollary 3.4.4. This attractor is finite-dimensional due to Theorem 3.4.5.

To prove the regularity properties in (6.3.54) and (6.3.55), we can use the inequality in (6.3.36) and the same idea as in the proof of Theorem 3.4.19. Indeed, let $\gamma = \{u(t) : t \in \mathbb{R}\}$ be a full trajectory of the system. This means that

$$(S_t u^s)(\theta) = u(t + s + \theta) \text{ for } \theta \in [-h, 0], \ s \in \mathbb{R}, \ t \ge 0.$$

Assume that $u^t \in \mathfrak{A}$ for all $t \in \mathbb{R}$. Consider the difference of this trajectory and its small shift $\gamma_\varepsilon = \{u(t + \varepsilon) : t \in \mathbb{R}\}$ and apply the inequality in (6.3.36) with the starting point at $s \in \mathbb{R}$:

$$\|u_t(t + \varepsilon) - u_t(t)\|^2 + \|A^{1/2}(u(t + \varepsilon) - u(t))\|^2$$

$$\le C_1(R)e^{-\tilde{\lambda}(t-s)}|u^{s+\varepsilon} - u^s|_W^2 + C_2(R)\max_{\xi \in [s,t]} \|A^{1/2-\delta}(u(\xi + \varepsilon) - u(\xi))\|^2.$$

Since $u^s \in \mathfrak{A}$ for all $s \in \mathbb{R}$, in the limit $s \to -\infty$ we obtain that

$$\|u_t(t + \varepsilon) - u_t(t)\|^2 + \|A^{1/2}(u(t + \varepsilon) - u(t))\|^2$$

$$\le C_2(R) \sup_{\xi \in [-\infty, t]} \|A^{1/2-\delta}(u(\xi + \varepsilon) - u(\xi))\|^2.$$

Now in the same way as in the proof of Theorem 3.4.19 we can conclude that

$$\frac{1}{\varepsilon^2}\left[||u_t(t+\varepsilon)-u_t(t)||^2 + ||A^{1/2}(u(t+\varepsilon)-u(t))||^2\right]$$

is uniformly bounded in $\varepsilon \in (0,1]$ and $t \in \mathbb{R}$. This implies (passing to the limit $\varepsilon \to 0$) that

$$||u_{tt}(t)||^2 + ||A^{1/2}u_t(t)||^2 \le C_R.$$

Now using equation (6.3.1) we conclude that $||Au(t)||^2 \le C_R$. This gives (6.3.54) and (6.3.55).

This completes the proof of Theorem 6.3.16.   □

Remark 6.3.17. One can show (see CHUESHOV/REZOUNENKO [66]) that in the situation considered in Remark 6.3.7 the global attractor $\mathfrak{A}$ is a bounded set on the manifold $\mathscr{L}$ given by (6.3.19).   ∎

Now we present a result on fractal exponential attractors. We recall (see Definition 3.4.6) that a compact set $\mathfrak{A}_{exp} \subset W$ is said to be a (generalized) fractal exponential attractor for the dynamical system (W, S_t) iff $\mathfrak{A}_{exp}$ is a positively invariant set whose fractal dimension is finite (in some extended space $\mathscr{W} \supset W$) and for every bounded set $D \subset W$ there exist positive constants t_D, C_D, and γ_D such that

$$d_W\{S_tD \,|\, \mathfrak{A}_{exp}\} \equiv \sup_{x\in D} \text{dist}_W(S_tx, \mathfrak{A}_{exp}) \le C_D \cdot e^{-\gamma_D(t-t_D)}, \quad t \ge t_D. \qquad (6.3.56)$$

Using the quasi-stability property and Theorem 3.4.7 we can construct fractal exponential attractors for the system considered.

Theorem 6.3.18 (Exponential attractor). *Let the hypotheses of Theorem 6.3.16 be in force. Then the dynamical system (W, S_t) possesses a (generalized) fractal exponential attractor whose dimension is finite in the space*

$$\mathscr{W} \equiv C([-h, 0]; H_{1/2-\delta}) \cap C^1([-h, 0]; H_{-\delta}), \quad \forall\, \delta > 0.$$

Proof. Using (6.3.1) we can see that $||u_{tt}(t)||_{-1} < C_R$ for all $t \in \mathbb{R}_+$ and for every solution from an absorbing ball. This allows us to show that $S_t\varphi$ is Hölder continuous in t in the space $\mathscr{W}$, i.e.,

$$|S_{t_1}\varphi - S_{t_2}\varphi|_{\mathscr{W}} \le C_{\mathscr{B}}|t_1 - t_2|^\gamma, \quad t_1, t_2 \in \mathbb{R}_+, \, y \in \mathscr{B},$$

for some positive $\gamma > 0$. Indeed, using the interpolation inequality (4.1.2) with an appropriate choice of parameters, we have that

$$||u(t+h)-u(t)||_{1/2-\delta} \le \left[||u(t+h)||_{1/2} + ||u(t)||_{1/2}\right]^{1-2\delta} ||u(t+h)-u(t)||^{2\delta}$$

for $o < \delta \le 1/2$. Since

$$\|u(t+h) - u(t)\| \le \left| \int_t^{t+h} \|u_t(\tau)\| d\tau \right| \le C_B |h|,$$

the former inequality implies that $u(t)$ is 2δ-Hölder in $H_{1/2-\delta}$, and similarly for $u_t(t)$. Thus the result follows from Theorem 3.4.7. $\square$

In conclusion, we note that using quasi-stability property (6.3.36) we can also establish some other asymptotic properties of the system (W, S_t). For instance, using the same idea as in Theorem 3.4.20 we can suggest criteria which guarantee the existence of a finite number of determining functionals.

The question of whether it is possible to avoid the assumption of large damping in the case of critical nonlinearities in Theorem 6.3.18 is still open.

6.3.5   Remark on models with structural damping

In this section we briefly discuss the model in (6.3.1) with structural damping of the form $2\varkappa \cdot A^{1/2} u_t$ instead of viscous damping $k \cdot u_t$. Namely, we consider the equation

$$u_{tt}(t) + 2\varkappa A^{1/2} u_t(t) + Au(t) + B(u(t)) + M(u^t) = 0, \quad t > 0. \tag{6.3.57}$$

The presence of structural damping leads to additional a priori estimates. This makes it possible to relax the conditions concerning the terms $B(\cdot)$ and $M(u^t)$. Moreover for the model in (6.3.57) we can apply the parabolic theory developed in Section 6.1. Indeed, equation (6.3.57) can be treated as a nonlinear delay perturbation of a linear model of the form

$$u_{tt}(t) + 2\varkappa A^{1/2} u_t(t) + Au(t) = 0, \quad t > 0.$$

Using the idea presented in LASIECKA/TRIGGIANI [144] we can rewrite (6.3.57) as a semilinear parabolic model. For instance, in the case when[11] $\varkappa > 1$ we can use the variables

$$y_1(t) = -\frac{1}{2\beta} \left(u_t(t) + (\varkappa - \beta) A^{1/2} u(t) \right)$$

and

$$y_2(t) = \frac{1}{2\beta} \left(u_t(t) + (\varkappa + \beta) A^{1/2} u(t) \right),$$

---

[11]The case $0 < \varkappa < 1$ can be considered in a similar way, but it requires a complexification procedure. See LASIECKA/TRIGGIANI [144] and also CHUESHOV/LASIECKA [58, Chapter 13].

where $\beta = \sqrt{\varkappa^2 - 1}$. The inverse transformation has the form

$$u = A^{-1/2}(y_1 + y_2), \quad u_t = -\mu(y_1 + y_2) - \beta(y_1 - y_2).$$

Using these variables y_1 and y_2, equation (6.3.57) can be written in the form

$$\frac{d}{dt}Y + \mathbb{A}Y = \mathbb{B}(Y^t), \tag{6.3.58}$$

where

$$Y = \begin{pmatrix} y_1 \\ y_2 \end{pmatrix}, \quad \mathbb{A} = \begin{pmatrix} (\varkappa + \beta)A^{1/2} & 0 \\ 0 & (\varkappa - \beta)A^{1/2} \end{pmatrix},$$

and

$$\mathbb{B}(Y^t) = \frac{1}{2\beta} \begin{pmatrix} B(A^{-1/2}(y_1(t) + y_2(t))) + M(A^{-1/2}(y_1^t + y_2^t)) \\ -B(A^{-1/2}(y_1(t) + y_2(t))) - M(A^{-1/2}(y_1^t + y_2^t)) \end{pmatrix}.$$

It is clear that $\mathbb{A}$ is a positive operator in $H \times H$ with a discrete spectrum. Thus problem (6.3.58) has the same form as (6.1.1), and thus we can apply the methods presented in Sections 6.1 and 6.2. In particular, this makes it possible to obtain a local existence result without Lipschitz continuity hypotheses and establish additional compactness properties for the system generated by (6.3.57).

6.3.6 Applications: plate and wave models

In this section we consider several possible applications of the results above.

6.3.6.1 Plate models

Our main applications are related to nonlinear plate models.

Let $\Omega \subset \mathbb{R}^2$ be a bounded smooth domain. In the space $H = L_2(\Omega)$ we consider the following problem:

$$u_{tt}(t, x) + ku_t(t, x) + \Delta^2 u(t, x)$$
$$+ [f(u(t, \cdot))](x) + u(t - \tau[u^t], x) = 0, \; x \in \Omega, \; t > 0, \tag{6.3.59a}$$

$$u = \frac{\partial u}{\partial n} = 0 \; \text{ on } \partial\Omega, \; u(\theta) = \varphi(\theta) \; \text{ for } \theta \in [-h, 0]. \tag{6.3.59b}$$

We assume that

$$\tau \ : \ C([-h,0]; H_0^2(\Omega)) \cap C^1([-h,0]; L_2(\Omega)) \mapsto [0,h]$$

is a Lipschitz continuous mapping. As was already mentioned, the delay term in (6.3.59a) models the foundation reaction with a delayed (state-dependent) response.

The model in (6.3.59) can be written in the abstract form (6.3.1) in the space $H = L_2(\Omega)$ with $A = \Delta^2$ defined on the domain $\mathscr{D}(A) = (H^4 \cap H_0^2)(\Omega)$. Here and below $H^s(\Omega)$ is the Sobolev space of order s and $H_0^s(\Omega)$ is the closure of $C_0^\infty(\Omega)$ in $H^s(\Omega)$.

As the simplest example of a state-dependent delay satisfying all hypotheses in Theorems 6.3.16 and 6.3.18 we can consider

$$\tau[u^t] = g(Q[u^t]), \tag{6.3.60}$$

where g is a smooth mapping from $\mathbb{R}$ into $[0,h]$ and

$$Q[u^t] = \sum_{i=1}^{N} c_i u(t - \sigma_i, a_i).$$

Here $c_i \in \mathbb{R}$, $\sigma_i \in [0,h]$, $a_i \in \Omega$ are arbitrary elements. We could also consider the term Q with the Stieltjes integral over delay interval $[-h,0]$ instead of the sum. Another possibility is to consider a combination of averages like

$$Q[u^t] = \sum_{i=1}^{N} \int_{\Omega} u(t - \sigma_i, x)\xi_i(x)dx, \tag{6.3.61}$$

where $\sigma_i \in [0,h]$ and $\{\xi_i\}$ are functions from $L_2(\Omega)$. We can also consider linear combinations of these Q's as well as their powers and products. The corresponding calculations are simple and related to the fact that the space $\mathscr{D}(A^{1/2})$ is an algebra belonging to $C(\overline{\Omega})$.

The nonlinearities f satisfying all requirements of Theorems 6.3.16 and 6.3.18 are the same as in CHUESHOV/LASIECKA[56, 58], and the delay perturbations of the models considered in these sources in the case of linear damping provide us with a series of examples. Here we only mention three of them.

Simplified Kirchhoff model: In this case $f(u) = f_0(u) - h(x)$, where $h \in L_2(\Omega)$, and

$$f_0 \in \mathrm{Lip_{loc}}(\mathbb{R}) \quad \text{satisfies} \quad \liminf_{|s|\to\infty} f_0(s)s^{-1} = \infty. \tag{6.3.62}$$

This is a subcritical case (relation (6.3.32) holds with $\eta > 0$). The growth condition in (6.3.62) is needed to satisfy Assumption 6.3.9.

The following two examples are critical ((6.3.32) holds with $\eta = 0$).

Von Karman model: In this model (see, e.g., CHUESHOV/LASIECKA [58] or LIONS [151])

$$f(u) = -[u, v(u) + F_0] - h(x),$$

where $F_0 \in H^4(\Omega)$ and $h \in L_2(\Omega)$ are given functions,

$$[u, v] = \partial^2_{x_1} u \cdot \partial^2_{x_2} v + \partial^2_{x_2} u \cdot \partial^2_{x_1} v - 2 \cdot \partial_{x_1 x_2} u \cdot \partial_{x_1 x_2} v,$$

and the function $v(u)$ satisfies the equations:

$$\Delta^2 v(u) + [u, u] = 0 \text{ in } \Omega, \quad \frac{\partial v(u)}{\partial n} = v(u) = 0 \text{ on } \partial\Omega.$$

For details we refer to CHUESHOV/LASIECKA [56, 58].

Berger model: In this case $f(u) = \Pi'(u)$, where

$$\Pi(u) = \frac{\kappa}{4} \left[\int_\Omega |\nabla u|^2 dx \right]^2 - \frac{\mu}{2} \int_\Omega |\nabla u|^2 dx - \int_\Omega u(x)h(x)dx,$$

where $\kappa > 0$ and $\mu \in \mathbb{R}$ are parameters, $h \in L_2(\Omega)$. We refer to the analyses presented in CHUESHOV [39, Chapter 4] and CHUESHOV/LASIECKA [56, Chapter 7].

In all these models we can also include a non-conservative non-delay force of the form

$$f^*(u) = (a_1 \partial_{x_1} + a_2 \partial_{x_2})u, \quad \text{where } (a_1; a_2) \in \mathbb{R}^2.$$

These kinds of models arise in some aero-elastic problems; see, e.g., CHUESHOV/ LASIECKA [58].

6.3.6.2 Wave model

Let $\Omega \subset \mathbb{R}^d$, $d = 2, 3$, be a bounded domain with a sufficiently smooth boundary Γ. The exterior normal on Γ is denoted by n. We consider the following wave equation:

$$u_{tt} - \Delta u + ku_t + f(u) + u(t - \tau[u^t]) = 0 \text{ in } Q = [0, \infty) \times \Omega$$

subject to boundary conditions either of Dirichlet type

$$u = 0 \text{ on } \Sigma \equiv [0, \infty) \times \Gamma, \tag{6.3.63}$$

or else of Robin type

$$\frac{\partial u}{\partial n} + u = 0 \quad \text{on} \quad \Sigma. \tag{6.3.64}$$

The initial conditions are given by $u(\theta) = \varphi(\theta)$, $\theta \in [-h, 0]$. In this case $H = L_2(\Omega)$ and A is $-\Delta$ with either the Dirichlet (6.3.63) or the Robin (6.3.64) boundary conditions. So $\mathscr{D}(A^{1/2})$ is either $H_0^1(\Omega)$ or $H^1(\Omega)$ in this case.

We assume that k is a positive parameter and the function $f \in C^2(\mathbb{R})$ satisfies the following polynomial growth condition: there exists a positive constant $M > 0$ such that

$$|f''(s)| \leq M(1 + |s|^{q-1}),$$

where $q \leq 2$ when $d = 3$ and $q < \infty$ when $d = 2$. Moreover, we assume the same lower growth condition as in (6.3.62). One can see that the hypotheses in Theorems 6.3.16 and 6.3.18 are satisfied (see [56, Chapter 5] for a detailed discussion). Moreover we have the subcritical case if $d = 2$ or $d = 3$ and $q < 2$. The case $d = 3$ and $q = 2$ is critical.

As for the delay term $u(t - \tau[u^t])$, we can assume that, as in the plate models above, $\tau[u^t]$ has the form (6.3.60) with $Q[u_t]$ given by (6.3.61). Moreover, instead of the averaging we can consider an arbitrary family of linear functionals on $H^{1-\delta}(\Omega)$ for some $\delta > 0$; i.e., we can take

$$Q[u^t] = \sum_{i=1}^{N} c_i l_i[u(t - \sigma_i)],$$

where $c_i \in \mathbb{R}$, $\sigma_i \in [0, h]$ and $l_i \in [H^{1-\delta}(\Omega)]'$ are arbitrary elements.

Appendix A
Auxiliary Facts

In this appendix we start with a discussion of various issues related to solvability of finite-dimensional ODEs. Then we discuss some issues related to Gronwall's lemma which are important for parabolic problems. We also consider properties of measurable functions with values in infinite-dimensional spaces, some approximations of the identity operator on Banach and Hilbert spaces, and elements of differential calculus on these spaces. We also show that uniqueness for semilinear parabolic problems is a generic property and discuss the monotonicity method for 2D hydrodynamical problems.

A.1 Generation of continuous systems by ODEs

In this section we review several classical results on the generation of dynamical systems by ordinary differential equations (ODEs).

Let $X = \mathbb{R}^d$ with the (Euclidean) norm $\| \cdot \|$ and $f : X \mapsto X$ be a continuous function. We consider the following Cauchy problem in X:

$$\dot{u} = f(u), \quad t > t_0, \quad u(t_0) = u_0 \in X. \tag{A.1.1}$$

Definition A.1.1. A function $u(t)$ with values in X is said to be a solution to problem (A.1.1) on a (semi-open) interval $[t_0, t_0 + T)$ if

$$u \in C([t_0, t_0 + T); X) \cap C^1((t_0, t_0 + T); X)$$

and satisfies (A.1.1) on the open interval $(t_0, t_0 + T)$. We can similarly define a solution on closed intervals of the form $[t_0, t_0 + T]$. The notation $C([a, b]; X)$ and $C^1((a, b); X)$ has an obvious meaning (see Section A.3.1 below). ∎

© Springer International Publishing Switzerland 2015
I. Chueshov, *Dynamics of Quasi-Stable Dissipative Systems*, Universitext,
DOI 10.1007/978-3-319-22903-4

The following result is standard and can be found in many books on ODEs (see, e.g., CODDINGTON/LEVINSON [75] or HARTMAN [120]). It is usually attributed to the contributions of G. Peano (1858–1932) and C. Carathéodory (1873–1950).

Theorem A.1.2 (Carathéodory-Peano). *Let f be a continuous function from $X = \mathbb{R}^d$ into itself. Then for any $u_0 \in X$ and $t_0 \in \mathbb{R}$ there exists $T \leq \infty$ such that problem (A.1.1) has a solution on the interval $[t_0, t_0 + T)$. Moreover,*

- *every solution can be extended to an (a maximal) interval $[t_0, t_0 + \tilde{T})$ possessing the property: if $\tilde{T} < \infty$ then the solution blows up, i.e.,*

$$\limsup_{t \nearrow \tilde{T}} \|u(t)\| = +\infty;$$

- *if we assume in addition that f is locally Lipschitz in the sense that for every $R > 0$ there is L_R such that*

$$\|f(u_1) - f(u_2)\| \leq L_R \|u_1 - u_2\| \quad \text{for all } \|u_i\| \leq R, \ i = 1, 2, \qquad \text{(A.1.2)}$$

then the solution is unique.

Sketch of the proof. We present the main steps only. For details we refer to the classical sources CODDINGTON/LEVINSON [75] or HARTMAN [120].

We first note that the Cauchy problem in (A.1.1) is equivalent to the problem: find $u \in C([t_0, t_0 + T); X)$ satisfying the integral relation

$$u(t) = u_0 + \int_{t_0}^{t} f(u(\tau))d\tau \quad \text{for all } t \in [t_0, t_0 + T). \qquad \text{(A.1.3)}$$

Next we define approximate solutions. For $T_* > 0$ and for every $n = 1, 2, \ldots$ we consider a sequence of functions $v_n(t)$ on $[t_0, t_0 + T_*]$ satisfying the relation

$$v_n(t) = \begin{cases} u_0, & \text{for } t \in [t_0, t_0 + T_*/n]; \\ u_0 + \displaystyle\int_{t_0}^{t-T_*/n} f(v_n(\tau))d\tau, & \text{for } t \in [t_0 + T_*/n, t_0 + T_*], \end{cases} \qquad \text{(A.1.4)}$$

for $n = 1, 2, \ldots$. It is clear that $v_1(t) = u_0$ and the delayed character of the integral above allows us to define $v_n(t)$ by the step-by-step procedure starting from the initial interval $[t_0, t_0 + T_*/n]$. One can show that each function v_n is continuous on $[t_0, t_0 + T_*]$.

Now we fix $R > 0$ and consider the ball $B_R(u_0) = \{u \in X : \|u - u_0\| \leq R\}$. Let

$$M_R(u_0) = \sup\{\|f(u)\| : u \in B_R(u_0)\} \quad \text{and} \quad T_* \leq M_R(u_0)^{-1}R.$$

Using the step-by-step method it is easy to see that

$$\max\{\|v_n(t) - u_0\| \ : \ t \in [t_0, t_0 + T_*]\} \le R.$$

This implies that

$$\|v_n(t_1) - v_n(t_2)\| \le M_R(u_0)|t_1 - t_2| \quad \text{for all} \quad t_1, t_2 \in [t_0, t_0 + T_*].$$

Thus $\{v_n(t)\}$ is a uniformly bounded and equicontinuous sequence on $[t_0, t_0 + T_*]$.

Since X is finite-dimensional, by the Arzelà-Ascoli theorem (see, e.g., DIEUDONNÉ [85] and also Lemma A.3.5) there exists a convergent subsequence $\{v_{n_m}(t)\}$:

$$\max_{t \in [t_0, t_0 + T_*]} \|v_{n_m}(t) - v(t)\| \to 0 \quad \text{as} \quad m \to \infty$$

for some function $v \in C([t_0, t_0 + T_*], X)$. It follows from (A.1.4) that

$$v_{n_m}(t) = u_0 + \int_{t_0}^t f(v_{n_m}(\tau))d\tau - \int_{t - T_*/n_m}^t f(v_{n_m}(\tau))d\tau$$

for $t_0 + T_*/n_m \le t \le t_0 + T_*$ and $m = 1, 2, \ldots$. This yields that $v(t)$ satisfies (A.1.3).

Thus the existence of solutions to problem (A.1.1) on a "small" interval $[t_0, t_0 + T_*]$ is proved. Taking initial data $u(t_0 + T_*)$ at the time $t_0 + T_*$ we can extend the solution to a greater interval. Hence we can construct a solution on some semi-open interval $[t_0, t_0 + \tilde{T})$ which is maximal in the sense that we cannot extend the solution beyond $t_0 + \tilde{T}$.

Let $[t_0, t_0 + T)$ be an interval of existence for a solution $u(t)$ and $T < \infty$. If we assume that $u(t)$ is bounded as $t \nearrow T$, then by (A.1.3) we have that $u(t)$ is continuous at $t = t_0 + T$. Thus we can extend the solution beyond $t_0 + T$. This implies that every solution can be extended to an interval $[t_0, t_0 + \tilde{T})$ with the property: if $\tilde{T} < \infty$, then $\limsup_{t \nearrow \tilde{T}} \|u(t)\| = +\infty$.

The statement on uniqueness of solutions follows via a Gronwall's-type argument.

Let f be locally Lipschitz and assume that $u_1(t)$ and $u_2(t)$ are two solutions to (A.1.1) on some joint interval $[t_0, t_0 + T]$. Using representation (A.1.3) one can show that

$$\|u_1(t) - u_2(t)\| \le L_T(u_1, u_2) \int_{t_0}^t \|u_1(\tau) - u_2(\tau)\|d\tau \quad \text{for all} \quad t \in [t_0, t_0 + T],$$

where $L_T(u_1, u_2) = L_R$ with $R = \max_{t \in [t_0, t_0 + T]}\{\|u_1(t)\| + \|u_2(t)\|\}$. Thus the standard Gronwall lemma implies the uniqueness.

This completes the proof of Theorem A.1.2.   □

We mention that Example 1.7.15 in Section 1.7 shows that the uniqueness statement in Theorem A.1.2 cannot be true without the Lipschitz assumption for f.

The following result provides a criterion for global existence.

Theorem A.1.3. *Let* $X = \mathbb{R}^d$ *and* $f : X \mapsto X$ *be a continuous mapping. Assume that* f *satisfies the following dissipativity condition: there exist* $k_1 \geq 0$ *and* $k_2 \geq 0$ *such that*

$$(f(u), u) \leq k_1 \|u\|^2 + k_2 \quad \text{for all } u \in X. \tag{A.1.5}$$

Then for any $u_0 \in X$ *and* $t_0 \in \mathbb{R}$ *problem* (A.1.1) *has a solution on the semi-axis* $[t_0, +\infty)$. *Moreover, this solution admits the estimate*

$$\|u(t)\|^2 \leq e^{2k_1(t-t_0)} \|u_0\|^2 + \frac{k_2}{k_1} \left(e^{2k_1(t-t_0)} - 1 \right) \quad \text{for all } t \geq t_0. \tag{A.1.6}$$

If we assume in addition that f *is locally Lipschitz (see* (A.1.2)), *then the solution is unique and depends continuously on initial data. Furthermore, for any two solutions* $u^1(t)$ *and* $u^2(t)$ *with initial data* u_0^1 *and* u_0^2, *from the ball* $B_\rho = \{v \in X : \|v\| \leq \rho\}$ *we have the following estimate:*

$$\|u^1(t) - u^2(t)\| \leq C_{T,\rho} \|u_0^1 - u_0^2\| \quad \text{for all } t \in [t_0, t_0 + T], \tag{A.1.7}$$

where $C_{T,\rho} = \exp\{T L_R\}$ *with* $R = \sqrt{\rho^2 + k_2/k_1} \exp\{k_1 T\}$, *with the same* L_R *as in* (A.1.2).

Proof. The existence of a local solution $u(t)$ on some interval $[t_0, t_0 + \tau)$ follows from Theorem A.1.2.

If we substitute this solution $u(t)$ into (A.1.1) and multiply (A.1.1) by $u(t)$, then we obtain

$$\frac{1}{2} \frac{d}{dt} \|u(t)\|^2 = (f(u(t)), u(t)) \leq k_1 \|u(t)\|^2 + k_2, \quad t \in [t_0, t_0 + \tau).$$

Via a Gronwall's-type argument this implies the estimate in (A.1.6) on the existence interval. In particular this means that the solution cannot blow up at the end of this interval, and thus by Theorem A.1.2 the solution can be extended on the semi-axis $[t_0, +\infty)$. Obviously estimate (A.1.6) remains true on this semi-axis.

To prove (A.1.7) we note that by (A.1.6) we have

$$\|u^i(t)\|^2 \leq R^2 \equiv e^{2k_1 T}(\rho^2 + k_2/k_1), \quad t \in [t_0, t_0 + T], \ i = 1, 2.$$

Therefore using (A.1.1) and the Lipschitz condition in (A.1.2) we can conclude that $u(t) = u^1(t) - u^2(t)$ satisfies the relation

$$\frac{1}{2}\frac{d}{dt}\|u(t)\|^2 = (f(u(t)^1) - f(u(t)^2), u(t)) \le L_R\|u(t)\|^2, \quad t \in [t_0, t_0 + T].$$

By the Gronwall lemma this implies (A.1.7). □

The Lipschitz condition in Theorem A.1.3 can be relaxed if we assume some monotonicity of $f(u)$. More precisely, the following assertion holds.

Theorem A.1.4. *Let $X = \mathbb{R}^d$ and $f : X \mapsto X$ be a continuous mapping. Assume that f satisfies the following one-sided Lipschitz condition: there exists $k_0 \ge 0$ such that*

$$(f(u^1) - f(u^2), u^1 - u^2) \le k_0\|u^1 - u^2\|^2 \quad \text{for all } u^1, u^2 \in X. \tag{A.1.8}$$

Then for any $u_0 \in X$ and $t_0 \in \mathbb{R}$ problem (A.1.1) has a unique solution on the semi-axis $[t_0, +\infty)$. Moreover, this solution admits the estimate (A.1.6) with $k_1 = k_0 + 1/2$ and $k_2 = \|f(0)\|/2$, and for the difference of two solutions $u^1(t)$ and $u^2(t)$ with initial data u_0^1 and u_0^2 we have the estimate

$$\|u^1(t) - u^2(t)\| \le e^{k_0(t-t_0)}\|u_0^1 - u_0^2\| \quad \text{for all } t \ge t_0. \tag{A.1.9}$$

Proof. The statement follows from the argument given in the proof of Theorem A.1.3. We leave it to the reader to provide all the details. □

The results in Theorems A.1.3 and A.1.4 give us conditions for the generation of continuous time dynamical systems in $\mathbb{R}^d$ by ODEs. The evolution operator S_t is given by $S_t u_0 = u(t)$, where $u(t)$ is a solution to (A.1.1) with $t_0 = 0$.

A.2   Two Gronwall-type lemmas

In this section we prove two assertions which provide us with some nonstandard extensions of the classic (see, e.g., HARTMAN [120]) Gronwall's inequality.

Lemma A.2.1. *Let $\psi(t)$ and $g(t)$ be given scalar functions from $L_1^{loc}(\mathbb{R}_+)$. Assume that a continuous function $h(t)$ defined on $\mathbb{R}_+$ satisfies the inequality*

$$h(t) + \int_s^t \psi(\tau)h(\tau)\,d\tau \le h(s) + \int_s^t g(\tau)\,d\tau \tag{A.2.1}$$

for all $t \ge s \ge 0$. Then

$$h(t) \le h(s)\exp\left\{-\int_s^t \psi(\sigma)\,d\sigma\right\} + \int_s^t g(\tau)\exp\left\{-\int_\tau^t \psi(\sigma)d\sigma\right\}d\tau \tag{A.2.2}$$

for all $t \ge s \ge 0$.

Proof. The main difficulty in the proof of this lemma is that it is not assumed that h is a C^1 function and/or ψ and g have fixed signs. For instance, in the case when $h \in C^1(\mathbb{R}_+)$ we can choose $t = s + \Delta s$ in (A.2.1) and after the limit transition $\Delta s \to +0$ show that

$$h'(s) + \psi(s)h(s) \le g(s) \quad \text{for almost all} \ \ s \in \mathbb{R}_+.$$

Therefore using the multiplier $e(s) = \exp\left\{\int_0^s \psi(\sigma)\,d\sigma\right\}$ we can see that

$$\frac{d}{ds}\big[h(s)e(s)\big] \le g(s)e(s) \quad \text{for almost all} \ \ s \in \mathbb{R}_+.$$

Thus after integration we can obtain (A.2.2).

To overcome the difficulty of the insufficient smoothness of the function $h(s)$ we first prove the following assertion.

Lemma A.2.2. *Assume that $f(t)$ is a continuous function on an interval $[a, b]$ such that*

$$\liminf_{\delta \to 0, \delta < 0} \frac{1}{|\delta|}\, [f(t + \delta) - f(t)] \ge -m(t) \tag{A.2.3}$$

for almost all $t \in (a, b)$, where $m(t) \in L^1(a, b)$. Then

$$f(t_2) - f(t_1) \le \int_{t_1}^{t_2} m(\tau)\,d\tau \quad \text{for all} \quad a \le t_1 < t_2 \le b. \tag{A.2.4}$$

Proof. It is clear that

$$M(t) \equiv f(t) - \int_a^t m(\tau)\,d\tau \in C[a, b]$$

and satisfies the relation

$$\liminf_{\delta \to 0, \delta < 0} \frac{1}{|\delta|}\, [M(t + \delta) - M(t)] \ge 0 \tag{A.2.5}$$

for all $t \in \mathscr{B}$, where $\mathscr{B}$ is a measurable set of full measure in (a, b). To obtain (A.2.4) we should prove that $M(t)$ is a non-increasing function on $[a, b]$. It is sufficient to prove that the function $\Phi(t) = M(t) - \gamma t$ is non-increasing for any $\gamma > 0$. From (A.2.5) we have

$$\liminf_{\delta \to 0, \delta < 0} \frac{1}{|\delta|}\, [\Phi(t + \delta) - \Phi(t)] \ge \gamma > 0, \quad t \in \mathscr{B}.$$

This implies that for every $t \in \mathscr{B}$ there exists $\sigma(t) > 0$ such that

$$\Phi(t - \tau) \geq \Phi(t), \quad 0 \leq \tau < \sigma(t), \ t \in \mathscr{B} . \tag{A.2.6}$$

Let $t_1 < t_2$ be points from $\mathscr{B}$. Consider the covering of the segment $[t_1, t_2)$ by intervals $(t - \min\{\sigma(t_2), \sigma(t)\}, t)$, where $t \in \mathscr{B}$. It is clear that there exists a finite subcovering. Moreover we can choose the points $\tau_1 < \tau_2 < \ldots < \tau_N$ from $\mathscr{B} \cap (t_1, t_2)$ such that

$$t_1 \in (\tau_1 - \sigma(\tau_1), \tau_1), \quad \tau_N \in (t_2 - \sigma(t_2), t_2)$$

and

$$\tau_k \in (\tau_{k+1} - \sigma(\tau_{k+1}), \tau_{k+1}), \quad k = 1, \ldots N - 1 .$$

Therefore from (A.2.6) we have

$$\Phi(t_1) \geq \Phi(\tau_1); \quad \Phi(\tau_k) \geq \Phi(\tau_{k+1}), \ k = 1, \ldots N - 1; \quad \Phi(\tau_N) \geq \Phi(t_2) .$$

This implies that $\Phi(t_1) \geq \Phi(t_2)$.   □

We apply Lemma A.2.2 to the function

$$f(t) = h(t) \exp \left\{ \int_0^t \psi(\sigma) \, d\sigma \right\} \equiv h(t) e(t).$$

It follows from (A.2.1) with $s = t + \delta, \delta < 0$, that

$$\begin{aligned} f(t + \delta) - f(t) &= [h(t + \delta) - h(t)] e(t + \delta) + h(t) [e(t + \delta) - e(t)] \\ &\geq \left[\int_{t+\delta}^t \psi(\tau) h(\tau) d\tau - \int_{t+\delta}^t g(\tau) \, d\tau \right] e(t + \delta) \\ &\quad + h(t) [e(t + \delta) - e(t)] . \end{aligned}$$

This relation implies (A.2.3) with $m(t) = g(t) e(t)$. Therefore the application of Lemma A.2.2 yields the inequality in (A.2.2).   □

Our second Gronwall-type lemma provides an important tool for parabolic problems. The corresponding modification was suggested in HENRY [123].

Lemma A.2.3 (Henry-Gronwall). *Let $u(t)$ be a measurable locally bounded non-negative function on $\mathbb{R}_+$ satisfying the inequality*

$$u(t) \leq a + b \int_0^t \frac{u(\tau) d\tau}{(t - \tau)^\alpha} \quad \text{for almost all } t \geq 0, \tag{A.2.7}$$

where $a, b > 0$ and $\alpha \in [0, 1)$ are constants. Then

$$u(t) \leq 2a \exp\{c_\alpha b^{1/(1-\alpha)} t\} \text{ for almost all } t \geq 0, \qquad (A.2.8)$$

for some constant $c_\alpha > 0$ depending on α only.

Proof. The original argument in HENRY [123] relies on an iteration procedure. The idea of the proof presented here is borrowed from ROBINSON [196, p. 132]; see also CARVALHO/LANGA/ROBINSON [26].

We start with the case when $u(t)$ is continuous. In this case (A.2.7) is satisfied for *all* $t \geq 0$. First we prove a comparison principle for this case. We claim that if a non-negative continuous function $y(t)$ satisfies the inequality

$$y(t) \geq a^* + b \int_0^t \frac{y(\tau)d\tau}{(t-\tau)^\alpha}, \quad t \geq 0, \qquad (A.2.9)$$

for some $a^* > a$, then $u(t) < y(t)$ for all $t \in \mathbb{R}_+$. Indeed, if this is not true, then for $z(t) = y(t) - u(t)$ we have that $z(0) = a^* - a > 0$ and thus there is $t_* > 0$ such that $z(t) > 0$ for all $t \in [0, t_*)$ and $z(t_*) = 0$. On the other hand, if we subtract from (A.2.9) relation (A.2.7), then we obtain that

$$z(t_*) \geq a^* - a + b \int_0^{t_*} \frac{y(\tau)d\tau}{(t-\tau)^\alpha} > a^* - a > 0.$$

This gives a contradiction.

Next we construct a comparison function for continuous u satisfying (A.2.7).

Rescaling $u(t) \mapsto a^{-1}u(t)$, if necessary, we can assume that $a = 1$ in (A.2.7). We take $a^* = 3/2$ and look for a parameter N such that the function $y_0(t) = 2e^{Nt}$ satisfies (A.2.9). We have

$$\int_0^t \frac{y_0(\tau)d\tau}{(t-\tau)^\alpha} = 2 \int_0^t \frac{1}{(t-\tau)^\alpha} e^{N\tau} d\tau = 2e^{Nt} \int_0^t \frac{1}{(t-\tau)^\alpha} e^{-N(t-\tau)} d\tau$$

$$= y_0(t) \int_0^t s^{-\alpha} e^{-Ns} ds \leq y_0(t) \int_0^\infty s^{-\alpha} e^{-Ns} ds.$$

Introducing the variable $\xi = Ns$ in the integral above we obtain

$$\int_0^t \frac{y_0(\tau)d\tau}{(t-\tau)^\alpha} \leq y_0(t) \frac{\varkappa_\alpha}{N^{1-\alpha}} \quad \text{with } \varkappa_\alpha \equiv \int_0^\infty \xi^{-\alpha} e^{-N\xi} d\xi.$$

Thus

$$\frac{3}{2} + b \int_0^t \frac{y_0(\tau)d\tau}{(t-\tau)^\alpha} \leq \frac{3}{2} + y_0(t) \frac{b\varkappa_\alpha}{N^{1-\alpha}} \leq \left[\frac{3}{4} + \frac{b\varkappa_\alpha}{N^{1-\alpha}}\right] y_0(t).$$

Consequently $y_0(t)$ satisfies (A.2.9) with $a^* = 3/2$ provided we choose N such that $b\varkappa_\alpha N^{-(1-\alpha)} = 1/4$. By the comparison principle this implies the inequality in (A.2.8) in the class of continuous functions.

Now we assume that $u(t)$ is a measurable locally bounded function. In this case,

$$w(t) = B[u](t) \equiv a + b \int_0^t \frac{u(\tau)d\tau}{(t-\tau)^\alpha}$$

is a continuous function. Since the mapping B is order preserving, we can apply it to the inequality in (A.2.7) and show that $w(t)$ satisfies the inequality

$$w(t) \leq a + b \int_0^t \frac{w(\tau)d\tau}{(t-\tau)^\alpha}.$$

Thus applying Lemma A.2.3 for continuous functions, we obtain for $w(t) = B[u](t)$ the same bound as in (A.2.8). Since $u(t) \leq w(t)$ for almost all $t \geq 0$, we obtain the desired conclusion. □

A.3   Vector-valued functions and compactness theorems

Now we describe some properties of vector-valued functions and discuss several compactness theorems, which we use in the main text.

A.3.1   Continuous vector-valued functions

Let I be an interval in $\mathbb{R}$ and X be a Banach space. We denote by $C(I;X)$ the vector space of all continuous functions $f : I \mapsto X$. If I is a closed bounded interval, then $C(I;X)$ is a Banach space with the norm

$$\|f\|_{C(I;X)} = \max_{t\in I} \|f(t)\|_X.$$

For $k \in \mathbb{N}$ we denote by $C^k(I;X)$ the space of all k-times differentiable functions with continuous k-th derivative. Here we understand derivatives in the strong sense. For instance, the first derivative of f at a point $t \in I$ is an element f' in X such that

$$\lim_{h\to 0} \left\| \frac{f(t+h)-f(t)}{h} - f' \right\| = 0,$$

and similarly for higher derivatives. We also set $C^\infty(I;X) = \cap_{k\geq 1} C^k(I;X)$.

Sometimes it is convenient to use the Hölder spaces $C^\nu(I;X)$ which are analogs for $C^k(I;X)$ for a non-integer smoothness index. These spaces can be introduced in the following way.

Let I be a closed bounded interval and $\nu \in \mathbb{R}_+$. We suppose that $k = [\nu]$ and $\alpha = \nu - [\nu]$ are the integer and fractional parts of ν. We define the space $C^\nu(I;X)$ by the formula

$$C^\nu(I;X) = \left\{ \psi \in C^k(I;X) \; : \; \mathrm{Hld}_\alpha[\psi^{(k)}] < \infty \right\},$$

where $\psi^{(k)}$ denotes the derivative of ψ of order k and

$$\mathrm{Hld}_\alpha[\phi] = \sup\left\{ \frac{\|\phi(t_1) - \phi(t_2)\|_X}{|t_1 - t_2|^\alpha} \; : \; t_1, t_2 \in I, \; t_1 \neq t_2 \right\}.$$

This space $C^\nu(I;X)$ is Banach with the norm given by

$$|\psi|_{C^\nu(I;X)} = \sum_{m=0}^{k} \max_{t \in I} \|\psi^{(m)}(t)\|_X + \mathrm{Hld}_\alpha[\psi^{(k)}],$$

where $k = [\nu]$ and $\alpha = \nu - [\nu]$.

A.3.2   Bochner integral and vector-valued L_p functions

We start with a brief introduction to Bochner integration of vector-valued functions. For more details we refer to DUNFORD/SCHWARTZ [88, Chapters 3,4] or YOSIDA [229, Chapter 5].

Assume that X is a Banach space and $I = (a,b)$ is an interval (bounded or unbounded) in $\mathbb{R}$. Let $f : I \mapsto X$ be a vector-valued function.

The function f is called *simple* if it can be represented as

$$f = \sum_{k=1}^{n} x_k \chi_{\Delta_k}(t)$$

for elements $x_k \in X$ and measurable bounded subsets $\Delta_k \subset I$, where $\chi_\Delta(t)$ is the characteristic function of the set Δ. For a simple function f we define its integral by

$$\int_I f(s)ds := \sum_{k=1}^{n} x_k \mu(\Delta_k),$$

where $\mu(\Delta_k)$ denotes the Lebesgue measure of Δ_k on $\mathbb{R}$.

If a function $g : I \mapsto X$ can be approximated pointwise by simple functions; i.e., if there exists a sequence $\{f_n\}$ of simple functions on I such that

$$\lim_{n \to \infty} \|g(t) - f_n(t)\|_X = 0 \ \text{ for almost all } t \in I,$$

then we call f (strongly) measurable.

If g is measurable and there exists a sequence $\{f_n\}$ of simple functions on I such that

$$\lim_{n \to \infty} \int_I \|g(t) - f_n(t)\|_X dt = 0,$$

then g is said to be (Bochner) integrable. For an integrable function f we define its integral by

$$\int_I g(t) dt := \lim_{n \to \infty} \int_I f_n(t) dt.$$

We list some elementary properties of measurable functions. See DUNFORD/ SCHWARTZ [88] or YOSIDA [229] for the proofs.

- If $\{f_n\}$ is a sequence of measurable functions on I converging to f strongly for almost all $t \in I$, then f is measurable as well.
- If f is continuous, then it is measurable.
- If f is measurable, then f is integrable if and only if

$$\int_I \|f(t)\|_X dt < \infty.$$

- If f is integrable, then

$$\left\| \int_I f(t) dt \right\|_X \leq \int_I \|f(t)\|_X dt.$$

- If X is a separable space, then f is measurable if and only if it is weakly measurable; i.e., $l(f(t))$ is a (scalar) measurable function for every $l \in X^*$ (Pettis' theorem).

For $1 \leq p < +\infty$ we define the spaces

$$L_p(I;X) := \left\{ f : I \mapsto X \ \middle| \ \begin{array}{l} f \text{ is Bochner measurable,} \\[2mm] \|f\|_{L_p(I;X)} \equiv \left[\int_I \|f(t)\|_X^p dt \right]^{1/p} < \infty \end{array} \right\}$$

If $p = +\infty$ we suppose that

$$L_\infty(I;X) := \left\{ f : I \mapsto X \; \middle| \; \begin{array}{l} f \text{ is Bochner measurable,} \\ \\ \|f\|_{L_\infty(I;X)} \equiv \operatorname{esssup}\{\|f(t)\| \, : \, t \in I\} < \infty \end{array} \right\}$$

The spaces $L_p(I;X)$ are Banach for all $1 \le p \le +\infty$.

Our main point of interest is the case when X is a Hilbert space. In this case the spaces $L_p(I,X)$ are reflexive for $1 < p < +\infty$, and

$$\left[L_p(I;X)\right]^* = L_q(I;X) \text{ for } \frac{1}{p} + \frac{1}{q} = 1.$$

Any linear functional F on $L_p(I;X)$ has the form

$$\langle F, u \rangle = \int_I (f(t), u(t))_X dt, \quad \forall \, u \in L_p(I;X),$$

where f is some element from $L_q(I;X)$. We also have that

$$L_\infty(I;X) = \left[L_1(I;X)\right]^*.$$

The space $L_2(I;X)$ is Hilbert with the inner product

$$(f,g)_{L_2(I;X)} := \int_I (f(t), g(t))_X dt.$$

We also note that *-weak convergence of a sequence $\{f_n\}$ to an element f in $L_p(I,X)$ with $1 < p \le +\infty$ means that

$$\lim_{n \to \infty} \int_I (f_n(t), u(t))_X dt = \int_I (f(t), u(t))_X dt \text{ for every } u \in L_q(I;X),$$

where $q^{-1} + p^{-1} = 1$. In the reflexive case ($p \ne \infty$) this convergence is called weak. A well-known fact based on the Banach-Alaoglu theorem (see, e.g., RUDIN [199, Chapter 3]) states that any bounded set in $L_p(I;X)$ with $1 < p \le +\infty$ is *-weakly relatively compact in the sense that any sequence in this set contains a *-weakly convergent subsequence. We use this fact and also the compactness theorem stated in Section A.3.3 to perform limit transitions in approximate solutions in Chapters 5 and 6.

To guarantee weak continuity of solutions to evolution equations in a smoother space, we need the following lemma (see LIONS/MAGENES [152, Section 3.8.4]).

Lemma A.3.1 (Lions' lemma). *Let X and Y be two Banach spaces such that X is continuously embedded in Y. Assume that a function $f \in L_\infty(a,b;X)$ is weakly continuous in Y, i.e., the function $t \mapsto l(f(t))$ is continuous for every $l \in Y^*$. Then $f(t)$ is also weakly continuous in X.*

The following assertion deals with generalized derivatives of functions with values in Banach spaces (see, e.g., LIONS/MAGENES [152] or TEMAM [215, 216]).

Proposition A.3.2. *Let u and v be integrable functions on* $[a, b]$ *with values in a Banach space X. Then the following three conditions are equivalent:*

- *there exists* $\xi \in X$ *such that*

$$u(t) = \xi + \int_a^t v(\tau)d\tau \quad \text{for almost all } t \in [a, b];$$

- *for every test function* $\phi \in C_0^\infty((a, b); \mathbb{R})$ *we have*

$$\int_a^b \phi_t(t)u(t)dt = -\int_a^b \phi(t)v(t)dt \quad (\phi_t = (d/dt)\phi);$$

- *for every* $l \in X^*$ *we have*

$$\frac{d}{dt}l(u(t)) = l(v(t)) \quad \text{in the sense of distributions on } (a, b).$$

The function $v := \partial_t u \equiv u_t$ *is called a derivative of u in the distributional sense.*

Let A be a positive self-adjoint operator on a Hilbert space H and $V = \mathscr{D}(A)$ endowed with the graph norm $\|u\|_1 = \|Au\|$. We denote by V_{-1} the completion of H with respect to the norm $\|u\|_{-1} = \|A^{-1}u\|$. One can see that $V_{-1} = V^*$ and thus the triple $V \subset H \subset V^*$ of embedding spaces arises. The spaces V and V^* become Hilbert if we define inner products as $(u, v)_1 = (Au, Av)$ and $(u, v)_{-1} = (A^{-1}u, A^{-1}v)$. Using this triple we can introduce the space

$$W(a, b) = \{u \in L_2(a, b; V) : u_t \in L_2(a, b; V_{-1})\},$$

where u_t is the derivative of u in the distributional sense (see Proposition A.3.2). We equip the space $W(a, b)$ with the norm

$$\|u\|_{W(a,b)}^2 = \|u\|_{L_2(a,b;V)}^2 + \|u_t\|_{L_2(a,b;V_{-1})}^2,$$

which provides $W(a, b)$ with the Hilbert structure. The following assertion can be found in LIONS/MAGENES [152] or TEMAM [215].

Proposition A.3.3 (Continuous embedding). *The space* $W(a, b)$ *is continuously embedded into* $C([a, b]; H)$ *and*

$$\exists C > 0 : \quad \|u(t)\|_{C([a,b];H)} \leq \|u(t)\|_{W(a,b)}, \quad \forall u \in W(a, b),$$

and

$$\|u(t)\|^2 = \|u(a)\|^2 + 2 \int_a^t (u(\tau), u_t(\tau))d\tau, \ \ t \in [a, b], \ \ \forall\, u \in W(a, b).$$

We note that instead of the pair $\{V; V^*\}$ we can consider pairs $\{H_\alpha; H_{-\alpha}\}$, where the spaces H_s are defined in Section 4.1 with the help of a positive self-adjoint operator.

A.3.3   Compactness theorems for vector-valued functions

We first recall the definition and some criteria for compactness of sets in Banach spaces (see, e.g., DUNFORD/SCHWARTZ [88], YOSIDA [229], or ZEIDLER [231]).

A set F in a Banach space X is said to be *compact* if for every family of open sets covering F there exists a finite subfamily covering F. A set is *relatively compact* if its closure is compact. Given $\varepsilon > 0$, a set $C \subset X$ is said to be ε-*net* for a set $M \subset X$ if $M \subset \cup_{a \in C}\{x : \|x - a\|_X \le \varepsilon\}$.

The following equivalent conditions for compactness are well known (see, e.g., DUNFORD/SCHWARTZ [88, Chapter 5]).

Proposition A.3.4. *Let K be a set in a Banach space X. Then the following statements are equivalent.*

- *The set K is relatively compact.*
- *For every $\varepsilon > 0$ there exists a finite ε-net for K.*
- *For every $\varepsilon > 0$ there exists a relatively compact ε-net for K.*
- *Any sequence of elements from K contains a subsequence that converges to some element of X.*

We use the following compactness criteria applicable to vector-valued continuous functions.

Lemma A.3.5 (Arzelà-Ascoli theorem, I). *Let X be a Banach space. A set $F \subset C(a, b; X)$ is relatively compact if and only if*

(i) *$F(t) := \{f(t) : f \in F\}$ is relatively compact in X for each $t \in [a, b]$.*
(ii) *F is equicontinuous; that is, for any $\varepsilon > 0$ there exists $\delta > 0$ such that*

$$\|f(t) - f(s)\|_X \le \varepsilon \quad \text{for any } f \in F \text{ and } t, s \in [a, b] \text{ such that } |t - s| \le \delta.$$

The following version of the Arzelà-Ascoli theorem is a relaxed form of the previous lemma. Instead of assuming compactness for every fixed t, only the integral form of the compactness property is required.

Lemma A.3.6 (Arzelà-Ascoli theorem, II). *Let X be a Banach space. A set $F \subset C(a, b; X)$ is relatively compact if and only if*

(i) $\int_{t_1}^{t_2} F := \left\{ \int_{t_1}^{t_2} f(t)dt \ : \ f \in F \right\}$ *is relatively compact in X for each* $t_1, t_2 \in [a, b]$.

(ii) *F is equicontinuous.*

We also state the following result on compactness of vector functions, which plays a key role in many situations considered in Chapters 4–6.

Theorem A.3.7 (Aubin-Dubinskii-Lions). *Assume that* $X \subset Y \subset Z$ *is a triple of Banach spaces such that X is compactly embedded in Y.*

- *Let F be a bounded set in* $L_p(a, b; X)$ *for some* $1 \leq p < \infty$ *such that the set* $\partial_t F := \{\partial_t f \ : \ f \in F\}$ *is bounded in* $L_q(a, b; Z)$ *for some* $q \geq 1$. *Here* $\partial_t f$ *is the derivative in the distributional sense. Then F is relatively compact in* $L_p(a, b; Y)$. *If* $q > 1$, *then F is also relatively compact in* $C(a, b; Z)$.
- *If F is a bounded set in* $L_\infty(a, b; X)$ *and* $\partial_t F$ *is bounded in* $L_r(a, b; Z)$ *for some* $r > 1$, *then F is relatively compact in* $C(a, b; Y)$.

Particular cases[1] of Theorem A.3.7 can be found in AUBIN [4], DUBINSKII [87], and LIONS [151, Chapter 1, Section 5]). In almost the same form as above, Theorem A.3.7 is stated in SIMON [213, Corollary 4]. For the proof we refer to SIMON [213]. See also the Appendix in CHUESHOV/LASIECKA [58], where a concise argument based on the Arzelà-Ascoli theorem stated in Lemma A.3.6 is given in the case when $q > 1$.

A.4 Approximation of identity by uniformly bounded mappings

In well-posedness results related to transitions from the case of globally Lipschitz nonlinearities to the local case, we use a truncation procedure based on the following assertion.

Lemma A.4.1. *Let X be a Banach space. We define a mapping* π_R *on X by the formula*

$$\pi_R(x) = \begin{cases} x, & \text{if } \|x\| \leq R; \\ R x \|x\|^{-1}, & \text{if } \|x\| > R. \end{cases} \tag{A.4.1}$$

Then for each $R > 0$ *the mapping* π_R *is globally Lipschitz and we have that*

$$\|\pi_R(u) - \pi_R(v)\| \leq \mathrm{Lip}(\pi_R)\|u - v\| \quad \text{for every } u, v \in X,$$

where $\mathrm{Lip}(\pi_R) \leq 2$. *If X is a Hilbert space and* $\| \cdot \|$ *is the norm generated by its scalar product, then* $\mathrm{Lip}(\pi_R) = 1$.

---

[1]The result is named after Aubin, Dubinskii, and Lions. In the Western literature Theorem A.3.7 is usually referred to as the Aubin or Aubin-Lions result. In the Russian literature it is often called the Dubinskii theorem.

Proof. We distinguish three cases.

(i) If $\|u\|, \|v\| \le R$, the conclusion is obvious.

(ii) In the case $\|u\| \le R$ and $\|v\| > R$ we have

$$\|\pi_R(u) - \pi_R(v)\| = \left\| u - R\frac{v}{\|v\|} \right\| = \frac{\|u\|v\| - Rv\|}{\|v\|}.$$

Thus

$$\|\pi_R(u) - \pi_R(v)\| \le \frac{\|(u-v)\|v\| + (\|v\| - R)v\|}{\|v\|} \le \|u - v\| + (\|v\| - R)$$

$$\le \|u - v\| + (\|v\| - \|u\|) \le 2\|u - v\|.$$

(iii) In the case $\|u\| > R$, $\|v\| > R$ we have

$$\|\pi_R(u) - \pi_R(v)\| = R\left\| \frac{v}{\|v\|} - \frac{v}{\|v\|} \right\| = R\frac{\|u\|v\| - v\|u\|\|}{\|u\|\|v\|}.$$

Hence

$$\|\pi_R(u) - \pi_R(v)\| \le R\frac{\|(u-v)\|v\| + (\|v\| - \|u\|)v\|}{\|u\|\|v\|}$$

$$\le \|u - v\| + \big|\|v\| - \|u\|\big| \le 2\|u - v\|.$$

This completes the proof of Lemma A.4.1 in the Banach case.

In the case when X is a Hilbert space we can use the same argument as in CHUE-SHOV/ELLER/LASIECKA [47]; see also CHUESHOV/LASIECKA [58, Chapter 2]. □

A.5   Elements of differential calculus in Banach spaces

In this section we quote several notions and results related to the smoothness of mappings F between two Banach spaces X and Y. Our presentation relies on CARTAN [22].

We start with the following definition.

Definition A.5.1 (Fréchet derivative). Let $\mathscr{O}$ be an open set in X. A mapping $F : \mathscr{O} \mapsto Y$ is said to be *Fréchet differentiable* on $\mathscr{O}$ if for any $u \in \mathscr{O}$ there exists a bounded linear operator $F'(u)$ from X into Y such that

$$\frac{\|F(v) - F(u) - F'(u)(v - u)\|_Y}{\|v - u\|_X} \to 0 \quad \text{as} \quad \|v - u\|_X \to 0. \tag{A.5.1}$$

The operator $F'(u)$ is called the *(Fréchet) derivative* of F at the point $u \in \mathcal{O}$. The relation in (A.5.1) means that for every $u \in \mathcal{O}$ there exist $\delta > 0$ and a scalar function $\gamma(s)$ on $[0, \delta]$ such that $\gamma(s) \to 0$ as $s \to 0$ and

$$\|F(v) - F(u) - F'(u)(v - u)\|_Y \leq \gamma(\|v - u\|_X)\|v - u\|_X.$$

If the number δ and the function $\gamma(s)$ do not depend on u, then the mapping F is said to be *uniformly (Fréchet) differentiable* on $\mathcal{O}$. One says that $F : \mathcal{O} \mapsto Y$ is *continuously differentiable* on $\mathcal{O}$ or of the class C^1 on $\mathcal{O}$ if:

- F is differentiable at every point of $\mathcal{O}$;
- the derivative F' is a continuous mapping from $\mathcal{O}$ into $\mathcal{L}(X \mapsto Y)$, where $\mathcal{L}(X \mapsto Y)$ denotes the space of linear operators from X into Y equipped with the operator norm.

∎

The chain rule stated in the following assertion is important in many calculations.

Proposition A.5.2 (Derivative of a compound function). *Let X, Y, Z be three Banach spaces, $\mathcal{O}$ an open set in X, and $\mathcal{V}$ an open set in Y. Consider two continuous mappings*

$$F : \mathcal{O} \mapsto \mathcal{V} \subset Y, \quad G : \mathcal{V} \mapsto Z.$$

On the set $\mathcal{O}$ we consider the compound mapping $G \circ F : \quad \mathcal{O} \mapsto Z$. If F is differentiable at a point $a \in \mathcal{O}$ and G is differentiable at the point $b = F(a)$, then $H = G \circ F$ is differentiable at the point a and

$$H'(a) = G'(b)F'(a).$$

In other words, the linear mapping $H'(a) : X \mapsto Z$ is the composition of the linear mapping $F'(a) : X \mapsto Y$ and the linear mapping $G'(a) : Y \mapsto Z$.

Proof. See CARTAN [22].   □

This proposition implies that if F is continuously differentiable on an open set $\mathcal{O}$ and the interval $[u, v] := \{\lambda u + (1 - \lambda)v : 0 \leq \lambda \leq 1\}$ lies in $\mathcal{O}$ for some $u, v \in \mathcal{O}$, then

$$F(v) - F(u) = \int_0^1 F'(\lambda v + (1 - \lambda)u)(v - u)d\lambda. \tag{A.5.2}$$

Consider now a particular case of mappings Π from a Banach space X into the real axis $\mathbb{R}$. In this case the derivative $\Pi'(u)$ is a linear mapping from X into $\mathbb{R}$. Thus $\Pi'(u)$ can be treated as an element of adjoint space X^*. So the mapping $u \mapsto \Pi'(u)$ from X into X^* arises. This observation leads to the following definition.

A mapping $B : X \mapsto X^*$ is said to be *potential* on X if there exists a Frechét differentiable functional $\Pi(u)$ on X such that $B(u) = \Pi'(u)$, i.e.,

$$\lim_{\|v\|_X \to 0} \frac{1}{\|v\|_X} \left| \Pi(u+v) - \Pi(u) - \langle B(u), v \rangle \right| = 0, \tag{A.5.3}$$

where $\langle f, v \rangle$ denotes the value of functional f on v.

Applying Proposition A.5.2, one can show that in the situation above for every $u \in C^1([a, b]; X)$ we have that $t \mapsto \Pi(u(t))$ is a C^1 scalar function on $[a, b]$ and

$$\frac{d}{dt} \Pi(u(t)) = \langle B(u(t)), u_t(t) \rangle, \quad t \in [a, b]. \tag{A.5.4}$$

Taking now $u(t) = u + tv$, one can see that

$$\Pi(u+v) - \Pi(u) = \int_0^1 \langle B(u + \lambda v), v \rangle d\lambda \text{ for every } u, v \in X. \tag{A.5.5}$$

Now we introduce the derivatives of higher order.

Let $F : \mathcal{O} \subset X \mapsto Y$ be a differentiable mapping on $\mathcal{O}$. Then the derivative F' can be seen as a mapping from $\mathcal{O}$ into the Banach space $\mathscr{L}(X \mapsto Y)$. Thus we can define the second derivative $F''(u)$ of F as the first derivative of the mapping

$$F' : \mathcal{O} \mapsto \mathscr{L}(X \mapsto Y).$$

In this case $F''(a)$ is a linear mapping from X into $\mathscr{L}(X \mapsto Y)$, i.e.,

$$F''(a) \in \mathscr{L}(X \mapsto \mathscr{L}(X \mapsto Y)) \text{ for each } a \in \mathcal{O}.$$

One can see that the following isomorphism takes place:

$$\mathscr{L}(X \mapsto \mathscr{L}(X \mapsto Y)) \equiv \mathscr{L}(X \times X \mapsto Y) := \mathscr{L}(X^{\times 2} \mapsto Y),$$

where $\mathscr{L}(X^{\times n} \mapsto Y)$ denotes the space of all linear mappings from $X \times \ldots \times X$ into Y. It is clear that $\mathscr{L}(X^{\times n} \mapsto Y)$ is a Banach space.

Thus we can define by induction the n-th order derivative $F^{(n)}$ on $\mathcal{O}$ as the first derivative of the mapping

$$F^{(n-1)} : \mathcal{O} \mapsto \mathscr{L}(X^{\times (n-1)} \mapsto Y).$$

In this case $F^{(n)}(a)$ is an element from $\mathscr{L}(X^{\times n} \mapsto Y)$.

We accept the following definition. A mapping $F \; : \; \mathcal{O} \subset X \mapsto Y$ is of the class C^n on $\mathcal{O}$ (or F is n times continuously differentiable on $\mathcal{O}$) if there exist all derivatives $F^{(k)}$ with the order $k \leq n$ on $\mathcal{O}$ and if the mapping

$$F^{(n)} \; : \; \mathcal{O} \mapsto \mathscr{L}(X^{\times n} \mapsto Y)$$

is continuous. The following assertion provides us with the symmetry properties of the higher derivatives.

Proposition A.5.3 (Symmetry). *If $F : X \mapsto Y$ is n times differentiable on $\mathcal{O}$, then for every $a \in \mathcal{O}$ the derivative $F^{(n)}(a) \in \mathscr{L}(X^{\times n} \mapsto Y)$ is a multilinear symmetric mapping from the product space $X \times \ldots \times X$ into Y. This means that*

$$F^{(k)}(a)[v_1, \ldots, v_k] = F^{(k)}(a)[v_{\sigma(1)}, \ldots, v_{\sigma(k)}]$$

for every permutation σ of $\{1, 2, \ldots, k\}$. Here $F^{(k)}(u)[v_1, \ldots, v_k]$ denotes the value of the derivative $F^{(k)}(u)$ on elements $v_1, \ldots, v_k$.

We note that in the case when $Y = \mathbb{R}$, the Fréchet derivatives $\Pi^{(k)}(u)$ of the functional Π are *symmetric k-linear continuous (scalar) forms on X.*

In conclusion we recall Taylor's formula (see CARTAN [22] for the proof).

Theorem A.5.4 (Taylor's formula with integral remainder). *Let $F \; : \; \mathcal{O} \mapsto Y$ be a mapping of the class C^{n+1}. As previously, $\mathcal{O}$ denotes an open set of a Banach space X, and Y a Banach space. If the interval $[u, u+v] = \{x = u+\tau v \; : \; \tau \in [0, 1]\}$ is contained in $\mathcal{O}$, then*

$$F(u + v) = F(u) + \sum_{k=1}^{n} \frac{1}{k!} F^{(k)}(u)[v, \ldots, v] + \int_0^1 \frac{(1 - \tau)^n}{n!} F^{(n+1)}(u + \tau v)[v, \ldots, v] d\tau,$$

$$(\text{A.5.6})$$

where $F^{(k)}(u)[v_1, \ldots, v_k]$ denotes the value of the derivative $F^{(k)}(u)$ on elements $v_1, \ldots, v_k$.

A.6   The Orlicz theorem on uniqueness for parabolic problems

Our goal in this section is to show that the uniqueness of solutions to the parabolic-type problem considered in Chapter 4 with a continuous (non-Lipschitz, in general) nonlinearity is a *generic* property (in the Baire category sense). According to YUDOVICH [230], a similar result for ordinary differential equations seems to go back to the earlier paper ORLICZ [176].

Recall some definitions (see, e.g., BOURBAKI [16]).

Let $\mathscr{L}$ be a complete metric space. A set K is said to be *nowhere dense* if its closure contains no open sets. A subset D of $\mathscr{L}$ is said to be *meager* (or a first category set in the Baire sense), if it is contained in a countable union of closed nowhere dense subsets of $\mathscr{L}$. The complement of a meager set is called *residual*. By the Baire categories theorem (see, e.g., BOURBAKI [16]) any residual set is dense. A property $\mathscr{P}$ of elements of $\mathscr{L}$ is said to be *generic* in $\mathscr{L}$ if $\mathscr{P}$ holds in some residual set.

We consider the problem

$$u_t + Au = B(u), \quad u\big|_{t=0} = u_0 \in H, \tag{A.6.1}$$

under the following set of hypotheses.

Assumption A.6.1 (Basic hypotheses). We assume that H is a separable Hilbert space and

(A) A is a positive self-adjoint operator with a discrete spectrum on H (see Definition 4.1.1);

(B) B is a (nonlinear) continuous mapping from $H_\alpha = \mathscr{D}(A^\alpha)$ into H with $0 \leq \alpha < 1$ and bounded on every bounded set in H_α, i.e.,

$$\forall R > 0, \ \exists C_R : \ \|B(v)\| \leq C_R \ \text{ for all } \ \|v\|_\alpha \leq R.$$

Let $\mathscr{L}$ be a family of nonlinearities B satisfying Assumption A.6.1(B). We assume that $\mathscr{L}$ is a complete metric space $\mathscr{L}$ with respect to some metric $\varrho_{\mathscr{L}}$ such that (a) for every sequence $\{B_n\}$ we have that

$$\varrho_{\mathscr{L}}(B_n, B) \to 0 \ \text{ implies } \ \forall R > 0, \quad \sup_{\|v\|_\alpha \leq R} \|B_n(v) - B(v)\| \to 0 \tag{A.6.2}$$

as $n \to \infty$, and (b) every element of $\mathscr{L}$ can be approximated in $\mathscr{L}$ by a sequence $\{B_n\}$ of Lipschitz point continuous mappings. A mapping $B : H_\alpha \mapsto H$ is called *Lipschitz point continuous* if for every point $u \in H_\alpha$ there exists an open set $\mathcal{O}(u) \ni u$ and a constant $L(u)$ such that

$$\|B(u_1) - B(u_2)\| \leq L(u)\|A^\alpha(u_1 - u_2)\|, \ \forall u_i \in \mathcal{O}(u), \ i = 1, 2. \tag{A.6.3}$$

As an example of the class $\mathscr{L}$ we can take the space of *all* continuous mappings $B : H_\alpha \mapsto H$ satisfying Assumption A.6.1(B) and endowed with the metric

$$\varrho_{\mathscr{L}}(B_1, B_2) = \sum_{k=1}^{\infty} 2^{-k} \frac{d_k(B_1 - B_2)}{1 + d_k(B_1 - B_2)},$$

where $d_n(B) = \sup_{\|u\|_\alpha \leq n} \|B(u)\|$. It is easy to see that the property in (A.6.2) is valid. To show that Lipschitz point continuous mappings are dense in $\mathscr{L}$ we can apply the approximation lemma established in LASOTA/YORKE [146].

We recall that an initial datum $u_* \in \mathcal{D}(A^\alpha)$ is said to be a *uniqueness point* if there is a vicinity $\mathcal{O}$ of u_* in H_α such that problem (A.6.1) with $u_0 \in \mathcal{O}$ has a unique (local) mild solution[2] in this vicinity $\mathcal{O}$. We also say that problem (A.6.1) possesses a local uniqueness property for a given B if all initial data are uniqueness points for this B.

The following theorem states that the local uniqueness property is generic.

Theorem A.6.2 (Orlicz). *Let Assumption A.6.1 be in force. Then for every initial data $u_0 \in \mathcal{D}(A^\alpha)$ problem (A.6.1) has a (local) mild solution $u(t)$. Moreover the set $\mathcal{M}$ of all elements $B \in \mathcal{L}$ such that (A.6.1) has at least one non-uniqueness point is a meager set, i.e., $\mathcal{M}$ is a countable union of closed nowhere dense sets. Thus the set of all problems of the form (A.6.1) with the local uniqueness property is residual.*

Proof. The existence of solutions for continuous $B(u)$ follows from Proposition 6.1.3; see also Remark 6.1.4.

To prove the statement concerning uniqueness, we rely on the idea due to ORLICZ [176] in the form presented in YUDOVICH [230] for the ODE case.

Let $u_0 \in H_\alpha$ and τ, a be positive numbers. We denote by $\mathcal{M}(u_0, \tau, a)$ the set of nonlinearities B in $\mathcal{L}$ such that there exist

$$\tilde{u}_0 \in \mathcal{B}_1(u_0) = \{v \in H_\alpha \; : \; \|v - u_0\|_\alpha < 1\}$$

and two mild solutions $u(t)$ and $\tilde{u}(t)$ to problem (A.6.1) on the interval $[0, \tau)$ such that

- $u(t) \in \mathcal{B}_1(u_0)$ and $\tilde{u}(t) \in \mathcal{B}_1(u_0)$ for all $t \in [0, \tau)$,
- $u(0) = \tilde{u}_0$ and $\tilde{u}(0) = \tilde{u}_0$,

and

$$\sup_{t \in [\tau/2, \tau]} \|u(t) - \tilde{u}(t)\|_\alpha \geq a. \tag{A.6.4}$$

We claim that the set $\mathcal{M}(u_0, \tau, a)$ is closed in $\mathcal{L}$. Indeed, let $\{B_n\} \subset \mathcal{M}(u_0, \tau, a) \subset \mathcal{L}$ and $B_n \to B$ in $\mathcal{L}$. This means that there exist a sequence $\{u_0^n\}$ and solutions $u^n(t)$ and $\tilde{u}^n(t)$ such that

- $u^n(t) \in \mathcal{B}_1(u_0)$ and $\tilde{u}^n(t) \in \mathcal{B}_1(u_0)$ for all $t \in [0, \tau)$;
- $u^n(0) = u_0^n$ and $\tilde{u}(0) = u_0^n$;

and also

$$\sup_{t \in [\tau/2, \tau]} \|u^n(t) - \tilde{u}^n(t)\|_\alpha \geq a. \tag{A.6.5}$$

---

[2] We note that in this case a solution starting in $\mathcal{O}$ can be *globally* non-unique. See Remark 1.7.17.

We can assume that $u_0^n \to u_*$ in H_α weakly for some $u_* \in H_\alpha$. By the compactness statement in Proposition 4.1.6 this implies that

$$\sup_{t\in[0,\tau]} \|e^{-tA}u_0^n - e^{-tA}u_*\|_\theta + \sup_{t\in[\epsilon,\tau]} \|e^{-tA}u_0^n - e^{-tA}u_*\|_\alpha \to 0, \quad n \to \infty,$$

for every positive ϵ and $\theta < \alpha$. By the definition of mild solutions (see Definition 4.2.2) we have

$$u^n(t) = e^{-tA}u_0^n + \int_0^t e^{-(t-s)A}B_n(u^n(s))ds$$

and

$$\tilde{u}^n(t) = e^{-tA}u_0^n + \int_0^t e^{-(t-s)A}B_n(\tilde{u}^n(s))ds.$$

Since $\varrho_{\mathscr{L}}(B_n, B) \to 0$, we also have that

$$\sup_{|v|_\alpha \leq 1} \|B_n(u_0 + v) - B(u_0 + v)\| \to 0, \quad n \to \infty.$$

Thus $B_n(v)$ is uniformly bounded on $\mathscr{B}_1(u_0)$. This implies that

$$\sup_{t\in[\epsilon,\tau]} \left[\|\tilde{u}^n(t)\|_\gamma + \|u^n(t)\|_\gamma\right] \leq C_\epsilon, \quad n = 1, 2, \ldots$$

for every positive ϵ with $\gamma \in (\alpha, 1)$. As in the proof of Proposition 6.1.3 we can show that the sequences $\{u^n(t)\}$ and $\{\tilde{u}^n(t)\}$ are uniformly Hölder on $[\epsilon, \tau]$ in the sense that

$$|u^n|_{H_\alpha^\beta(\epsilon,\tau)} + |\tilde{u}^n|_{H_\alpha^\beta(\epsilon,\tau)} \leq C_{\epsilon,\tau}$$

for every $0 \leq \beta < 1 - \alpha$ and $\epsilon < \tau$, where

$$|v|_{H_\alpha^\beta(a,b)} \equiv \sup \left\{ \frac{\|v(t_1) - v(t_2)\|_\alpha}{|t_1 - t_2|^\beta} \ \middle| \ \begin{matrix} t_1, t_2 \in [a, b] \\ |t_1 - t_2| \leq 1 \end{matrix} \right\}.$$

By the Arzelà-Ascoli theorem, see Lemma A.3.5, we can conclude that there exist $u(t)$ and $\tilde{u}(t)$ in $C((0, \tau]; H_\alpha)$ such that along a subsequence we have

$$\sup_{t\in[\epsilon,\tau]} \left[\|\tilde{u}^n(t) - \tilde{u}(t)\|_\alpha + \|u^n(t) - u(t)\|_\alpha\right] \to 0, \quad n \to \infty.$$

and

$$\sup_{t\in[0,\tau]} [\|\tilde{u}^n(t) - \tilde{u}(t)\|_\theta + \|u^n(t) - u(t)\|_\theta] \to 0, \quad n \to \infty, \ \theta < \alpha.$$

These properties allow us to show that $u(t)$ and $\tilde{u}(t)$ are mild solutions for (A.6.1) with initial data u_*. It is also clear that (A.6.4) is valid for these solutions. Thus the set $\mathscr{M}(u_0, \tau, a)$ is closed in $\mathscr{L}$.

This set is nowhere dense. If this is not true, then we can find a Lipschitz point continuous function near B for which we have the uniqueness property.

We complete the argument in the same way as in YUDOVICH [230]. Let $\{w_k\}_{k\in\mathbb{N}}$ be a countable dense set in H_α and $\mathbb{Q}_+$ be positive rational numbers. Then the set

$$\mathscr{M} = \cup\{\mathscr{M}(w_k, \tau, a) : k \in \mathbb{N}, \ \tau \in \mathbb{Q}_+, \ a \in \mathbb{Q}_+\}$$

is meager. It is clear that every continuous nonlinearity B with non-uniqueness point belongs to some set $\mathscr{M}(w_k, \tau, a)$ with rational τ and a. This completes the proof. □

As an application of Theorem A.6.2 in a bounded smooth domain $\Omega \subset \mathbb{R}^3$, we consider the heat equation

$$u_t(x, t) - \Delta u(x, t) + f(u(x, t)) = 0 \tag{A.6.6}$$

endowed with boundary and initial conditions

$$u|_{\partial\Omega} = 0, \quad u|_{t=0} = u_0. \tag{A.6.7}$$

We assume that $f : \mathbb{R} \mapsto \mathbb{R}$ is a continuous function. In the 3D case we have the Sobolev embedding (see TRIEBEL [220]): $H^{3/2+\delta}(\Omega) \subset C(\overline{\Omega})$ for $\delta > 0$. This implies that the Nemytskii operator $u \mapsto f(u)$ is continuous from $H^{3/2+\delta}(\Omega)$ into $L_2(\Omega)$ for every continuous f.

In this example we take $\mathscr{L} = C(\mathbb{R})$ endowed with the norm

$$\varrho_{C(\mathbb{R})}(f_1, f_2) = \sum_{k=1}^\infty 2^{-k} \frac{\max_{|s|\le k} |f_1(s) - f_2(s)|}{1 + \max_{|s|\le k} |f_1(s) - f_2(s)|}.$$

The problem in (A.6.6) and (A.6.7) can be written in the form (A.6.1) in the space $H = L_2(\Omega)$ with $A = -\Delta$ defined on the domain $\mathscr{D}(A) = H^2(\Omega) \cap H_0^1(\Omega)$. In this case $\mathscr{D}(A^{3/4+\delta}) = (H^{3/2+2\delta} \cap H_0^1)(\Omega)$ for $\delta \in (0, 1/4)$ and thus we can apply Theorem A.6.2 with $\alpha = 3/4 + \delta$ and $\mathscr{L} = C(\mathbb{R})$. The property in (A.6.2) is obvious in this case. Thus we arrive at the following result.

Theorem A.6.3. *Let* $\Omega \subset \mathbb{R}^3$ *be a bounded smooth domain. Assume that initial data* u_0 *lies in* $(H^{3/4+\delta} \cap H_0^1)(\Omega)$. *Then for every continuous function* f *the problem in* (A.6.6) *and* (A.6.7) *has a (local) mild solution* $u(t)$ *which lies in the space*

$C([0, T); H_{3/4+\delta})$ with some $T = T(u_0, f) > 0$. Moreover the set of all continuous functions $f \in C(\mathbb{R})$ for which the problem has at least one non-uniqueness point is a meager set in $C(\mathbb{R})$ with respect to uniform convergence on finite intervals. Thus the uniqueness is a generic property in the space of all continuous reaction terms f.

A.7   Monotonicity method for well-posedness of 2D hydrodynamical systems

The goal of this section is to prove the well-posedness of the abstract 2D hydro-dynamical problem in (4.4.1) without[3] assuming that the operator A possesses a discrete spectrum. For this we use the Galerkin method in combination with the monotonicity-type argument suggested in MENALDI/SRITHARAN [165].

Thus we consider

$$u_t(t) + Au(t) + B(u(t), u(t)) + Ku(t) = f, \quad u\big|_{t=0} = u_0, \qquad (A.7.1)$$

in a separable Hilbert space H with the norm $\| \cdot \|$ and the inner product $(.,.)$. We assume that A is an (unbounded) self-adjoint positive linear operator on H and consider the triple $V \subset H \subset V'$, where

- $V = H_{1/2} = \mathscr{D}(A^{1/2})$ with $\|v\|_V \equiv \|A^{1/2}v\|$ for $v \in V$;
- $V' = H_{-1/2}$ is the dual of V with respect to $(.,.)$, $\|v\|_{V'} \equiv \|A^{-1/2}v\|$ for $v \in V'$.

In contrast with Section 4.4, there is no compact embedding in the pair $V \subset H$.

Concerning B, K, and f, we assume the same properties as in Assumption 4.4.1:

- $B : V \times V \to V'$ is a bilinear continuous mapping.
- The trilinear form $b(u_1, u_2, u_3) = (B(u_1, u_2), u_3)$ possesses the following skew-symmetry property:

$$(B(u_1, u_2), u_3) = -(B(u_1, u_3), u_2) \quad \text{for } u_i \in V, i = 1, 2, 3.$$

- There exists a Banach (interpolation) space $\mathscr{H}$ such that $V \subset \mathscr{H} \subset H$ and

$$\exists a_0 > 0 : \quad \|v\|_{\mathscr{H}}^2 \le a_0 \|v\| \, \|v\|_V \quad \text{for any } v \in V.$$

- There exists a constant $C > 0$ such that

$$|(B(u_1, u_2), u_3)| \le C \|u_1\|_{\mathscr{H}} \|u_2\|_V \|u_3\|_{\mathscr{H}}, \quad \text{for } u_i \in V, \ i = 1, 2, 3. \qquad (A.7.2)$$

- $K : H \mapsto H$ is globally Lipschitz, $f \in V'$.

---

[3]This corresponds to the case of 2D hydrodynamical flows in unbounded domains.

We recall that a function $u \in L_2(0, T; V)$ is said to be a weak solution on $[0, T]$ if (A.7.1) is satisfied in the sense of distributions (see Definition 4.4.5).

The following assertion is an analog of Theorem 4.4.7 for the noncompact case.

Theorem A.7.1 (Well-posedness). *Let the hypotheses stated above be in force. Then for any $u_0 \in H$ problem (A.7.1) has a unique weak solution u on any interval $[0, T]$. This solution possesses the property*

$$u \in C(0, T; H) \cap L_2(0, T; V), \quad u_t \in L_2(0, T; V'),$$

satisfies the energy balance relation in (4.4.14), and admits the estimates stated in (4.4.15) and (4.4.16).

Proof. As in the proof of Theorem 4.4.7, using a smooth orthonormal basis $\{\varphi_i\}$ in H we can construct approximate solutions $u^N(t)$ which solve the equations

$$(u_t^N - F(u^N), \varphi_k) = 0, k = 1, \ldots, N, \quad \text{and} \quad u^N\big|_{t=0} = P_N u_0 \in H, \qquad (A.7.3)$$

with

$$F(u) = -Au - B(u, u) - K(u) + f, \quad \forall u \in V,$$

where P_N is the orthogonal projector in H on $H_N = \mathrm{Span}\{\varphi_1, \cdots, \varphi_N\}$. These solutions satisfy the balance relation in (4.4.14) with u^N instead of u and admit the a priori estimate

$$\|u^N(t)\|^2 + \int_0^t \left[\|u_t^N(\tau)\|_{V'}^2 + \|u^N(\tau)\|_V^2 + \|u^N(\tau)\|_{\mathcal{H}}^4 \right] d\tau \leq C_T(f, u_0)$$

for every $t \in [0, T]$. As in the proof of Theorem 4.4.7, this estimate implies the existence of the function

$$u(t) \in L_\infty(0, T; H) \cap L_2(0, T; V) \cap L_4(0, T; \mathcal{H}) \quad \text{with} \quad u_t \in L_2(0, T; V'),$$

such that along a subsequence we have

(i) $u^N \to u$ weakly in $L^2(0, T; V)$ and in $L^4(0, T; \mathcal{H})$, *-weakly in $L_\infty(0, T; H)$.

In addition we can assume that

(ii) $u^N(T) \to \tilde{u}(T)$ weakly in H (for some $\tilde{u}(T)$),
(iii) $u_t^N \to u_t$ and $F(u^N) \to F$ weakly in $L^2(0, T; V')$ for some F.

We use this information to show, via a monotonicity argument, that $u(t)$ is a weak solution to (A.7.1). The main issue is to show that $F = F(u)$.

Let $f \in C^1([0, T])$ be such that $\|f\|_\infty = 1, f(0) = 1$. For any integer $j \geq 1$ we set $h_j(t) = f(t)\varphi_j$, where $\{\varphi_j\}_{j\geq1}$ is the previously chosen orthonormal basis for H.

Due to the convergences above, we obtain in the limit from (A.7.3) that

$$(\tilde{u}(T), \varphi_j) f(T) = (u_0, \varphi_j) + \int_0^T (u(s), \varphi_j) f'(s) ds + \int_0^T (F(s), h_j(s)) ds.$$
$$(A.7.4)$$

Let $t \in (0, T)$ be fixed. Take $0 \leq f_k \leq 1$ such that $f_k(s) = 1$ for $s \leq t - 1/k$ and $f_k(s) = 0$ for $s \geq t$. Set $f = f_k$ in (A.7.4). In the limit $k \to \infty$ we obtain that

$$0 = (u_0 - u(t), \varphi_j) + \int_0^t (F(s), \varphi_j) ds \quad \text{for almost all } t \in [0, T] \text{ and } j = 1, 2, \ldots.$$

This implies that

$$u(t) = u_0 + \int_0^t F(s) ds \in H \quad \text{for almost all } t \in [0, T]. \tag{A.7.5}$$

This means that we can suppose that $u(t)$ is continuous with values in V'. Moreover, taking $f \equiv 1$ in (A.7.4) we obtain

$$\tilde{u}(T) = u_0 + \int_0^T F(s) ds.$$

Thus $\tilde{u}(T) = u(T)$. Since $u_t \in L_2(0, T; V')$, by Proposition A.3.3, $u(t)$ is continuous in H.

To establish that $F(s) = F(u(s))$ we use the monotonicity-type argument presented in MENALDI/SRITHARAN [165]. We first note that inequality (A.7.2) implies (4.4.8). Therefore for any $\eta > 0$ there exists $C_\eta > 0$ such that for $u, v \in V$,

$$(F(u) - F(v), u - v) \leq -(1 - \eta) \|u - v\|_V^2 + (R_1 + C_\eta \|v\|_{\mathscr{H}}^4) \|u - v\|^2, \tag{A.7.6}$$

where R_1 is the Lipschitz constant of K and C_η is the constant from (4.4.8).

Let

$$v \in \mathscr{X} = L^4(0, T; \mathscr{H}) \cap L^\infty(0, T; H)) \cap L^2(0, T; V).$$

For every $t \in [0, T]$, we set

$$r(t) = \int_0^t \left[2R_1 + 2C_\eta \|v(s)\|_{\mathscr{H}}^4 \right] ds, \tag{A.7.7}$$

where C_η is a function of η such that (A.7.6) holds. We have that $0 \leq r(t) < \infty$ for all $t \in [0, T]$ and also

$$r \in L_\infty(0; T), \ e^{-r} \in L^\infty(0, T), \ r' \in L^1(0, T), \ r' e^{-r} \in L^1(0, T). \tag{A.7.8}$$

The weak convergence $u^N(T) \to u(T)$ and the property $P_N u_0 \to u_0$ in H imply that

$$\|u(T)\|^2 \, e^{-r(T)} - \|u_0\|^2 \leq \liminf_N \left[\|u^N(T)\|^2 \, e^{-r(T)} - \|P_N u_0\|^2 \right]. \qquad (A.7.9)$$

Since $u \in L_2(0, T; V)$ and $u_t \in L_2(0, T; V')$, by Proposition A.3.3,

$$\|u(t)\|^2 - \|u(0)\|^2 = 2 \int_0^t (u(s), u_t(s)) ds.$$

This implies the relation

$$\|u(T)\|^2 \, e^{-r(T)} - \|u(0)\|^2 = 2 \int_0^T e^{-r(s)} (u(s), u_t(s)) ds - \int_0^T r'(s) e^{-r(s)} \|u(s)\|^2 ds,$$

which can be justified due to (A.7.8). The same relation is definitely true with u^N instead of u. Using (A.7.5) and letting $u = v + (u - v)$ after simplification, from (A.7.9) we obtain

$$\int_0^T e^{-r(s)} \left[-r'(s) \{ \|u(s) - v(s)\|^2 + 2\big(u(s) - v(s),\, v(s)\big) \} \right.$$
$$\left. + 2\big(F(s), u(s)\big) \right] ds \leq \liminf_N X_N, \qquad (A.7.10)$$

where

$$X_N = \int_0^T e^{-r(s)} \left[-r'(s) \{ \|u^N(s) - v(s)\|^2 + 2\big(u^N(s) - v(s),\, v(s)\big) \} \right.$$
$$\left. + 2(F(u^N(s)), u^N(s)) \right] ds.$$

Relation (A.7.6) with $0 < \eta < 1$ and Schwarz's inequality imply that

$$Y_N := \int_0^T e^{-r(s)} \left[-r'(s) \|u^N(s) - v(s)\|^2 \right.$$
$$\left. + 2(F(u^N(s)) - F(v(s)), u^N(s) - v(s)) \right] ds \leq 0.$$

Furthermore, $X_N = Y_N + Z_N$ with

$$Z_N = \int_0^T e^{-r(s)} \left[-2r'(s)(u^N(s) - v(s), v(s)) + 2(F(u^N(s)), v(s)) \right.$$
$$\left. + 2\langle F(v(s)), u^N(s) \rangle - 2(F(v(s)), v(s)) \right] ds.$$

The weak convergence properties imply that $Z_N \to Z$ as $N \to \infty$, where

$$Z = \int_0^T e^{-r(s)}\big[-2r'(s)\big(u(s) - v(s), v(s)\big) + 2\langle F(s), v(s)\rangle + 2\langle F(v(s)), u(s)\rangle$$
$$- 2\langle F(v(s)), v(s)\rangle \big]ds.$$

Thus, (A.7.10) yields

$$\int_0^T e^{-r(s)}\Big\{ -r'(s)\|u(s) - v(s)\|^2 + 2(F(s) - F(v(s)), u(s) - v(s))\Big\}ds \leq 0,$$
$$\text{(A.7.11)}$$

for any $v \in \mathscr{X}$. For $\lambda \in \mathbb{R}$ and $\tilde{v} \in L^\infty(0, T, V)$, we set $v_\lambda = u - \lambda \tilde{v}$. It is clear that $v_\lambda \in \mathscr{X}$. Applying (A.7.11) to $v := v_\lambda$ yields

$$\int_0^T e^{-r_\lambda(s)}\Big[-\lambda^2 r'_\lambda(s)\|\tilde{v}(s)\|^2 + 2\lambda(F(s) - F(v_\lambda(s)), \tilde{v}(s))\Big]ds \leq 0, \quad \text{(A.7.12)}$$

where $r_\lambda(s)$ is given by (A.7.7) with v_λ instead of v. The same argument as for (A.7.6) yields

$$\big|(F(v_\lambda(s)) - F(u(s)), \tilde{v}(s))\big| \leq C|\lambda|\big[\|\tilde{v}(s)\|_V^2 + \|\tilde{v}(s)\|^2\|u(s)\|_{\mathscr{H}}^4\big]$$

for $\lambda \neq 0$ and $s \in [0, T]$. Thus in the limit $\lambda \to 0$ we deduce that

$$\int_0^T e^{-r_\lambda(s)}(F(s) - F(v_\lambda(s)), \tilde{v}(s))ds \to \int_0^T e^{-r_0(s)}(F(s) - F(u(s)), \tilde{v}(s))ds.$$

Now dividing (A.7.12) by $\lambda > 0$ (resp. $\lambda < 0$) and letting $\lambda \to 0$ we obtain that for every $\tilde{v} \in L^\infty(0, T; V)$, which is a dense subset of $L^2(0, T; V)$,

$$\int_0^T e^{-r_0(s)}\Big[(F(s) - F(u(s)), \tilde{v}(s))\Big]ds = 0.$$

Hence (A.7.5) can be rewritten as

$$u(t) = u_0 + \int_0^t F(u(s))ds, \quad t \in [0, T].$$

This means that $u(t)$ is a weak solution satisfying (4.4.15).

Other statements of Theorem A.7.1 follow from the argument given in the proof of Theorem 4.4.7.   □

References

1. R. Adams, *Sobolev Spaces*, Academic Press, New York, 1975.
2. R. Akhmerov, M. Kamenskii, A. Potapov, A. Rodkina, and B. Sadovskii, *Measures of Noncompactness and Condensing Operators*, Birkhäuser, Basel, 1992.
3. E. Aragão-Costa, T. Caraballo, A. Carvalho, and J. Langa, Stability of gradient semigroups under perturbations, *Nonlinearity*, 24 (2011), 2099.
4. J.-P. Aubin, Une théorè de compacité, *C.R. Acad. Sci. Paris*, 256 (1963), 5042–5044.
5. J.-P. Aubin, *Approximation of Elliptic Boundary-Value Problems*, Wiley, New York, 1972.
6. G. Autuori, P. Pucci and M.C. Salvatori, Asymptotic stability for nonlinear Kirchhoff systems, *Nonlinear Anal., RWA*, 10 (2009), 889–909.
7. A. Babin, Global Attractors in PDE. In: B. Hasselblatt and A. Katok (Eds.), Handbook of Dynamical Systems, vol. 1B, Elsevier, Amsterdam, 2006, 983–1085.
8. A. Babin and M. Vishik, Unstable invariant sets of semigroups of nonlinear operators and their perturbations, *Russian Math. Surveys*, 41(4) 1986, 1–41.
9. A. Babin and M. Vishik, *Attractors of Evolution Equations*, North-Holland, Amsterdam, 1992.
10. D. Barbato, M. Barsanti, H. Bessaih and F. Flandoli, Some rigorous results on a stochastic Goy model, *Journal of Statistical Physics*, 125 (2006) 677–716.
11. N.N. Bautin and E.A. Leontovich, *Methods and Examples of the Qualitative Analysis of Dynamical Systems in a Plane*, Nauka, Moscow, 1990 (in Russian).
12. S. Bernstein, Sur une classe d'équations fonctionelles aux dérivées partielles, *Bull. Acad. Sciences de l'URSS, Ser. Math.* 4 (1940), 17–26.
13. J. Billoti and J. LaSalle, Periodic dissipative processes, *Bull. Amer. Math. Soc.*, 6 (1971), 1082–1089.
14. G.D. Birkhoff, *Dynamical Systems*, AMS Colloquium Publications, vol. 9, AMS, Providence, RI, 1927.
15. V.A. Boichenko, G.A. Leonov and V. Reitmann, *Dimension Theory for Ordinary Differential Equations*, Teubner, Wiesbaden, 2005.
16. N. Bourbaki, *General Topology: Chapters 5–10*, Berlin, Springer, 1998.
17. L. Boutet de Monvel, I. Chueshov and A. Rezounenko, Long-time behaviour of strong solutions of retarded nonlinear P.D.E.s, *Commun. Partial Diff. Eqs.*, 22 (1997), 1453–1474.
18. N.F. Britton, Spatial structures and periodic travelling waves in an integro-differential reaction-diffusion population model, *SIAM. J. Appl. Math.*, 50 (1990), 1663–1688.
19. F. Bucci and I. Chueshov, Long-time dynamics of a coupled system of nonlinear wave and thermoelastic plate equations, *Discr. Cont. Dyn. Sys.*, 22 (2008), 557–586.

© Springer International Publishing Switzerland 2015

I. Chueshov, *Dynamics of Quasi-Stable Dissipative Systems*, Universitext,
DOI 10.1007/978-3-319-22903-4

20. F. Bucci, I. Chueshov and I. Lasiecka, Global attractor for a composite system of nonlinear wave and plate equations, *Commun. Pure Appl. Anal.*, 6 (2007), 113–140.

21. A. Busenberg, D. Fisher and M. Martelli, Better bounds for periodic solutions of differential equations in Banach spaces. *Proc. Amer. Math. Soc.*, 98 (1986), 376–378.

22. H. Cartan, *Calculus Différentielles*, Hermann, Paris, 1967.

23. A. Carvalho and J. Cholewa, Attractors for strongly damped wave equations with critical nonlinearities, *Pacific J. Math.*, 207 (2002), 287–310.

24. A. Carvalho, J. Cholewa and T. Dlotko, Strongly damped wave problems: bootstrapping and regularity of solutions. *J. Differential Equations*, 244 (2008), 2310–2333.

25. A. Carvalho and J. Langa, An extension of the concept of gradient semigroups which is stable under perturbation, *J. Diff. Equations*, 246 (2009), 2646–2668.

26. A. Carvalho, J. Langa and J. Robinson, *Attractors for Infinite-Dimensional Non-Autonomous Dynamical Systems*, New York, Springer, 2013.

27. M.M. Cavalcanti, V.N.D. Cavalcanti, J.S.P. Filho and J.A. Soriano, Existence and exponential decay for a Kirchhoff–Carrier model with viscosity, *J. Math. Anal. Appl.*, 226 (1998), 20–40.

28. S. Ceron and O. Lopes, α-contractions and attractors for dissipative semilinear hyperbolic equations and systems, *Ann. Math. Pura Appl. IV*, 160 (1991), 193–206.

29. V. Chepyzhov and A. Ilyin, On the fractal dimension of invariant sets: applications to the Navier-Stokes equations, *Discrete Contin. Dyn. Syst.*, 10 (2004), 117–136.

30. V. Chepyzhov, E. Titi and M. Vishik, On the convergence of solutions of the Leray-α model to the trajectory attractor of the 3D Navier-Stokes system, *Discrete Contin. Dyn. Syst.*, 17 (2007), 481–500.

31. V.V. Chepyzhov and M.I. Vishik, *Attractors for Equations of Mathematical Physics*, AMS, Providence, RI, 2002.

32. A. Cheskidov, Global attractors of evolutionary systems, *J. Dyn. Dif. Equations*, 21 (2009), 249–268.

33. A. Cheskidov and C. Foias, On global attractors of the 3D Navier-Stokes equations, *J. Dif. Equations*, 231 (2006), 714–754.

34. A. Cheskidov, D. Holm, E. Olson and E. Titi, On a Leray-α model of turbulence, *Proc. R. Soc. Lond. Ser. A*, 461 (2005), 629–649.

35. J.W. Cholewa and T. Dlotko, Strongly damped wave equation in uniform spaces, *Nonlinear Anal., TMA*, 64 (2006) 174–187.

36. I. Chueshov, On a system of equations with delay that arises in aero-elasticity, *J. Soviet Math.*, 58 (1992), no. 4, 385–390.

37. I. Chueshov, On the finiteness of the number of determining elements for von Karman evolution equations, *Math. Meth. Appl. Sci.*, 20 (1997), 855–865.

38. I. Chueshov, Theory of functionals that uniquely determine asymptotic dynamics of infinite-dimensional dissipative systems, *Russian Math. Surv.*, 53 (1998), 731–776.

39. I. Chueshov, *Introduction to the Theory of Infinite-Dimensional Dissipative Systems*, Acta, Kharkov, 1999, in Russian; English translation: Acta, Kharkov, 2002; see also http://www.emis.de/monographs/Chueshov/

40. I. Chueshov, Global attractors for a class of Kirchhoff wave models with a structural nonlinear damping, *J. Abstract Diff. Equations and Applications*, 1 (2010), 86–106.

41. I. Chueshov, A global attractor for a fluid-plate interaction model accounting only for longitudinal deformations of the plate, *Math. Meth. Appl. Sci.*, 34 (2011), 1801–1812.

42. I. Chueshov, Long-time dynamics of Kirchhoff wave models with strong nonlinear damping, *J. Diff. Equations*, 252 (2012), 1229–1262.

43. I. Chueshov, Quantum Zakharov model in a bounded domain, *Zeitschrift Angew. Math. Phys.*, 64 (2013), 967–989.

44. I. Chueshov, Discrete data assimilation via Ladyzhenskaya squeezing property in the 3D viscous primitive equations, Preprint arXiv:1308.1570 (August 2013).

45. I. Chueshov, Dynamics of a nonlinear elastic plate interacting with a linearized compressible viscous fluid, *Nonlinear Anal., TMA*, 95 (2014), 650–665.

46. I. Chueshov, Interaction of an elastic plate with a linearized inviscid incompressible fluid, *Commun. Pure Appl. Anal.*, 13 (2014), 1759–1778.

47. I. Chueshov, M. Eller and I. Lasiecka, On the attractor for a semilinear wave equation with critical exponent and nonlinear boundary dissipation, *Commun. Partial Diff. Eqs.*, 27 (2002), 1901–1951.

48. I. Chueshov, M. Eller and I. Lasiecka, Finite dimensionality of the attractor for a semilinear wave equation with nonlinear boundary dissipation, *Commun. Partial Diff. Eqs.*, 29 (2004), 1847–1976.

49. I. Chueshov and S. Kolbasin, Plate models with state-dependent damping coefficient and their quasi-static limits, *Nonlinear Anal., TMA*, 73 (2010), 1626–1644.

50. I. Chueshov and S. Kolbasin, Long-time dynamics in plate models with strong nonlinear damping, *Commun. Pure Appl. Anal.*, 11 (2012), 659–674.

51. I. Chueshov and I. Lasiecka, Attractors for second order evolution equations, *J. Dynam. Diff. Eqs.*, 16 (2004), 469–512.

52. I. Chueshov and I. Lasiecka, Global attractors for von Karman evolutions with a nonlinear boundary dissipation, *J. Diff. Equations*, 198 (2004), 196–221.

53. I. Chueshov and I. Lasiecka, Kolmogorov's ε-entropy for a class of invariant sets and dimension of global attractors for second order in time evolution equations with nonlinear damping. In: *Control Theory of Partial Differential Equations*, O. Imanuvilov et al. (Eds.), A Series of Lectures in Pure and Applied Mathematics, vol. 242, Chapman & Hall/CRC, Boca Raton, FL, 2005, 51–69.

54. I. Chueshov and I. Lasiecka, Global attractors for Mindlin–Timoshenko plates and for their Kirchhoff limits, *Milan J. Math.*, 74 (2006), 117–138.

55. I. Chueshov and I. Lasiecka, Long time dynamics of semilinear wave equation with nonlinear interior-boundary damping and sources of critical exponents. In *Control Methods in PDE - Dynamical Systems*, F. Ancona et al. (Eds.), Contemporary Mathematics, vol. 426, AMS, Providence, RI, 2007, 153–193.

56. I. Chueshov and I. Lasiecka, *Long-Time Behavior of Second Order Evolution Equations with Nonlinear Damping*, Memoirs of AMS, vol. 195, no. 912, AMS, Providence, RI, 2008.

57. I. Chueshov and I. Lasiecka, Attractors and long time behavior of von Karman thermoelastic plates, *Appl. Math. Optim.*, 58 (2008), 195–241.

58. I. Chueshov and I. Lasiecka, *Von Karman Evolution Equations*, Springer, New York, 2010.

59. I. Chueshov and I. Lasiecka, On global attractor for 2D Kirchhoff-Boussinesq model with supercritical nonlinearity, *Commun. Partial Dif. Eqs.*, 36 (2011), 67–99.

60. I. Chueshov and I. Lasiecka, Well-posedness and long time behavior in nonlinear dissipative hyperbolic-like evolutions with critical exponents. In: *Nonlinear Hyperbolic PDEs, Dispersive and Transport Equations*, HCDTE Lecture Notes, Part I, G. Alberti et al. (Eds.), AIMS on Applied Mathematics, vol. 6, AIMS, Springfield, 2013, 1–96.

61. I. Chueshov, I. Lasiecka and D. Toundykov, Long-term dynamics of semilinear wave equation with nonlinear localized interior damping and a source term of critical exponent, *Discr. Cont. Dyn. Sys.*, 20 (2008), 459–509.

62. I. Chueshov, I. Lasiecka and D. Toundykov, Global attractor for a wave equation with nonlinear localized boundary damping and a source term of critical exponent, *J. Dynam. Diff. Eqs.*, 21 (2009), 269–314.

63. I. Chueshov, I. Lasiecka and J.T. Webster, Attractors for delayed, non-rotational von Karman plates with applications to flow-structure interactions without any damping, *Commun. Partial Dif. Eqs.*, 39, (2014), 1965–1997.

64. I. Chueshov, I. Lasiecka and J.T. Webster, Flow-plate interactions: well-posedness and long-time behavior, *Discrete Continuous Dynamical Systems Ser. S*, 7 (2014), 925–965.

65. I. Chueshov and A. Rezounenko, Global attractors for a class of retarded quasilinear partial differential equations, *C. R. Acad. Sci. Paris, Ser. I*, 321 (1995), 607–612.

66. I. Chueshov and A. Rezounenko, Dynamics of second order in time evolution equations with state-dependent delay, *Nonlinear Anal. TMA*, 123–124 (2015) 126–149.

67. I. Chueshov and A. Rezounenko, Finite-dimensional global attractors for parabolic nonlinear equations with state-dependent delay. *Commun. Pure Appl. Anal.*, 14 (2015), 1685–1704.

68. I. Chueshov and I. Ryzhkova, A global attractor for a fluid-plate interaction model, *Commun. Pure Appl. Anal.*, 12 (2013), 1635–1656.

69. I. Chueshov and I. Ryzhkova, Unsteady interaction of a viscous fluid with an elastic shell modeled by full von Karman equations. *J. Diff. Equations*, 254 (2013), 1833–1862.

70. I. Chueshov and A. Shcherbina, On 2D Zakharov system in a bounded domain, *Diff. Integral Eqs.*, 18 (2005), 781–812.

71. I. Chueshov and A. Shcherbina, Semi-weak well-posedness and attractors for 2D Schrö-dinger-Boussinesq equations, *Evolution Equations and Control Theory*, 1 (2012), 57–80.

72. P. Ciarlet, *Mathematical Elasticity, Vol. III: Theory of Shells*, North-Holland, Amsterdam, 2000.

73. B. Cockburn, D.A. Jones and E. Titi, Determining degrees of freedom for nonlinear dissipative systems, *C.R. Acad. Sci. Paris, Ser. I*, 321 (1995), 563–568.

74. B. Cockburn, D.A. Jones and E. Titi, Estimating the number of asymptotic degrees of freedom for nonlinear dissipative systems, *Math. Comp.*, 66 (1997), 1073–1087.

75. E.A. Coddington and N. Levinson, *Theory of Ordinary Differential Equations,* McGraw-Hill, New York, 1955.

76. P. Constantin, C. Doering and E. Titi, Rigorous estimates of small scales in turbulent flows, *J. Math. Phys.*, (1996), 6152–6156.

77. E Constantin and C. Foias, Global Lyapunov exponents, Kaplan-Yorke formulas and the dimension of the attractors for 2D Navier-Stokes equations, *Comm. Pure Appl. Math.*, 38 (1985), 1–27.

78. P. Constantin and C. Foias, *Navier-Stokes Equations*, University of Chicago Press, Chicago, 1988.

79. P. Constantin, C. Foias and R. Temam, *Attractors Representing Turbulent Flows*, Memoirs of AMS, vol. 53, no. 314, AMS, Providence, RI, 1985.

80. P. Constantin, B. Levant and E.S. Titi, Analytic study of the shell model of turbulence, *Physica D* 219 (2006), 120–141.

81. K.L. Cooke and Z. Grossman, Discrete delay, distributed delay and stability switches, *J. Math. Anal. Appl.*, 86 (1982), 592–627.

82. *Data Assimilation. Making Sense of Observations* (Eds: W. Lahoz, B. Khattatov, R. Ménard), Springer, New York, 2010.

83. L. De, The critical forms of the attractors for semigroups and the existence of critical attractors, *Proc. Royal Society of London, Ser. A*, 454 (1998), 2157–2171.

84. O. Diekmann, S. van Gils, S. Verduyn Lunel and H.-O. Walther, Delay Equations: Functional, Complex, and Nonlinear Analysis, Springer-Verlag, New York, 1995.

85. J. Dieudonné, *Foundations of Modern Analysis*, Academic Press, New York, 1960.

86. R.D. Driver, A two-body problem of classical electrodynamics: the one-dimensional case, *Ann. Physics*, 21 (1963), 122–142.

87. Yu. A. Dubinskiĭ, Weak convergence in nonlinear elliptic and parabolic equtions, *Math. USSR Sbornik* 67(4) (1965), 609–642.

88. N. Dunford and J. Schwartz, *Linear Operators, Part I: General Theory*, Interscience, New York, 1958.

89. N. Dunford and J. Schwartz, *Linear Operators, Part II: Spectral Theory*, Interscience, New York, 1963.

90. G. Duvaut and J.L. Lions, Inéquations en thermoélasticité et magnéto hydrodynamique, *Arch. Rational Mech. Anal.* 46 (1972), 241–279.

91. M. Efendiev, A. Miranville and S. Zelik, Exponential attractors for nonlinear reaction-diffusion systems in $\mathbb{R}^n$, *C.R. Acad. Sci. Paris, Ser. I* 330 (2000), 713–718.

92. A. Eden, C. Foias, B. Nicolaenko and R. Temam, *Exponential Attractors for Dissipative Evolution Equations*, Research in Appl. Math. 37, Masson, Paris, 1994.

93. K. Engel and R.Nagel, *One-Parameter Semigroups for Linear Evolution Equations*, Springer, New York, 2000.

94. P. Fabrie, C. Galusinski, A. Miranville and S. Zelik, Uniform exponential attractors for singularly perturbed damped wave equation, *Discr. Cont. Dyn. Syst.*, 10 (2004), 211–238.

95. K. Falconer, *Fractal Geometry: Mathematical Foundations and Applications*, Wiley, Chichester, 1990.

96. X. Fan and S. Zhou, Kernel sections for non-autonomous strongly damped wave equations of non-degenerate Kirchhoff-type, *Appl. Math. Computation*, 158 (2004), 253–266.

97. M. Farkas, *Periodic Motions*, Springer, New York, 1994.

98. T. Fastovska, Upper semicontinuous attractor for 2D Mindlin–Timoshenko thermoelastic model with memory, *Commun. Pure Appl. Anal.*, 6 (2007), 83–101.

99. T. Fastovska, Upper semicontinuous attractors for a 2D Mindlin–Timoshenko thermoviscoelastic model with memory, *Nonlinear Anal., TMA*, 71 (2009) 4833–4851.

100. E. Feireisl and D. Pražák, *Asymptotic Behavior of Dynamical Systems in Fluid Mechanics*, Springfield, AIMS, 2010.

101. W.E. Fitzgibbon, Semilinear functional differential equations in Banach space, *J. Differential Equations*, 29 (1978) 1–14.

102. C. Foias and E. Olson, Finite fractal dimension and Hölder–Lipschitz parametrization, *Indiana Univ. Math. J.*, 45 (1996), 603–616.

103. C. Foias, O. Manley and R. Temam, Attractors for the Bénard problem: existence and physical bounds on their fractal dimension. *Nonlinear Analysis*, 11 (1987), 939–967.

104. C. Foias, O. Manley, R. Temam and Y.M. Treve, Asymptotic analysis of the Navier–Stokes equations, *Physica D*, 9 (1983), 157–188.

105. C. Foias and G. Prodi, Sur le comportement global des solutions non stationnaires des equations de Navier-Stokes en dimension deux, *Rend. Sem. Mat. Univ. Padova*, 36 (1967), 1–34.

106. C. Foias, R. Rosa, and R. Temam, Topological properties of the weak global attractor of the three-dimensional Navier-Stokes equations, *Discrete Contin. Dyn. Syst.*, 27 (2010), 611–1631.

107. C. Foias and R. Temam, Determination of solutions of the Navier–Stokes equations by a set of nodal values, *Math. Comp.*, 43 (1984), 117–133.

108. C. Foias and E.S. Titi, Determining nodes, finite difference schemes and inertial manifolds, *Nonlinearity*, 4 (1991), 135–153.

109. G.P. Galdi and M. Padula, A new approach to energy theory in the stability of fluid motion, *Arch. Rational Mech. Anal.*, 110 (1990), 187–286.

110. M.J. Garrido-Atienza and J. Real, Existence and uniqueness of solutions for delay evolution equations of second order in time, J. Math. Anal. Appl., 283 (2003), 582–609.

111. S. Gatti and V. Pata, A one-dimensional wave equation with nonlinear damping, *Glasgow Math. J.*, 48 (2006), 419–430.

112. M. Ghisi, Global solutions for dissipative Kirchhoff strings with non-Lipschitz nonlinear term, *J. Differential Equations*, 230 (2006), 128–139.

113. S. Gourley, J. So and J. Wu, Non-locality of reaction–diffusion equations induced by delay: biological modeling and nonlinear dynamics. In: D.V. Anosov, A. Skubachevskii (Eds.), Contemporary Mathematics, Thematic Surveys, Kluwer, Plenum, Dordrecht, New York, 2003, 84–120.

114. J. Guckenheimer and P. Holmes, *Nonlinear Oscillations, Dynamical Systems, and Bifurcations of Vector Fields*, Springer, New York, 1983.

115. J.K. Hale, *Theory of Functional Differential Equations*, 2nd ed., Springer, New York, 1977.

116. J.K. Hale, *Asymptotic Behavior of Dissipative Systems*, AMS, Providence, RI, 1988.

117. J. Hale and H. Kocak. *Dynamics and Bifurcations*. Springer, New York, 1991.

118. A. Haraux, Two remarks on dissipative hyperbolic problems. In: *Research Notes in Mathematics*, Pitman, 1985, pp. 161–179.

119. A. Haraux, *Semilinear Hyperbolic Problems in Bounded Domains*, Mathematical Reports, vol. 3, Harwood Gordon Breach, New York, 1987.

120. P. Hartman, *Ordinary Differential Equations*, 2nd ed., SIAM, Philadelphia, 2002.

121. F. Hartung, T. Krisztin, H.-O. Walther and J. Wu, Functional differential equations with state-dependent delays: Theory and applications. In: A. Canada, P. Drabek, and A. Fonda (Eds.) Handbook of Differential Equations, Ordinary Differential Equations, vol. 3, Elsevier, North Holland, 2006, pp. 435–545.

122. K. Hayden, E. Olson and E.S. Titi, Discrete data assimilation in the Lorenz and 2D Navier-Stokes equations, *Physica* D, 240 (2011), 1416–1425.

123. D. Henry, *Geometric Theory of Semilinear Parabolic Equations*, Springer, New York, 1981.

124. L. Hoang, E. Olson and J. Robinson, On the continuity of global attractors, Preprint ArXiv:1407.3306 (July 2014).

125. B.R. Hunt and V.Y. Kaloshin, Regularity of embeddings of infinite-dimensional fractal sets into finite-dimensional spaces. *Nonlinearity*, 12 (1999), 1263–1275.

126. D.A. Jones and E.S. Titi, Determination of the solutions of the Navier-Stokes equations by finite volume elements, *Physica D*, 60 (1992), 165–174.

127. D.A. Jones and E.S. Titi, Upper bounds on the number of determining modes, nodes and volume elements for the Navier-Stokes equations, *Indiana Univ. Math. J.*, 42 (1993), 875–887.

128. V. Kalantarov and S. Zelik, Finite-dimensional attractors for the quasi-linear strongly-damped wave equation, *J. Differential Equations*, 247 (2009), 1120–1155.

129. E. Kalnay, *Atmospheric Modeling, Data Assimilation and Predictability*. Cambridge University Press, Cambridge, UK, 2003.

130. L.V. Kapitansky and I.N. Kostin, Attractors of nonlinear evolution equations and their approximations,*Leningrad Math. J.*, 2 (1991), 97–117.

131. A.G. Kartsatos and L.P. Markov, An L_2-approach to second-order nonlinear functional evolutions involving *m*-accretive operators in Banach spaces, *Differential Integral Equations*, 14 (2001), 833–866.

132. A. Katok and B. Hasselblatt, *Introduction to the Modern Theory of Dynamical Systems*, Cambridge University Press, Cambridge, UK, 1996.

133. N.H. Katz and N. Pavlović, Finite time blow-up for a dyadic model of the Euler equations, *Trans. Amer. Math. Soc.* 357 (2005), 695–708.

134. A.K. Khanmamedov, Global attractors for von Karman equations with nonlinear dissipation, *J. Math. Anal. Appl.*, 318 (2006), 92–101.

135. P. Kloeden and M. Rasmussen, *Nonautonomous Dynamical Systems*. AMS, Providence, RI, 2011.

136. N.N. Krasovskii, *Stability of Motion: Applications of Lyapunov's Second Method to Differential Systems and Equations with Delay*. Stanford University Press, 1963.

137. T. Krisztin and O. Arino, The two-dimensional attractor of a differential equation with state-dependent delay, *J. Dynam. Diff. Eqs.*, 13 (2001) 453–522.

138. K. Kunisch and W. Schappacher, Necessary conditions for partial differential equations with delay to generate C_0-semigroups, *J. Differential Equations*, 50 (1983), 49–79.

139. Yu. Kuznetsov, *Elements of Applied Bifurcation Theory*, Springer, New York, 1998.

140. O. Ladyzhenskaya, A dynamical system generated by the Navier–Stokes equations, *J. Soviet Math.*, 3(4) (1975), 458–479.

141. O. Ladyzhenskaya, Estimates for the fractal dimension and number of deterministic modes for invariant sets of dynamical systems, *J. Soviet Math.*, 49 (1990), 1186–1201.

142. O. Ladyzhenskaya, *Attractors for Semigroups and Evolution Equations*, Cambridge University Press, Cambridge, UK, 1991.

143. O. Ladyzhenskaya and V. Solonnikov, Solution of some nonstationary magnetohydrodynamical problems for incompressible fluid, *Trudy Steklov Math. Inst.* 59 (1960), 115–173; in Russian.

144. I. Lasiecka and R. Triggiani, Exact null controllability of structurally damped and thermoelastic parabolic models, *Atti Accad. Naz. Lincei Cl. Sci. Fis. Mat. Natur. Rend. Lincei, Mat. Appl.*, **9** (1998), 43–69.

145. I. Lasiecka and R. Triggiani, *Control Theory for Partial Differential Equations*, Cambridge University Press, Cambridge, UK, 2000.

146. A. Lasota and J. Yorke. The generic property of existence of solutions of differential equations in Banach space, *J. Diff. Eqs.*, 13 (1973), 1–12.

147. P. Lazo, Global solutions for a nonlinear wave equation, *Appl. Math. Computation*, 200 (2008), 596–601.

148. S. Lefschetz, *Differential Equations: Geometric Theory*, Dover New York, 1977.

149. J. Leray, Essai sur le mouvement d'un fluide visqueux emplissant l'espace, *Acta Math.*, 63 (1934), 193–248.

150. N. Levinson, Transformation theory of non-linear differential equations of the second order, *Annals of Mathematics*, 45 (1944), 723–737.

151. J.L. Lions, *Quelques Méthodes de Résolution des Problèmes aux Limites Non Linéaires*, Dunod, Paris, 1969.

152. J.L. Lions and E. Magenes, *Non-Homogeneous Boundary Value Problems and Applications, vol. 1.* Springer, New York, 1972.

153. J.L. Lions, On some questions in boundary value problems in mathematical physics. In: International Symposium on Continuum Mechanics and Partial Differential Equations, Rio de Janeiro, 1977. North-Holland, Amsterdam, 1978.

154. A. Lunardi, *Analytic Semigroups and Optimal Regularity in Parabolic Problems.* Birkhäuser, Basel, 1995.

155. V.S. Lvov, E. Podivilov, A. Pomyalov, I. Procaccia, and D. Vandembroucq, Improved shell model of turbulence, *Physical Review E*, 58 (1998), 1811–1822.

156. Q. Ma, S. Wang and C. Zhong, Necessary and sufficient conditions for the existence of global attractors for semigroups and applications, *Indiana Univ. Math. J.*, 51 (2002), 1541–1559.

157. J. Málek and J. Nečas, A finite dimensional attractor for three dimensional flow of incompressible fluids, *J. Differential Equations*, 127 (1996), 498–518.

158. J. Málek and D. Pražák, Large time behavior via the method of *l*-trajectories, *J. Differential Equations*, 181 (2002), 243–279.

159. J. Mallet-Paret, Negatively invariant sets of compact maps and an extension of a theorem of Cartwright, *J. Differential Equations*, 22 (1976), 331–348.

160. J. Mallet-Paret, R.D. Nussbaum and P. Paraskevopoulos, Periodic solutions for functional-differential equations with multiple state-dependent time lags, Topol. Methods Nonlinear Anal., 3 (1994), 101–162.

161. R. Mañé, On the dimension of the compact invariant sets of certain nonlinear maps, Dynamical Systems and Turbulence, Warwick 1980 (Coventry, 1979/1980), *Lecture Notes in Math.*, vol. 898, Springer, Berlin (1981), 230–242.

162. T. Matsuyama and R. Ikehata, On global solution and energy decay for the wave equation of Kirchhoff type with nonlinear damping term, *J. Math. Anal. Appl.*, 204 (1996), 729–753.

163. L.A. Medeiros, J.L. Ferrel and S.B. de Menezes, Vibration of elastic strings: Mathematical Aspects, Part One, *J. Comp. Analysis Appl.*, 4 (2002), 91–127.

164. L.A. Medeiros and M. Milla Miranda, On a nonlinear wave equation with damping, *Revista Mat. Univ. Complutense Madrid*, 3 (1990), 213–231.

165. J.-L. Menaldi and S. Sritharan, Stochastic 2-D Navier-Stokes equation, *Appl. Math. Optimization*, 46 (2002), 31–54.

166. A. Miranville and S. Zelik, Attractors for dissipative partial differential equations in bounded and unbounded domains. In: C.M. Dafermos, and M. Pokorny (Eds.), Handbook of Differential Equations: Evolutionary Equations, vol. 4, Elsevier, Amsterdam, 2008, pp. 103–200.

167. R. Moreau, *Magnetohydrodynamics*, Kluwer, Dordrecht, 1990.

168. O. Naboka, On synchronization of oscillations of two coupled Berger plates with nonlinear interior damping, *Commun. Pure Appl. Anal.*, 8 (2009), 1933–1956.

169. M. Nakao, An attractor for a nonlinear dissipative wave equation of Kirchhoff type, *J. Math. Anal. Appl.*, 353 (2009), 652–659.

170. M. Nakao and Y. Zhijian, Global attractors for some quasi-linear wave equations with a strong dissipation, *Adv. Math. Sci. Appl.*, 17 (2007), 89–105.

171. V.V. Nemytskii and V.V. Stepanov, *Qualitative Theory of Differential Equations*, Princeton University Press, NJ, 1960.

172. M. Nieuwenhuis, J. Robinson, and S. Steinerberger, Minimal periods for ordinary differential equations in strictly convex Banach spaces and explicit bounds for some l_p-spaces, *J. Differential Equations*, 256 (2014), 2846–2857.

173. K. Ohkitani and M. Yamada, Temporal intermittency in the energy cascade process and local Lyapunov analysis in fully developed model of turbulence, *Prog. Theor. Phys.*, 89 (1989), 329–341.

174. K. Ono, Global existence, decay, and blow up of solutions for some mildly degenerate nonlinear Kirchhoff strings, *J. Differential Equations*, 137 (1997), 273–301.

175. K. Ono, On global existence, asymptotic stability and blowing up of solutions for some degenerate non-linear wave equations of Kirchhoff type with a strong dissipation, *Math. Methods Appl. Sci.*, 20 (1997), 151–177.

176. W. Orlicz, Zur Theorie der Differentialgleichung $y' = f(t, y)$, *Bull. de Acad. Pol. des Sciences*, Ser. A, 1932, 221–228.

177. V. Pata and S. Zelik, Smooth attractors for strongly damped wave equations, *Nonlinearity*, 19 (2006), 1495–1506.

178. V. Pata and S. Zelik, Global and exponential attractors for 3-D wave equations with displacement dependent damping, *Math. Meth. Appl. Sci.*, 29 (2006), 1291–1306.

179. V. Pata and S. Zelik, A result on the existence of global attractors for semigroups of closed operators, *Commun. Pure. Appl. Anal.*, 6 (2007) 481–486.

180. V. Pata and S. Zelik, Attractors and their regularity for 2-D wave equations with nonlinear damping, *Adv. Math. Sci. Appl.*, 17 (2007), 225–237.

181. A. Pazy, *Semigroups of Linear Operators and Applications to Partial Differential Equations*, Springer, New York, 1986.

182. V.A. Pliss, *Nonlocal Problems of the Theory of Oscillations*, Academic Press, New York, 1966.

183. V.A. Pliss, *Integral Sets of Periodic Systems of Differential Equations*, Nauka, Moscow, 1977 (in Russian).

184. S.I. Pohozhaev, On a class of quasilinear hyperbolic equations, *Math. USSR, Sbornik*, 25 (1975), no. 1, 145–158.

185. E Poláček, Parabolic equations: Asymptotic behavior and dynamics on invariant manifolds. In: B. Fiedler (Ed.), Handbook of Dynamical Systems, vol. 2, Elsevier, Amsterdam, 2002, 835–883.

186. M. Potomkin, Asymptotic behavior of thermoviscoelastic Berger plate, *Commun. Pure Appl. Anal.*, 9 (2010), 161–192.

187. D. Pražák, On finite fractal dimension of the global attractor for the wave equation with nonlinear damping, *J. Dyn. Diff. Eqs.*, 14 (2002), 764–776.

188. G. Raugel, Global attractors in partial differential equations. In: B. Fiedler (Ed.), Handbook of Dynamical Systems, vol. 2, Elsevier, Amsterdam, 2002, 885–982.

189. R. Reissing, G. Sansone and R. Conti, *Qualitative Theory of Ordinary Differential Equations*, Nauka, Moscow, 1974 (in Russian).

190. A.V. Rezounenko, Partial differential equations with discrete distributed state-dependent delays, *J. Math Anal. and Appl.*, 326 (2007), 1031–1045.

191. A.V. Rezounenko, Differential equations with discrete state-dependent delay: uniqueness and well-posedness in the space of continuous functions, *Nonlinear Anal., TMA*, 70 (2009), 3978–3986.

192. A.V. Rezounenko, Non-linear partial differential equations with discrete state-dependent delays in a metric space, *Nonlinear Anal., TMA*, 73 (2010), 1707–1714.

193. A.V. Rezounenko, A condition on delay for differential equations with discrete state-dependent delay, *J. Math Anal. and Appl.*, 385 (2012), 506–516.

194. A.V. Rezounenko and P. Zagalak, Non-local PDEs with discrete state-dependent delays: well-posedness in a metric space, *Discrete Contin. Dyn. Syst.*, 33 (2013), 819–835.

195. J. Robinson, *Infinite-Dimensional Dynamical Systems: An Introduction to Dissipative Parabolic PDEs and the Theory of Global Attractors*, Cambridge University Press, 2001.

196. J. Robinson, *Dimensions, Embeddings, and Attractors*, Cambridge University Press, 2011.

197. J. Robinson and A. Vidal-López, Minimal periods of semilinear evolution equations with Lipschitz nonlinearity, *J. Differential Equations*, 220 (2006), 396–406.

198. J. Robinson and A. Vidal-López, Minimal periods of semilinear evolution equations with Lipschitz nonlinearity revisited, *J. Differential Equations*, 254 (2013), 4279–4289.

199. W. Rudin, *Functional Analysis*, McGraw-Hill, Inc., New York, 1991.

200. W.M. Ruess, Existence of solutions to partial differential equations with delay. In: Theory and Applications of Nonlinear Operators of Accretive Monotone Type, Lecture Notes Pure Appl. Math. 178 (1996), 259–288.

201. I. Ryzhkova, On a retarded PDE system for a von Karman plate with thermal effects in the flow of gas, *Matem. Fizika, Analiz, Geometrija*, 12 (2005), 173–186.

202. A. Savostianov and S. Zelik, Recent progress in attractors for quintic wave equations, *Mathematica Bohemica*, 139 (2014), 657–665.

203. A. Savostianov and S. Zelik, Smooth attractors for the quintic wave equations with fractional damping, *Asymptotic Analysis*, 87 (2014), 191–221.

204. A.P.S. Selvadurai, *Elastic Analysis of Soil Foundation Interaction*, Elsevier, Amsterdam, 1979.

205. M. Sermange and R. Temam, Some mathematical questions related to MHD equations, *Commun. Pure Appl. Math.* 36 (1983), 635–664.

206. G.R. Sell and Y. You, *Dynamics of Evolutionary Equations*, Springer, New York, 2002.

207. A. Sharkovsky, Ideal turbulence, *Nonlinear Dynamics*, 44 (2006), 15–27.

208. A. Sharkovsky, S. Kolyada, A. Siwak and V. Fedorenko, *Dynamics of One-Dimensional Maps*, Naukova Dumka, Kiev, 1989 (in Russian).

209. A. Sharkovsky, Yu. Maistrenko and E. Romanenko, *Difference Equations and Their Applications*, Naukova Dumka, Kiev, 1986 (in Russian).

210. R. Showalter, *Monotone Operators in Banach Spaces and Nonlinear Partial Differential Equations*, AMS, Providence, RI, 1997.

211. V.I. Shubov, On subsets of a Hilbert space which have a finite Hausdorff dimension,*Zapiski Nauchnyh Seminarov LOMI*, 163 (1987), 154–165; in Russian.

212. K.S. Sibirsky, *Introduction to Topological Dynamics*, Noordhoff, Leyden, 1975.

213. J. Simon, Compact sets in the space $L^p(0, T; B)$, *Annali Mat. Pura Appl.*, 148 (1987), 65–96.

214. T. Taniguchi, Existence and asymptotic behaviour of solutions to weakly damped wave equations of Kirchhoff type with nonlinear damping and source terms, *J. Math. Anal. Appl.*, 361 (2010), 566–578.

215. R. Temam, *Navier-Stokes Equations and Nonlinear Functional Analysis*, 2nd edition, SIAM, Philadelphia, 1995.

216. R. Temam, *Infinite Dimensional Dynamical Systems in Mechanics and Physics*, 2nd edition, Springer, New York, 1997.

217. G. Teschl, *Ordinary Differential Equations and Dynamical Systems*, AMS, Providence, RI, 2012.

218. C.C. Travis and G.F. Webb, Existence and stability for partial functional differential equations, *Transactions of AMS*, 200 (1974), 395–418.

219. C.C. Travis and G.F. Webb, Existence, stability, and compactness in the α-norm for partial functional differential equations, *Transactions of AMS*, 240 (1978), 129–143.

220. H. Triebel, *Interpolation Theory, Functional Spaces and Differential Operators*, North-Holland, Amsterdam, 1978.

221. V.Z. Vlasov and U.N. Leontiev, *Beams, Plates, and Shells on Elastic Foundation, Israel Program for Scientific Translations*, Jerusalem, 1966 (translated from Russian).

222. H.-O. Walther, The solution manifold and C^1-smoothness for differential equations with state-dependent delay, *J. Differential Equations*, 195 (2003), 46–65.

223. H.-O. Walther, On Poisson's state-dependent delay, *Discrete Contin. Dyn. Syst.*, 33 (2013), 365–379.

224. J. Wu, *Theory and Applications of Partial Functional Differential Equations*, Springer, New York, 1996.

225. Z.-J. Yang, Longtime behavior of the Kirchhoff type equation with strong damping in $\mathbf{R}^N$, *J. Differential Equations*, 242 (2007), 269–286.

226. Z.-J. Yang, Y.-Q. Wang, Global attractor for the Kirchhoff type equation with a strong dissipation *J. Differential Equations*, 249 (2010), 3258–3278.

227. Z. Yang, X. Li, Finite dimensional attractors for the Kirchhoff equation with a strong dissipation, *J. Math. Anal. Appl.*, **375** (2011), 579–593.

228. J.A. Yorke, Periods of periodic solutions and the Lipschitz constant. *Proc. AMS*, 22 (1969), 509–512.

229. K. Yosida, *Functional Analysis*, 4th ed., Springer, Berlin, 1974.

230. V.I. Yudovich, *Mathematical Models of Natural Sciences*, Vuzovckaya kniga, 2009; in Russian.

231. E. Zeidler, *Nonlinear Functional Analysis and Its Applications, vol.I-IV*, Springer, Berlin, 1986–1995.

232. S. Zelik, The attractor for a nonlinear reaction-diffusion system with a supercritical nonlinearity and its dimension, *Rend. Accad. Naz. Sci. XL Mem. Mat. Appl.*, 24 (2000) 1–25.

Index

© Springer International Publishing Switzerland 2015
I. Chueshov, *Dynamics of Quasi-Stable Dissipative Systems*, Universitext,
DOI 10.1007/978-3-319-22903-4

Printed in the United States
By Bookmasters